O9-BTO-148

biology

THE CORE

ERIC J. SIMON

New England College

PEARSON

Boston   Columbus   Indianapolis   New York   San Francisco   Upper Saddle River
Amsterdam   Cape Town   Dubai   London   Madrid   Milan   Munich   Paris   Montreal   Toronto
Delhi   Mexico City   São Paulo   Sydney   Hong Kong   Seoul   Singapore   Taipei   Tokyo

Editor-in-Chief: Beth Wilbur
Executive Director of Development: Deborah Gale
Acquisitions Editor: Alison Rodal
Development Editors: Nora Lally-Graves, Mary Ann Murray
Program Manager: Anna Amato
Editorial Assistant: Libby Reiser
Permissions Specialist: Michael Farmer
Executive Managing Editor: Erin Gregg
Managing Editor: Michael Early
Senior Project Manager: Shannon Tozier
Production Project Manager: Michael Penne
Production Management and Composition: S4Carlisle Publishing Services
Illustrations: Precision Graphics
Copyeditor: Lorretta Palagi
Editorial Proofreader: Julie Lewis
Art Coordination: S4Carlisle Publishing Services
Design Manager: Marilyn Perry
Interior Design: Dorling Kindersley Limited
Cover Design: Tandem Creative, Inc.
Photo Permissions Management: Donna Kalal
Photo Researcher: Kristen Piljay of Wanderlust Photos
Director of Marketing: Christy Lesko
Executive Marketing Manager: Amee Mosley
Senior Market Development Manager: Michelle Cadden
Manufacturing Buyer: Stacey Weinberger
Text Printer: Courier Kendallville
Cover Printer: Lehigh-Phoenix
Cover Photo Credit: Getty Images/Adrian Samson

Credits and acknowledgments for materials borrowed from other sources and reproduced, with permission, in this textbook appear on pages C-1–C-6.

Copyright © 2015 Pearson Education, Inc. All rights reserved. Manufactured in the United States of America. This publication is protected by Copyright, and permission should be obtained from the publisher prior to any prohibited reproduction, storage in a retrieval system, or transmission in any form or by any means, electronic, mechanical, photocopying, recording, or likewise. To obtain permission(s) to use material from this work, please submit a written request to Pearson Education, Inc., Permissions Department, 1900 E. Lake Ave., Glenview, IL 60025. For information regarding permissions, call (847) 486-2635.

Many of the designations used by manufacturers and sellers to distinguish their products are claimed as trademarks. Where those designations appear in this book, and the publisher was aware of a trademark claim, the designations have been printed in initial caps or all caps.

MasteringBiology® & BioFlix® is a trademark, in the U.S. and/or other countries, of Pearson Education, Inc. or its affffiliates.

Library of Congress Cataloging in Publication Data
Simon, Eric J.
    Biology : the core / Eric J. Simon, New England College.—First edition.
       pages cm
    ISBN 978-0-321-73586-7
  1. Biology.   I. Title.
    QH308.2.S56 2013
    570—dc22

                                                      2013022480

1 2 3 4 5 6 7 8 9 10—CRK—16 15 14 13 12

www.pearsonhighered.com

ISBN 10: 0-321-73586-2
ISBN 13: 978-0-321-73586-7 (Student edition)

About the Author

ERIC J. SIMON is a professor in the Department of Biology and Health Science at New England College in Henniker, New Hampshire. He teaches introductory biology to science majors and non-science majors, as well as upper-level courses in tropical marine biology and careers in science. Dr. Simon received a B.A. in biology and computer science and an M.A. in biology from Wesleyan University and a Ph.D. in biochemistry from Harvard University. His research focuses on innovative ways to use technology to improve teaching and learning in the science classroom, particularly for non-science majors. He lives in rural New Hampshire with his wife, two boys, two Cavalier King Charles Spaniels, a few dozen chickens, a dwarf hamster, and a leopard gecko. Dr. Simon is the lead author of the introductory non-majors biology textbooks *Campbell Essential Biology* (5th ed.) and *Campbell Essential Biology with Physiology* (4th ed.), and a co-author of the introductory biology textbook *Campbell Biology: Concepts and Connections* (8th ed.), all published by Pearson Benjamin Cummings.

> " *I dedicate this book to those who form the core of my life: Amanda, a partner of unwavering patience and kindness, and my boys Reed and Forest.* "

Preface

To the Student,

Being a college student today means juggling many priorities: work, school, extracurricular activities, family. I imagine that, if you're reading this book, you are enrolled in your first college science course, and it may be the only one you'll ever take. When it seems like there are so many priorities all competing for your attention, you may be unsure about how to fit studying biology into your busy life. Good news: This book is written specifically for you!

Over the years, I've observed my own students strive to succeed in their biology course in as efficient a manner as possible. *Biology: The Core* has been designed from the ground up to help you study effectively and succeed. Only the most important and relevant information—the core of biology content—is included. Biological concepts are displayed in highly visual and approachable two-page modules that guide you along a clear learning path, allowing your study time to be as efficient and effective as possible.

You might also be wondering how this course—and biology generally—applies to your own life. Luckily, this is easy to address, since issues like nutrition, cancer, reproductive health, and exercise physiology directly affect you and those you love. *The Core* is paired with a robust online library, *MasteringBiology*, that contains videos, current events, and interactive tutorials that help you draw connections between the course material and the world around you. Questions you might have about many topics will be addressed in this online complement to your textbook.

I hope that *Biology: The Core* meshes with your goals and your priorities, acting as a helpful guide for this course and addressing questions you run into in your broader life. Please feel free to drop me a line to tell me about your experience with *Biology: The Core* or to provide feedback regarding the text or online resources. Best wishes for a successful semester—and enjoy the big adventure of biology! It's not only in the pages of this book, but all around you.

ERIC J. SIMON, Ph.D.
SimonBiology@gmail.com

To the Instructor,

In a world with so many options for non-major biology textbooks, why write a new one? The answer is simple: today's students. We've all watched our non-science-major students struggle with the depth of material and relating biology to their lives. Which concepts do non-science students *need to know* in order to understand the relevance of biology? If we pare down the content and focus on the most important take-home lessons—the information that we hope students will remember 10 years after graduation—what remains is the core: a set of essential biological concepts that presents the big picture, providing students with a scientific basis for the issues they will confront throughout their lives.

Biology: The Core is a new kind of textbook, one that presents information in small chunks using a nonlinear, engaging, visual style. The book contains only the most essential content for each topic. All information is presented in stand-alone two-page modules that fully integrate narrative and art into a single teaching tool. Each module is complete with a topic statement, introductory paragraph, all text and graphics needed to explain the topic, a summary, and a self-quiz. In addition to a consistent pedagogical structure, each module is designed to stand on its own. Modules can be read in any order, allowing you the flexibility to assign topics in whatever sequence best suits your course.

The printed text is paired with *MasteringBiology,* an online tutorial platform that allows you to reinforce the book content and expand on the basic concepts presented in each module as needed. The activities and resources in *MasteringBiology* also offer you the flexibility to incorporate a wide variety of applications and current issues into your teaching. Unlimited by the particular set of examples printed in a static textbook, a rich collection of online resources—including Current Topic PowerPoint presentations, news videos from ABC News and the BBC, *New York Times* articles, and interactive tutorials—enables you to connect the core content to interesting, relevant, and timely applications and issues that are important to you and your students.

I hope that the aims of *Biology: The Core* resonate with the teaching and learning goals of your non-major introductory biology course. Feel free to drop me a line to tell me about your course and your students, to provide feedback regarding the text or the online resources, or just to chat about the non-major course in general—it's my favorite topic of conversation!

Best wishes for a successful semester,

ERIC J. SIMON, Ph.D.
SimonBiology@gmail.com

A New Biology Learning Program Built for Today's Students

> *A brief textbook focused on only the **core** content that students need to learn for a non-majors course.*

MasteringBiology®

is an online homework and tutoring system that delivers self-paced activities designed to help students arrive to class prepared. Instructors can efficiently maximize class time with easy-to-assign and automatically graded assessments.

A Modern Teaching Program Supports Innovation and Active Learning

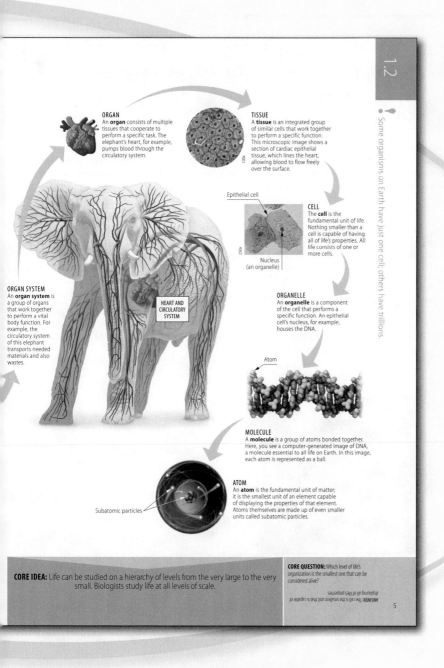

1.2

Some organisms on Earth have just one cell; others have trillions.

ORGAN
An **organ** consists of multiple tissues that cooperate to perform a specific task. The elephant's heart, for example, pumps blood through the circulatory system.

TISSUE
A **tissue** is an integrated group of similar cells that work together to perform a specific function. This microscopic image shows a section of cardiac epithelial tissue, which lines the heart, allowing blood to flow freely over the surface.

100X

Epithelial cell

CELL
The **cell** is the fundamental unit of life. Nothing smaller than a cell is capable of having all of life's properties. All life consists of one or more cells.

250X

Nucleus
(an organelle)

ORGAN SYSTEM
An **organ system** is a group of organs that work together to perform a vital body function. For example, the circulatory system of this elephant transports materials needed and also wastes.

HEART AND
CIRCULATORY
SYSTEM

ORGANELLE
An **organelle** is a component of the cell that performs a specific function. An epithelial cell's nucleus, for example, houses the DNA.

Atom

MOLECULE
A **molecule** is a group of atoms bonded together. Here, you see a computer-generated image of DNA, a molecule essential to all life on Earth. In this image, each atom is represented as a ball.

ATOM
An **atom** is the fundamental unit of matter; it is the smallest unit of an element capable of displaying the properties of that element. Atoms themselves are made up of even smaller units called subatomic particles.

Subatomic particles

CORE IDEA: Life can be studied on a hierarchy of levels from the very large to the very small. Biologists study life at all levels of scale.

CORE QUESTION: Which level of life's organization is the smallest one that can be considered alive?

ANSWER: The cell is the smallest unit that is capable of displaying all of life's properties.

5

A modular format and dynamic set of applications give instructors flexibility in teaching the course.

MasteringBiology® offers:

▶ **Interactive online activities** to help students apply and relate biological concepts to real life.

▶ **Unique teaching materials and resources** to assist instructors in preparing an innovative and effective course.

See the Big Picture

Biology: The Core is designed to help you efficiently learn the material and see the big picture. Begin studying each **concise module** by reading the **concept statement** which summarizes the key biological concept presented below.

▶ Next, the **narrative** introduces you to the key concept. The prose in each module is brief and works together with the illustrations to convey only the most core information so you never get lost in a sea of details.

4.4

Photosynthesis occurs in two linked stages

The overall equation for photosynthesis is relatively straightforward: carbon dioxide and water, along with the energy in sunlight, are used to produce sugar, releasing oxygen gas as a by-product. This process actually occurs in two stages. The **light reactions** capture sunlight and provide high-energy molecules to the **Calvin cycle**, which uses the high-energy molecules to produce sugar from carbon dioxide (CO_2).

Stage 1
THE LIGHT REACTIONS: CAPTURING ENERGY

Within thylakoids, energy from sunlight is absorbed by molecules of chlorophyll. This energy is used to split water, producing O_2 and high-energy electrons, which are stored by converting molecules of the electron carrier $NADP^+$ to NADPH. The energy from sunlight is also used to produce high-energy ATP molecules. To summarize, the light reactions capture the energy in sunlight and store it within high-energy molecules of ATP and NADPH.

$$6CO_2 + 6H_2O$$

ENERGY FROM SUNLIGHT

H_2O

ENERGY SHUTTLE

ATP (higher energy)

ELECTRON SHUTTLE

NADPH (higher energy)

O_2

LINKING THE STAGES: The Energy and Electron Shuttles

The light reactions and the Calvin cycle are linked by energy and electron shuttles. **NADPH** is a molecule that acts as a high-energy electron shuttle. It is produced from a lower-energy form called $NADP^+$. **ATP** (adenosine triphosphate) is a high-energy molecule that acts as an energy shuttle. It is produced from a lower-energy molecule called ADP (adenosine diphosphate).

66

▼ Then, read through each module looking at both the text and the illustrations. **Figures and narrative work together** to convey concepts and help you understand the material. Everything you need to study a core concept is at your fingertips.

MasteringBiology®

Guided Video Tours walk you through key concepts in each module and let you check your understanding of the core ideas.

4.4

Most of the mass of a plant ultimately derives from CO_2 in the air.

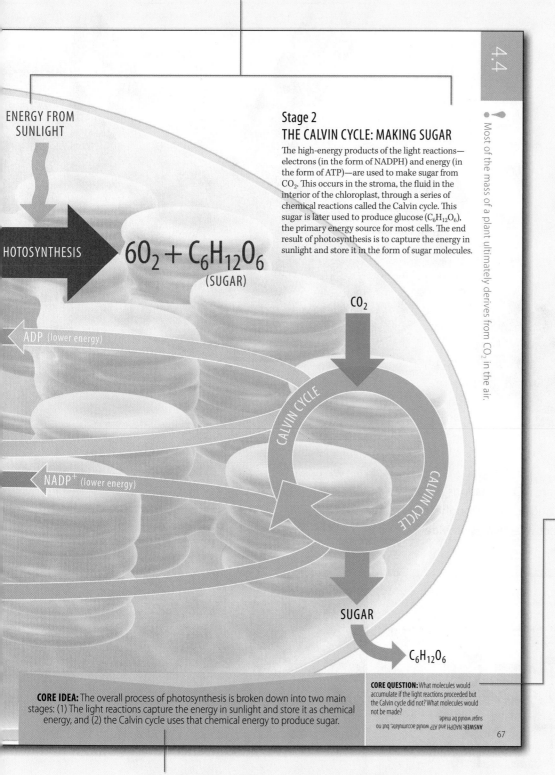

ENERGY FROM SUNLIGHT

PHOTOSYNTHESIS

$6O_2 + C_6H_{12}O_6$
(SUGAR)

ADP (lower energy)

NADP⁺ (lower energy)

CO_2

CALVIN CYCLE

CALVIN CYCLE

SUGAR

$C_6H_{12}O_6$

Stage 2
THE CALVIN CYCLE: MAKING SUGAR

The high-energy products of the light reactions—electrons (in the form of NADPH) and energy (in the form of ATP)—are used to make sugar from CO_2. This occurs in the stroma, the fluid in the interior of the chloroplast, through a series of chemical reactions called the Calvin cycle. This sugar is later used to produce glucose ($C_6H_{12}O_6$), the primary energy source for most cells. The end result of photosynthesis is to capture the energy in sunlight and store it in the form of sugar molecules.

◀ After you have studied the entire module, the **Core Questions** allow you to apply your knowledge and check your understanding of the concepts.

CORE IDEA: The overall process of photosynthesis is broken down into two main stages: (1) The light reactions capture the energy in sunlight and store it as chemical energy, and (2) the Calvin cycle uses that chemical energy to produce sugar.

CORE QUESTION: What molecules would accumulate if the light reactions proceeded but the Calvin cycle did not? What molecules would not be made?

ANSWER: NADPH and ATP would accumulate, but no sugar would be made.

67

▲ Review the **core idea** with a summary of the module to reinforce what you just learned.

Dynamic Visuals Explain Each Concept

▶ **Vibrant illustrations** take center stage with narrative integrated seamlessly to help you learn each concept. You never have to flip back and forth between pages or between text and visuals to grasp a concept.

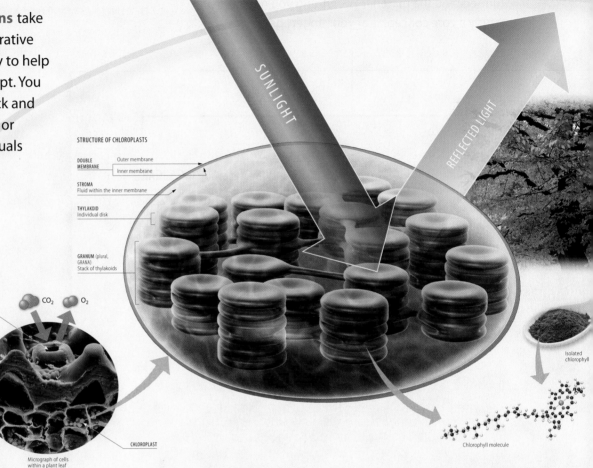

STRUCTURE OF CHLOROPLASTS

DOUBLE MEMBRANE — Outer membrane / Inner membrane

STROMA — Fluid within the inner membrane

THYLAKOID — Individual disk

GRANUM (plural, GRANA) — Stack of thylakoids

SUNLIGHT

REFLECTED LIGHT

STOMA (plural, STOMATA)

CO₂   O₂

Size 5752x

CHLOROPLAST

Micrograph of cells within a plant leaf

H₂O

Isolated chlorophyll

Chlorophyll molecule

▶ **BioFlix Animations** are 3-D animations that help you visualize and learn the toughest topics in biology. Interactive activities provide coaching and feedback.

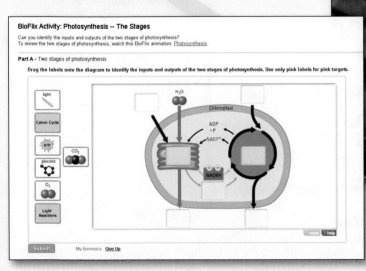

BioFlix Activity: Photosynthesis -- The Stages

Can you identify the inputs and outputs of the two stages of photosynthesis?
To review the two stages of photosynthesis, watch this BioFlix animation: Photosynthesis.

Part A - Two stages of photosynthesis

Drag the labels onto the diagram to identify the inputs and outputs of the two stages of photosynthesis. Use only pink labels for pink targets.

light

Calvin Cycle

ATP

glucose

O₂

Light Reactions

CO₂

H₂O

Chloroplast

ADP +P

NADP⁺

NADPH

reset   help

Submit   My Answers   Give Up

MasteringBiology® **Brings Concepts to Life**

MasteringBiology® is designed to help you practice the course concepts and apply them to current topics. Interactive activities and individualized coaching help you arrive to class prepared and offer study tools for you to use wherever you are!

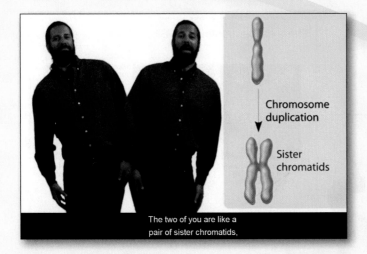

The two of you are like a pair of sister chromatids,

◀ **Video Tutor Sessions and MP3 Tutor Sessions.** Get help with key concepts with on-the-go tutorials hosted by author Eric Simon. Each video or audio session will help you build on the basic knowledge presented in your textbook.

Eric Simon's Video Tutor Sessions include:

- DNA Profiling
- DNA Structure
- Mitosis vs. Meiosis
- Phylogenetic Trees
- Sex-Linked Pedigrees
- Survey of Biodiversity

▶ **Interpreting Data and You Decide activities.** Receive coaching on how to analyze data and graphs, and learn how you can use data to make informed decisions in everyday life.

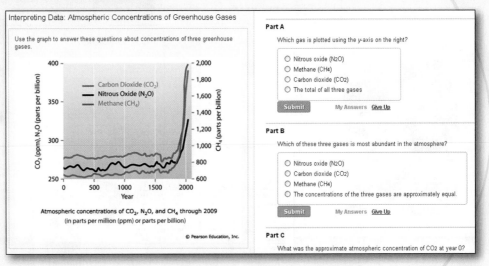

◀ **ABC News videos** and **current events** from the *New York Times* cover a wide range of biological topics to show you how science connects to your everyday life.

Teach the Course Your Way

Flexible and innovative instructor resources make it easy to prepare engaging classes that hook students into learning about biology-related topics such as agriculture, athletic cheating, cancer, food & nutrition, global climate change, evolution, and more.

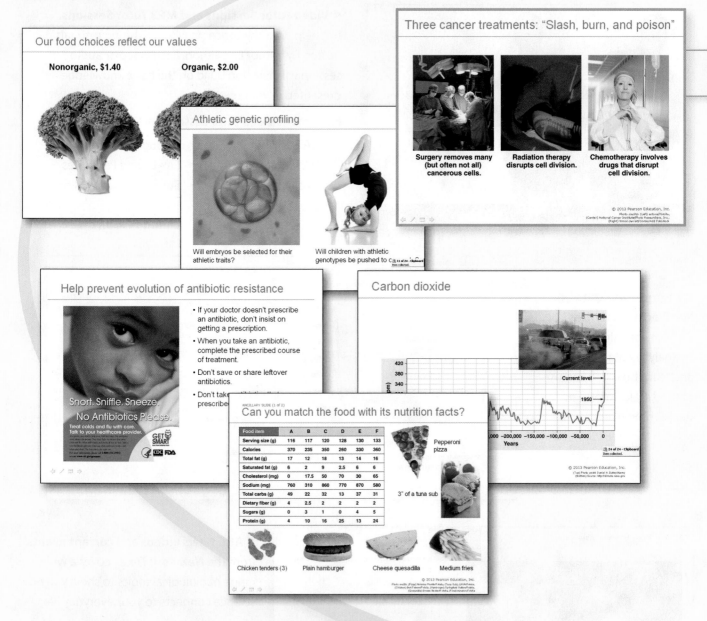

▶ Design high interest lectures and save time. Each application is assembled into a **learning unit** to support instructors with a detailed **topic guide** developed by author Eric Simon, along with a customizable **PowerPoint Lecture Presentation** and a related **MasteringBiology® Current Topic pre-built assignment**.

▶ Share ideas for easy to use in-class activities with **The Instructor Exchange**. This resource offers active lecture options for classes of any size and can be accessed through MasteringBiology®.

Flexible Resources Support Your Course Goals

Assign just the right amount of content with self-contained modules that cover only the core concepts you need to support your course syllabus and teaching goals.

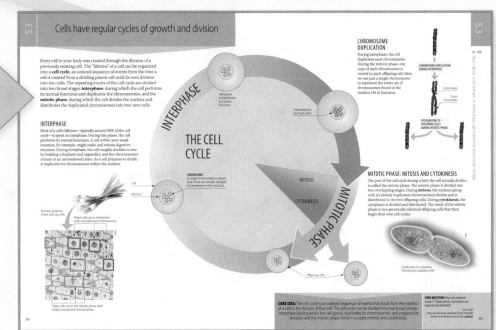

▶ Jump between topics with ease. **The modular organization** of the textbook gives you the freedom to teach concepts in your preferred order throughout the course and helps your students maximize their study time with modules that cover just what they need to know.

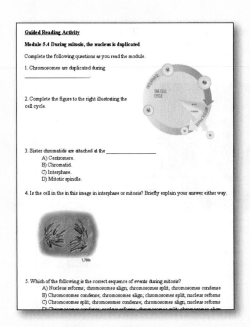

▲ Encourage Active Reading with **Guided reading activities** to help students navigate the text and allow them to practice their understanding of every module.

▼ Ensure your students arrive to class prepared. Assign automatically graded activities, animations, and reading quizzes in **MasteringBiology®** to encourage students to practice basic biological concepts outside of class.

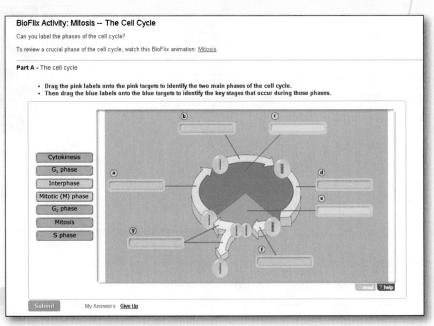

Evaluate Student Understanding and Boost Success

Help students practice foundational biology principles outside of class and ensure they arrive to lecture prepared. MasteringBiology® activities allow students to assess their understanding and receive personalized feedback designed to help them grasp key concepts.

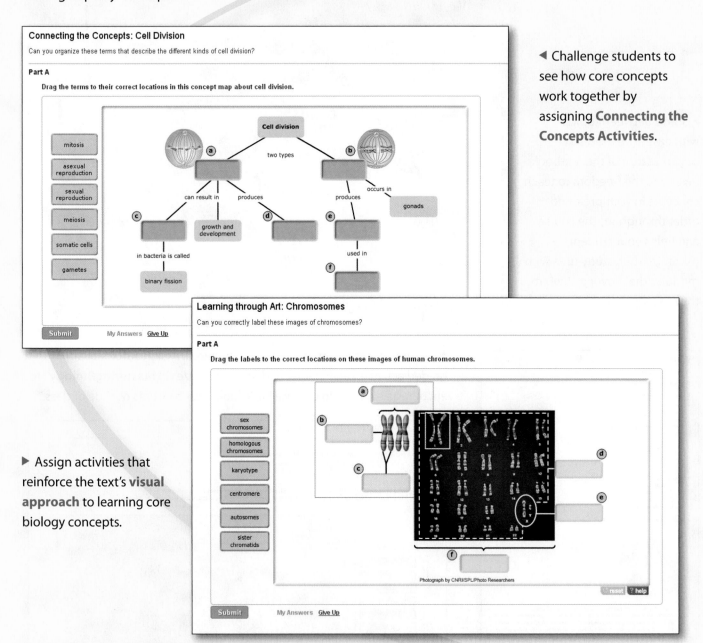

◀ Challenge students to see how core concepts work together by assigning **Connecting the Concepts Activities**.

▶ Assign activities that reinforce the text's **visual approach** to learning core biology concepts.

▼ Provide students with **personalized wrong-answer feedback** by assigning coaching activities in MasteringBiology®.

Try Again
You labeled 3 of 6 targets incorrectly. For target (f), what term describes this photographic inventory of a complete set of chromosomes from one cell? This one shows all 46 chromosomes in a human somatic cell.

▶ **Learning Catalytics** is a "bring your own device" student engagement and assessment educational tool, allowing you to actively engage each student during lecture and access rich data to assess student understanding. Students can use any web-enabled device and instructors can create their own assessment items or select from an existing set of items.

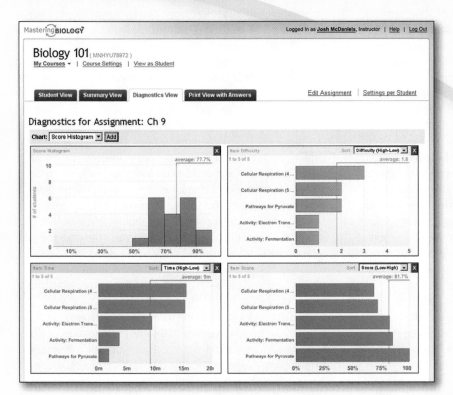

◀ Learn what your students do and don't understand before class using **one-click diagnostics**, so that you can make the most of valuable time in class.

▶ The **Mastering gradebook** provides easy-to-interpret insight into student performance. Every assignment is automatically graded and shades of red highlight challenging assignments.

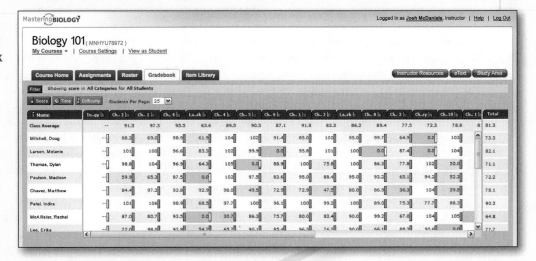

Answer Stats:	Students	% Correct	% Unfinished	% Req'd Solution	Wrong/student	Hints/student
System Average	5548	98%	1.4%	0.6%	0.6	0
This Course (MBDEMOGRADES)	25	100%	0%	0%	0.5	0

◀ Learn which topics are challenging for your students and **compare your class data** to the national average.

Acknowledgments

After five years of working on *Biology: The Core*, there are many people to thank and acknowledge for their contributions. One of the most gratifying aspects of writing a book is the opportunity it presents to interact with so many skilled professionals at Pearson Education and so many talented biology colleagues from around the country. While any problems or mistakes within this book or related to it are solely my responsibility, the successful aspects are due to the efforts of dozens of people.

A few people deserve special mention for their deep contributions; these people form the core of *The Core*. First and foremost is Nora Lally-Graves—Developmental Editor, Project Manager, and all-around problem-solver—whose passion for communicating undergraduate science can be seen on literally every page of this book. Chalon Bridges used her unbounded positivity, unbridled enthusiasm, and a deep empathy toward students and teachers to help launch this project, leaving strong echoes of her talents even after she moved on to pursue other opportunities. Alison Rodal, as Acquisitions Editor, worked with me and the rest of the team nearly every day, bringing a broad array of talents to bear on this project.

Rounding out the original Pearson team is Beth Wilbur, Editor-in-Chief, whose vision, understanding, and unwavering high standards have benefited every project since she brought me into the Pearson family over a decade ago. Editorial Assistants Rachel Brickner and Libby Reiser provided never-ending help coordinating reviews and supplements and solving countless problems, always with a buoying positive energy. Deborah Gale, Director of Development, provided her masterful oversight of all matters editorial and gave invaluable help when it was most needed. Thanks are also due to President of Science, Business and Technology Paul Corey and former President of Pearson Science and Math, Linda Davis, whose vision and insights were vital to launching this project.

I love my editors! All members of the Pearson editorial team bring decades of experience and talent to every project they take on. Developmental editors Mary Ann Murray and Evelyn Dahlgren provided eagle-eyed editorial help during the late stages of the book. For saving me from countless embarrassments large and small, I thank Copyeditor Lorretta Palagi and Editorial Proofreader Julie Lewis. Thank you to Indexer Robert Swanson for making the book much more user-friendly (see "Indexer, gratitude for"). Anna Amato, Program Manager, stepped in during the later stages of production to aid in ushering this text to publication.

If you like the look of this book—the layout, photographs, and graphics—then you can thank the amazing design team at DK Publishing (Dorling Kindersley). This team was a key part of *The Core* from literally the first day. Stuart Jackman and Sophie Mitchell helped manage the London-based crew and kindly turned their keen senses of style to this project. Anthony Limerick designed every spread in *The Core* and is therefore the one person most responsible for its beautiful look.

Once the text was written and the art developed, the production team took over, turning countless individual scraps of information into a coherent book. Led by Senior Project Manager Shannon Tozier, this group graciously implemented ideas, turning them into the reality of the book you see before you. For this we thank Production Project Manager Michael Penne, Managing Editor Michael Early, photo researcher Kristin Piljay of Wanderlust Photos, Pearson Image Lead Donna Kalal, and permissions specialist Michael Farmer. Thank you also to Amanda Waldo for her assistance in compiling the art manuscript and glossary. Jon Ballard began media production, and Daniel Ross capably continued that effort. I thank Yvo Riezebos for his design of the book cover, and Design Manager Marilyn Perry for her help in producing it. Roxanne Klaas at S4Carlisle provided expert composition, and Kristina Seymour, Heidi Richter, Megan Stewart, and Jan Troutt of Precision Graphics lent their talents to art development and art creation. Some of the truly special three-dimensional art in the book was developed by Martin Hale, Terri Hamer, and Craig Vrankovich of Animated Biomedical Productions Studio—thank you all for your special artistic contributions!

I am deeply indebted to the many talented thinkers who lent their expertise in the form of reviews of early stages of the text. Two reviewers stand out in particular for their willingness to read every chapter and for the quality of their feedback: Jim Newcomb of New England College and Amanda Marsh. Special thanks also go to Jay Withgott, Jean Dickey, Marshall Simon, Jamey Barone, Paula Marsh, and Terry Austin for providing key advice when it was most needed. For technical advice on specific topics, I thank David S. Hibbett of Clark University, Kim T. Fredricks of the Upper Midwest Environmental Science Center, Maria Colby, and the zoology class at New England College. I am also indebted to the many faculty colleagues around the country who served on our Faculty Advisory Board or reviewed chapters of *The Core*, each of whom used his or her own teaching and research experience to provide countless pieces of advice that helped steer the project through big decisions—such as its content and title—right down to the smallest details. You can find a complete list of all those who have made contributions on pages xvii–xviii; thanks go to every name on that list. I would also like to thank all of my colleagues at New England College—especially Lori Bergeron, Deb Dunlop, Sachie Howard, Tod Ramseyer, Aaron Daniels, Mark Watman, and Michelle Perkins—for providing support of various kinds during this project.

Of equal importance to the book itself are the many multimedia supplements that accompany it. Many teaching colleagues provided help with writing supplements for *The Core*. In particular, I thank test bank author Brandon Foster of Wake Tech Community College, instructor's guide author Brenda Leady of University of Toledo, PowerPoint presentation author Wendy Kuntz of Kapiolani Community College, active learning question author Heather Miller of Front Range Community College and Kaplan University, and Active Reading Guide author Dana Kurpius of Elgin Community College. Special thanks go to Cindy Klevickis of James Madison University for writing the reading questions that supplement the book in *MasteringBiology*. It is especially gratifying to thank Mackie Glashow (a former student of mine at New England College) for her contributions to *The Core* multimedia. I'd like to extend particular thanks to the many instructors and students who class-tested *The Core* prior to publication, helping to ensure its usefulness and effectiveness.

Many people at Pearson also helped produce the media that accompanies this text. In particular, Director of *MasteringBiology* Editorial Content Tania Mlawer and *MasteringBiology* Developmental Editor Juliana Tringali provided much assistance in preparing this vital part of the program. From the engineering side of *MasteringBiology*, I thank Katie Foley, Caroline Ross, and Taylor Merck. I also thank Senior Supplements Project Manager Susan Berge and Supplements Project Manager Kim Wimpsett.

After a book is written, the marketing team steps in to ensure that the complete story is told to all who might benefit. For this we thank Executive Marketing Manager Amee Mosley, Market Development Manager Michelle Cadden, and Market Development Coordinator Kait Nagi. Acting as the final facilitators of the long journey from author to student, I thank the entire Pearson Education sales team including all the sales managers and publisher's representatives who work tirelessly every semester to help students learn and instructors teach.

In closing, I beg forgiveness from those who lent their unique talents to this book but who I failed to mention—I hope you will forgive my oversight and know that you have earned my gratitude.

With deepest, sincerest, and humblest thanks and respect to all who contributed their talents to *Biology: The Core*,

ERIC J. SIMON, Ph.D.
NEW ENGLAND COLLEGE, HENNIKER, NH

Reviewers

Shamili Ajgaonkar,
Sandiford College of DuPage

Penny Amy,
University of Nevada, Las Vegas

Kim Atwood,
Cumberland University

David Ballard,
Southwest Texas Junior College

Marilyn Banta,
Texas State University

Patricia Barg,
Pace University

David Belt,
Metropolitan Community College, Penn Valley

Anna Bess Sorin,
University of Memphis

Andrea Bixler,
Clarke University

Susan Bornstein-Forst,
Marian University

Randy Brewton,
University of Tennessee, Knoxville

Peggy Brickman,
University of Georgia

Steven Brumbaugh,
Green River Community College

Stephanie Burdett,
Brigham Young University

Greg Dahlem,
Northern Kentucky University

Mary Dettman,
Seminole State College of Florida

Eden L. Effert,
Eastern Illinois University

Jose Egremy,
Northwest Vista College

Hilary Engebretson,
Whatcom Community College

Brian Forster,
St. Joseph's University

Brandon Foster,
Wake Technical Community College

Thomas Gehring,
Central Michigan University

Larry Gomoll,
Stone Child College

Tammy Goulet,
University of Mississippi

Eileen Gregory,
Rollins College

David Grise,
Texas A&M University–Corpus Christi

Tom Hinckley,
Landmark College

Kelly Hogan,
University of North Carolina at Chapel Hill

Christopher Jones,
Moravian College

Jacob Krans,
Central Connecticut State University

Pramod Kumar,
Northwest Vista College

Wendy Kuntz,
University of Hawai'i

Dana Kurpius,
Elgin Community College

Brenda Leady,
University of Toledo

Maureen Leupold,
Genesee Community College

Mark Manteuffel,
St. Louis Community College

Debra McLaughlin,
University of Maryland University College

Heather Miller,
Front Range Community College and Kaplan University

Lisa Misquitta,
Quinebaug Valley Community College

Pamela Monaco,
Molloy College

Ulrike Muller,
California State University, Fresno

Lori Nicholas,
New York University

Monica Parker,
Florida State College at Jacksonville

Don Plantz,
Mohave Community College

Gregory Podgorski,
Utah State University

Robyn A. Puffenbarger,
Bridgewater College

Kayla Rihani,
Northeastern Illinois University

Nancy Risner,
Ivy Tech Community College

Bill Rogers,
Ball State University

David Rohrbach,
Northwest Vista College

Chris Romero,
Front Range Community College, Larimer Campus

Checo Rorie,
North Carolina Agricultural and Technical State University

Amanda Rosenzweig,
Delgado Community College

Kim Sadler,
Middle Tennessee State University

Steve Schwartz,
Bridgewater State University

Tara Scully,
George Washington University

Cara Shillington,
Eastern Michigan University

Stephen Sumithran,
Eastern Kentucky University

Suzanne Wakim,
Butte Community College

Frances Weaver,
Widener University

Susan Whitehead,
Becker College

Jennifer Wiatrowski,
Pasco-Hernando Community College

Matthew Wund,
The College of New Jersey

Class Testers and Interview Participants

Leo Alves,
Manhattan College

Tonya Bates,
University of North Carolina at Charlotte

Brian Baumgartner,
Trinity Valley Community College

Lisa Blumke,
Georgia Highlands College

TJ Boyle,
Blinn College

Michelle Brewer,
Central Carolina Technical College

Melissa Caspary,
Georgia Gwinnett College

Krista Clark,
University of Cincinnati, Clermont

Merry Clark,
Georgia Highlands College

Reggie Cobb,
Nash Community College

Angela Costanzo,
Hawai'i Pacific University, Loa Campus

Evelyn Cox,
University of Hawai'i, West Oahu

Hattie Dambrowski,
Normandale Community College

Kelsey Deus,
Casper College

Lisa Delissio,
Salem State University

Dani Ducharme,
Waubonsee Community College

Jennifer Ellie,
Wichita State University

Sachie Etherington,
University of Hawai'i, Manoa

Christy Fleishacker,
University of Mary

Brandon Foster,
Wake Technical Community College

Valerie Franck,
Hawai'i Pacific University

Jennifer Fritz,
University of Texas at Austin

Kathy Galluci,
Elon University

Chunlei Gao,
Middlesex Community College

Mary Gobbett,
University of Indianapolis

Erin Goergen,
St. Petersburg College Clearwater

Marla Gomez,
Nicholls State University

Larry Gomoll,
Stone Child College

David Grise,
Texas A&M University Corpus Christi

Mellissa Gutierrez,
University of Southern Mississippi

Barbara Hass Jacobus,
Indiana University-Purdue University, Columbus

Debra Hautau,
Alpena Community College

Jon Hoekstra,
Heartland Community College

Tina Hopper,
Missouri State University

Joseph Husband,
Florida State College at Jacksonville

John Jenkin,
Blinn College

Jamie Jensen,
Brigham Young University

Julie Johns,
Cincinnati State Community College

Anta'Sha Jones,
Albany State University

Ambrose (Trey) Kidd,
University of Missouri, St. Louis

Manju Kishore,
Heartland Community College

Cindy Klevickis,
James Madison University

Tatyana Kliorina,
Trinity University

Karen Koster,
University of South Dakota

Barbara Kuehner,
University of Hawai'i, West Hawai'i

Dana Kurpius,
Elgin Community College

Jennifer Landin,
North Carolina State University

Grace Lasker,
Lake Washington Institute of Technology

Brenda Leady,
University of Toledo

Sharon Lee-Bond,
Northhampton Community College

Ernest May,
Kansas City Kansas Community College

MaryAnn Menvielle,
California State University, Fullerton

Kim Metera,
Wake Technical Community College

Heather Miller,
Front Range Community College and Kaplan University

Pamela Monaco,
Molloy College

Punya Nachappa,
Indiana University-Purdue University, Fort Wayne

Kathryn Nette,
Cuyamaca College

Betsy Ott,
Tyler Junior College

Mary O'Sullivan,
Elgin Community College

Dianne Purves,
Crafton Hills College

Peggy Rolfsen,
Cincinnati State Community College

Checo Rorie,
North Carolina A&T State University

Brian Sailer,
Central New Mexico Community College

Daita Serghi,
University of Hawai'i, Manoa

Vishal Shah,
Dowling College

David Smith,
Lock Haven

Patti Smith,
Valencia Community College, East Campus

Adrienne Smyth,
Worcester State

Wendy Stankovich,
University of Wisconsin at Platteville

Frank Stanton,
Leeward Community College

Olga Steinberg,
Hostos Community College

Fengjie Sun,
Georgia Gwinnett College

Ed Tall,
Seton Hall University

Lavon Tonga,
Longview Community College

Marie Trone,
Valencia College, Osceola Campus

Dan Trubovitz,
San Diego Miramar College

Encarni Trueba,
Community College of Baltimore County

Larchinee Turner,
Central Carolina Technical College

Marty Vaughan,
Indiana University-Purdue University, Indianapolis

Justin Walguarnery,
University of Hawai'i, Manoa

Jim Wallis,
St. Petersburg College, Tarpon Springs Campus

Rebekah Ward,
Georgia Gwinnett College

Jamie Welling,
South Suburban College

Clay White,
Lone Star College

Leslie Winemiller,
Texas A&M University

Focus Group Participants

Christine Andrews,
Lane Community College

Nickolas Butkevich,
Schoolcraft College

Susan Finazzo,
Georgia Perimeter College

Kristy Halverson,
University of Southern Mississippi

Wendy Jamison,
Chadron State College

Jennifer Kneafsey,
Tulsa Community College

Ruben Murcia,
Rose State College

Lisa Maranto,
Prince Georges Community College

Cassandra Moore-Crawford,
Prince Georges Community College

Christina Weir,
Eastern New Mexico University, Roswell

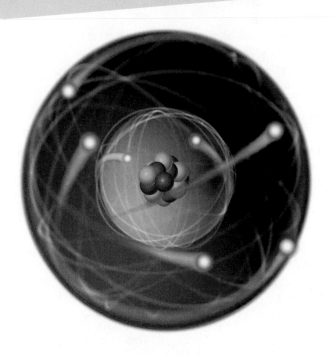

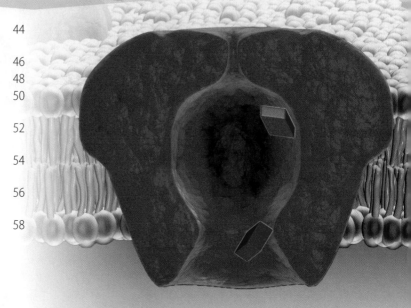

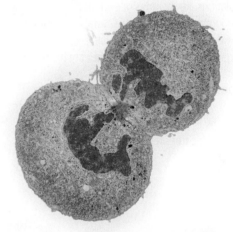

8 Biodiversity 1: Microscopic Organisms 178

11 Human Body Systems 248

12 Ecology 296

biology THE CORE

All living organisms share certain properties

Biology is the scientific study of life. While the definition of biology is very straightforward, it does raise some important questions. Perhaps the most obvious is: What is life? How do we distinguish living organisms from nonliving matter? How do we know that an elephant is alive, but a boulder is not? Biologists recognize **life** through a series of characteristics shared by all living things. We define life through the properties that living things display. An object is alive if and only if it displays all of these properties simultaneously.

THE PROPERTIES OF LIFE

REPRODUCTION
Like begets like; all organisms reproduce their own kind. Thus, elephants reproduce only elephants— never zebras or lions.

GROWTH AND DEVELOPMENT
Information carried by genes controls the pattern of growth in all organisms. For example, male elephants grow tusks as they age.

ENERGY USE
Every organism takes in energy, converts it to useful forms, and expels energy. This elephant is taking in energy by eating a plant. It can use that energy to move. It also releases energy as heat.

ORDER
Each living thing has a complex but well-ordered structure, as seen in the elephant's eye.

A VIRUS IS NOT ALIVE

We know that a virus is not alive because it does not display all of the properties of life. For example, a virus is not composed of cells, and it cannot reproduce on its own. While nonliving matter may display some of life's properties (a virus has order, for example), it never displays all of life's properties simultaneously.

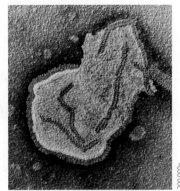

200,000x

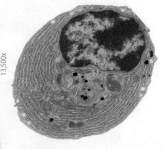

13,500x

CELLS

All living organisms consist of cells. Some living organisms have just one cell, but others (such as an elephant) have trillions.

RESPONSE TO THE ENVIRONMENT

All organisms respond to changes in the environment. Many of these responses help to keep an organism's internal environment within narrow limits even when the external environment changes a lot. This elephant is responding to the heat of the day by taking a bath, which helps keep its body temperature steady.

EVOLUTION

Individuals with traits that help them survive and reproduce pass the genes for those traits to offspring. Over many generations, such adaptations drive the evolution of populations. For example, modern elephants and woolly mammoths evolved from a common ancestor species that lived until about 5 million years ago.

CORE IDEA: Biology is the scientific study of life. All living things display a shared set of characteristics. Nonliving matter never displays all of these characteristics of life simultaneously.

CORE QUESTION: Which properties of life does a car display? Which does it not?

ANSWER: A car uses energy, is ordered, and responds to the environment. A car does not reproduce, grow or develop, or evolve, and a car is not made of cells.

Life can be studied at many levels

The study of life encompasses a very broad range of scales, from the microscopic world of cells to the vast scope of Earth's ecosystems. This figure summarizes some of the levels at which biologists study life on Earth, starting at the upper end of the scale.

THE LEVELS OF BIOLOGICAL ORGANIZATION

BIOSPHERE
The **biosphere** consists of all life on Earth and all of the environments that support life, from the deepest oceans to high in the atmosphere.

ECOSYSTEM
An **ecosystem** includes all the living organisms in one particular area (such as this African savannah) as well as the nonliving components that affect life, such as soil, air, and sunlight.

COMMUNITY
A **community** consists of all the interacting populations of organisms occupying an ecosystem. This community includes plants, animals, and even microscopic organisms.

ORGANISM
An **organism** is an individual living being, such as one African savannah elephant (*Loxodonta africana*).

POPULATION
A **population** is a group of interacting individuals of one species, such as the African savannah elephants shown here.

ORGAN

An **organ** consists of multiple tissues that cooperate to perform a specific task. The elephant's heart, for example, pumps blood through the circulatory system.

TISSUE

A **tissue** is an integrated group of similar cells that work together to perform a specific function. This microscopic image shows a section of cardiac epithelial tissue, which lines the heart, allowing blood to flow freely over the surface.

100x

Epithelial cell

CELL

The **cell** is the fundamental unit of life. Nothing smaller than a cell is capable of having all of life's properties. All life consists of one or more cells.

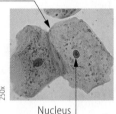
250x

Nucleus
(an organelle)

ORGAN SYSTEM

An **organ system** is a group of organs that work together to perform a vital body function. For example, the circulatory system of this elephant transports needed materials and also wastes.

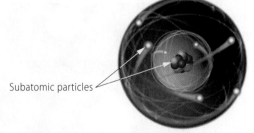

HEART AND
CIRCULATORY
SYSTEM

ORGANELLE

An **organelle** is a component of the cell that performs a specific function. An epithelial cell's nucleus, for example, houses the DNA.

Atom

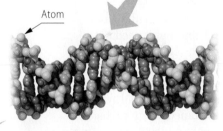

MOLECULE

A **molecule** is a group of atoms bonded together. Here, you see a computer-generated image of DNA, a molecule essential to all life on Earth. In this image, each atom is represented as a ball.

ATOM

An **atom** is the fundamental unit of matter; it is the smallest unit of an element capable of displaying the properties of that element. Atoms themselves are made up of even smaller units called subatomic particles.

Subatomic particles

CORE IDEA: Life can be studied on a hierarchy of levels from the very large to the very small. Biologists study life at all levels of scale.

CORE QUESTION: Which level of life's organization is the smallest one that can be considered alive?

ANSWER: The cell is the smallest unit that is capable of displaying all of life's properties.

Scientists use well-established methods to investigate the natural world

Biologists use the process of science to study life. That raises obvious questions: What is science? What distinguishes scientific thinking from other ways of investigating nature? There are several important principles that underlie scientific investigations.

THE SCIENTIFIC METHOD

Science always begins by observing the world. Such observations inevitably lead to questions about why the world is the way it is. How can you uncover an explanation for an observed phenomenon? The **scientific method** is a rough "recipe" for discovery, a series of steps that, if followed, may help a scientist understand an observation. The scientific method is simply a way of formalizing how we usually try to solve problems.

Scientists use the scientific method as a guideline, but it need not be followed rigidly. During a particular investigation, for example, a scientist might investigate multiple hypotheses simultaneously, or perhaps fail to make a specific prediction.

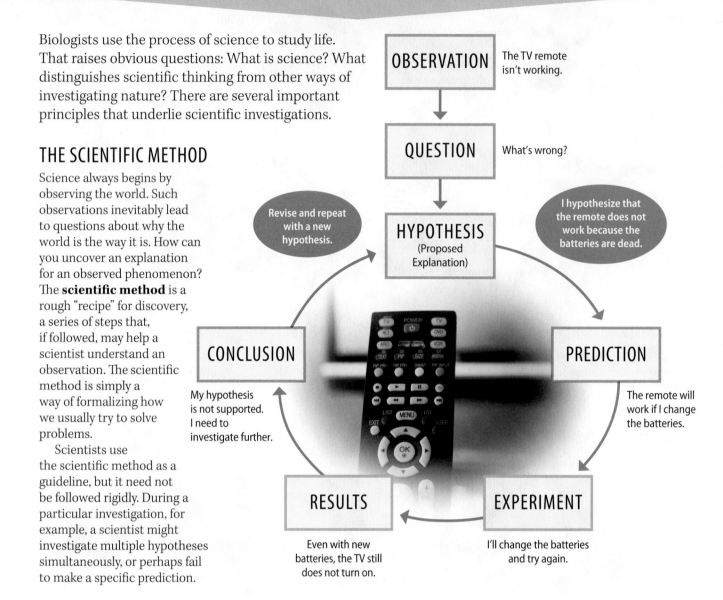

OBSERVATION
The TV remote isn't working.

QUESTION
What's wrong?

Revise and repeat with a new hypothesis.

HYPOTHESIS (Proposed Explanation)

I hypothesize that the remote does not work because the batteries are dead.

CONCLUSION
My hypothesis is not supported. I need to investigate further.

PREDICTION
The remote will work if I change the batteries.

RESULTS
Even with new batteries, the TV still does not turn on.

EXPERIMENT
I'll change the batteries and try again.

DISCOVERY SCIENCE

When scientists make verifiable observations, take careful measurements, and gather data—even in the absence of a hypothesis—they are performing **discovery science**. Discovery science provides data that can be used to describe the natural world. The researcher shown here, for example, is gathering information on the insect community living in a rainforest canopy. The data gathered via discovery science can prompt questions and guide the scientific method. For example, Charles Darwin's careful descriptions of the plants and animals he observed during his journeys led to hypotheses about how organisms evolve. Both methods of investigation—hypothesis-driven science and discovery science—allow scientists to investigate the natural world.

HYPOTHESES AND THEORIES

When discussing scientific ideas, it is important to distinguish between hypotheses and theories. Notice that a scientist uses the word "theory" differently than we tend to use it in everyday speech. In common usage, the word "theory" often means "a guess," which is not how scientists use the word.

HYPOTHESIS	THEORY
The scientific method depends on the development of hypotheses. A **hypothesis** is a proposed explanation for an observation. A valid hypothesis must be testable, and the results of such tests will either support or refute the hypothesis. For example, the endosymbiotic hypothesis proposes that some cellular components (such as the chloroplast and mitochondrion visible in the cell here) were once free-living organisms that were long ago incorporated into a larger cell.	A **theory** is much broader in scope than a hypothesis. It is much more comprehensive, it has not been shown false, and it already explains a great many observations. Theories are supported by a large and growing body of evidence. For example, the Cell Theory states that every living organism consists of cells that arose from preexisting cells. Theories can be used to devise specific hypotheses to be tested.

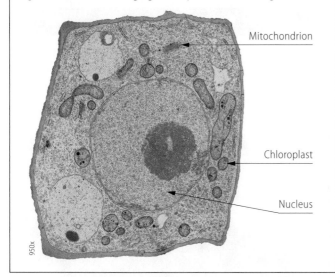

Mitochondrion

Chloroplast

Nucleus

950x

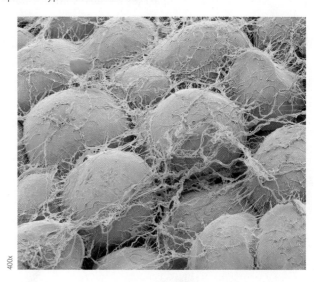

400x

CONTROLLED EXPERIMENTS

To investigate a hypothesis, a scientist may choose to perform a **controlled experiment** in which a test is run multiple times with one variable changing—and, ideally, all other variables held constant. The use of a controlled experiment allows a scientist to draw conclusions about the effect of the one variable that did change. The photos shown here represent a controlled experiment on the effect of using cake flour versus all-purpose flour on the thickness of chocolate chip cookies.

This batch of cookies was baked using cake flour.

This batch of cookies was baked using all-purpose flour. Other than the kind of flour, the recipe was exactly the same. This controlled experiment supports the hypothesis that cake flour increases the height of these chocolate chip cookies.

CORE IDEA: Scientific investigations may be hypothesis-driven (using the scientific method) or discovery-based. Careful observations and controlled experiments allow scientists to investigate hypotheses and develop theories.

CORE QUESTION: If you observe squirrels, come up with a tentative explanation for their behavior, and then test your idea, what method of inquiry are you performing?

ANSWER: The scientific method.

Cells, the fundamental units of life, contain DNA

The cell is the fundamental unit of life: Every living organism is composed of one or more cells, and nothing smaller is capable of performing all of the activities required for life. Some living creatures (such as microscopic bacteria) are unicellular, composed of just a single cell. Others (such as you) are composed of trillions of cells. Within each cell, one or more molecules of DNA act as the hereditary material.

TWO KINDS OF CELLS

All life on Earth is composed of one of two types of cells: prokaryotic or eukaryotic. These two types of cells have many similarities but also some fundamental differences.

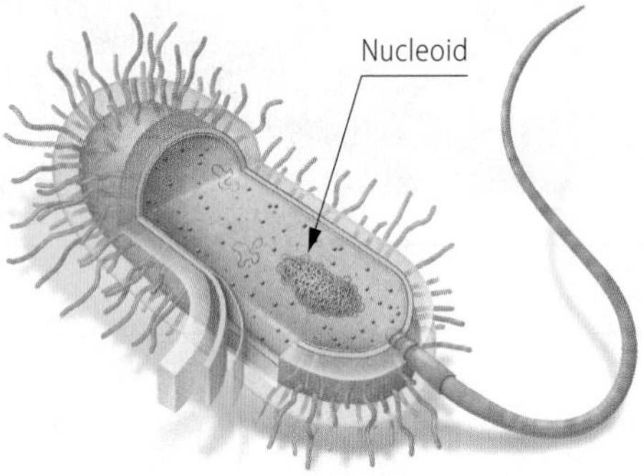

	PROKARYOTIC CELL	EUKARYOTIC CELL
	Nucleoid A typical prokaryotic bacterial cell	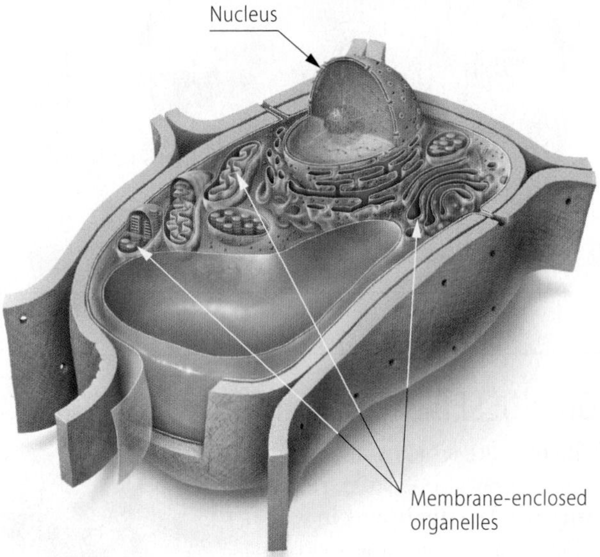 Nucleus Membrane-enclosed organelles A typical eukaryotic plant cell
Size	Smaller	Typically 10–100× larger
Complexity	Simpler	More structurally complex
Organelles	No membrane-enclosed organelles	Has membrane-enclosed organelles
Evolution	First appeared approximately 3.5 billion years ago	Evolved from prokaryotes approximately 2.1 billion years ago
DNA	Not contained within any cellular structure	Housed in membrane-enclosed nucleus
Number of cells	Unicellular	Unicellular or multicellular
Examples	Bacteria and archaea	Plants, animals, fungi, and protists

DNA IS THE MOLECULE OF LIFE

Although cells vary widely in many respects, they share certain features. For example, all cells contain **genes**—units of hereditary information—made from the same molecule: DNA. Although different organisms may have different genes, the genes of all organisms are coded by DNA molecules that contain the same four building blocks, abbreviated A, T, C, and G.

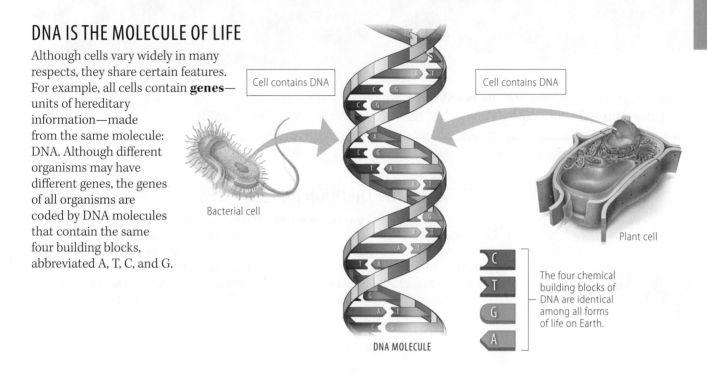

Cell contains DNA

Cell contains DNA

Bacterial cell

Plant cell

The four chemical building blocks of DNA are identical among all forms of life on Earth.

DNA MOLECULE

DNA CAN BE MIXED AND MATCHED

Because the hereditary information of all life is written in the identical chemical language of DNA, a gene from one species may be cut and pasted into the DNA of a different species. Genetic engineers have produced many such organisms. For example, biologists have made mice (as well as fish, cats, and monkeys) that contain a glow gene from a jellyfish, and many of our food crops are genetically modified to contain genes from other organisms.

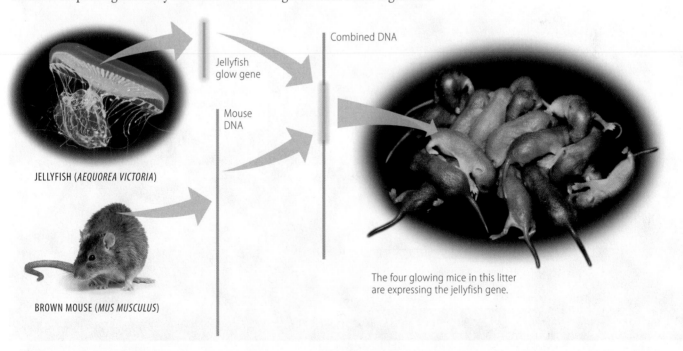

Combined DNA

Jellyfish glow gene

Mouse DNA

JELLYFISH (*AEQUOREA VICTORIA*)

BROWN MOUSE (*MUS MUSCULUS*)

The four glowing mice in this litter are expressing the jellyfish gene.

CORE IDEA: All life on Earth is composed of either small, simple prokaryotic cells, or larger, complex, organelle-containing eukaryotic cells. All cells contain genes made from the same four building blocks of DNA.

CORE QUESTION: Both eukaryotic and prokaryotic cells contain DNA. How do they differ in terms of where the DNA is stored?

ANSWER: In eukaryotic cells, the DNA is stored within a membrane-enclosed nucleus. In prokaryotic cells, the DNA is not enclosed.

All organisms interact with their ecosystems

Every organism interacts with its **ecosystem**, which includes other living organisms as well as nonliving physical factors such as air, sunlight, wind, and water. The dynamics of every ecosystem depend on two main processes: the flow of energy and the recycling of nutrients.

ENERGY FLOW THROUGH AN ECOSYSTEM

Every ecosystem is powered by a continuous flow of energy, usually from sunlight. Solar energy is captured and converted to chemical energy by producers (photosynthetic organisms such as plants), which are then eaten by consumers (organisms, such as animals, that eat other organisms). Meanwhile, all organisms release energy out of the system as heat. Maintaining this flow—and therefore maintaining life on Earth—requires a constant input of energy from the sun.

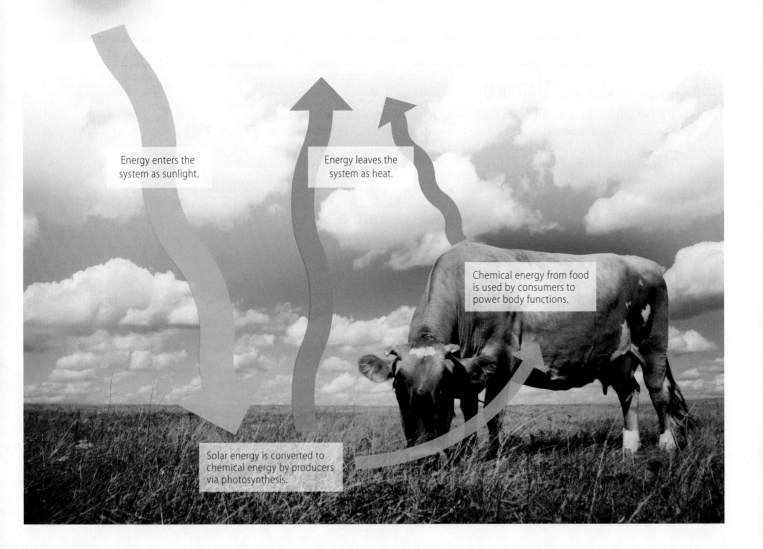

SUN

Energy enters the system as sunlight.

Energy leaves the system as heat.

Chemical energy from food is used by consumers to power body functions.

Solar energy is converted to chemical energy by producers via photosynthesis.

CHEMICAL RECYCLING WITHIN AN ECOSYSTEM

Chemical nutrients cycle within an ecosystem. Elements are absorbed from the environment by producers. For example, CO_2 from the air is used to produce sugars in plants via photosynthesis. These elements are transferred to consumers when the plants are eaten. Within the cells of both producers and consumers, sugars are broken down to provide cellular energy. This process releases CO_2 back into the atmosphere.

CO_2

Within the cells of producers and consumers, sugars are broken down, providing cellular energy and releasing CO_2.

Sugars are absorbed by consumers that eat plants.

CO_2 is converted to sugars and other molecules by producers via photosynthesis.

CORE IDEA: Every organism is part of an ecosystem through which energy flows and within which chemical nutrients are recycled.

CORE QUESTION: What is the primary difference between the movement of energy and the movement of nutrients in an ecosystem?

ANSWER: Energy flows into and out of an ecosystem, whereas nutrients are recycled within an ecosystem.

Biologists organize species into groups

There is a tremendous diversity of life on Earth. Biologists have identified and named nearly 2 million species. How many other species remain undiscovered or unnamed is a matter of pure speculation, but it surely numbers in the many millions, with new species discovered on a daily basis. To help keep track of and organize so many species, biologists have devised an organizational system. The broadest units of biological classification are the domains and kingdoms.

THE DOMAINS OF LIFE

The most comprehensive unit into which life on Earth is classified is the **domain**. Biologists recognize three domains, organized by the type of cell found in each. The first two—Bacteria and Archaea—consist of single-celled organisms with relatively small and simple prokaryotic cells (and so are called **prokaryotes**). The third domain—Eukarya—consists of single-celled and multicelled organisms that have relatively large and structurally sophisticated eukaryotic cells (and so are called **eukaryotes**).

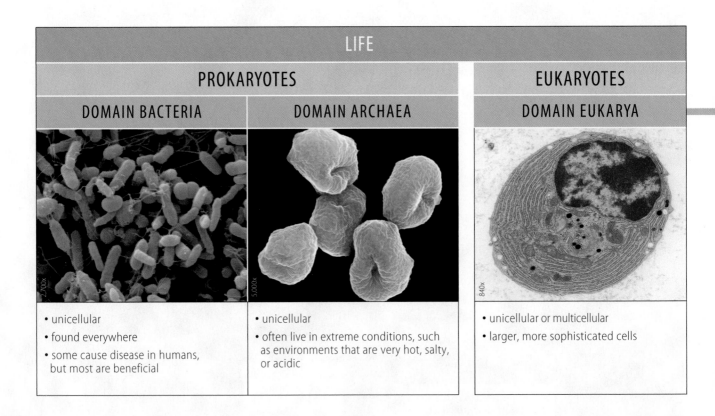

LIFE		
PROKARYOTES		**EUKARYOTES**
DOMAIN BACTERIA	**DOMAIN ARCHAEA**	**DOMAIN EUKARYA**

- unicellular
- found everywhere
- some cause disease in humans, but most are beneficial

- unicellular
- often live in extreme conditions, such as environments that are very hot, salty, or acidic

- unicellular or multicellular
- larger, more sophisticated cells

THE KINGDOMS OF EUKARYOTES

All eukaryotes are placed within the domain Eukarya. This domain is further subdivided into smaller groups called **kingdoms**. Three kingdoms (plants, fungi, and animals) are recognized by how their members obtain dietary energy. All eukaryotes that do not fit into these three kingdoms are grouped into a catch-all category called the **protists**. Most ecosystems, such as the woodland shown below, contain tremendous variety from among all the eukaryotic kingdoms.

DOMAIN EUKARYA			
Kingdom Plantae	Kingdom Fungi	Kingdom Animalia	Protists
• multicellular • use sunlight to produce sugars via photosynthesis	• single-celled or multicellular • decompose and digest dead organisms	• multicellular • eat and digest other organisms	• single-celled or multicellular • catch-all category for all remaining eukaryotes • includes many kingdoms

Fungus: Emetic russula (*Russula emetica*)

Protist: various species found in soil

Animal: European hedgehog (*Erinaceus europaeus*)

Plant: Delicate Fern Moss (*Thuidium delicatulum*)

CORE IDEA: Biologists organize life on Earth into three domains based on types of cells. The domain Eukarya is further divided into three kingdoms (plants, animals, and fungi) and the protists.

CORE QUESTION: To which domain do you belong? To which kingdom?

Evolution by natural selection is biology's unifying theme

Evolution is the core theme of biology. Every level of biological organization, from molecules to ecosystems, can be studied and understood through the lens of evolution. This is because all life on Earth is connected through a shared evolutionary history that stretches back over 3 billion years. This evolutionary view of life was formalized in 1859, when Charles Darwin published *The Origin of Species*. Any study of biology should therefore start with an understanding of Darwin's ideas.

NATURAL SELECTION: UNEQUAL REPRODUCTIVE SUCCESS

In *The Origin of Species*, Darwin laid out a very easy-to-understand logical argument for evolution by natural selection. Darwin argued that a few readily verifiable observations lead to a profound conclusion.

OBSERVATION

EXPONENTIAL GROWTH
Every species has the potential to increase its numbers very rapidly.

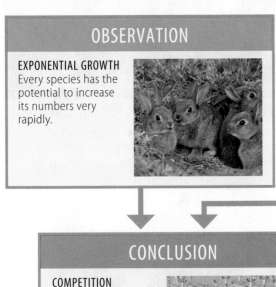

OBSERVATION

RESOURCES
The environment has an essentially fixed amount of resources. The amount of food, shelter, sunlight, etc., does not significantly increase over time.

CONCLUSION

COMPETITION
The overproduction of offspring relative to availability of environmental resources leads to competition among individuals for those limited resources.

OBSERVATION

VARIATION
Individuals of every species vary in many inherited traits. No two individuals are exactly alike.

CONCLUSION

NATURAL SELECTION
In the struggle to survive, which individuals gain available resources is not random: Those with traits better suited to compete in the local environment will, on average, survive and produce more offspring, passing their successful traits to the next generation. Darwin called such nonrandom unequal reproductive success **natural selection**.

Natural selection: nonrandom unequal reproductive success

Under a microscope, your cells are indistinguishable from those of an elephant.

EVOLUTION: DESCENT WITH MODIFICATION

In *The Origin of Species*, Darwin made two main points. First, he proposed **natural selection**: Individuals with traits that make them well suited to compete for available resources will have, on average, more offspring. Second, Darwin proposed that over many generations, continuous natural selection within a population will result in "descent with modification," Darwin's term for **evolution**. Darwin realized that traits that enhance survival will be represented in greater numbers in successive generations. The result is **adaptation**, the accumulation of favorable traits in a population over time. We see such adaptations throughout the natural world. For example, over many generations, the selective pressure for stealthiness while hunting favored the evolution of stripes in the tiger population.

THE DIVERSITY AND UNITY OF LIFE

An understanding of evolution can account for the fact that life on Earth displays both tremendous diversity and underlying unity. Life is highly diverse in terms of the broad range of species. But all life shares a common cellular and molecular basis. Descent from common ancestors, with modifications over time, produces both unity and diversity. For example, although this tropical scene displays a broad variety of species, every organism here has cells with many common features, and all use chemically similar DNA as their genetic material. Why? Because all of these species evolved from a common ancestor.

1,000x

A representative plant cell

900x

A representative animal cell

All cells have DNA made from the same chemical building blocks.

CORE IDEA: Darwin reasoned that competition for resources leads to unequal reproductive success. Over time, natural selection results in evolutionary adaptations, accounting for both the diversity and unity of life on Earth.

CORE QUESTION: Can an individual organism evolve?

ANSWER: No. Evolution requires changes to accumulate during descent through many generations.

Evolution affects our daily lives

As biology's unifying theme, evolution helps to explain a vast range of biological phenomena, from the striking unity among the cells of all living creatures to the tremendous biodiversity we see all around us. In fact, if you stop to think about it, evolution affects your own life in many important ways.

ARTIFICIAL SELECTION

In *The Origin of Species*, Darwin noted that for millennia humans had been substituting our desires for the effects of the natural environment, thereby enforcing unequal reproductive success on species with which we interact. He called this **artificial selection**, the selective breeding of domesticated crops and animals. By choosing to mate individuals with desirable traits (such as large fruit, desirable behaviors, or plentiful meat), humans have enforced our own version of "descent with modification," Darwin's term for evolution. The resulting domesticated breeds, customized to suit our needs, may bear little resemblance to their wild ancestors.

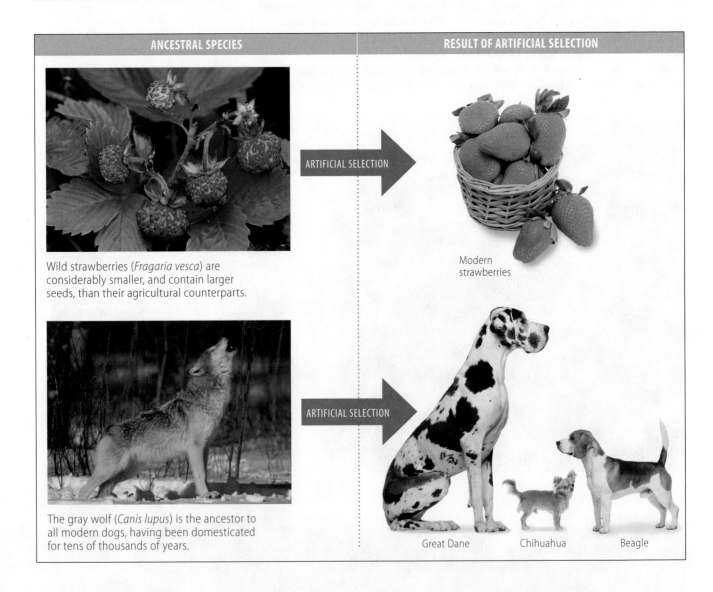

ANCESTRAL SPECIES	RESULT OF ARTIFICIAL SELECTION

ARTIFICIAL SELECTION

Wild strawberries (*Fragaria vesca*) are considerably smaller, and contain larger seeds, than their agricultural counterparts.

Modern strawberries

ARTIFICIAL SELECTION

The gray wolf (*Canis lupus*) is the ancestor to all modern dogs, having been domesticated for tens of thousands of years.

Great Dane　　　Chihuahua　　　Beagle

NATURAL SELECTION IN ACTION

In the more than 150 years since Darwin published his theory of natural selection, countless examples of this process in action have been documented. From bacteria to apple trees to insects to people, evolution is one of biology's best-demonstrated and longest-lasting theories. It is important for you to realize that evolution is not just an abstract concept, it is a force that affects your life in many ways every day.

When first discovered, **antibiotics** (naturally occurring drugs that kill bacteria) were thought to be a cure-all. But in the decades following their introduction, some antibiotics (such as penicillin) have become virtually useless because bacteria have evolved resistance to them. Methicillin-resistant *Staphylococcus aureus* (MRSA) is a particularly dangerous strain of bacteria that resists several antibiotics.

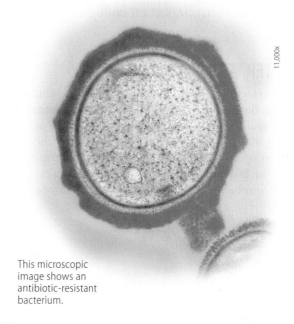

This microscopic image shows an antibiotic-resistant bacterium.

The disease malaria, which kills millions of people in tropical areas, is caused by an infection transmitted by mosquitoes. For decades, organizations such as the United Nations World Health Organization tried to eradicate the host mosquitoes using toxic pesticides such as DDT. Their efforts have been thwarted by the ability of mosquito populations to evolve faster than the pesticides can act. Due to concerns about its effects on human health and the environment, DDT is rarely used today.

Female *Anopheles gambiae* mosquito feeding on human blood

1950s bottle of a DDT insecticide

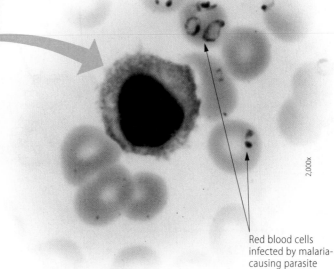

Red blood cells infected by malaria-causing parasite

CORE IDEA: You can witness examples of evolution in action all around you: domesticated pets, foods created via artificial selection, and the effect of natural selection on disease-causing organisms.

CORE QUESTION: Why is the domestication of animals considered "artificial" selection?

ANSWER: Because humans, rather than the environment, decide the reproductive success of individuals.

17

All life is made of molecules, which are made of atoms

Every living organism is, at a fundamental level, a chemical system. Interactions among basic chemical ingredients drive all of life's processes. Viewed this way, you are just a big watery bag of chemicals undergoing countless reactions. Therefore, knowledge of basic **chemistry**, the scientific study of matter, is necessary to understand many important biological processes.

MATTER, ATOMS, AND MOLECULES

Every object in the universe—all the "stuff" around you—is composed of **matter**, defined as anything that occupies space and has mass (substance). On Earth, matter can be in the form of a gas, liquid, or solid, referred to as the three phases of matter. All matter consists of **atoms**, the smallest units that retain all of the properties of their type of matter. In nature, atoms are not often found in isolation. Instead, atoms are usually bonded to each other to form **molecules**. You are probably already familiar with many kinds of molecules.

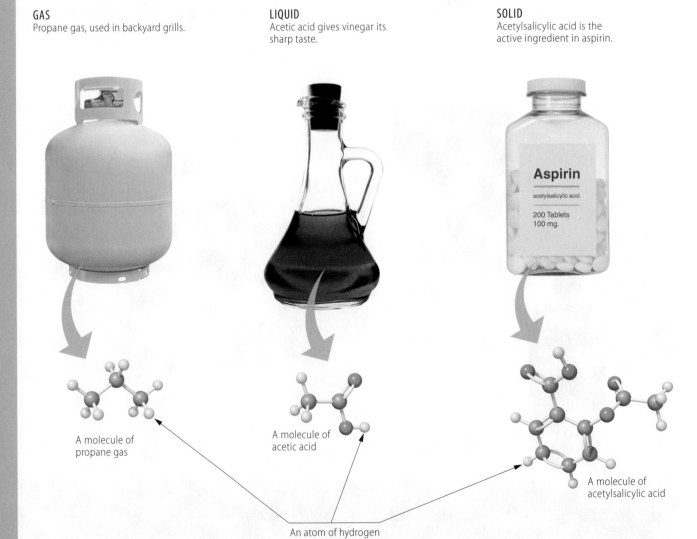

GAS
Propane gas, used in backyard grills.

LIQUID
Acetic acid gives vinegar its sharp taste.

SOLID
Acetylsalicylic acid is the active ingredient in aspirin.

Aspirin
acetylsalicylic acid
200 Tablets
100 mg.

A molecule of propane gas

A molecule of acetic acid

A molecule of acetylsalicylic acid

An atom of hydrogen

It would take over a million carbon atoms to stretch across the period at the end of this sentence.

ELEMENTS AND COMPOUNDS

All matter is composed of individual **elements**, substances that cannot be broken down into other substances by chemical reactions. Examples of elements include hydrogen, carbon, and gold. Although elements are occasionally found by themselves—neon gas that lights up a sign, for example, is just a collection of atoms of the element neon—they more often combine to form **compounds**, substances with two or more elements in a fixed ratio.

SODIUM CHLORIDE

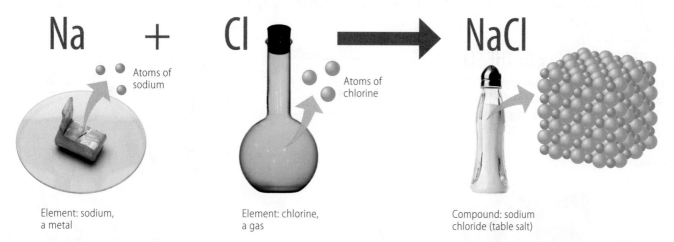

Atoms of sodium

Atoms of chlorine

Element: sodium, a metal

Element: chlorine, a gas

Compound: sodium chloride (table salt)

CHEMICAL REACTIONS

The composition of living matter is constantly changing through **chemical reactions**. During a chemical reaction atoms remain whole, but they are swapped as molecules are broken down and built up. Chemical reactions are written with the **reactants** (starting matter) on the left and the **products** (ending matter) on the right. The arrow between the reactants and the products represents one or more chemical reactions. Notice that chemical reactions rearrange atoms, but atoms are never created nor destroyed. (Go ahead and count them yourself to verify!) This particular series of chemical reactions, called cellular respiration, uses oxygen and a sugar called glucose to provide energy to your living cells.

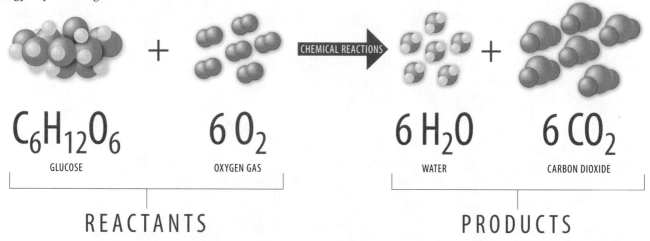

CHEMICAL REACTIONS

$C_6H_{12}O_6$

GLUCOSE

$6\,O_2$

OXYGEN GAS

$6\,H_2O$

WATER

$6\,CO_2$

CARBON DIOXIDE

REACTANTS

PRODUCTS

CORE IDEA: All matter consists of atoms, which are often bonded together into molecules. Individual elements combine to form compounds. Reactants are transformed into products through chemical reactions.

CORE QUESTION: When two molecules of glucose ($C_6H_{12}O_6$) combine, they produce one molecule of water and one maltose. Remembering that atoms are neither created nor destroyed during a chemical reaction, write this chemical reaction.

ANSWER: $C_6H_{12}O_6 + C_6H_{12}O_6 \rightarrow H_2O + C_{12}H_{22}O_{11}$.

All matter consists of chemical elements

All living organisms are composed of matter, and all matter is composed of **elements**, substances that cannot be broken down into other substances by chemical reactions. There are 92 naturally occurring elements on Earth, and several others have been artificially created in the laboratory. Here, you will see examples of familiar elements and learn their importance in living systems.

THE PERIODIC TABLE OF THE ELEMENTS

The **periodic table of the elements** lists all of the chemical elements, ordered by atomic number. Each entry in the table provides important information about one element. In this table, we use colors to highlight the elements that are essential to life.

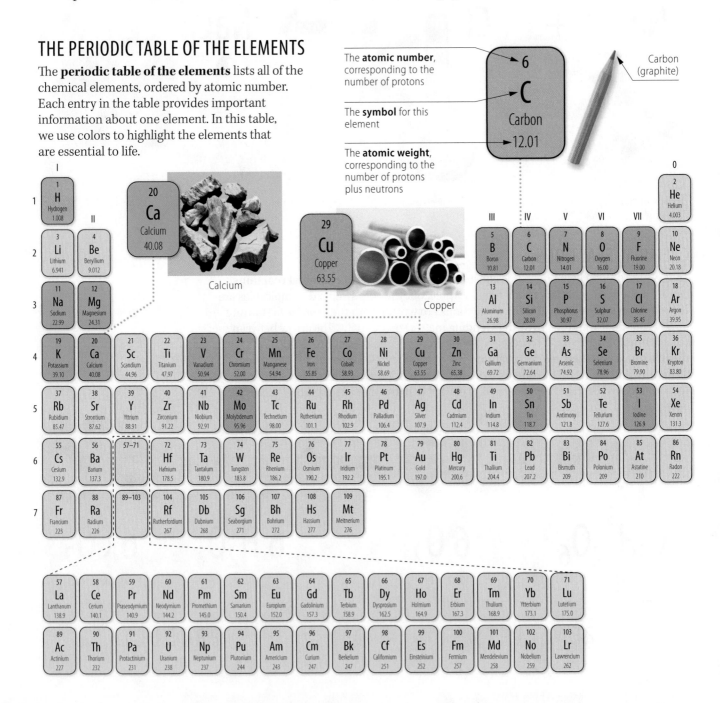

The **atomic number**, corresponding to the number of protons

The **symbol** for this element

The **atomic weight**, corresponding to the number of protons plus neutrons

Carbon (graphite)

Calcium

Copper

ELEMENTS ESSENTIAL TO LIFE

Four of the 92 naturally occurring elements make up the vast majority of matter within living organisms. Another 7 elements account for much of the remaining mass. Finally, 14 **trace elements** are present in very tiny amounts, but cells cannot survive without them.

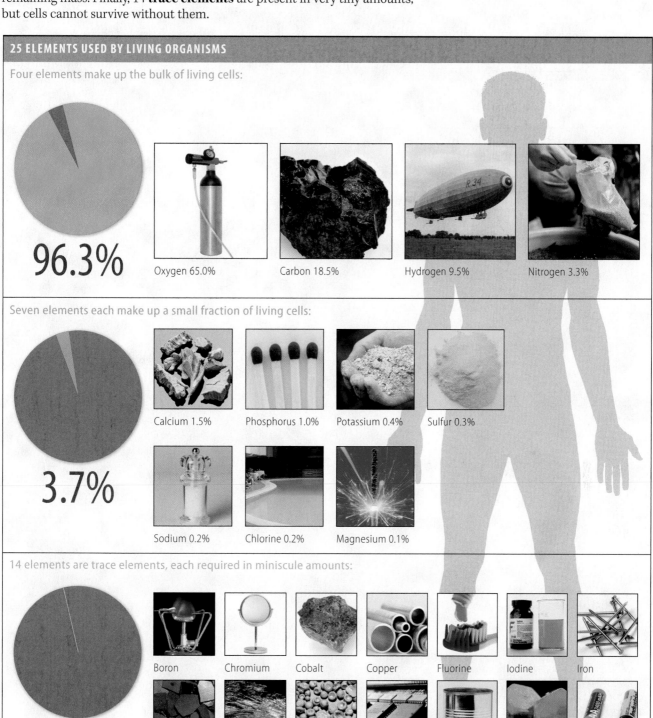

25 ELEMENTS USED BY LIVING ORGANISMS

Four elements make up the bulk of living cells:

96.3%

Oxygen 65.0% Carbon 18.5% Hydrogen 9.5% Nitrogen 3.3%

Seven elements each make up a small fraction of living cells:

3.7%

Calcium 1.5% Phosphorus 1.0% Potassium 0.4% Sulfur 0.3%

Sodium 0.2% Chlorine 0.2% Magnesium 0.1%

14 elements are trace elements, each required in miniscule amounts:

0.1%

Boron Chromium Cobalt Copper Fluorine Iodine Iron

Manganese Molybdenum Selenium Silicon Tin Vanadium Zinc

The iron in fortified breakfast cereal is identical to that in iron nails.

CORE IDEA: All matter is composed of elements. Of the 92 natural elements, only 25 are used by living cells. Of these, four make up the bulk of the cell, seven are required in small quantities, and 14 are required in only tiny amounts.

CORE QUESTION: Is a dietary supplement that contains gold beneficial to your body?

ANSWER: No. Gold is not one of the 25 elements used by living cells.

Atoms are composed of subatomic particles

All matter is composed of atoms. But atoms themselves are composed of even smaller **subatomic particles**. Two of these particles (**neutrons** and **protons**) have about equal mass and are located in the **nucleus** at the center of the atom. The number of protons in an atom determines its chemical element. Particles of the third type (**electrons**) have very little mass and orbit the nucleus at high speeds. Electrons are not found just anywhere around the nucleus; they orbit at specific locations called electron shells.

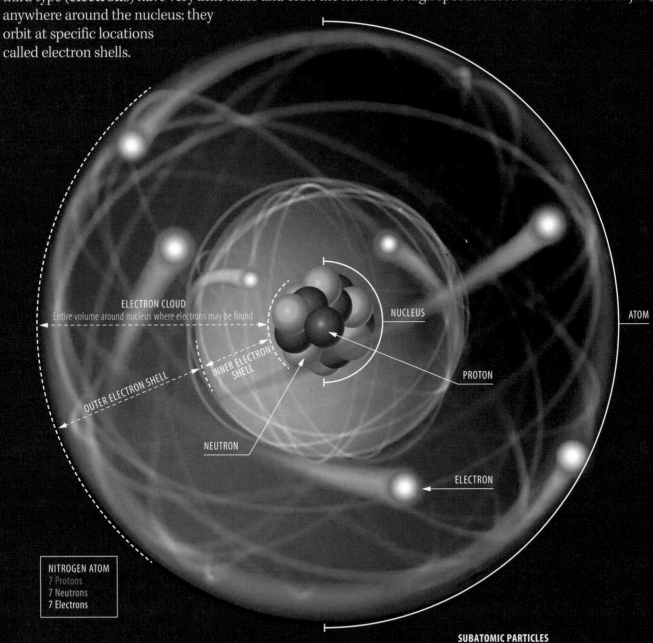

ELECTRON CLOUD
Entire volume around nucleus where electrons may be found

NUCLEUS

ATOM

INNER ELECTRON SHELL

OUTER ELECTRON SHELL

PROTON

NEUTRON

ELECTRON

NITROGEN ATOM
7 Protons
7 Neutrons
7 Electrons

ATOM OF NITROGEN

This model shows the subatomic structure of an atom of nitrogen, with protons and neutrons in the central nucleus and individual fast-moving electrons forming an electron cloud around the nucleus. Electrons orbit the nucleus at specific diameters. Notice that this atom has two electrons in the inner shell and five electrons in the outer shell. The electrons in the outer shell may interact with the electrons of other atoms, determining the chemical properties of the atom.

SUBATOMIC PARTICLES

PARTICLE	MASS	CHAR
PROTON *Protons determine element.*	1	+1
NEUTRON *Neutrons determine isotope.*	1	0
ELECTRON *Electrons determine ion.*	0	−1

An electron has only about 1/1,000th the mass of a proton or neutron.

NITROGEN-15 ISOTOPE

This model shows the subatomic structure of the most common isotope of nitrogen. **Isotopes** vary in the number of neutrons in the nucleus. The "15" in the name of the isotope refers to the atomic weight (protons plus neutrons, in this case 7 protons plus 8 neutrons); the name of this isotope may also be written as "N-15" or "^{15}N." Because protons determine the identity of the element, isotopes of the same element vary in the number of neutrons.

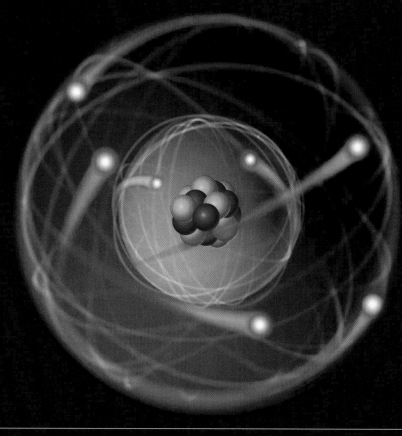

NITROGEN-15 ISOTOPE
7 Protons
8 Neutrons
7 Electrons

NITROGEN ION

This model shows the subatomic structure of a common ion of nitrogen: N^{3-}. **Ions** vary in the number of electrons in the electron cloud. Atoms are neutrally charged (with the same number of protons as electrons), but ions carry a charge. The name of the ion indicates its total charge, with one extra negative charge for each extra electron, or one extra positive charge for each missing electron.

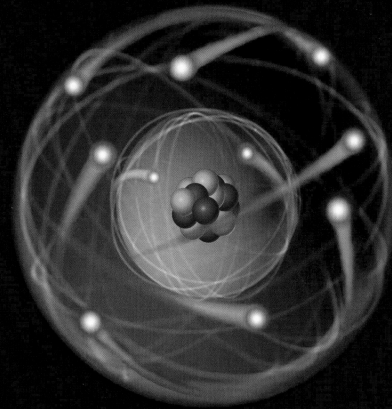

NITROGEN ION
7 Protons
7 Neutrons
10 Electrons

CORE IDEA: All atoms are composed of protons, neutrons, and electrons. The number of protons determines the element; the number of neutrons determines the isotope; and the number of electrons determines the ion state and chemical reactivity.

CORE QUESTION: Carbon always has six protons. How many neutrons and electrons are there in a C-14 isotope?

ANSWER: Eight neutrons and six electrons.

Atoms are held together by chemical bonds

Within every living cell, countless chemical reactions between molecules drive the processes of life. During a chemical reaction, atoms gain, release, or share electrons with other atoms. As they do, the atoms involved may become attracted to each other and be held together by **chemical bonds**. Here, you can survey the different kinds of chemical bonds to see how they are formed.

CHEMICAL BONDS

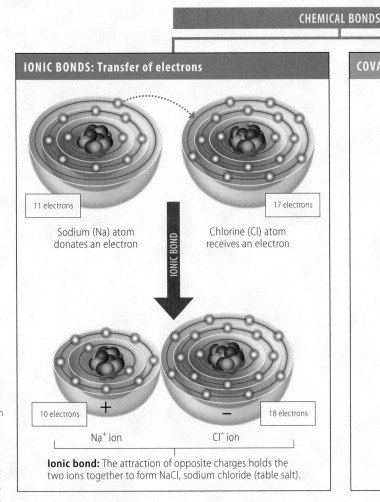

IONIC BONDS: Transfer of electrons

11 electrons

17 electrons

Sodium (Na) atom donates an electron

Chlorine (Cl) atom receives an electron

IONIC BOND

10 electrons

+

Na^+ ion

−

18 electrons

Cl^- ion

Ionic bond: The attraction of opposite charges holds the two ions together to form NaCl, sodium chloride (table salt).

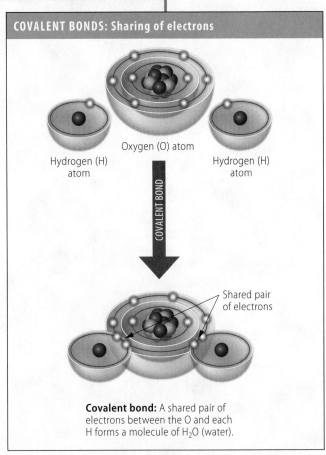

COVALENT BONDS: Sharing of electrons

Hydrogen (H) atom

Oxygen (O) atom

Hydrogen (H) atom

COVALENT BOND

Shared pair of electrons

Covalent bond: A shared pair of electrons between the O and each H forms a molecule of H_2O (water).

IONIC BONDS

Ionic bonds involve the transfer of one or more electrons from one atom to another. Because electrons are negatively charged, the atom that receives the electron(s) becomes negatively charged, while the atom that donates the electron(s) becomes positively charged. Both atoms are now **ions**, an atom or group of atoms that has acquired a charge by the gain or loss of electrons. The two ions in an ionic bond are held together by the attraction of their opposite charges, in a manner similar to the way magnets stick together.

COVALENT BONDS

Covalent bonds involve the sharing of one or more electrons between atoms. In forming a covalent bond, each atom shares one electron with the other. The bond therefore consists of a pair of shared electrons, one from each atom involved. Covalent bonds hold atoms together in molecules. A single shared pair of electrons is called a **single bond** and is represented in a chemical structure by a solid line. If atoms share two pairs of electrons, they form a **double bond**, shown as a pair of solid lines, as in hydrogen gas (H_2): H=H. Three pairs of electrons are shared in triple bonds, which are relatively rare.

Coal and diamonds are both made from pure carbon, but they differ in how the atoms are bonded together.

POLAR VERSUS NONPOLAR BONDS

All covalent bonds involve shared pairs of electrons, but the sharing may be equal or unequal. If the electrons are shared equally between the two atoms in the bond, the result is a **nonpolar bond**. If electrons are more strongly attracted to one atom than the other, the result is a **polar bond**. Because the electrons are no longer evenly spaced between the two atoms, a polar bond has one slightly positive pole (end) and one slightly negative pole, although the molecule overall remains neutrally charged.

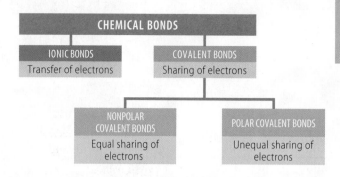

NONPOLAR COVALENT BONDS: Equal sharing of electrons

CH₄: methane (natural gas)

Nonpolar bond: equal sharing of electrons between C and H atoms

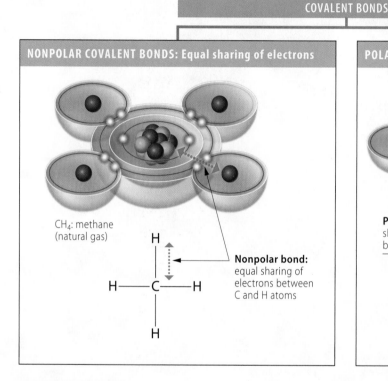

POLAR COVALENT BONDS: Unequal sharing of electrons

Due to its chemical nature, the oxygen nucleus more strongly attracts the electrons than the hydrogen nucleus does.

H₂O: water

Polar bond: unequal sharing of electrons between O and H atoms

The oxygen atom carries a slight negative charge.

Polar bond

The hydrogens each carry a slight positive charge.

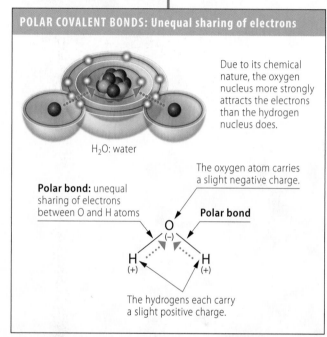

HYDROGEN BONDS

A molecule of water contains two polar covalent bonds. Because the electrons in these bonds are more attracted to the oxygen nucleus than the hydrogen nucleus, the oxygen atom carries a slightly negative charge, while the hydrogen atoms each carry slightly positive charges. Water molecules tend to align themselves so that a negatively charged oxygen faces a positively charged hydrogen from another water molecule. The resulting bond, which is called a **hydrogen bond**, is fairly weak by itself. However, one water molecule may participate in several hydrogen bonds, and the resulting networks of hydrogen bonds are strong enough to form liquid water. There are many other examples of hydrogen bonds, sometimes occurring between molecules (as in water) and sometimes occurring between atoms within the same molecule (as in many proteins).

HYDROGEN BONDS: Attractions between opposite charges

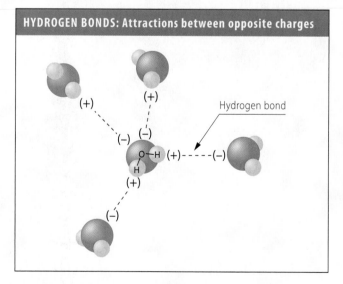

Hydrogen bond

CORE IDEA: Chemical bonds can be ionic (via electron transfer) or covalent (via electron sharing). Covalent bonds can be nonpolar (equal sharing) or polar (unequal sharing). Hydrogen bonds are important in the structure of water.

CORE QUESTION: What kind of bond holds the atoms together *within* a molecule of water? What kind of bond acts *between* molecules of water to hold them together?

ANSWER: Polar covalent bonds; hydrogen bonds.

The structure of water gives it unique properties

All life depends on water. In fact, life first appeared in water and evolved there for billions of years before moving on to land. Additionally, nearly all cells are, by weight, mostly water. What is it about water that makes it so central to life's processes? Water has special properties that are a consequence of its unique chemical structure.

HYDROGEN BONDING IN WATER

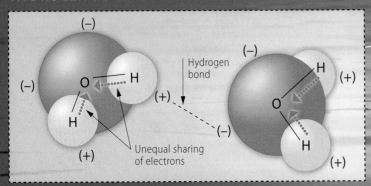

For each O—H bond in a molecule of water, one pair of electrons is shared between the two atoms. But this sharing is unequal: The electrons are more attracted to the O nucleus than they are to the H nucleus. As a result, water is a polar (unequally charged) molecule, with a slightly negative charge on its oxygen and slightly positive charges on its hydrogens. This allows a molecule of water to form hydrogen bonds with one or several other water molecules. Liquid water contains extensive networks of hydrogen bonds between the individual molecules.

ICE FLOATING

Unlike nearly every other liquid, when water molecules freeze, they move apart, forming a rigid network of hydrogen bonds. It is as if each water molecule holds all others at "arm's length." Thus, water expands when it freezes, so ice is less dense than water. As a result, ice floats. This is extremely biologically relevant: During winter, a thin layer of floating ice insulates the water below the surface, allowing life to survive until the spring thaw.

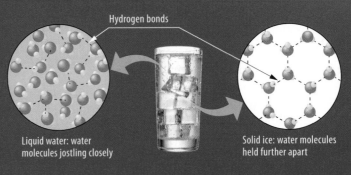

Hydrogen bonds

Liquid water: water molecules jostling closely

Solid ice: water molecules held further apart

WATER AS A SOLVENT

Many liquids can act as a **solvent**, a dissolving agent, to form a mixture called a **solution**. Due to its polar structure, water is an extremely effective solvent, able to dissolve just about anything. Within cells, water's polar nature allows it to hold many substances in solution, making them available for life's processes.

TEMPERATURE REGULATION

Liquid water readily absorbs and releases heat. Water therefore resists temperature changes more than most substances. The presence of water can thereby act to moderate temperatures. On a global scale, the oceans help moderate the temperature of Earth's surface, keeping it within a livable range. On a more personal scale, sweating helps moderate your temperature by cooling off your skin.

COHESION

Due to hydrogen bonding, water molecules have a strong tendency to stick to each other, a property called **cohesion**. The cumulative effect of all these hydrogen bonds is to create surface tension, a film-like surface on which items (such as this wolf spider) can be suspended.

CORE IDEA: The polar nature of water molecules allows them to form networks of hydrogen bonds. This special ability endows water with many life-supporting properties.

CORE QUESTION: What force allows a "water walker" insect to stay on top of a pond?

ANSWER: A network of hydrogen bonds produces cohesion between water molecules within liquid water.

pH is a measure of the acidity of a solution

Most of the chemical reactions that maintain life occur in water. An **aqueous solution** is one that contains a substance dissolved in water. Within any aqueous solution, a small percentage of the water molecules break apart into hydrogen ions (H^+) and hydroxide ions (OH^-). The concentration of H^+ ions in an aqueous solution determines its pH. The **pH scale** runs from 0 (most acidic) to 14 (most basic), with 7 as neutral. Each number in the pH scale represents a tenfold change in H^+ ion concentration.

pH SCALE

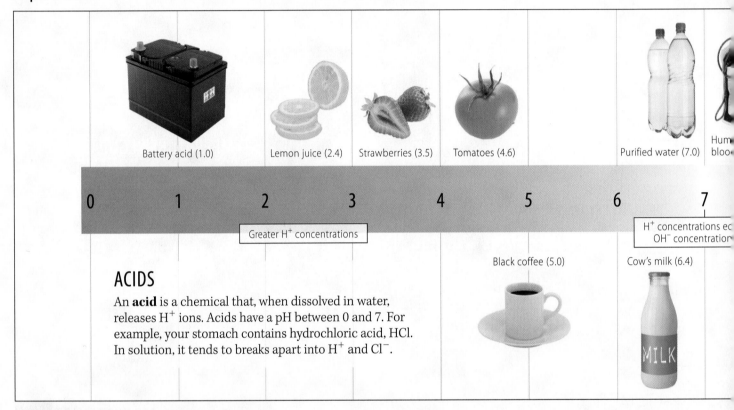

Battery acid (1.0) · Lemon juice (2.4) · Strawberries (3.5) · Tomatoes (4.6) · Purified water (7.0) · Hum blood

0 1 2 3 4 5 6 7

Greater H^+ concentrations

H^+ concentrations equ OH$^-$ concentration

Black coffee (5.0) · Cow's milk (6.4)

ACIDS

An **acid** is a chemical that, when dissolved in water, releases H^+ ions. Acids have a pH between 0 and 7. For example, your stomach contains hydrochloric acid, HCl. In solution, it tends to breaks apart into H^+ and Cl^-.

ACID PRECIPITATION

Burning fossil fuels releases chemicals that react with water in the air to form strong acids. As they fall back to Earth in the form of snow, rain, or fog, this acid precipitation damages lakes, streams, forests, and soil. The U.S. Clean Air Act has been effective at reducing acid precipitation.

BUFFERS

Most cells regulate their pH through the use of **buffers**, chemicals that minimize changes in pH by accepting H^+ ions when they are present in excess and donating H^+ ions when they are in short supply. For example, within human blood and other fluids, there are several different types of buffers that keep the body's solutions at a nearly neutral pH, despite changes in the concentration of H^+. For instance, buffers within your blood counteract a drop in pH that occurs whenever you exercise.

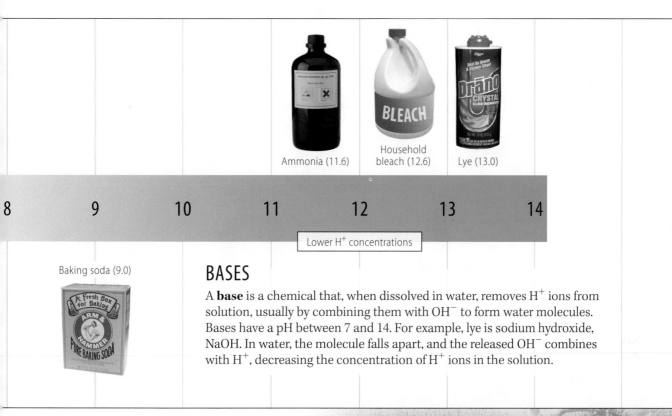

Ammonia (11.6)

Household bleach (12.6)

Lye (13.0)

| 8 | 9 | 10 | 11 | 12 | 13 | 14 |

Lower H^+ concentrations

Baking soda (9.0)

BASES

A **base** is a chemical that, when dissolved in water, removes H^+ ions from solution, usually by combining them with OH^- to form water molecules. Bases have a pH between 7 and 14. For example, lye is sodium hydroxide, NaOH. In water, the molecule falls apart, and the released OH^- combines with H^+, decreasing the concentration of H^+ ions in the solution.

OCEAN ACIDIFICATION

As CO_2 levels rise in the atmosphere, about 25% of the excess CO_2 is absorbed by the oceans. As it dissolves, CO_2 undergoes a chemical reaction that lowers the pH of the ocean. This in turn damages coral reefs and other ecosystems by limiting the ability of organisms to perform the chemical reactions used to build their skeletons or shells.

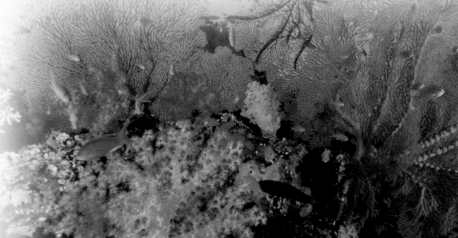

CORE IDEA: The concentration of H^+ ions in an aqueous solution determines the pH, from acidic (0 to 7) to basic (7 to 14). Buffers can help reduce changes in pH. pH changes in the environment can affect the health of ecosystems.

CORE QUESTION: How does the pH of lemonade compare to the pH of water? How does the concentration of H^+ ions compare between the two?

ANSWER: Lemonade is more acidic and so will have a lower pH, meaning that more free H^+ ions are dissolved in lemonade than in water.

All life on Earth is based on carbon

Perhaps you have heard that all organisms on Earth are "carbon-based life-forms." That is because, besides water, most of the molecules that make up living matter are **organic compounds**, molecules that contain carbon bonded to other elements. Why is life carbon-based and not, say, nitrogen-based? Unlike other atoms, because it can bond with as many as four other atoms, carbon is able to form large, highly branched, diverse chains that can serve as the basic skeletons for a wide variety of chemical compounds. Here, you can see a survey of some of the many carbon compounds that are found in living matter.

CARBON SKELETONS

Every organic compound contains a skeleton of carbon atoms. Notice that each carbon atom forms four bonds to other atoms. Carbon skeletons vary in length and branching pattern, and some fuse together to form rings.

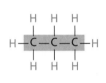

3-carbon skeleton

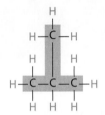

Branched carbon skeleton

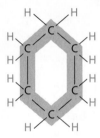

Ring carbon skeleton

FUNCTIONAL GROUPS

Every organic compound has a carbon skeleton, but most also contain one or more **functional groups**, sets of atoms that are attached to the carbon skeleton. Because functional groups participate in chemical reactions, they often determine the overall properties of an organic compound. Here, you can see some of the biologically important functional groups—highlighted in color—and some common molecules in which they are found.

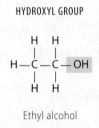

HYDROXYL GROUP

Ethyl alcohol

AMINO GROUP

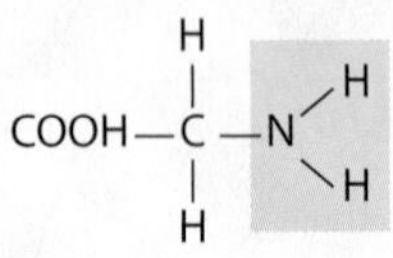

Amino acid

PHOSPHATE GROUP

Adenosine triphosphate (ATP)

Cellulose is the most abundant chemical compound on Earth.

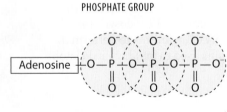

BIOLOGICALLY IMPORTANT ORGANIC COMPOUNDS

Organic compounds—all with carbon skeletons, and most with functional groups attached—form a huge variety of biologically important molecules. There are four classes of large organic molecules—carbohydrates, lipids, proteins, and nucleic acids—that are particularly important to life on Earth.

CLASS	EXAMPLES	
CARBOHYDRATES	**Cellulose** is a large, complex carbohydrate that forms much of the structure of a plant.	**Glucose** is a sugar (a simple carbohydrate) that acts as an energy source for all living cells.
LIPIDS	**Coconut oil** is a lipid that is rich in fat and serves as an important dietary staple in much of the tropical world.	**Cholesterol** is a lipid that circulates in the bloodstream and acts as a molecular ingredient to make steroid hormones.
PROTEINS	**Hexokinase** is an enzyme, a protein that helps drive a chemical reaction, found in most living cells.	**Keratin** is a structural protein found in hair, nails, and skin.
NUCLEIC ACIDS	**DNA** is a nucleic acid that serves as the hereditary material of all life on Earth.	**RNA** is a nucleic acid that acts as a messenger between DNA and other parts of the cell. It is found in all types of cells.

CORE IDEA: All organisms have an abundance of organic compounds, consisting of carbon skeletons that may have functional groups attached. There are four classes of large organic compounds that are particularly important to life.

CORE QUESTION: What property of carbon makes it well suited to form the basis of life's molecules?

ANSWER: Each atom of carbon forms bonds with four other atoms, allowing it to be the base of complex molecules.

Most biological macromolecules are polymers

The majority of your body weight is water. Most of the rest consists of **macromolecules**, large molecules that can have complex structures. Despite their complexity, the structures of most macromolecules are fairly straightforward since they are made from smaller building blocks. Although the different classes of macromolecules vary in structure and function, they are all built up and broken down via similar chemical reactions.

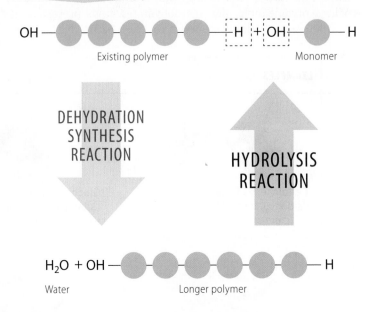

OH — ⬤ — ⬤ — ⬤ — ⬤ — ⬤ — [H] + [OH] — ⬤ — H

Existing polymer   Monomer

DEHYDRATION SYNTHESIS REACTION

HYDROLYSIS REACTION

H_2O + OH — ⬤ — ⬤ — ⬤ — ⬤ — ⬤ — H

Water   Longer polymer

HYDROLYSIS REACTIONS

Most of the organic macromolecules in living cells are **polymers**, large molecules made by joining many smaller molecules called **monomers**. Polymers can be broken down into the monomers that make them up via **hydrolysis reactions**. During hydrolysis, a water molecule is split, and its atoms are used to separate a monomer from the rest of the chain (see the dashed boxes in the figure above).

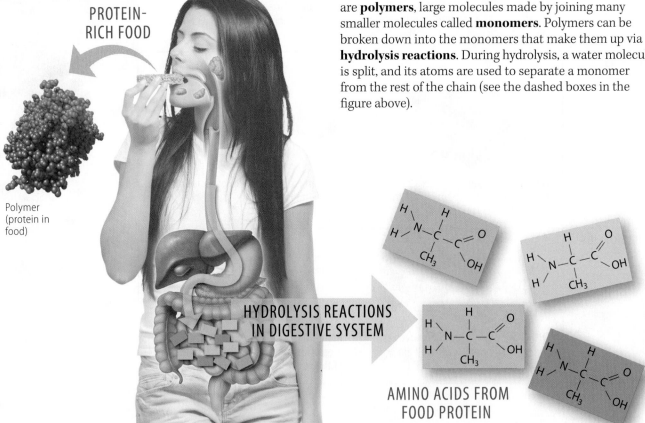

PROTEIN-RICH FOOD

Polymer (protein in food)

HYDROLYSIS REACTIONS IN DIGESTIVE SYSTEM

AMINO ACIDS FROM FOOD PROTEIN

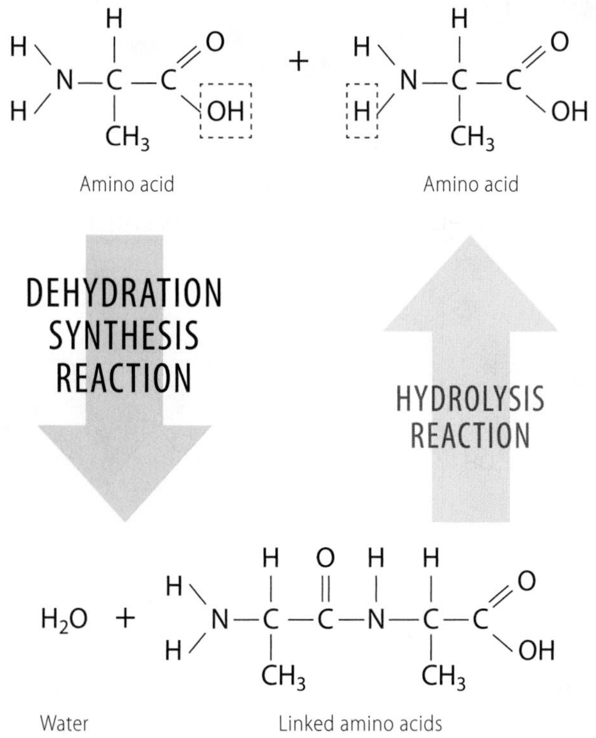

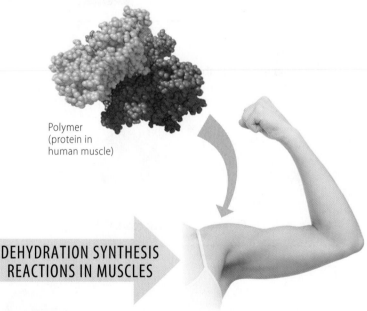

H—N—C—C + H—N—C—C

(Amino acid)   +   (Amino acid)

Amino acid   Amino acid

DEHYDRATION
SYNTHESIS
REACTION

HYDROLYSIS
REACTION

H_2O +

Water   Linked amino acids

DEHYDRATION SYNTHESIS REACTIONS

Monomer building blocks are linked together to form larger polymers through a chemical reaction called a **dehydration synthesis reaction**. As each new monomer is added to a chain, a hydrogen atom (H) from one monomer and a hydroxyl group (OH) from another monomer (see the dashed boxes in the figure to the left) are removed, creating a new chemical bond between the two monomers and releasing a molecule of water (H_2O). One molecule of water is released for each monomer added to the chain. Notice that a dehydration synthesis reaction (which builds up polymers) is the opposite of a hydrolysis reaction (which breaks down polymers).

METABOLISM

Your **metabolism** is the sum total of all the chemical reactions that take place in your body. Many important metabolic reactions involve the breaking down and building up of polymers. Your digestive system breaks down the macromolecules you eat (protein in peanut butter, for example) into the monomers that make them up (in this case amino acids, the building blocks of all proteins). Your cells then use these monomer building blocks to construct new polymers (by building new muscle proteins, for example).

Polymer
(protein in
human muscle)

AMINO ACIDS

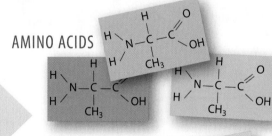

DEHYDRATION SYNTHESIS
REACTIONS IN MUSCLES

MONOMERS TRANSPORTED
THROUGH BODY VIA
CIRCULATORY SYSTEM

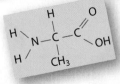

CORE IDEA: Macromolecules (large molecules) are often polymers made by joining together monomers via dehydration synthesis reactions. Polymers can be broken down into the monomers that make them up via hydrolysis reactions.

CORE QUESTION: Why are the building-up reactions called dehydration reactions?

ANSWER: Because a molecule of water is removed (dehydrated) from the molecules involved.

33

Carbohydrates are composed of monosaccharides

Carbohydrates—commonly known as "carbs"—include sugars and large molecules made from sugars. All carbohydrates are molecules constructed from one or more monosaccharides (simple sugars). Carbohydrates are a common source of dietary energy for animals and a structural component of plants.

MONOSACCHARIDES

Monosaccharides are the building blocks of carbohydrates. Every carbohydrate consists of one or more monosaccharides. Notice that two of the most common monosaccharides—glucose and fructose—are **isomers**, meaning they have the same numbers and kinds of atoms but vary in the arrangement of atoms.

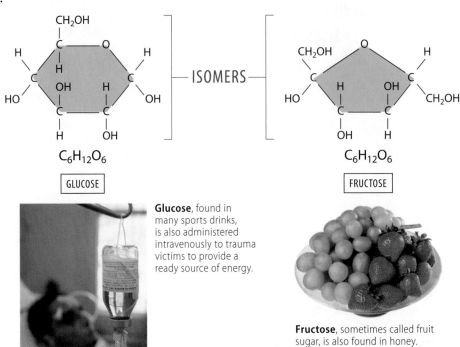

ISOMERS

$C_6H_{12}O_6$

GLUCOSE

$C_6H_{12}O_6$

FRUCTOSE

Glucose, found in many sports drinks, is also administered intravenously to trauma victims to provide a ready source of energy.

Fructose, sometimes called fruit sugar, is also found in honey.

DISACCHARIDES

A **disaccharide** is a double sugar formed by joining two monosaccharides. Notice that many sugars have names that end in the suffix "-ose," making them easy to recognize on a food label.

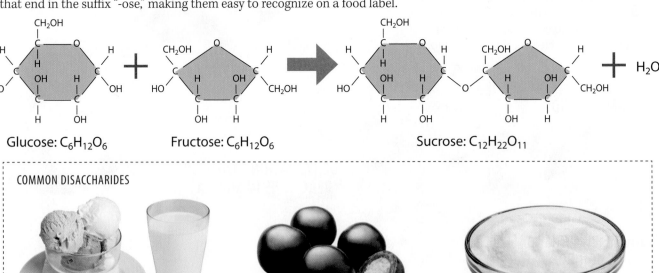

Glucose: $C_6H_{12}O_6$ Fructose: $C_6H_{12}O_6$ Sucrose: $C_{12}H_{22}O_{11}$ H_2O

COMMON DISACCHARIDES

Lactose: also known as milk sugar

Maltose: used for brewing and in malted milk candy

Sucrose: common table sugar

POLYSACCHARIDES

A **polysaccharide** is a complex carbohydrate made by joining many monosaccharides together into a long chain. Here you can see three familiar polysaccharides, all of which are made from long chains of glucose.

STARCH

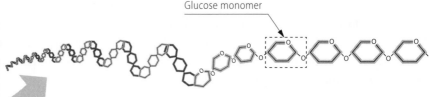

Glucose monomer

Plants store excess sugar as the polysaccharide **starch**, consisting of long, twisted, unbranched chains of glucose molecules. Many animals consume starch as a source of dietary energy.

CELLULOSE

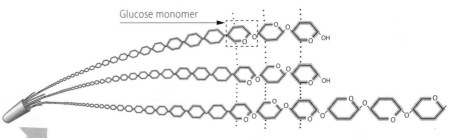

Glucose monomer

Cellulose makes up much of a plant's body in the form of cable-like fibers in the walls of plant cells. A molecule of cellulose contains many long straight chains of glucose, with bonds joining the chains. Cellulose is the major component of wood, and it is the "fiber" in your diet.

GLYCOGEN

Glucose monomer

Glycogen consists of branched chains of glucose molecules. In most animals, excess sugar is stored as glycogen granules in liver and muscle cells, where it is available if needed for about 24 hours.

The majority of sugar used in the United States comes not from sugar cane but from sugar beets.

CORE IDEA: Carbohydrates consist of one or more monosaccharides joined together. Simple sugars (monosaccharides) include glucose and fructose; disaccharides include sucrose; and polysaccharides include starch and cellulose.

CORE QUESTION: What do starch and cellulose have in common in terms of their structure? How are they different?

ANSWER: They are both large chains made by joining many glucose molecules together, but how the molecules are joined differs.

Lipids are a diverse group of hydrophobic molecules

Lipids are a very diverse group of organic compounds, but they all share one important property: All lipids are **hydrophobic** ("water-fearing"), meaning they do not mix with water. You can see this chemical behavior yourself in the way that salad oil (a lipid) and vinegar (which is water based) stay separated. Even if you vigorously shake them together, the hydrophobic oil soon separates from the watery vinegar. There are several kinds of lipids, many of which are already familiar to you.

Oil and vinegar

PHOSPHOLIPIDS

Every living cell is surrounded by a membrane that helps regulate the passage of materials into and out of the cell. These membranes are **phospholipid bilayers**; that is, they are made by stacking two layers of a molecule called a phospholipid. Each **phospholipid** contains a phosphate group in its hydrophilic ("water-loving") head, and two long hydrophobic tails. Many proteins float within the phospholipid layers.

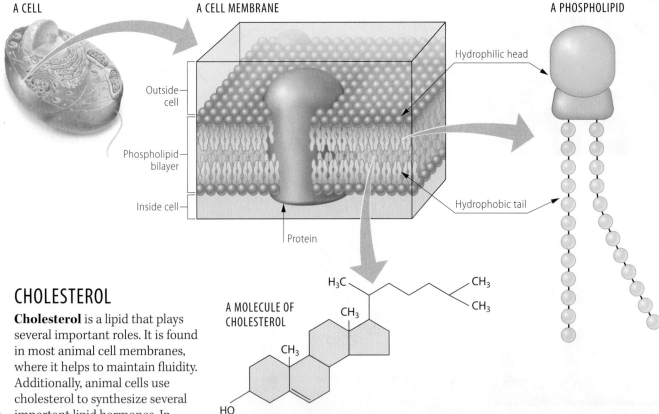

A CELL

A CELL MEMBRANE

A PHOSPHOLIPID

Outside cell

Phospholipid bilayer

Inside cell

Protein

Hydrophilic head

Hydrophobic tail

A MOLECULE OF CHOLESTEROL

CHOLESTEROL

Cholesterol is a lipid that plays several important roles. It is found in most animal cell membranes, where it helps to maintain fluidity. Additionally, animal cells use cholesterol to synthesize several important lipid hormones. In addition to being produced in your body, cholesterol is found in animal-derived foods such as eggs and red meat.

TYPES OF CHOLESTEROL			
Name	What it stands for	Common name	Health factors
LDL	Low-density lipoprotein	"Bad cholesterol"	Levels can be increased through poor diet
HDL	High-density lipoprotein	"Good cholesterol"	Levels can be increased through exercise

● One pound of body fat stores 3,500 calories.

TRIGLYCERIDES

Lipids include fats, and when you think of "fat" you probably think of your diet. A typical dietary fat consists of a molecule called a **triglyceride**, which is made from one molecule of glycerol joined to three fatty acid molecules. The carbon/hydrogen chains in the **fatty acid** tails store a lot of energy. Fatty foods therefore have many calories, and excess calories in the body are stored by adding triglycerides to adipose tissue, also called body fat.

A MOLECULE OF TRIGLYCERIDE

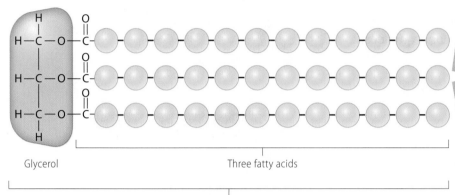

Glycerol

Three fatty acids

One molecule of triglyceride

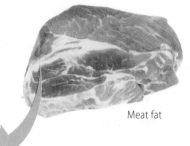

Meat fat

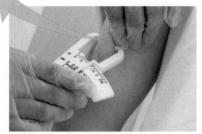

Human fatty tissue

STEROID HORMONES

Steroids are lipids that contain four fused chemical rings made primarily of carbon. Cholesterol is one familiar steroid. Within the body, cholesterol is used to produce a variety of steroid hormones, such as the sex hormones estrogen and testosterone. **Anabolic steroids** are synthetic variants of testosterone that mimic its effects, increasing body mass, but also causing potentially dangerous side effects. Many athletes, including former baseball player Mark McGwire and sprinter Marion Jones, have admitted to using these performance-enhancing drugs.

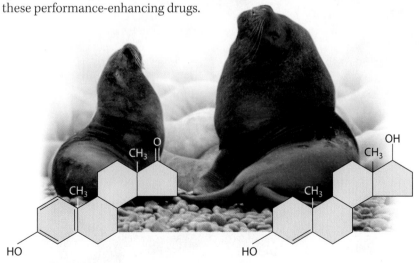

A type of estrogen

Testosterone

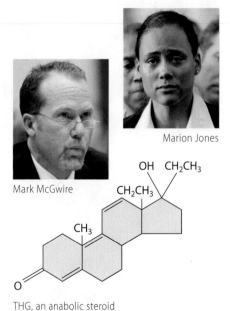

Marion Jones

Mark McGwire

THG, an anabolic steroid

CORE IDEA: Lipids are a diverse group of hydrophobic organic compounds. Lipids include phospholipids, cholesterol, triglycerides, and steroid hormones.

CORE QUESTION: What physical property is shared by dietary fats, cholesterol, and anabolic steroids?

ANSWER: They are all hydrophobic.

Your diet contains several different kinds of fats

Most of the lipids that you consume in your diet consist of molecules of triglyceride. Each triglyceride molecule contains three long fatty acid chains connected to a molecule of glycerol. Although the different kinds of dietary fats share structural features, there are important differences between the various types that have health consequences.

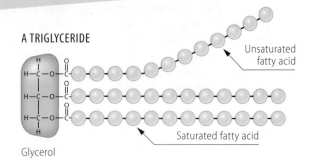

A TRIGLYCERIDE

Unsaturated fatty acid

Saturated fatty acid

Glycerol

SATURATED FATS

Dietary fats come in two basic varieties: saturated fats and unsaturated fats. **Saturated fats** have the maximum number of hydrogens along the fatty acid tail, which corresponds to all single chemical bonds in the chain. As a result, the fatty acid tails in saturated fats are straight. This shape allows them to easily stack together and form solids, so highly saturated fats tend to be solid at room temperature. Saturated fats are found in highest quantities (but not exclusively) in animal products.

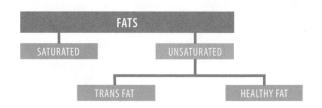

FATS

SATURATED | UNSATURATED

TRANS FAT | HEALTHY FAT

Single bond

- Maximum number of hydrogens in tail
- All single bonds
- Straight shape
- Forms solids at room temperature
- Found in higher amounts in animal products
- Less healthy

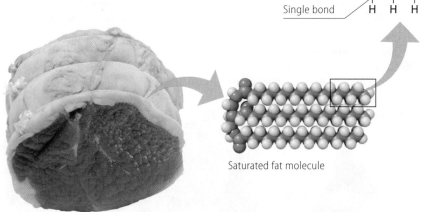

Saturated fat molecule

Beef

Butter

Butter contains a high percentage of saturated fat.

Unsaturated fat

Saturated fat

Cheese

UNSATURATED FATS

Unsaturated fats have one or more double bonds in the fatty acid tail, causing them to have fewer than the maximum number of hydrogens. As a result, the fatty acid tails of unsaturated fats have a bend or kink at each double bond, giving the molecule a twisted shape. This prevents them from stacking easily, so unsaturated fats tend to be liquid at room temperature. Unsaturated fats are found in greatest quantities (but not exclusively) in vegetable and fish oils.

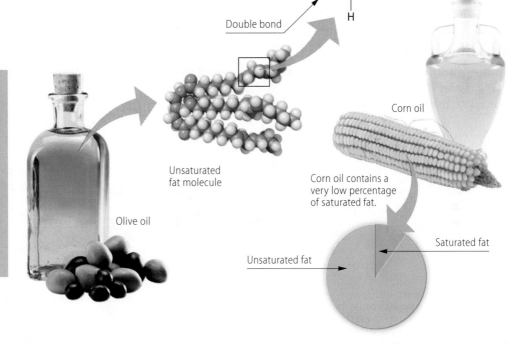

Double bond

Olive oil

Unsaturated fat molecule

Corn oil

Corn oil contains a very low percentage of saturated fat.

Saturated fat

Unsaturated fat

- Less than the maximum number of hydrogens in tail
- One or more double bonds
- Bent shape
- Stays liquid at room temperature
- Found in higher amounts in plant products
- More healthy

A pound of fat packs more than twice as much energy as a pound of carbohydrate.

TRANS FATS

An unsaturated fat can be rendered solid through a manufacturing process called **hydrogenation**. The hydrogenation process can produce **trans fat**, a type of unsaturated fat that contains an unusual bond that does not occur naturally. Trans fats are quite unhealthy, and they must be explicitly listed on food labels. Many food manufacturers are phasing out the use of trans fats.

Processed foods may contain trans fats.

HEALTHY FATS

Not all fats are unhealthy. In fact, some fats are essential to a healthy diet. For example, fats containing **omega-3 fatty acids** are known to reduce the risk of heart disease. Foods rich in essential fats should be part of a well-balanced diet.

Selection of omega-3–rich foods: fish, chicken, eggs, peanuts, and beans

CORE IDEA: Lipids include dietary fats: saturated fats and unsaturated fats (including trans fats and omega-3 fats). Unsaturated fats tend to be healthier than saturated fats, but trans unsaturated fats are particularly unhealthy.

CORE QUESTION: What is wrong with this statement: Animal fats are saturated, whereas plant fats are unsaturated.

ANSWER: Animal and plant fats each contain both saturated and unsaturated fats, but in different proportions.

Proteins perform many of life's functions

Of all the large biological molecules that play important roles in your body, **proteins** are the most diverse. Each kind of protein in a cell has a unique structure and shape that allows it to perform a specific function. Here, you can see the chemical composition of proteins and survey a variety of the functions they perform.

PROTEIN STRUCTURE

All proteins are polymers made by joining many **amino acid** monomers together. There are 20 different kinds of amino acids. The specific order of the amino acids within a protein determines the overall structure of that protein.

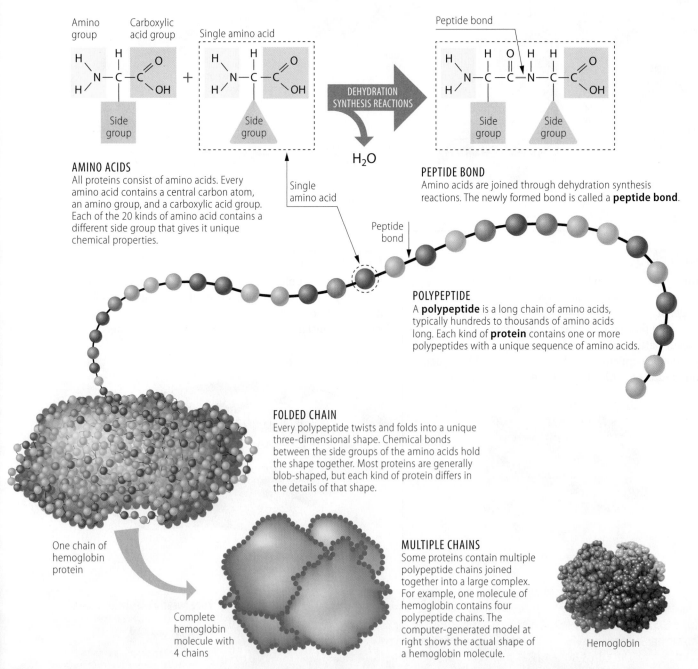

AMINO ACIDS
All proteins consist of amino acids. Every amino acid contains a central carbon atom, an amino group, and a carboxylic acid group. Each of the 20 kinds of amino acid contains a different side group that gives it unique chemical properties.

PEPTIDE BOND
Amino acids are joined through dehydration synthesis reactions. The newly formed bond is called a **peptide bond**.

POLYPEPTIDE
A **polypeptide** is a long chain of amino acids, typically hundreds to thousands of amino acids long. Each kind of **protein** contains one or more polypeptides with a unique sequence of amino acids.

FOLDED CHAIN
Every polypeptide twists and folds into a unique three-dimensional shape. Chemical bonds between the side groups of the amino acids hold the shape together. Most proteins are generally blob-shaped, but each kind of protein differs in the details of that shape.

MULTIPLE CHAINS
Some proteins contain multiple polypeptide chains joined together into a large complex. For example, one molecule of hemoglobin contains four polypeptide chains. The computer-generated model at right shows the actual shape of a hemoglobin molecule.

One chain of hemoglobin protein

Complete hemoglobin molecule with 4 chains

Hemoglobin

FUNCTIONS OF PROTEINS

Proteins perform a huge variety of tasks within your body and the bodies of all organisms. If something is getting done in your body, chances are there is a protein doing it. Here, you can see just some of the functions performed by proteins in animal bodies. Notice that each kind of protein has a unique shape that enables it to perform its unique function.

Structure: Keratin is an important component of hair, skin, nails, and fur.

Transport: Hemoglobin, found within red blood cells, carries oxygen as it is transported through the body.

Enzymes: Lactase is an enzyme within your digestive system that breaks down the milk sugar lactose.

Defense: Antibodies are proteins within your immune system that label foreign invaders for destruction.

Movement: Actin is one of the proteins that enables muscles to contract.

You have over 100,000 kinds of proteins in your body, each with a unique shape.

PROTEIN FORM AND FUNCTION

The precise amino acid sequence of a protein determines its overall shape and its function. Change the amino acid sequence, even a little, and you may alter the ability of the protein to perform its normal task. For example, changing just one of the 146 amino acids that make up one of the polypeptides (protein chains) in hemoglobin causes the protein to mis-fold. The altered protein cannot perform its function properly, leading to sickle-cell disease.

NORMAL

Normal hemoglobin

ONE AMINO ACID CHANGE

MUTATED

Mutated hemoglobin with slight change in shape

Normal red blood cells
1,700x

Sickled red blood cells
1,700x

CORE IDEA: Proteins are a diverse set of molecules made from amino acids joined by peptide bonds. Proteins perform most of the tasks required for life. Each kind of protein has a unique shape that determines its function.

CORE QUESTION: What is a polypeptide? What does the name mean?

ANSWER: A polypeptide is a chain of amino acids joined by many ("poly") peptide bonds.

Enzymes speed chemical reactions

Every living cell contains a huge variety of chemicals that constantly interact with each other through chemical reactions. The sum total of all the chemical reactions in an organism is called its metabolism. Almost no metabolic reaction occurs without the help of an **enzyme**, a protein that speeds up a chemical reaction without being changed itself. All living cells contain a huge variety of different enzymes, each of which promotes its own specific chemical reaction.

ENZYMES AND THEIR SUBSTRATES

Each enzyme recognizes one specific target molecule called its **substrate**. The substrate binds to the enzyme at a particular place on the enzyme called the **active site**. The active site has a shape that complements the shape of the substrate; the two fit together like a hand in a glove. Once the substrate binds, the enzyme catalyzes (promotes) a specific chemical reaction. For example, the enzyme lactase splits the milk sugar lactose into the simple sugars glucose and galactose. However, the enzyme itself is unchanged by the reaction and so can work again and again.

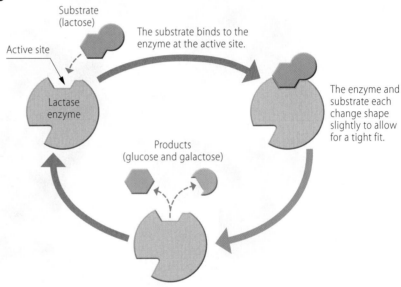

Substrate (lactose)

Active site

Lactase enzyme

The substrate binds to the enzyme at the active site.

The enzyme and substrate each change shape slightly to allow for a tight fit.

Products (glucose and galactose)

The substrate is converted to one or more products, which are released from the active site. The enzyme is then ready to accept another molecule of substrate.

ACTIVATION ENERGY

How are enzymes able to speed chemical reactions? The **activation energy** is the amount of energy required for a chemical reaction to proceed. An enzyme reduces the activation energy, which allows the reaction to proceed faster.

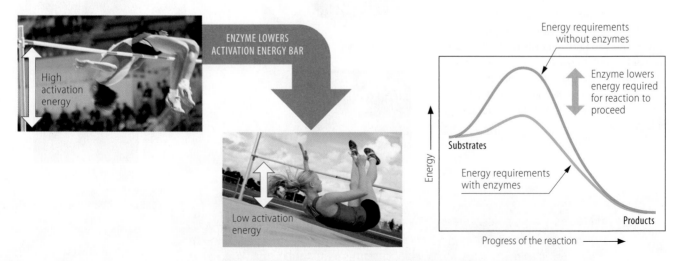

High activation energy

ENZYME LOWERS ACTIVATION ENERGY BAR

Low activation energy

Energy requirements without enzymes

Enzyme lowers energy required for reaction to proceed

Substrates

Energy requirements with enzymes

Products

Energy

Progress of the reaction

INHIBITORS

Within a living cell, enzymes can be turned off—that is, prevented from reacting with a substrate to produce products—when not needed. Certain molecules, called **inhibitors**, can bind to an enzyme and disrupt its function. Different enzyme inhibitors achieve this effect through different mechanisms.

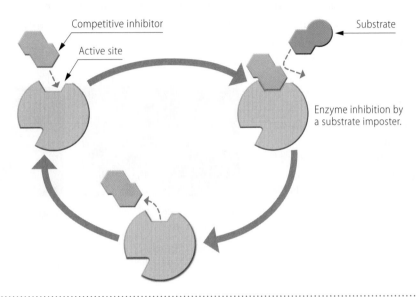

Competitive inhibitor

Active site

Substrate

Some inhibitors, called competitive inhibitors, work by binding to the active site, thereby "gumming up the works" and preventing the real substrate from binding there.

Enzyme inhibition by a substrate imposter.

Active site

Substrate

Noncompetitive inhibitor

Enzyme inhibition by a molecule that causes the active site to change shape.

Some inhibitors, called noncompetitive inhibitors, work by binding to a distant site on the enzyme. This binding causes the enzyme to change shape in such a way that the substrate can no longer bind to the active site.

FUNCTION FOLLOWS FORM

Enzymes are proteins. And, like all proteins, their function depends on their shape. Change the shape of an enzyme and you change its function. For example, this computer image shows the actual shape of the lactase enzyme. Certain mutations in the enzyme change its shape and prevent it from breaking down lactose. This can lead to lactose intolerance, the inability to properly digest milk sugar.

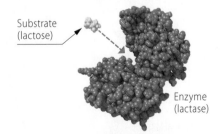

Substrate (lactose)

Enzyme (lactase)

CORE IDEA: Enzymes are proteins that speed chemical reactions by lowering activation energy. A substrate binds to the active site of an enzyme and is converted to one or more products. Inhibitors are molecules that prevent enzymes from working.

CORE QUESTION: If a sugar had the shape of lactose but could not be broken down, what effect would that have on the lactase enzyme?

ANSWER: The sugar would act as a competitive inhibitor, binding to the active site of the enzyme but preventing it from completing the reaction.

Cells are the fundamental units of life

All life is composed of cells. Some relatively simple organisms, such as bacteria and amoebas, consist of only a single cell. Other organisms, such as mammals, consist of trillions of cells. But whatever their source, all cells on Earth can be classified into two general kinds: prokaryotic cells and eukaryotic cells.

PROKARYOTIC CELLS

Two domains (large groups) of single-celled life—bacteria and archaea—are collectively called **prokaryotes**. All prokaryotes are relatively simple organisms consisting of a single prokaryotic cell. Prokaryotic fossils date back at least 3.5 billion years. Keep in mind that not all prokaryotes have all of the structures shown below.

2,700x

BACTERIA CELLS
This scanning electron micrograph shows *Pseudomonas aeruginosa*, a common disease-causing bacterium.

5,000x

ARCHAEA CELLS
This organism, called *Sulfolobus archaea*, grows only in very hot, acidic environments.

There are more bacteria living in your mouth right now than there are humans on Earth.

STRUCTURE OF AN IDEALIZED BACTERIUM
The structures labeled in blue are unique to prokaryotic cells.

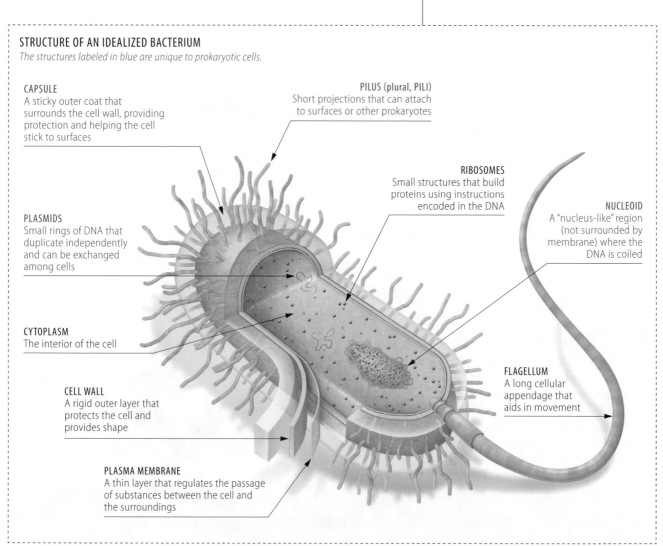

CAPSULE
A sticky outer coat that surrounds the cell wall, providing protection and helping the cell stick to surfaces

PILUS (plural, PILI)
Short projections that can attach to surfaces or other prokaryotes

RIBOSOMES
Small structures that build proteins using instructions encoded in the DNA

NUCLEOID
A "nucleus-like" region (not surrounded by membrane) where the DNA is coiled

PLASMIDS
Small rings of DNA that duplicate independently and can be exchanged among cells

CYTOPLASM
The interior of the cell

CELL WALL
A rigid outer layer that protects the cell and provides shape

FLAGELLUM
A long cellular appendage that aids in movement

PLASMA MEMBRANE
A thin layer that regulates the passage of substances between the cell and the surroundings

EUKARYOTIC CELLS

Eukaryotic cells are found in all non-prokaryotic forms of life (collectively called eukaryotes), including plants and animals. Compared to prokaryotic cells, eukaryotic cells are relatively large (about 10-fold bigger) and more complex. Eukaryotic cells contain **organelles** ("little organs"), membrane-enclosed structures that perform specific functions. Eukaryotes evolved from prokaryotes around 2 billion years ago.

ONE CELL   NUCLEUS   SPORES   CHLOROPLASTS

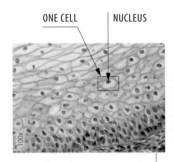

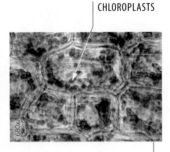

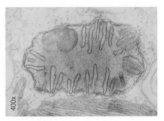

ANIMAL CELLS
Each small pink circle (many with visible nuclei) is a cell from the human reproductive tract.

FUNGUS CELLS
The blue structures at the end of these *Penicillium* fungus cells are reproductive spores.

PLANT CELLS
Chloroplasts are visible as green blobs within these cells from *Elodea*, an aquatic plant.

PROTIST CELL
This micrograph shows a *Euglena*, a single-celled protist.

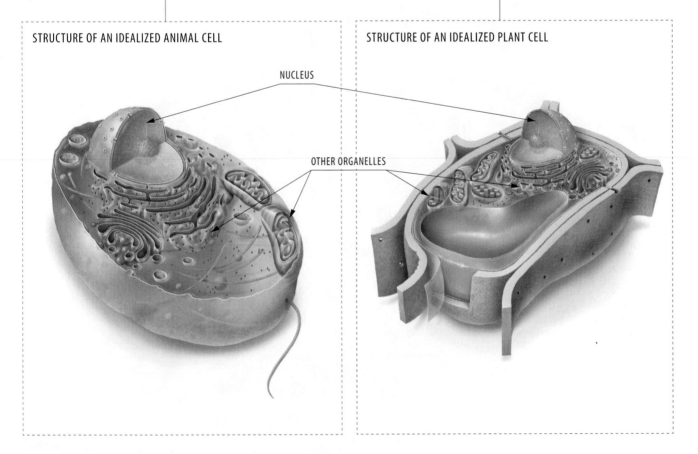

STRUCTURE OF AN IDEALIZED ANIMAL CELL

STRUCTURE OF AN IDEALIZED PLANT CELL

NUCLEUS

OTHER ORGANELLES

CORE IDEA: All cells are either relatively small, simple prokaryotic cells (bacteria and archaea) or larger, organelle-containing eukaryotic cells (animals, fungi, plants, and protists).

CORE QUESTION: Of the three kinds of cells shown here—bacterial, animal, plant—which kind(s) contain a nucleus? Plasmids?

ANSWER: Plant and animal cells contain nuclei; only prokaryotes such as bacteria contain plasmids.

Plant and animal cells have common and unique structures

All eukaryotic cells are fundamentally alike. All are surrounded by a **plasma membrane** and some, such as plant cells, also have an outer cell wall. The most prominent feature is the **nucleus**, a region surrounded by membrane where the DNA is housed. The **cytoplasm**, the liquidy region between the plasma membrane and the nucleus, contains many **organelles**—membrane-enclosed structures found only in eukaryotic cells—that perform specific functions. Most organelles are common to both animal and plant cells, while some are found in just one or the other.

STRUCTURE OF AN IDEALIZED ANIMAL CELL

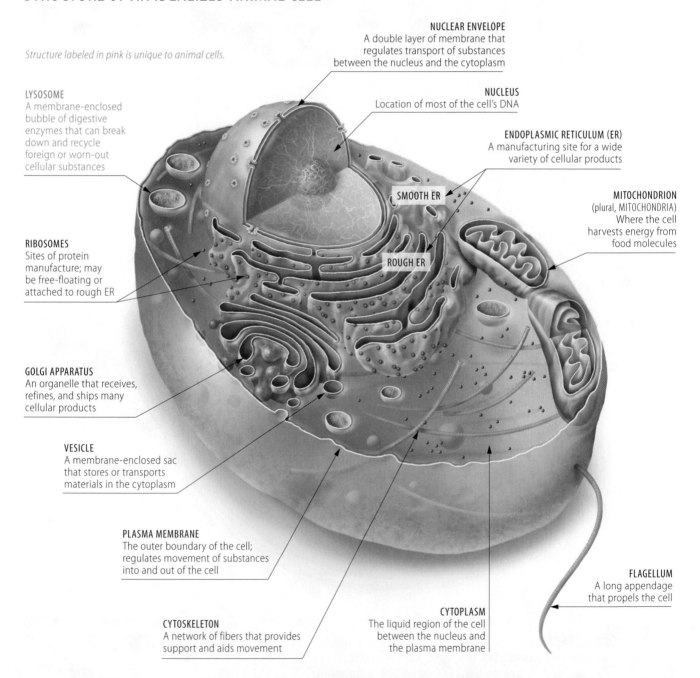

Structure labeled in pink is unique to animal cells.

NUCLEAR ENVELOPE
A double layer of membrane that regulates transport of substances between the nucleus and the cytoplasm

NUCLEUS
Location of most of the cell's DNA

LYSOSOME
A membrane-enclosed bubble of digestive enzymes that can break down and recycle foreign or worn-out cellular substances

ENDOPLASMIC RETICULUM (ER)
A manufacturing site for a wide variety of cellular products

SMOOTH ER

ROUGH ER

MITOCHONDRION
(plural, MITOCHONDRIA)
Where the cell harvests energy from food molecules

RIBOSOMES
Sites of protein manufacture; may be free-floating or attached to rough ER

GOLGI APPARATUS
An organelle that receives, refines, and ships many cellular products

VESICLE
A membrane-enclosed sac that stores or transports materials in the cytoplasm

PLASMA MEMBRANE
The outer boundary of the cell; regulates movement of substances into and out of the cell

FLAGELLUM
A long appendage that propels the cell

CYTOSKELETON
A network of fibers that provides support and aids movement

CYTOPLASM
The liquid region of the cell between the nucleus and the plasma membrane

STRUCTURE OF AN IDEALIZED PLANT CELL

Structures labeled in green are unique to plant cells.

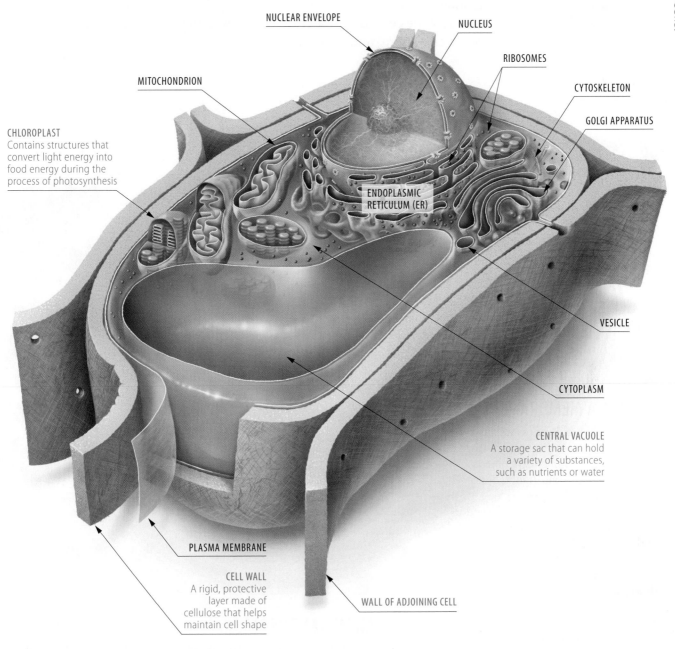

NUCLEAR ENVELOPE

NUCLEUS

RIBOSOMES

MITOCHONDRION

CYTOSKELETON

GOLGI APPARATUS

CHLOROPLAST
Contains structures that convert light energy into food energy during the process of photosynthesis

ENDOPLASMIC RETICULUM (ER)

VESICLE

CYTOPLASM

CENTRAL VACUOLE
A storage sac that can hold a variety of substances, such as nutrients or water

PLASMA MEMBRANE

CELL WALL
A rigid, protective layer made of cellulose that helps maintain cell shape

WALL OF ADJOINING CELL

CORE IDEA: Animal and plant cells share many organelles, such as the nucleus, ribosomes, and mitochondria. Only animal cells have lysosomes, while only plant cells have chloroplasts, cell walls, and central vacuoles.

CORE QUESTION: Which animal organelle(s) is/are involved in energy production? Which plant organelle(s)?

ANSWER: Mitochondria in animal cells; mitochondria and chloroplasts in plant cells.

Membranes are made from two layers of lipids

Although you may think of cholesterol as a dietary nuisance, it is a vital component of the membranes that surround every cell of your body.

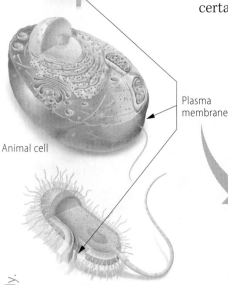

Plant cell

Animal cell

Plasma membrane

Bacterial cell

To survive, every cell must maintain an internal environment that is different than that of the outside world. Every cell (whether prokaryotic or eukaryotic) accomplishes this using a **plasma membrane**. This thin, flexible, oily sheet forms the boundary between each living cell and its surroundings. The plasma membrane has a relatively simple structure, but can perform a great many functions. Most importantly, membranes regulate the passage of materials into and out of the cell. Some materials are allowed to pass freely, others only under certain conditions, and some not at all.

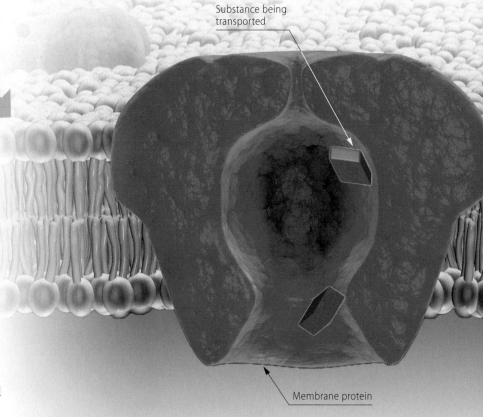

Substance being transported

Membrane protein

STRUCTURE OF A PLASMA MEMBRANE

The plasma membrane (as well as other membranes within the cell) are composed mostly of phospholipids and proteins. The phospholipids are organized into two layers, called a **phospholipid bilayer**. The bilayer and most of the proteins in it drift about in the plane of the membrane, causing the membrane to flex and undulate. Thus, a membrane is called a **fluid mosaic**—fluid because the molecules can move freely past one another, and a mosaic because of the diversity of proteins that float like icebergs in the phospholipid sea.

MEMBRANE PROTEINS

Most membranes have proteins embedded within them. Some of these proteins, such as the ones highlighted here, help regulate the passage of materials into and out of the cell. Others aid in communication between neighboring cells, facilitate enzymatic reactions, or help anchor the cell or its components. Some of these proteins are fixed in place, while others float about within the phospholipid bilayer.

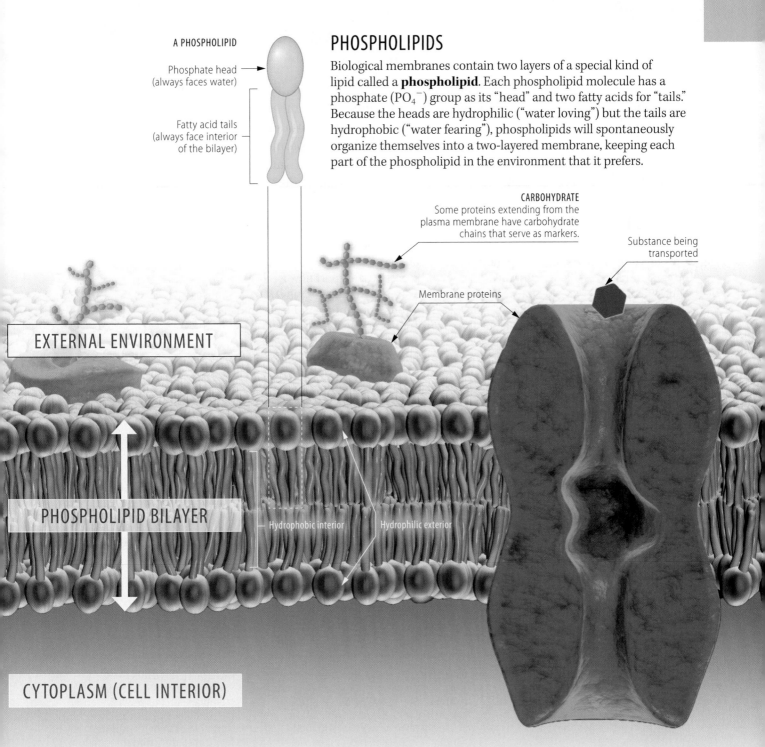

A PHOSPHOLIPID

Phosphate head
(always faces water)

Fatty acid tails
(always face interior
of the bilayer)

PHOSPHOLIPIDS

Biological membranes contain two layers of a special kind of lipid called a **phospholipid**. Each phospholipid molecule has a phosphate (PO_4^-) group as its "head" and two fatty acids for "tails." Because the heads are hydrophilic ("water loving") but the tails are hydrophobic ("water fearing"), phospholipids will spontaneously organize themselves into a two-layered membrane, keeping each part of the phospholipid in the environment that it prefers.

CARBOHYDRATE
Some proteins extending from the plasma membrane have carbohydrate chains that serve as markers.

Membrane proteins

Substance being transported

EXTERNAL ENVIRONMENT

PHOSPHOLIPID BILAYER

Hydrophobic interior   Hydrophilic exterior

CYTOPLASM (CELL INTERIOR)

THE CYTOPLASM

The interior of the cell is called the **cytoplasm**. It consists primarily of a watery liquid called the **cytosol**, various organelles, and dissolved molecules.

CORE IDEA: Every cell is surrounded by a plasma membrane made from two layers of phospholipids and integrated proteins. Membranes perform important functions such as regulating the passage of materials.

CORE QUESTION: Why is the structure of a membrane called a phospholipid bilayer?

ANSWER: Because it consists of two stacked layers of phospholipid molecules.

Membranes regulate the passage of materials

Every living cell is surrounded by a plasma membrane. Within eukaryotic cells, many organelles (such as the nucleus) are surrounded by their own membranes. And some organelles (such as mitochondria and chloroplasts) have internal membranes. The most important function of all these membranes is to regulate the flow of materials. Every membrane is selectively permeable, meaning it allows some substances to flow freely, others only under certain circumstances, and some not at all. Here, you can see some of the different mechanisms by which cells use membranes to control the flow of materials.

PASSIVE TRANSPORT

Passive transport occurs when a substance moves across a membrane from an area of higher concentration to an area of lower concentration. Such movement is said to occur along the concentration gradient. Substances that move by passive transport flow freely with no requirement for energy expenditure by the cell. For example, within your lungs, O_2 gas diffuses from the air into your blood, while CO_2 diffuses from your blood into your lungs. Both of these movements are passive, occurring along the concentration gradient for each gas.

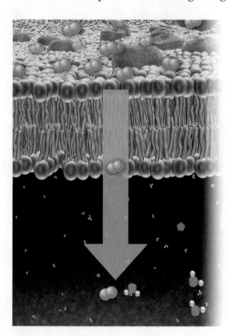

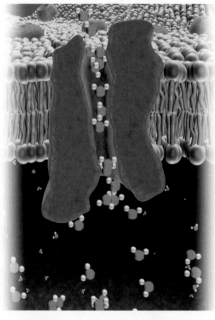

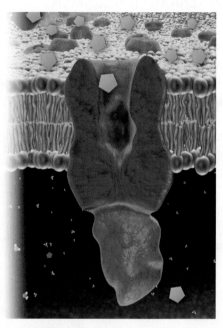

DIFFUSION
Because molecules are constantly in motion, they spread out to fill any available space. **Diffusion** is the movement of molecules from an area of higher concentration to an area of lower concentration, as when perfume molecules move out of a bottle and into a room, or when O_2 diffuses from the air sacs of your lungs into your blood.

OSMOSIS
The diffusion of water is called **osmosis**. Water always flows from an area of higher water concentration to an area of lower water concentration. For example, placing meat in salt will "cure" it by drawing out the water. Sometimes water can flow through a membrane itself; other times it requires a protein channel to pass through the membrane.

FACILITATED DIFFUSION
Some substances cannot cross a membrane on their own (because, for example, they are too large). Such substances can be transported via **facilitated diffusion** through specific transport proteins embedded in the membrane. These proteins act as selective channels. As in all forms of passive transport, substances always move down the concentration gradient.

3.4

All your senses depend on active transport to communicate their signals.

ACTIVE TRANSPORT

In contrast to passive transport, **active transport** involves moving a substance against its concentration gradient, from where it is less concentrated to where it is more so. This always requires an expenditure of energy on the part of the cell. Active transport is usually driven by a protein that sits within the membrane. Here, you can see a protein called the sodium-potassium (Na^+/K^+) pump moving three potassium ions into the cell. This movement of ions is essential in the formation of nerve signals.

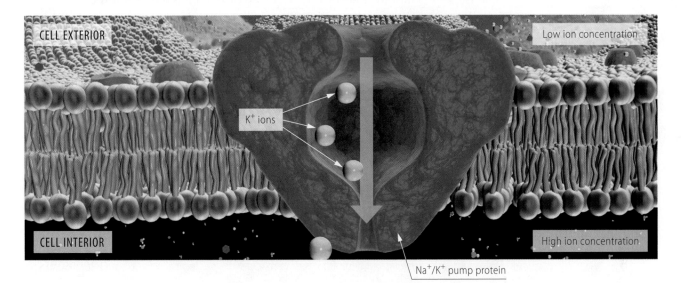

CELL EXTERIOR

Low ion concentration

K^+ ions

CELL INTERIOR

High ion concentration

Na^+/K^+ pump protein

ENDOCYTOSIS

The transport of large substances into the cell is usually accomplished via **endocytosis**. Substances to be ingested (such as food, fluids, or foreign substances to be destroyed) are packaged into vesicles—small bubbles surrounded by membranes—that bud inward from the plasma membrane. The vesicle then travels through the cytoplasm to its destination.

EXOCYTOSIS

Exocytosis is the export of large quantities of material from the cell. Vesicles containing the material to be exported fuse with the plasma membrane, dumping the contents into the environment around the cell. For example, cells in your tear glands use exocytosis to export salty tears.

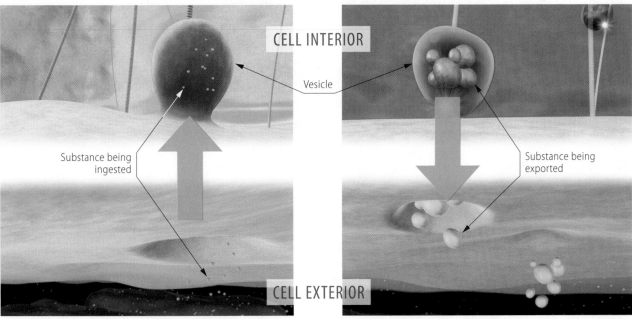

CELL INTERIOR

Vesicle

Substance being ingested

Substance being exported

CELL EXTERIOR

CORE IDEA: Membranes regulate the movement of substances. Passive transport requires no energy as substances move along the concentration gradient, while active transport requires energy to move substances against the gradient.

CORE QUESTION: What form of passive transport always requires membrane-bound proteins?

ANSWER: Facilitated diffusion. (Osmosis sometimes, but not always, requires protein channels.)

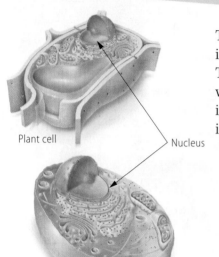

Plant cell

Nucleus

Animal cell

The most important difference between prokaryotic and eukaryotic cells is that eukaryotic cells contain organelles surrounded by membranes. The most prominent membrane-enclosed organelle is the **nucleus**, which houses most of the cell's DNA packaged as chromosomes. If a cell is like a factory, then the nucleus is the executive board room; orders are issued here that result in the performance of the cell's processes.

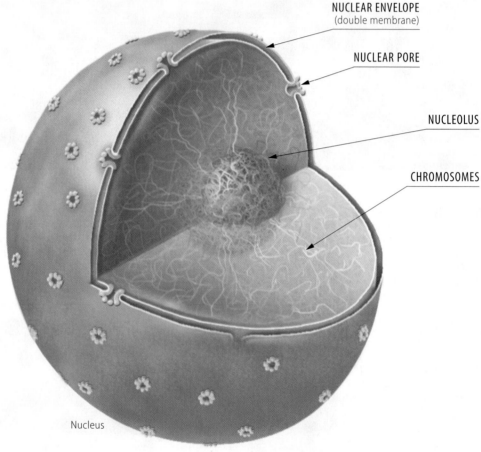

NUCLEAR ENVELOPE
(double membrane)

NUCLEAR PORE

NUCLEOLUS

CHROMOSOMES

Nucleus

THE NUCLEUS

Every eukaryotic cell (including plant and animal cells) contains a nucleus. The nucleus contains most of the cell's DNA stored in chromosomes, which control the cell's activities by directing protein production. The DNA is a bit like the executive director of a factory; while it doesn't do the hands-on work itself, the DNA issues orders (via RNA) that result in proteins being made to perform needed tasks.

An actual chromosome is less than 1/2,000th the size of the one shown on the right side of this page.

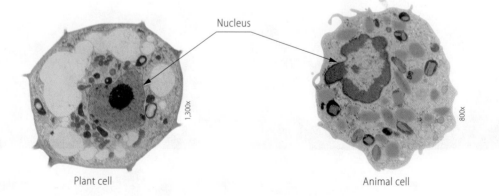

Nucleus

1,300x

800x

Plant cell

Animal cell

NUCLEAR ENVELOPE

The nucleus is surrounded by a double membrane called the **nuclear envelope** (also called the nuclear membrane). Each of the two layers has a structure similar to the cell's outer plasma membrane, with a phospholipid bilayer and associated proteins. The nuclear membrane controls the passage of materials into and out of the nucleus. Protein-lined **nuclear pores** allow certain molecules, such as RNA, to pass through.

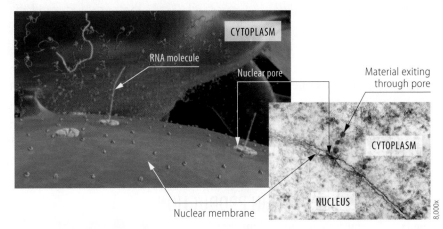

CYTOPLASM

RNA molecule

Nuclear pore

Material exiting through pore

CYTOPLASM

NUCLEUS

Nuclear membrane

8,000x

CHROMOSOMES

Within the nucleus, DNA molecules are wrapped around proteins to form fibers called **chromatin**. Each very long chromatin fiber twists and folds to form a **chromosome**. Each human body cell has 46 chromosomes in the nucleus. When a cell is not dividing, the chromatin is very loose, but it coils tightly during cell division.

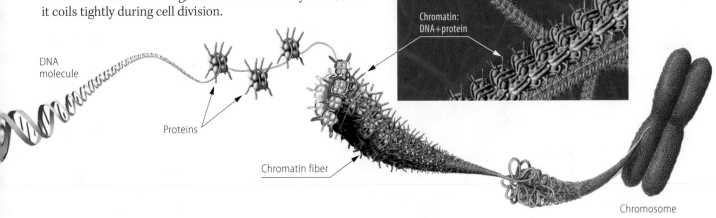

Section of a chromosome

Chromatin: DNA+protein

DNA molecule

Proteins

Chromatin fiber

Chromosome

NUCLEOLUS

The nucleolus is a location within the nucleus where DNA from multiple chromosomes directs the production of a kind of RNA called ribosomal RNA (rRNA). The rRNA subunits exit through the nuclear pores and enter the cytoplasm, where they can join together with protein molecules to form a ribosome, a cellular structure that manufactures proteins.

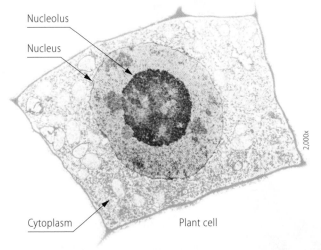

Nucleolus

Nucleus

Cytoplasm

Plant cell

2,000x

CORE IDEA: The nucleus, surrounded by an envelope and containing DNA, directs the activities of the cell. Materials, such as RNA, can pass out of the nucleus via protein-lined pores.

CORE QUESTION: If the DNA never leaves, why would it be detrimental for the cell's nucleus to be completely enclosed?

ANSWER: RNA must be able to exit the nucleus to carry messages from the DNA to the cytoplasm.

Several organelles participate in the production of proteins

Within the nucleus of eukaryotic cells, DNA acts as the executive director, controlling the functions of the cell. But DNA itself does not actually perform any of the cell's functions. Rather, it exerts control through the production of proteins, which act as the workers in the cellular factory. Several organelles within the cell participate in the important task of protein production.

AN OVERVIEW OF PROTEIN PRODUCTION

The DNA "executives" in the nucleus direct the "workers" in the cytoplasm through the production of proteins. DNA accomplishes this through two steps: **Transcription** in the nucleus results in the production of RNA from DNA. The RNA then carries the instructions for synthesizing proteins from the nucleus to the cytoplasm, where **translation** at the ribosomes results in the production of proteins. You can follow the process of protein synthesis in more detail through each organelle.

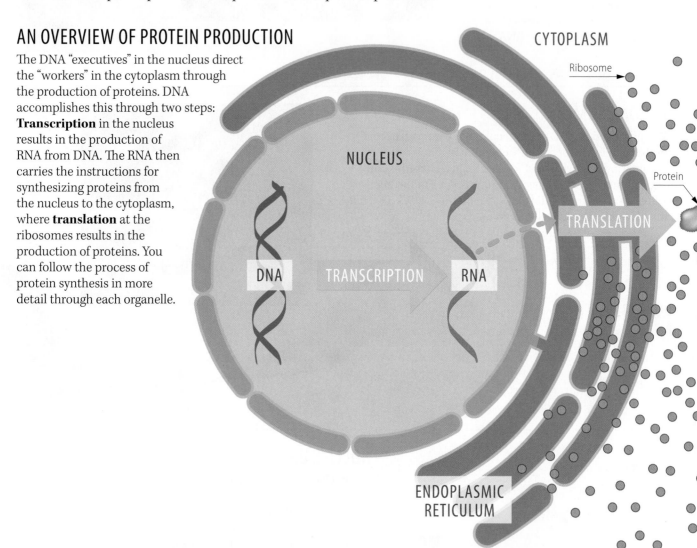

CYTOPLASM

Ribosome

Protein

NUCLEUS

DNA TRANSCRIPTION RNA

TRANSLATION

ENDOPLASMIC RETICULUM

ENDOPLASMIC RETICULUM

The **endoplasmic reticulum (ER)** is a network of membrane-enclosed passageways and sacs that touches the outside of the nuclear envelope and extends deep into the cytoplasm. The membrane of the **smooth ER** contains enzymes that produce lipids (such as steroid hormones). The **rough ER** contains ribosomes that produce many kinds of proteins.

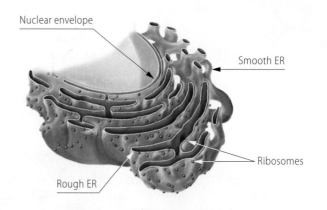

Nuclear envelope

Smooth ER

Ribosomes

Rough ER

The smooth endoplasmic reticulum of liver cells helps cleanse the blood of toxins such as drugs.

RIBOSOMES

Using information carried from the nucleus by a molecule of RNA, **ribosomes** are where proteins are made. Some ribosomes are bound to the membrane of the rough ER; others float freely in the cytoplasm. Through the process of translation, each RNA is used to create a specific protein.

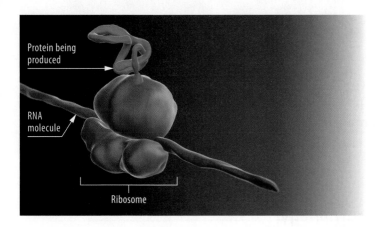

Protein being produced

RNA molecule

Ribosome

GOLGI APPARATUS

After being produced by the ribosomes, some proteins travel to the **Golgi apparatus**, where they are refined, stored, and distributed. On one of its sides, the Golgi receives vesicles (bubbles of membrane) with proteins to be processed. After modification by enzymes, the other side of the Golgi ships the refined products in new vesicles that may wind up inside the cell or be exported outside the cell.

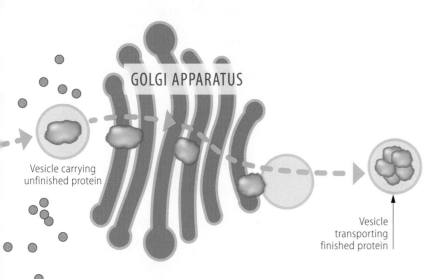

GOLGI APPARATUS

Vesicle carrying unfinished protein

Vesicle transporting finished protein

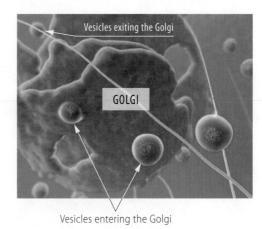

Vesicles exiting the Golgi

GOLGI

Vesicles entering the Golgi

VESICLES

Vesicles are small bubbles made of membrane that are used to transport materials through the cell. **Lysosomes** are vesicles containing digestive enzymes that can dissolve large food molecules, old cellular components, or invasive organisms such as bacteria.

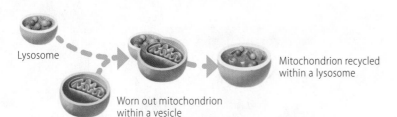

Lysosome

Worn out mitochondrion within a vesicle

Mitochondrion recycled within a lysosome

CORE IDEA: DNA directs a cell's activities through the production of proteins. RNA made in the nucleus travels to the ER and the ribosomes, where it is translated into proteins, which are finalized in the Golgi.

CORE QUESTION: List the organelles involved in protein production in order, starting where the DNA is.

ANSWER: Nucleus, ER, ribosome, Golgi, vesicles.

Chloroplasts and mitochondria provide energy to the cell

All cells require a continuous supply of energy to power life's processes. Two organelles help provide that energy: chloroplasts and mitochondria. Plant cells have chloroplasts, while both animal and plant cells have mitochondria. Both of these organelles contain deeply folded membranes that form fluid-filled spaces.

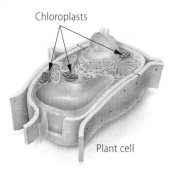

Chloroplasts

Plant cell

CHLOROPLASTS AND CHLOROPHYLL

Chloroplasts, found in all plant cells and the cells of some algae, are the organelles of photosynthesis. In **photosynthesis**, the energy of sunlight is captured and used to create molecules of sugar. To complete these reactions, chloroplasts require a steady supply of water (H_2O) and carbon dioxide (CO_2). During photosynthesis, chloroplasts release oxygen gas (O_2) as a waste product. The sugars produced by photosynthesis provide the energy that powers most ecosystems on Earth.

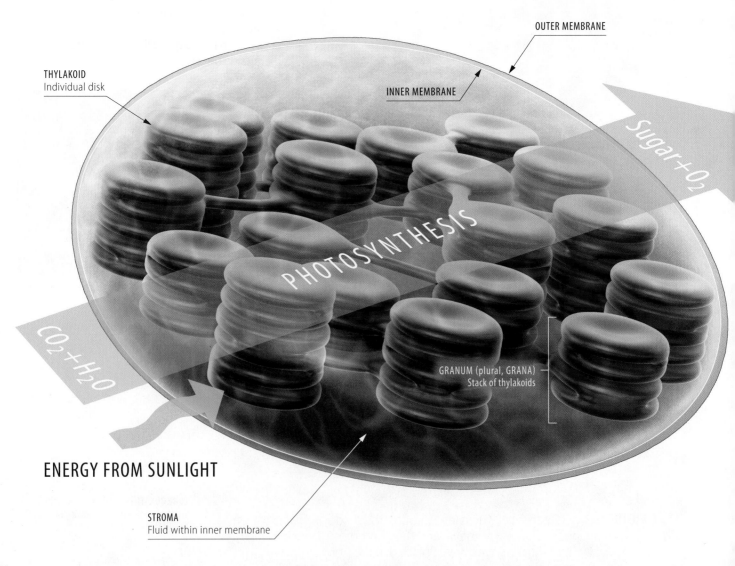

OUTER MEMBRANE

INNER MEMBRANE

THYLAKOID
Individual disk

Sugar + O₂

PHOTOSYNTHESIS

GRANUM (plural, GRANA)
Stack of thylakoids

CO₂ + H₂O

ENERGY FROM SUNLIGHT

STROMA
Fluid within inner membrane

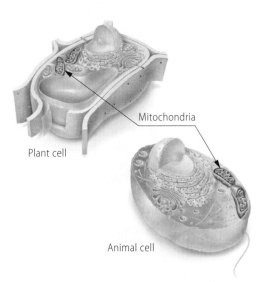

Plant cell

Mitochondria

Animal cell

MITOCHONDRIA

Mitochondria (singular, *mitochondrion*) provide energy to nearly all eukaryotic cells, including your own. Within mitochondria, a series of enzymes performs a sequence of chemical reactions called **cellular respiration**. This process uses oxygen (O_2) to harvest chemical energy from molecules of sugar (in particular, glucose). The harvested energy is stored as chemical energy in molecules of **ATP**, which can then be used to power many other cellular processes. Carbon dioxide and water are released as waste products.

A typical human muscle cell has a few hundred mitochondria.

INNER MEMBRANE
Highly folded, contains many of the enzymes that perform cellular respiration

INTERNAL FLUID

OUTER MEMBRANE

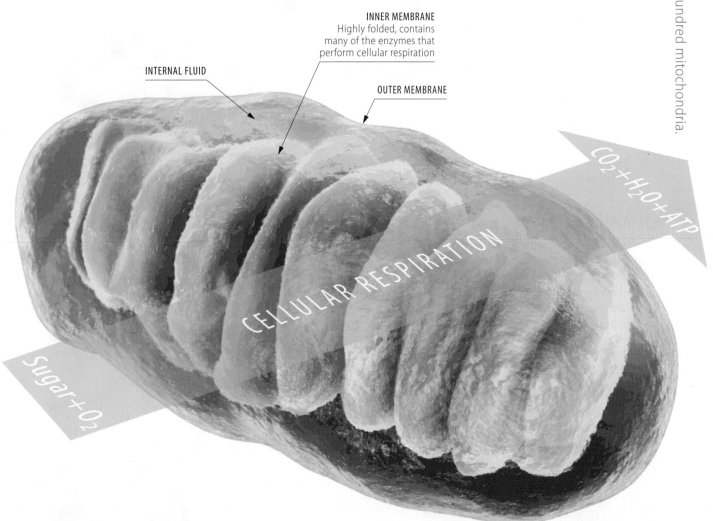

Sugar + O_2

CELLULAR RESPIRATION

$CO_2 + H_2O + ATP$

CORE IDEA: Chloroplasts use CO_2 and water to capture the energy of sunlight and store it as chemical energy in molecules of sugar. Mitochondria use oxygen to convert the chemical energy stored in molecules of sugar to ATP.

CORE QUESTION: Do you notice something interesting about the inputs and outputs of photosynthesis and cellular respiration?

ANSWER: The outputs of photosynthesis serve as the inputs for cellular respiration, and vice versa.

Other organelles provide cell shape, movement, storage, and recycling

Eukaryotic cells contain a wide variety of organelles that perform many functions important to life. Some of these structures are found in both plant and animal cells, while others are found in only one or the other.

VACUOLES

Vacuoles are intracellular sacs that come in a variety of sizes and perform a variety of functions. Some vacuoles store food, while others help pump water out of a cell. Many plant cells have a very large central vacuole that can store nutrients, pigments, toxins, and water.

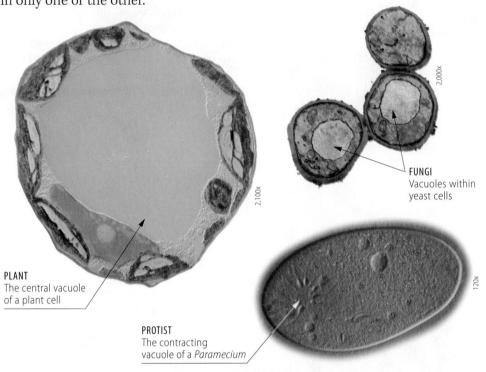

2,000x

FUNGI
Vacuoles within yeast cells

2,100x

PLANT
The central vacuole of a plant cell

120x

PROTIST
The contracting vacuole of a *Paramecium*

CILIA AND FLAGELLA

Some cells have moving appendages. A **flagellum** (plural, *flagella*) is a long extension that can propel the cell by whipping back and forth, like the familiar tail of a sperm. **Cilia** (singular, *cilium*), which are usually shorter and more numerous than flagella, move in a coordinated back-and-forth motion. For example, ciliated cells in your windpipe help keep your airway free of dust.

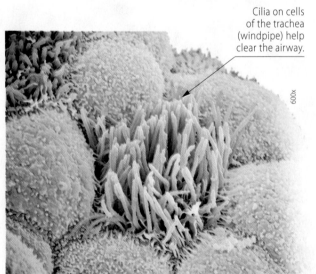

Cilia on cells of the trachea (windpipe) help clear the airway.

600x

10,000x

Helicobacter pylori (the bacteria that causes stomach ulcers) has numerous whip-like flagella.

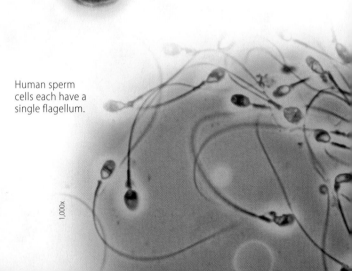

Human sperm cells each have a single flagellum.

1,000x

Tobacco smoke can destroy the cilia that line the windpipe, causing frequent coughing.

CELL WALLS

The cells of plants, fungi, and some protists have a rigid **cell wall** surrounding the plasma membrane. The walls of plant cells are made from fibers of a carbohydrate called cellulose; the cell walls of other types of cells are made from other molecules. The cell wall provides protection, helps maintain the shape of a cell, and provides the stiffness that allows thin leaves to stand upright.

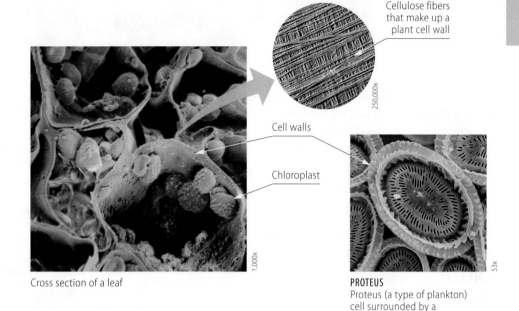

Cellulose fibers that make up a plant cell wall

250,000x

Cell walls

Chloroplast

Cross section of a leaf

1,000x

PROTEUS
Proteus (a type of plankton) cell surrounded by a protective cell wall

53x

CYTOSKELETON

Animal cells are able to maintain their shape because they have a **cytoskeleton**, a network of protein fibers that provides mechanical support, anchorage, and reinforcement. The cytoskeleton also provides a series of tracks along which vesicles move. As you can see here, the cytoskeleton forms an extensive network. But it can be quickly dismantled and reassembled, providing flexibility to a cell's shape.

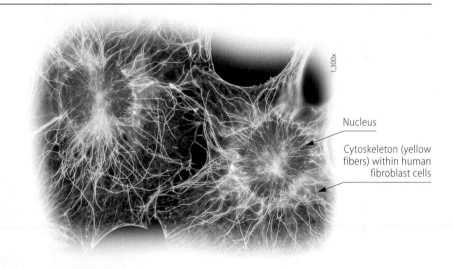

1,300x

Nucleus

Cytoskeleton (yellow fibers) within human fibroblast cells

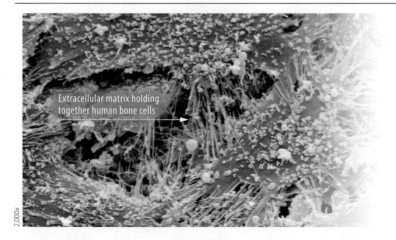

Extracellular matrix holding together human bone cells

2,000x

EXTRACELLULAR MATRIX

Animal cells produce a sticky **extracellular matrix** that helps hold cells together. For example, individual muscle cells are held together into a tissue by the extracellular matrix. Lying just outside the plasma membrane, the matrix consists primarily of an interwoven network of collagen protein. This network connects to proteins that span the plasma membrane, allowing signals to be transmitted into the cell.

CORE IDEA: Various structures within and outside of cells provide infrastructure, support, movement, storage, communication, and protection.

CORE QUESTION: Of the structures discussed here, which are on the surface of the cell? Which are in the interior?

ANSWER: Surface: cilia/flagella, extracellular matrix, cell wall. Interior: cytoskeleton, vacuole.

Energy can be converted from one form to another

Energy really does make the world go round. Every living organism metabolizes energy—taking it in from the environment, converting it to useful forms, and expending it. But what is energy? **Energy** is defined as the capacity to do work. **Work** is the movement of an object against an opposing force. Through the expenditure of energy (that is, through work), an object can be moved in a way that it would not move if left alone. All of life's processes are driven by converting energy from one form to another.

KINETIC ENERGY AND POTENTIAL ENERGY

Natural systems contain different forms of energy, and living systems continuously convert one form of energy to another. **Kinetic energy** is the energy of motion, such as when a child moves down a slide. **Potential energy** is the energy an object has due to its location or structure, such as the energy contained by a child poised at the top of the slide. Stored potential energy can be released and converted to other forms of energy. In the photo, the potential energy that a child has at the top of the slide is converted to kinetic energy during the ride down.

High potential energy:
top of the slide

The potential energy of the slider (due to his raised elevation) is converted to kinetic energy of motion.

Potential (chemical) energy in food is converted to kinetic energy of muscle movement (climbing the stairs).

Low potential energy:
bottom of the slide

CHEMICAL ENERGY

One form of energy that is particularly important to all biological systems is **chemical energy**, potential energy that is stored in the bonds that hold atoms together into molecules. Living organisms can break these bonds, releasing the stored energy. The released chemical energy can then be put to work in the performance of life's functions.

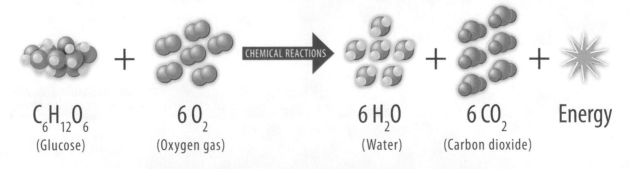

$C_6H_{12}O_6$ + $6\,O_2$ CHEMICAL REACTIONS → $6\,H_2O$ + $6\,CO_2$ +   Energy

(Glucose)   (Oxygen gas)   (Water)   (Carbon dioxide)

CONSERVATION OF ENERGY

Energy can be converted from one form to another (say, from potential energy to kinetic energy), but it cannot be created or destroyed. For example, a power plant does not create energy; instead, it converts energy from a less convenient form (the chemical energy stored in coal) to a more convenient form (electricity). If a biological system is studied carefully, many energy conversions become apparent. If we could account for every bit of energy that enters and leaves the system, we would see that the total amount of energy is constant.

Energy from sunlight warms the lizard.

Heat dissipates from the lizard into the environment.

Chemical energy stored in the lizard's muscles is converted to the kinetic energy of motion as its tongue flicks.

The chemical energy in the insect is captured by the lizard's digestive system and stored as other forms of chemical energy inside the lizard's body.

The chemical energy in one peanut can be converted into enough heat to boil two tablespoons of room temperature water.

ENTROPY

Within living systems, no energy conversion is perfectly efficient. There is always some waste expelled as **heat**, a form of kinetic energy stored in the random motions of atoms and molecules. Heat energy is very chaotic and is therefore hard to recapture in a biological system. Thus, every time energy is converted, the **entropy**, or amount of disorder within a system, increases. Increasing entropy (randomness) is a natural consequence of life's processes. To restore order, considerable energy must be expended (which releases heat, creating its own entropy). You've probably noticed in your own life how easy it is for disorder to be increased, while it requires considerable effort to decrease disorder.

CORE IDEA: Energy can be converted from one form to another, but it cannot be created or destroyed. In biological systems, chemical (potential) energy is often converted to kinetic (motion) energy, releasing heat energy.

CORE QUESTION: If you compress a spring and then release it, what energy conversions are taking place?

ANSWER: Compressing the spring converts kinetic energy into potential energy. Releasing the spring converts potential energy into kinetic energy.

Energy flows through an ecosystem

Nearly every organism on Earth ultimately derives its energy from the sun. Some organisms capture solar energy directly through the process of photosynthesis, converting the energy of sunlight to chemical energy stored in sugars. These organisms (and other organisms that eat them) can then use the process of cellular respiration to burn the energy stored in sugars to fuel life's processes. As energy is converted within and between organisms, heat is given off. This heat exits the ecosystem. Energy thus flows through ecosystems, entering as sunlight and leaving as heat. If you observe carefully, you can trace the flow of energy through every ecosystem in the world around you.

PRODUCERS

Most ecosystems receive a steady input of energy from sunlight. Organisms called **producers** can absorb the sun's energy and convert it to chemical energy stored in sugars and other organic molecules. To put it simply, these organisms can produce their own food. In addition to feeding themselves, producers are eaten by organisms that cannot use sunlight directly. Most ecosystems thereby depend on producers to feed the community. Producers include land plants, aquatic protists (such as seaweed, a form of algae), and certain bacteria.

SUNLIGHT

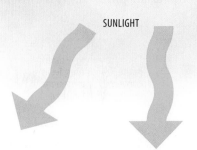

AQUATIC ALGAE

LAND PLANTS

PHOTOSYNTHETIC BACTERIA

$CO_2 + H_2O$

ENERGY FROM SUNLIGHT

PHOTOSYNTHESIS

Producers are able to capture the energy of sunlight and convert it to chemical energy through the process of **photosynthesis**. Notice in the figure that the chemical ingredients for photosynthesis are carbon dioxide (CO_2) and water (H_2O). In the cells of plants and algae, organelles called **chloroplasts** use light energy to rearrange the atoms of these ingredients, producing sugars and other organic molecules. A by-product of photosynthesis is oxygen gas (O_2), which is released into the environment.

PHOTOSYNTHESIS

CHLOROPLAST

SUGAR + O_2

CONSUMERS

Many organisms—including humans—cannot obtain energy directly from the environment via photosynthesis. Instead, we can obtain energy by eating producers, which gives us access to the sugars and other molecules they produced through photosynthesis. We are therefore **consumers**, organisms that obtain food by eating plants or by eating animals that have eaten plants (or animals that ate animals that ate plants, etc.). Consumers include a wide variety of organisms, such as animals, fungi, and some microscopic protists.

ANIMALS

FUNGI

SOME PROTISTS

CELLULAR RESPIRATION

CO₂ + H₂O

CELLULAR RESPIRATION

MITOCHONDRION

SUGAR + O₂

ATP MOLECULES

All organisms—both producers and consumers—are able to release the chemical energy stored in sugars through the process of **cellular respiration**. The chemical ingredients for cellular respiration are sugars (such as those produced by photosynthesis) and oxygen (O_2). In nearly every eukaryotic cell, enzymes in the cytoplasm and within organelles called **mitochondria** (singular, *mitochondrion*) break the chemical bonds in sugar. This releases the energy stored in those bonds, and that energy can be used to produce many copies of a molecule called ATP (adenosine triphosphate). Cells can then use ATP as a power source for many of their functions. During cellular respiration, CO_2 and H_2O are released into the environment as by-products, where they can then be used by producers as the ingredients for photosynthesis. Only producers perform photosynthesis, but both producers and consumers perform cellular respiration.

Because it uses sunlight to warm its body, a reptile requires only about 10% as much food energy as a mammal of equivalent size.

CORE IDEA: Energy enters most ecosystems as sunlight. Producers capture solar energy and use it to drive photosynthesis, producing sugars. Consumers then obtain energy by eating organisms. The cells of both producers and consumers obtain energy by breaking down sugars through cellular respiration.

CORE QUESTION: What is incorrect about this statement: Plants perform photosynthesis, but only animals perform cellular respiration.

ANSWER: Plants perform both photosynthesis and cellular respiration.

Within chloroplasts, the energy of sunlight is used to produce sugars

Through the process of **photosynthesis**, a producer (such as a plant) uses water (H_2O), carbon dioxide (CO_2), and the energy in sunlight to produce sugars. Oxygen gas (O_2) is released into the environment as a by-product. Within plant cells, the reactions of photosynthesis occur inside organelles called **chloroplasts**. In most plants, the interior cells of leaves have the most chloroplasts, but photosynthesis also occurs in other green parts of the plant body, such as the stems. By capturing sunlight and using that energy to produce sugar (which is a building block for the mass of a plant), photosynthesis is the process that provides food to nearly all life on Earth.

WHERE AND HOW PHOTOSYNTHESIS OCCURS

Most of the chemical reactions of photosynthesis take place within chloroplasts. A chloroplast has an extensive inner framework of membranes folded into stacks of disks, which gives the chloroplast a large surface area where these reactions can take place. The stacks are called **grana** (singular, *granum*); each individual disk is called a **thylakoid**. To proceed, photosynthesis requires water (H_2O) and carbon dioxide (CO_2). In most plants, water is absorbed by the roots and transported by veins to the cells where photosynthesis takes place. Carbon dioxide enters a leaf through tiny pores called *stomata* (singular, *stoma*). Oxygen gas (O_2) is given off as a by-product and exits through these same pores.

STRUCTURE OF CHLOROPLASTS

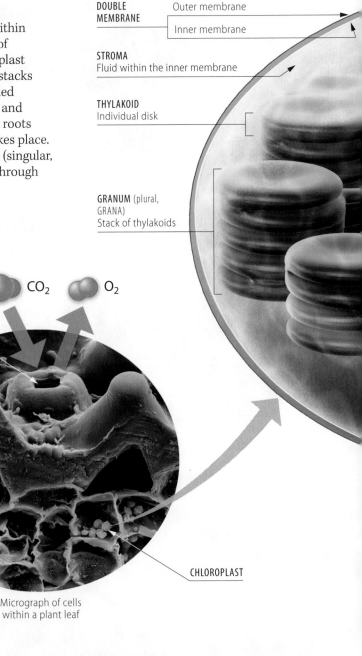

DOUBLE MEMBRANE
Outer membrane
Inner membrane

STROMA
Fluid within the inner membrane

THYLAKOID
Individual disk

GRANUM (plural, GRANA)
Stack of thylakoids

STOMA (plural, STOMATA)

CO_2 O_2

5,757x

CHLOROPLAST

Micrograph of cells within a plant leaf

H_2O

A typical oak leaf has about 10 billion chloroplasts.

CHLOROPHYLL

Chlorophyll is the primary pigment (light-absorbing molecule) in chloroplasts. Molecules of chlorophyll built into the thylakoid membranes selectively absorb light in the blue/violet and orange/red ranges. Reflected green/yellow light is what reaches our eye, and this is why plants appear green to us. Leaves also have other pigments; these become visible in autumn as the chlorophyll is broken down, producing the brilliant colors of fall foliage that you can see in the photo of a New England autumn.

SUNLIGHT

REFLECTED LIGHT

Isolated chlorophyll

Chlorophyll molecule

CORE IDEA: Photosynthesis uses CO_2 (obtained via stomata on leaves), H_2O (obtained via the roots), and light energy (absorbed by the pigment chlorophyll, found in the thylakoid membranes of chloroplasts) to produce sugars (used by the plant) and O_2 gas (released via stomata).

CORE QUESTION: Why do all photosynthesizing parts of a plant appear green?

ANSWER: Because chlorophyll found in chloroplasts absorbs other colors of light, reflecting yellow and green light to your eye.

Photosynthesis occurs in two linked stages

The overall equation for photosynthesis is relatively straightforward: carbon dioxide and water, along with the energy in sunlight, are used to produce sugar, releasing oxygen gas as a by-product. This process actually occurs in two stages. The **light reactions** capture sunlight and provide high-energy molecules to the **Calvin cycle**, which uses the high-energy molecules to produce sugar from carbon dioxide (CO_2).

Stage 1
THE LIGHT REACTIONS: CAPTURING ENERGY

Within thylakoids, energy from sunlight is absorbed by molecules of chlorophyll. This energy is used to split water, producing O_2 and high-energy electrons, which are stored by converting molecules of the electron carrier $NADP^+$ to NADPH. The energy from sunlight is also used to produce high-energy ATP molecules. To summarize, the light reactions capture the energy in sunlight and store it within high-energy molecules of ATP and NADPH.

$$6\,CO_2 + 6\,H_2O$$

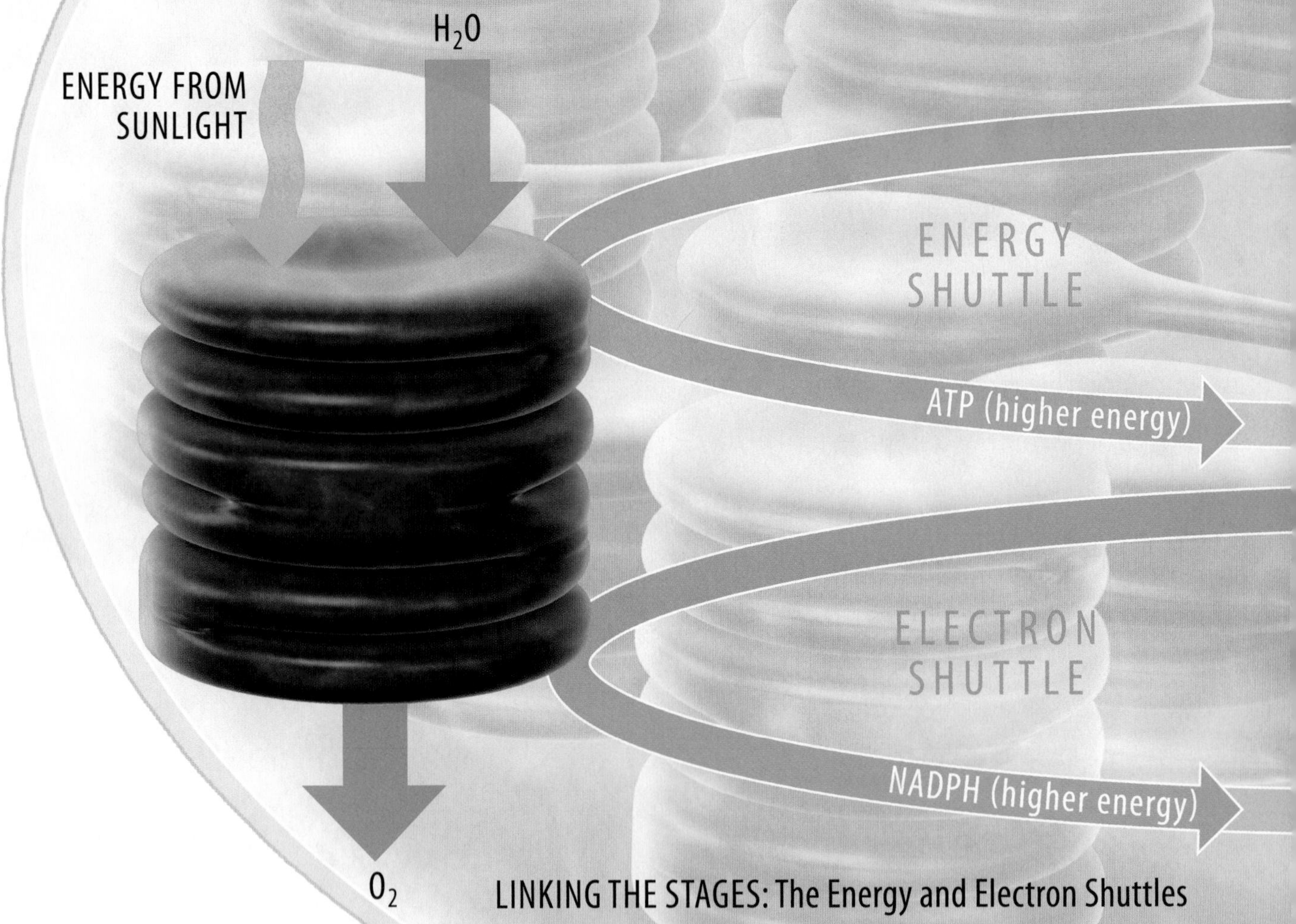

H_2O

ENERGY FROM SUNLIGHT

ENERGY SHUTTLE

ATP (higher energy)

ELECTRON SHUTTLE

NADPH (higher energy)

O_2

LINKING THE STAGES: The Energy and Electron Shuttles

The light reactions and the Calvin cycle are linked by energy and electron shuttles. **NADPH** is a molecule that acts as a high-energy electron shuttle. It is produced from a lower-energy form called $NADP^+$. **ATP** (adenosine triphosphate) is a high-energy molecule that acts as an energy shuttle. It is produced from a lower-energy molecule called ADP (adenosine diphosphate).

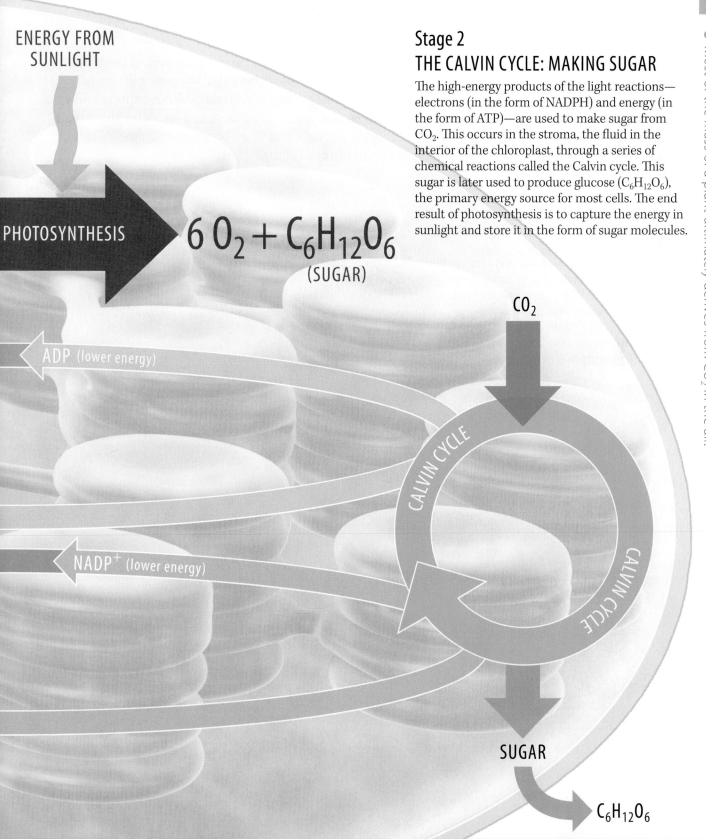

ENERGY FROM SUNLIGHT

PHOTOSYNTHESIS → $6\,O_2 + C_6H_{12}O_6$
(SUGAR)

ADP (lower energy)

NADP$^+$ (lower energy)

CO_2

CALVIN CYCLE

CALVIN CYCLE

SUGAR

$C_6H_{12}O_6$

Stage 2
THE CALVIN CYCLE: MAKING SUGAR

The high-energy products of the light reactions—electrons (in the form of NADPH) and energy (in the form of ATP)—are used to make sugar from CO_2. This occurs in the stroma, the fluid in the interior of the chloroplast, through a series of chemical reactions called the Calvin cycle. This sugar is later used to produce glucose ($C_6H_{12}O_6$), the primary energy source for most cells. The end result of photosynthesis is to capture the energy in sunlight and store it in the form of sugar molecules.

Most of the mass of a plant ultimately derives from CO_2 in the air.

CORE IDEA: The overall process of photosynthesis is broken down into two main stages: (1) The light reactions capture the energy in sunlight and store it as chemical energy, and (2) the Calvin cycle uses that chemical energy to produce sugar.

CORE QUESTION: What molecules would accumulate if the light reactions proceeded but the Calvin cycle did not? What molecules would not be made?

ANSWER: NADPH and ATP would accumulate, but no sugar would be made.

In the light reactions, the energy of sunlight is captured as chemical energy

Plants are able to capture the energy of sunlight and convert it to chemical energy stored in the bonds of sugar molecules. This is accomplished via the process of **photosynthesis** within organelles called chloroplasts. Photosynthesis proceeds through two stages: (1) the light reactions, which capture energy in sunlight and use it to produce high-energy molecules, and (2) the Calvin cycle, which uses these high-energy molecules to produce sugar. Here, we'll examine the light reactions. Next time you see a plant in the sunlight, try to imagine these reactions going on in the cells of the leaves.

PHOTOSYSTEMS

The cells of most plant leaves contain many chloroplasts, the organelles of photosynthesis. Each chloroplast contains grana, stacks of disk-like thylakoids. Each thylakoid is bound by a membrane. Within the thylakoid membranes are clusters of proteins and pigment molecules called **photosystems**. During the light reactions, pigment molecules within these photosystems capture the energy from sunlight and begin the process that converts the energy to forms that cells can use.

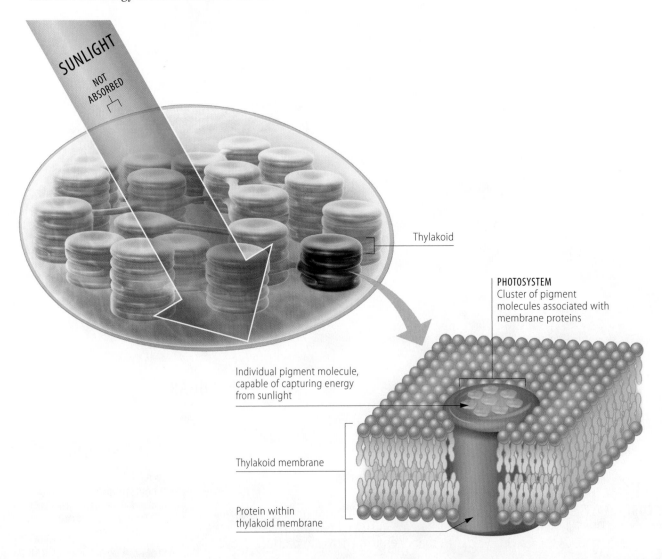

SUNLIGHT

NOT ABSORBED

Thylakoid

PHOTOSYSTEM
Cluster of pigment molecules associated with membrane proteins

Individual pigment molecule, capable of capturing energy from sunlight

Thylakoid membrane

Protein within thylakoid membrane

HOW A PHOTOSYSTEM WORKS

When light strikes a molecule of chlorophyll pigment within a photosystem, that energy is absorbed by an electron. This "excited" electron jumps to a high-energy state. Right after it does, the high-energy electron falls back down to its original (ground) state, giving up the energy that it had. This released energy is captured by a series of reactions called the electron transport chain and is used to convert molecules of ADP (adenosine diphosphate) into high-energy ATP (adenosine triphosphate) molecules. The electron is then reenergized in a second photosystem. As the electron returns to its ground state this time, the energy is used to produce molecules of NADPH. Notice that the final products of the light reactions—the high-energy molecules ATP and NAPDH—move to the Calvin cycle. Notice also that the electron involved in these reactions is supplied by splitting a molecule of water (H_2O), releasing a molecule of oxygen as a by-product.

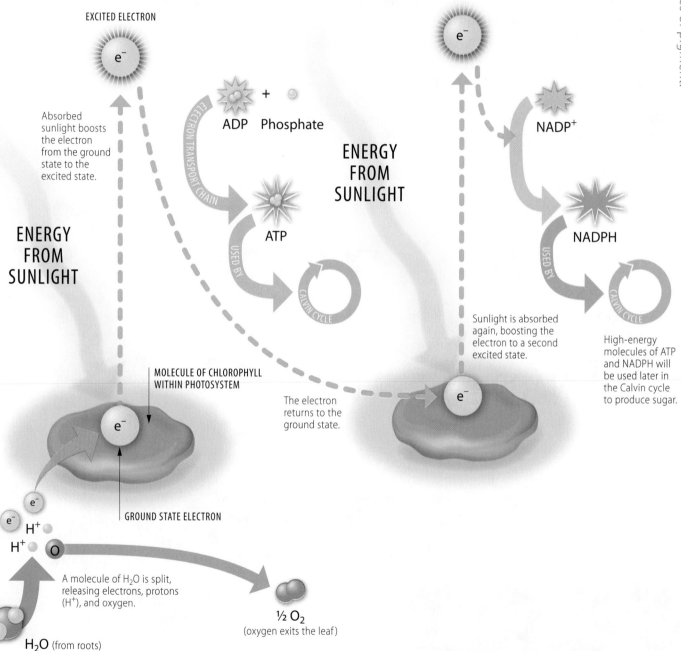

EXCITED ELECTRON

Absorbed sunlight boosts the electron from the ground state to the excited state.

ADP   Phosphate

ELECTRON TRANSPORT CHAIN

NADP⁺

ENERGY FROM SUNLIGHT

ATP

USED BY

NADPH

USED BY

CALVIN CYCLE

CALVIN CYCLE

ENERGY FROM SUNLIGHT

Sunlight is absorbed again, boosting the electron to a second excited state.

High-energy molecules of ATP and NADPH will be used later in the Calvin cycle to produce sugar.

MOLECULE OF CHLOROPHYLL WITHIN PHOTOSYSTEM

The electron returns to the ground state.

GROUND STATE ELECTRON

H⁺
H⁺   O

A molecule of H_2O is split, releasing electrons, protons (H⁺), and oxygen.

½ O₂
(oxygen exits the leaf)

H_2O (from roots)

CORE IDEA: Sunlight drives photosynthesis by exciting electrons, which then return to their original state. As they do, released energy is stored in the molecules ATP and NADPH, which are later used by the Calvin cycle to produce sugar.

CORE QUESTION: Why does a plant need water in order to perform photosynthesis?

ANSWER: Water molecules are split to provide the electrons that are used in the light reactions.

69

In the Calvin cycle, high-energy molecules are used to make sugar

Overall, the process of photosynthesis uses carbon dioxide (CO_2) and water (H_2O) to make sugars. This process is divided into two main stages. During the light reactions, energy from sunlight is used to create the high-energy molecules ATP and NADPH. In the Calvin cycle, these high-energy molecules are used to drive the production of sugar from CO_2, which the plants take in from the air via stomata (pores) in the leaves. The products of the Calvin cycle are vital to the lives of all plants and, since we consume plants as well as animals that consume plants, to our livelihood as well.

THE INPUTS OF THE CALVIN CYCLE

Within the stroma (fluid) of a chloroplast, the Calvin cycle uses the high-energy products of the light reactions (ATP and NADPH) and CO_2 from the air to drive the synthesis of sugars.

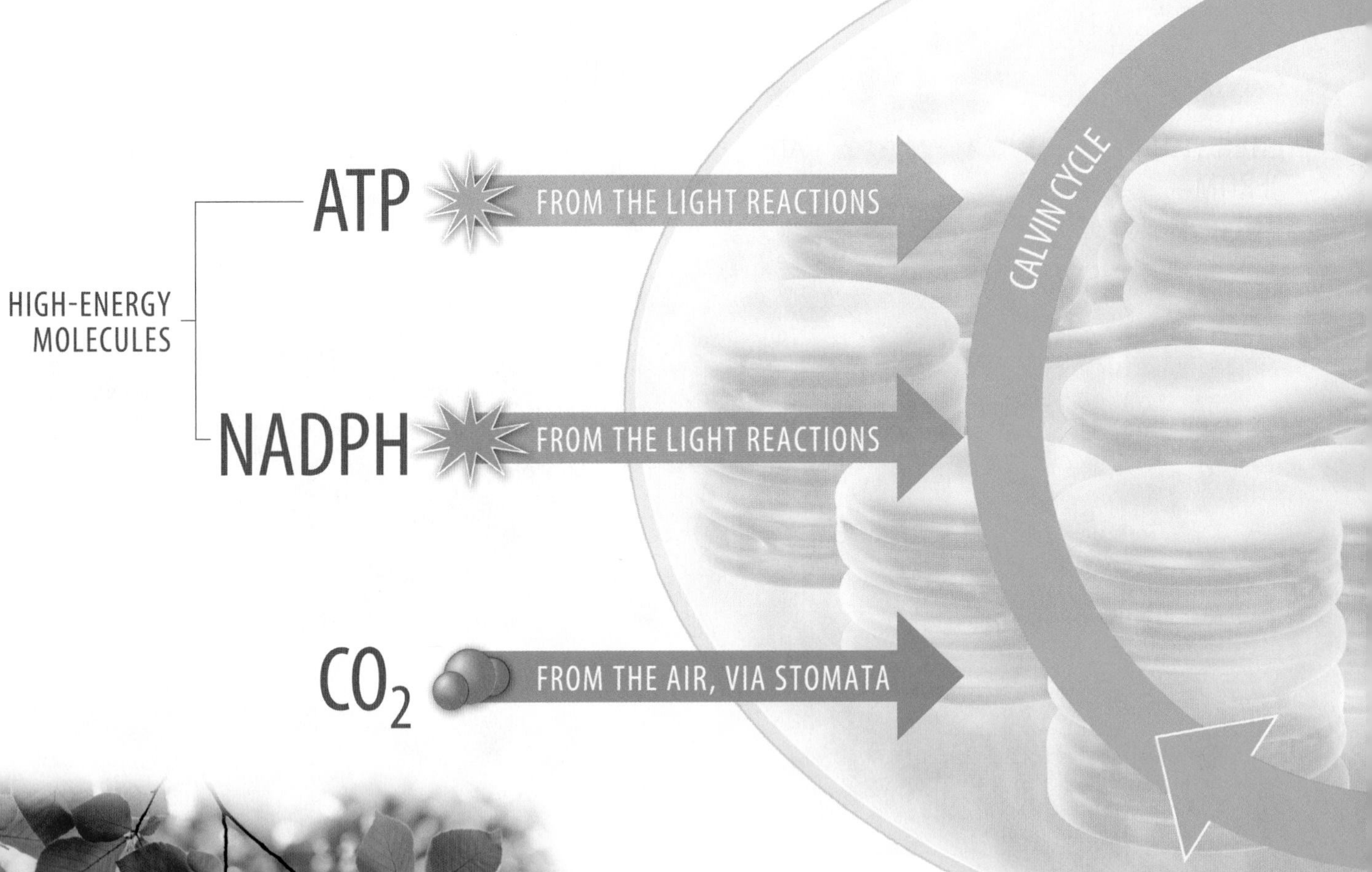

ATP   FROM THE LIGHT REACTIONS

HIGH-ENERGY MOLECULES

NADPH   FROM THE LIGHT REACTIONS

CO_2 FROM THE AIR, VIA STOMATA

CALVIN CYCLE

THE OUTPUTS OF THE CALVIN CYCLE

The Calvin cycle produces a 3-carbon sugar called glyceraldehyde 3-phosphate (G3P). After this sugar exits the chloroplast, the plant cell uses it to make glucose, which is then utilized by the plant in several ways.

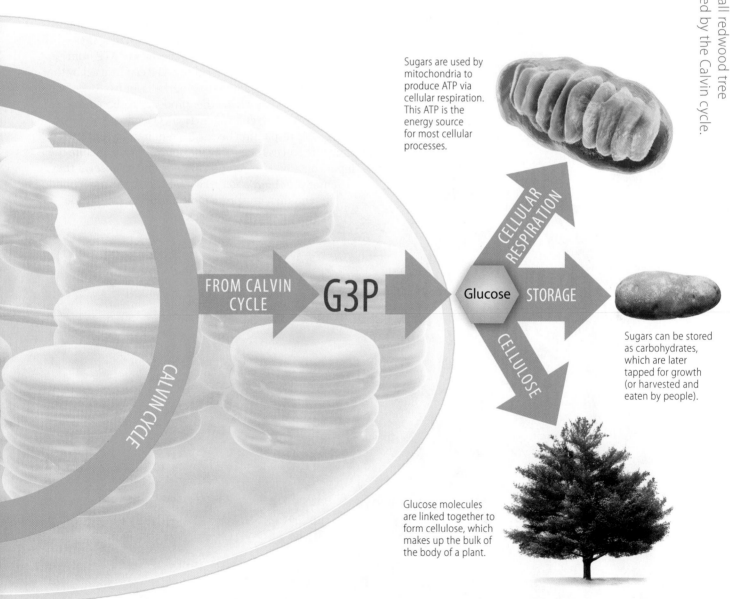

Sugars are used by mitochondria to produce ATP via cellular respiration. This ATP is the energy source for most cellular processes.

FROM CALVIN CYCLE

G3P

Glucose

CELLULAR RESPIRATION

STORAGE

CELLULOSE

CALVIN CYCLE

Sugars can be stored as carbohydrates, which are later tapped for growth (or harvested and eaten by people).

Glucose molecules are linked together to form cellulose, which makes up the bulk of the body of a plant.

CORE IDEA: The Calvin cycle uses high-energy molecules (provided by the light reactions) and CO_2 (from the air) to construct sugars, the ultimate product of photosynthesis. These sugars can then be used in a variety of ways by the plant.

CORE QUESTION: Can the Calvin cycle proceed in the dark? How?

ANSWER: Yes; a plant can create an excess of high-energy molecules during the daylight and then use those molecules to power the Calvin cycle in the dark.

In cellular respiration, oxygen is used to harvest energy stored in sugar

Why must we eat? Why do we have to constantly breathe? We eat and breathe because these processes provide the ingredients—glucose via the digestive system and oxygen via the respiratory and circulatory systems—required to power cellular respiration. Within the mitochondria of our cells, the enzymes of cellular respiration produce ATP, an energy storage molecule. The ATP produced by cellular respiration is used by all cells of the body to power life's processes. Without a steady supply of ATP via cellular respiration, our cells would quickly die.

THE INS AND OUTS OF RESPIRATION

AEROBIC RESPIRATION

Within the mitochondria of all eukaryotic cells (including consumers, such as animals, and producers, such as plants), the process of cellular respiration requires oxygen (O_2, obtained from the air) and glucose (obtained from food). Using these ingredients, our cells produce many molecules of ATP from each molecule of glucose. The resulting ATP can be used by the cell that made it or transported to other cells for use. Cellular respiration also produces two other products: carbon dioxide (CO_2, which is exhaled) and water (H_2O). Overall, this process is called aerobic respiration because it requires oxygen. Try to imagine this process occurring in your own body: Picture the inputs of cellular respiration (oxygen and glucose) as they enter your body and make their way to the mitochondria, and then imagine a molecule of carbon dioxide making its way back out.

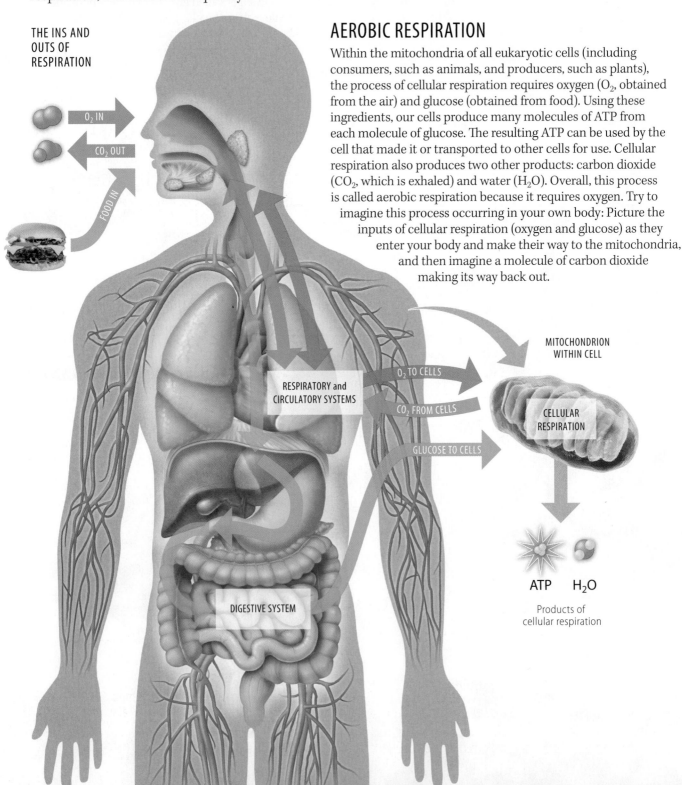

O_2 IN

CO_2 OUT

FOOD IN

RESPIRATORY and CIRCULATORY SYSTEMS

O_2 TO CELLS

CO_2 FROM CELLS

GLUCOSE TO CELLS

MITOCHONDRION WITHIN CELL

CELLULAR RESPIRATION

DIGESTIVE SYSTEM

ATP H_2O

Products of cellular respiration

ENERGY STORED IN ATP

Although the food you eat provides energy to your cells, it does not do so directly. The process of cellular respiration harvests the chemical energy stored in molecules of sugar and uses it to produce molecules of ATP from molecules of ADP. **ADP** stands for adenosine diphosphate, a molecule that contains two phosphate groups linked together. If another phosphate group is added to ADP (a process that requires energy to be spent), the result is ATP, or adenosine triphosphate, a molecule that contains three phosphate groups linked together. The third phosphate contains considerable potential energy that, when released, can be used by the cell for a variety of purposes. As shown in the figure, ATP acts like an energy shuttle, transferring chemical energy from bodily processes that provide energy (such as eating food) to ones that use energy (such as moving muscles).

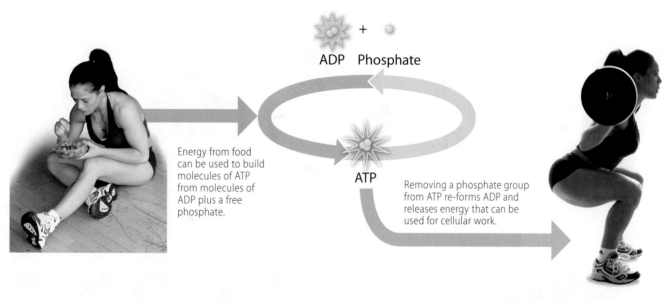

ADP   +   Phosphate

Energy from food can be used to build molecules of ATP from molecules of ADP plus a free phosphate.

ATP

Removing a phosphate group from ATP re-forms ADP and releases energy that can be used for cellular work.

KILOCALORIES

The amount of chemical energy stored in food is measured in kilocalories (kcal), which are thousands of calories. Kilocalories are often abbreviated as Calories (with a capital "C"). This same measure describes the amount of energy burned by your body as it performs any activity, such as exercise. Weight gain or loss is a simple matter of caloric arithmetic: If calories in from eating are greater than calories out from activity, you gain weight.

Energy in food	Walking = about 220 Calories per hour
1 hamburger (557 Calories)	2.2 hours
1 chocolate bar (220 Calories)	1 hour
1 orange (45 Calories)	0.2 hour

NUTRITION FACTS

Serving size: 1 bar (1.45 oz) (41g)

Amount Per Serving		
Calories 220	Calories from Fat 110	
		% Daily Value*
Total Fat 13 g		20%
Saturated Fat 8 g		40%
Trans Fat 0 g		
Cholesterol 10 mg		3%
Sodium 35 mg		1%
Total Carbohydrate 26 g		9%
Dietary Fiber 1 g		4%
Sugars 24 g		
Protein 3 g		
Vitamin A		0%
Vitamin C		0%
Calcium		8%
Iron		2%

Food labels contain information about how much energy is in each serving, expressed as Calories.

CORE IDEA: Cells use energy stored in molecules of ATP to perform the functions of life. ATP is assembled from ADP during the process of cellular respiration. Cellular respiration uses O_2 and sugar as inputs, and gives off CO_2 and H_2O as by-products.

CORE QUESTION: Trace a molecule of CO_2 through your body from its point of origin to where it exits.

ANSWER: CO_2 is released by mitochondria, exits the cell, enters the bloodstream, goes to the lungs, and is exhaled.

Cellular respiration is divided into three stages

Cellular respiration uses oxygen (O₂) to "burn" sugar (glucose, $C_6H_{12}O_6$), producing carbon dioxide (CO_2) and water (H_2O). This process releases a lot of energy, which is used to produce molecules of ATP. The overall equation for cellular respiration is fairly simple. Behind that equation, however, is a series of more than two dozen chemical reactions that form a pathway. This pathway can be divided into three broad stages: ❶ glycolysis, ❷ the citric acid cycle (sometimes called the Krebs cycle, after its discoverer), and ❸ the electron transport chain. Each of the three stages takes place in a different location within the cell.

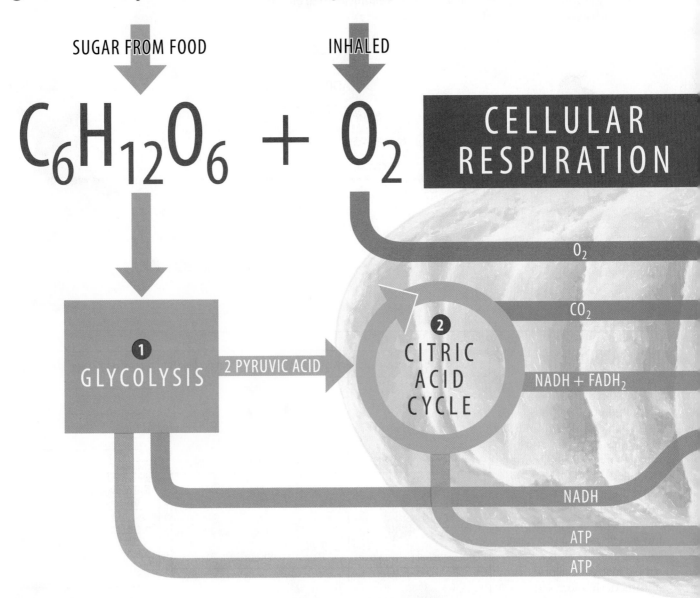

SUGAR FROM FOOD

INHALED

$$C_6H_{12}O_6 + O_2$$

CELLULAR RESPIRATION

O₂

CO₂

NADH + FADH₂

❶ GLYCOLYSIS

2 PYRUVIC ACID

❷ CITRIC ACID CYCLE

NADH

ATP

ATP

❶ GLYCOLYSIS

Glycolysis, which takes place in the cytoplasm, involves the splitting of one molecule of glucose into two molecules of pyruvic acid. This produces a bit of ATP and some high-energy electrons carried by a molecule called NADH.

❷ CITRIC ACID CYCLE

The **citric acid cycle**, which takes place in the fluid within mitochondria, completes the burning of glucose, breaking down pyruvic acid to CO_2, which is released from the cell. A bit of ATP is produced along with several high-energy electrons, which are carried by additional molecules of NADH as well as a molecule called FADH₂.

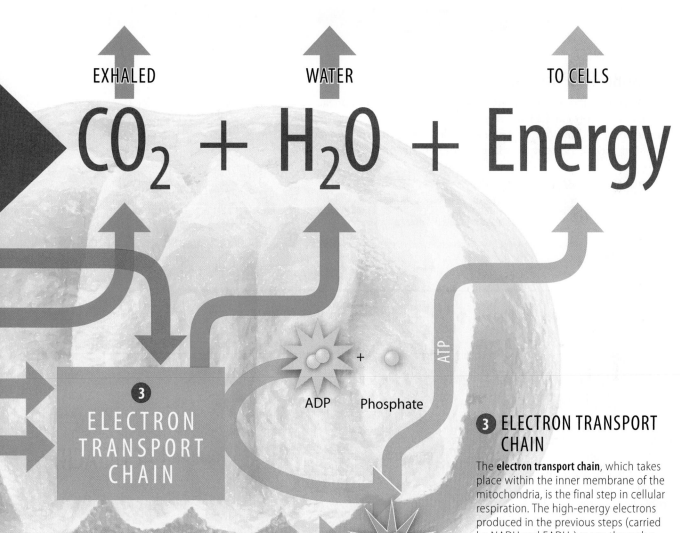

EXHALED WATER TO CELLS

$$CO_2 + H_2O + Energy$$

③
**ELECTRON
TRANSPORT
CHAIN**

ADP + Phosphate

ATP

ATP

③ ELECTRON TRANSPORT CHAIN

The **electron transport chain**, which takes place within the inner membrane of the mitochondria, is the final step in cellular respiration. The high-energy electrons produced in the previous steps (carried by NADH and $FADH_2$) move through a series of proteins. As they do, ATP is synthesized from ADP and phosphate. In the last step, the electrons are combined with O_2 to form H_2O, which is released as a by-product. Of the three steps shown here, the electron transport chain produces the most ATP by far.

Every molecule of CO_2 that you exhale can be traced back to the mitochondria of one of your cells.

CORE IDEA: Three stages of cellular respiration—glycolysis, citric acid cycle, and electron transport chain—use oxygen to gradually disassemble a molecule of glucose. The chemical energy released from the breakdown of glucose is used to form ATP.

CORE QUESTION: By the time that cellular respiration is complete, what has happened to the carbon atoms in the starting glucose molecule?

ANSWER: All of the carbon goes into molecules of carbon dioxide.

In fermentation, energy is harvested from sugar without oxygen

The process of cellular respiration uses oxygen (O_2) to break down glucose ($C_6H_{12}O_6$), producing carbon dioxide (CO_2), water (H_2O), and lots of ATP molecules that can be used by the cell as an energy source. This is why we need to breathe: O_2 helps ensure a steady supply of energy to all our working cells. The harvesting of food energy in the presence of oxygen, which is sustainable for as long as you keep breathing, is called aerobic cellular respiration:

$$C_6H_{12}O_6 + O_2 \xrightarrow{\text{CELLULAR RESPIRATION}} CO_2 + H_2O + \text{ENERGY}$$

ANAEROBIC RESPIRATION

Although the breakdown of glucose proceeds optimally in the presence of oxygen, food energy can be harvested in the absence of oxygen through an anaerobic ("without oxygen") process called **fermentation**. Anaerobic respiration plays important roles in the lives of many organisms.

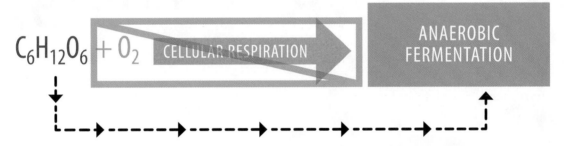

LACTIC ACID FERMENTATION IN HUMAN MUSCLE CELLS

If you exercise to the point of overexertion, you cannot breathe fast enough to supply your cells with the O_2 needed to maintain aerobic cellular respiration. During such anaerobic exercise, your muscles switch to an "emergency mode" in which glucose is fermented to produce lactic acid and a bit of ATP—much less ATP than is made via aerobic cellular respiration. This is not sustainable; after a few minutes of anaerobic exercise, your muscles will stop functioning (although no one is quite sure exactly why) and you will collapse.

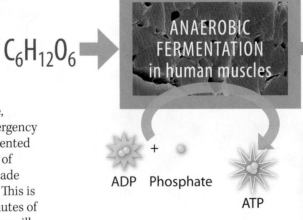

$$C_6H_{12}O_6 \rightarrow \boxed{\text{ANAEROBIC FERMENTATION in human muscles}} \rightarrow \text{LACTIC ACID}$$

ADP + Phosphate → ATP

LACTIC ACID FERMENTATION BY BACTERIA

Although your muscles cannot function under anaerobic conditions for very long, many microorganisms can use fermentation to provide for all their energy needs. Some bacteria, for example, can break down glucose in the absence of oxygen to produce lactic acid. Humans have domesticated such microorganisms to produce many familiar foods, most of which taste sharp due to the lactic acid.

Pickles

Olives

Yogurt

Pepperoni

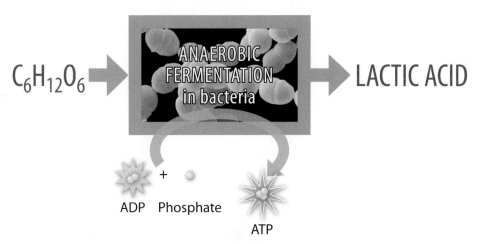

$C_6H_{12}O_6 \rightarrow$ ANAEROBIC FERMENTATION in bacteria $\rightarrow$ LACTIC ACID

ADP + Phosphate

ATP

ALCOHOL FERMENTATION BY YEAST

When placed in an anaerobic environment, yeast will ferment sugars, producing CO_2 and ethyl alcohol (ethanol). Beer and sparkling wines contain both of these products. When baking, CO_2 from fermentation causes bread to rise, but the alcohol is baked off.

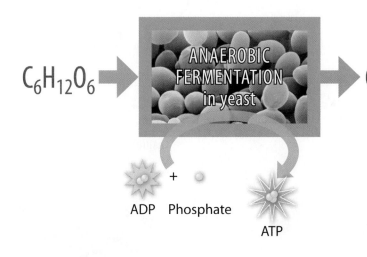

$C_6H_{12}O_6 \rightarrow$ ANAEROBIC FERMENTATION in yeast $\rightarrow CO_2 +$ ALCOHOL

ADP + Phosphate

ATP

Beer

Bread

Humans have been fermenting grain to produce beer for at least 5,570 years.

CORE IDEA: Fermentation is the anaerobic harvesting of energy from glucose. It produces far less ATP per molecule of glucose than cellular respiration. Human cells cannot run on fermentation alone for long, but some microbial cells can.

CORE QUESTION: What gas is contained within champagne bubbles? Where does it come from?

ANSWER: Champagne bubbles contain CO_2, produced through fermentation by yeast.

Cellular respiration is a central hub of many of life's metabolic processes

Although glucose is the primary fuel burned via cellular respiration, you don't actually eat much glucose. You obtain most of your food energy from other carbohydrates, fats, and proteins. Each of these food molecules can be used by cellular respiration, ultimately storing energy in the form of molecules of ATP, which can then power most cellular processes. Your **metabolism** is the sum total of all the chemical reactions that occur in your body. ATP is the central currency of your metabolism: All food energy leads to it, and all bodily work is powered by it.

VARIOUS FOOD MOLECULES CAN PROVIDE ENERGY

Although we have focused on glucose as the fuel for cellular respiration, many other food molecules also contribute to the body's production of ATP. Carbohydrates, fats, and proteins all feed into different points along the pathway of cellular respiration. In the end, the energy from all of them is used to make molecules of ATP.

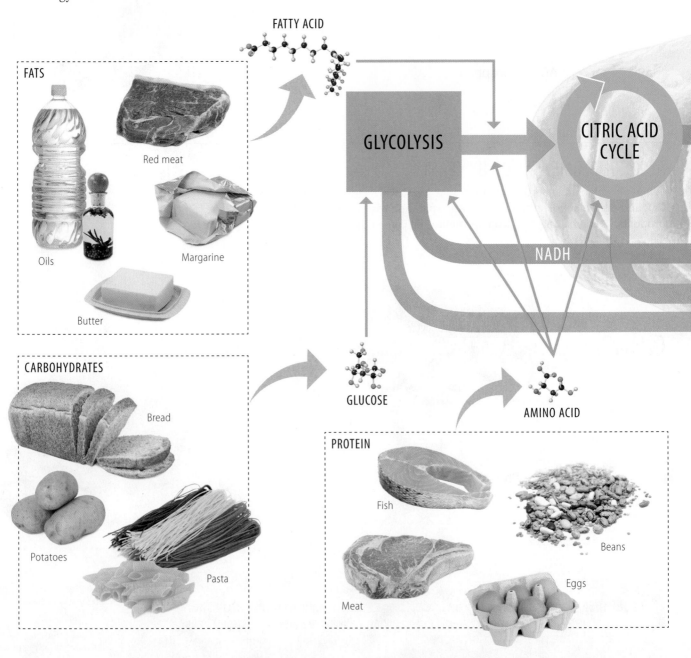

FATTY ACID

FATS
Red meat
Oils
Margarine
Butter

GLYCOLYSIS

CITRIC ACID CYCLE

NADH

CARBOHYDRATES
Bread
Potatoes
Pasta

GLUCOSE

AMINO ACID

PROTEIN
Fish
Beans
Meat
Eggs

An ounce of fat can be used to produce more than twice the ATP that an ounce of carbohydrate or protein can produce.

ENERGY USE BY ORGANISMS

The ATP produced by cellular respiration is used to power nearly all of life's processes. ATP is thus a key player in your metabolism and in the metabolism of all living organisms on Earth.

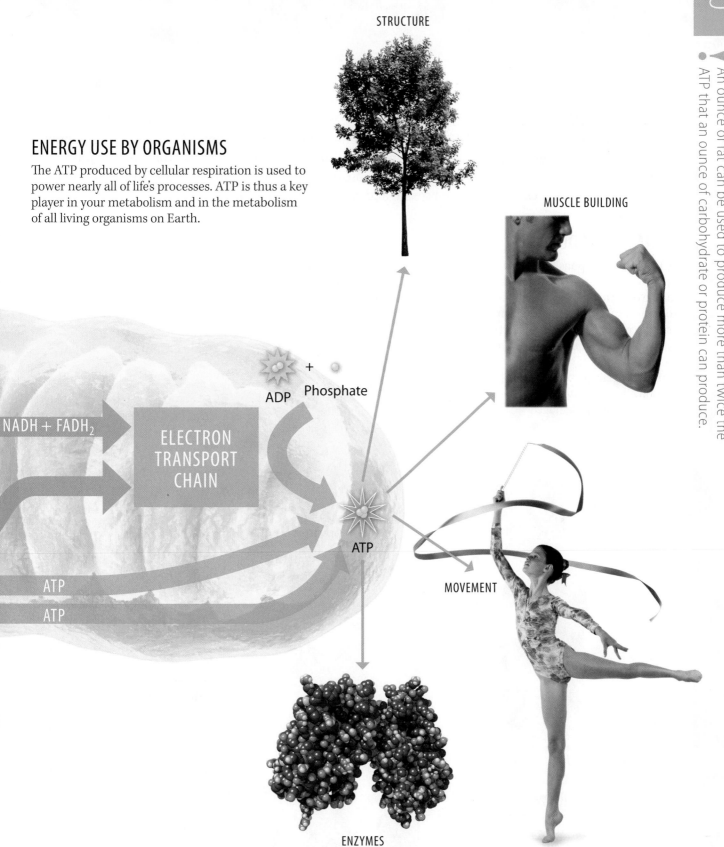

STRUCTURE

MUSCLE BUILDING

NADH + FADH$_2$

ELECTRON TRANSPORT CHAIN

ADP + Phosphate

ATP

ATP

ATP

MOVEMENT

ENZYMES

CORE IDEA: All the molecules in food that provide energy to your body do so by feeding into cellular respiration. The energy produced via cellular respiration and stored in molecules of ATP is used to power all of life's processes.

CORE QUESTION: What molecule serves as the central currency of energy in all your cells? From what molecule is it made?

ANSWER: ATP, ADP

Cell division provides for reproduction, growth, and repair

One of biology's most fundamental concepts is the **cell theory**, which states that all life is cellular, and that all cells arise from preexisting cells. The ability of cells to generate more cells through the process of **cell division** therefore underlies all of life's processes. When a cell undergoes cell division, the two offspring cells are genetically identical to each other and to the original parent cell. This is because, prior to cell division, the original cell replicates (copies) its chromosomes, the structures that contain most of the organism's DNA. During the process of cell division, one set of chromosomes is distributed to each offspring cell. This process is key to both sexual and asexual reproduction.

SEXUAL REPRODUCTION

Cell division is responsible for many important steps in the life cycle of sexually reproducing organisms. **Sexual reproduction** is the formation of genetically unique offspring. It begins with **fertilization**, the joining of gametes (sperm from the father and egg from the mother) to form a single cell called the **zygote**. Fertilization is followed by development of the embryo via many rounds of cell division. After birth, more cells produced via cell division allow the organism to grow and to repair damaged or old cells.

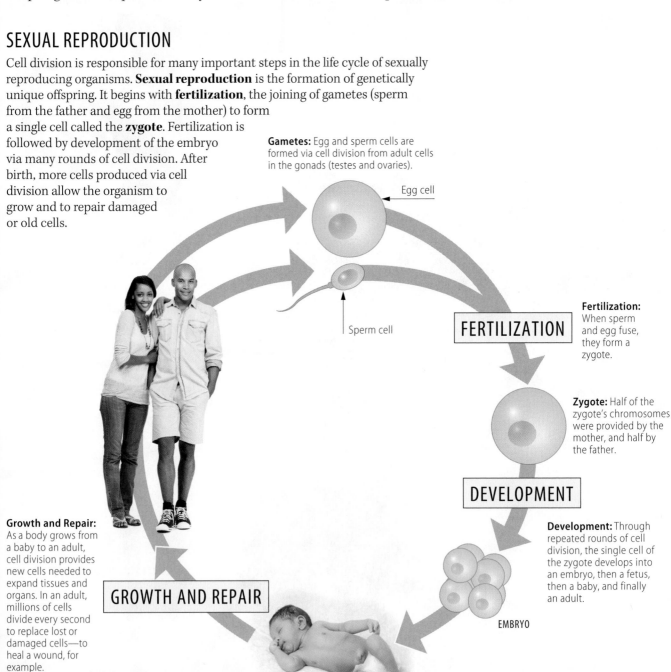

Gametes: Egg and sperm cells are formed via cell division from adult cells in the gonads (testes and ovaries).

Egg cell

Sperm cell

FERTILIZATION

Fertilization: When sperm and egg fuse, they form a zygote.

Zygote: Half of the zygote's chromosomes were provided by the mother, and half by the father.

DEVELOPMENT

Development: Through repeated rounds of cell division, the single cell of the zygote develops into an embryo, then a fetus, then a baby, and finally an adult.

EMBRYO

Growth and Repair: As a body grows from a baby to an adult, cell division provides new cells needed to expand tissues and organs. In an adult, millions of cells divide every second to replace lost or damaged cells—to heal a wound, for example.

GROWTH AND REPAIR

ASEXUAL REPRODUCTION

In addition to sexual reproduction via gametes, cell division can also result in **asexual reproduction**, the creation of a new individual by a lone parent without participation of sperm and egg. Because only one set of chromosomes is passed down, the offspring of asexual reproduction are genetically identical to each other and to the lone parent. Asexual reproduction plays an important role in the lives of many unicellular and multicellular organisms.

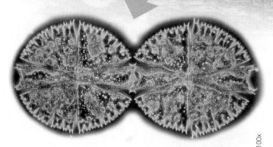

BINARY FISSION

Single-celled organisms can reproduce asexually by binary fission ("splitting into two"). In binary fission, an organism's single cell divides, creating two genetically identical offspring.

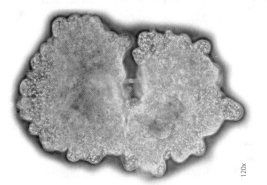

120x

100x

Amoeba proteus, a single-celled protist

Micrasterias, a single-celled green algae

PLANT REPRODUCTION

Most plants reproduce sexually, but many plants can also reproduce themselves asexually by sprouting (such as the eyes of potatoes) or by sending out runners. Humans take advantage of this ability by removing cuttings and using them to grow new plants.

Runner

Strawberry plant runners create new plants via asexual reproduction.

A cutting from a sunflower can be used to grow a new plant.

REGENERATION

Some animals can regenerate lost limbs, and some of them (such as the sea star shown here) can even regenerate an entire body starting from a severed limb.

This sea star is regrowing a lost arm.

This sea star is regrowing its whole body from one severed arm plus a bit of the central body.

Under ideal conditions, *E. coli* bacteria can duplicate themselves via binary fission every 20 minutes.

CORE IDEA: Cell division—the formation of cells from preexisting cells—underlies many important biological processes, including the formation of gametes, the development of an adult from an embryo, the growth and repair of bodily tissues, and asexual reproduction.

CORE QUESTION: From a cell division perspective, why are all of the trillions of cells in your body genetically identical?

ANSWER: Because they all descended via cell division from a single original cell (the zygote).

Chromosomes are associations of DNA and protein

All life on Earth uses **DNA** as the genetic material. One molecule of DNA may contain a great number of **genes**, the units of inheritance made from DNA that code for proteins. Almost all of the genes in a eukaryotic cell—about 21,000 in humans—are found on structures called **chromosomes**, located in the nucleus. (Here, we focus only on eukaryotes such as animals, plants, and fungi.) As you can see here, chromosome structure can be studied on several different levels.

CHROMOSOME COMPARISONS

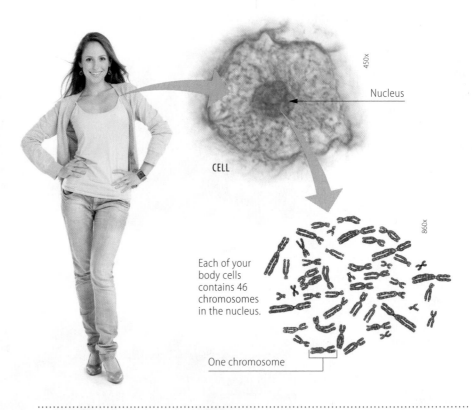

DIFFERENT SPECIES

78

56

46

40

20

8

0   25   50   75   100
NUMBER OF CHROMOSOMES IN A BODY CELL

CHROMOSOME NUMBER

All eukaryotic cells have a prominent nucleus where chromosomes are stored. Different species have different numbers of chromosomes in the nucleus. Every nonreproductive cell belonging to every organism of a particular species will have the same number of chromosomes. For example, all human body cells have 46 chromosomes. The chromosomes of members of the same species are nearly identical. Note from the graph that the number of chromosomes does not correspond to the size or complexity of an organism.

450x

Nucleus

CELL

860x

Each of your body cells contains 46 chromosomes in the nucleus.

One chromosome

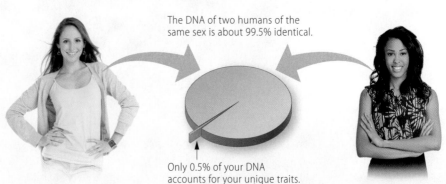

The DNA of two humans of the same sex is about 99.5% identical.

Only 0.5% of your DNA accounts for your unique traits.

If stretched out, the DNA in one of your cells would be taller than you.

CHROMOSOME STRUCTURE

Each chromosome is one very long piece of DNA, typically bearing thousands of genes, associated with proteins. Together, the DNA and the protein form a package called **chromatin**. The proteins in chromatin help keep the DNA tightly organized in the nucleus. Most of the time, the chromosomes are unraveled as thin chromatin fibers, which makes the genes accessible to cellular machinery. As a cell prepares to divide, it replicates (copies) its chromosomes and then compacts them into a dense pair called **sister chromatids**, which stay joined at the **centromere**. This compaction makes it easier to organize and divide the chromosomes. During cell division, the two genetically identical sister chromatids separate, one going to each offspring cell.

UNCONDENSED CHROMOSOMES

CONDENSED CHROMOSOMES

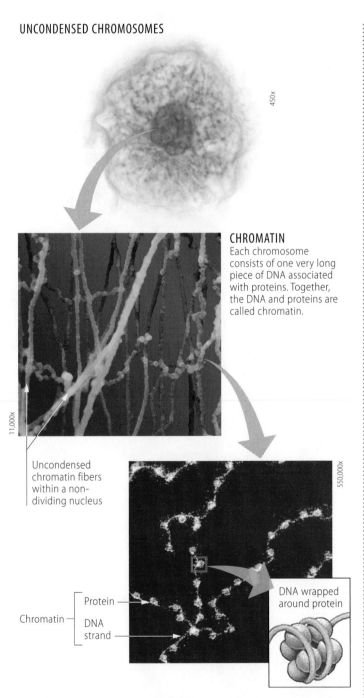

450x

11,000x

CHROMATIN
Each chromosome consists of one very long piece of DNA associated with proteins. Together, the DNA and proteins are called chromatin.

550,000x

Uncondensed chromatin fibers within a non-dividing nucleus

Chromatin ⎱ Protein
⎰ DNA strand

DNA wrapped around protein

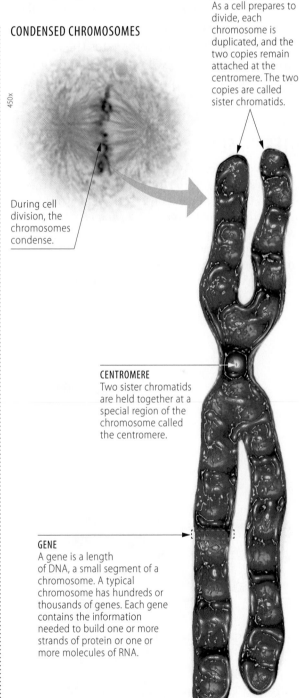

450x

SISTER CHROMATIDS
As a cell prepares to divide, each chromosome is duplicated, and the two copies remain attached at the centromere. The two copies are called sister chromatids.

During cell division, the chromosomes condense.

CENTROMERE
Two sister chromatids are held together at a special region of the chromosome called the centromere.

GENE
A gene is a length of DNA, a small segment of a chromosome. A typical chromosome has hundreds or thousands of genes. Each gene contains the information needed to build one or more strands of protein or one or more molecules of RNA.

CORE IDEA: Every eukaryotic cell contains chromosomes in the nucleus. Each chromosome consists of one long piece of DNA associated with proteins that help to compact and organize it. Together, the DNA and protein form chromatin.

CORE QUESTION: How many individual pieces of DNA are in the nucleus of each of your body cells?

ANSWER: Forty-six, one long piece of DNA for each chromosome.

Cells have regular cycles of growth and division

Every cell in your body was created through the division of a previously existing cell. The "lifetime" of a cell can be organized into a **cell cycle**, an ordered sequence of events from the time a cell is created from a dividing parent cell until its own division into two cells. The repeating events of the cell cycle are divided into two broad stages: **interphase**, during which the cell performs its normal functions and duplicates the chromosomes, and the **mitotic phase**, during which the cell divides the nucleus and distributes the duplicated chromosomes into two new cells.

INTERPHASE

Most of a cell's lifetime—typically around 90% of the cell cycle—is spent in interphase. During this phase, the cell performs its normal functions. A cell within your small intestine, for example, might make and release digestive enzymes. During interphase, the cell roughly doubles in size by building cytoplasm and organelles, and the chromosomes remain in an uncondensed state. As a cell prepares to divide, it duplicates its chromosomes within the nucleus.

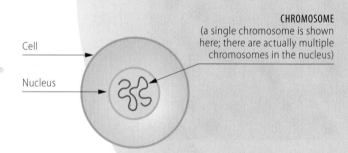

INTERPHASE

Cell grows and performs its normal functions

THE CELL CYCLE

CHROMOSOME
(a single chromosome is shown here; there are actually multiple chromosomes in the nucleus)

Cell

Nucleus

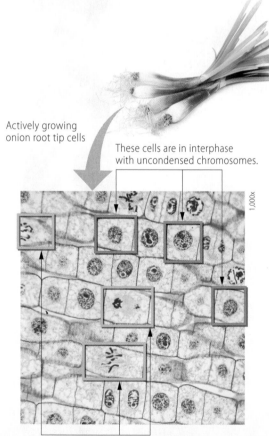

Actively growing onion root tip cells

These cells are in interphase with uncondensed chromosomes.

1,000x

These cells are in the mitotic phase with visibly compacted chromosomes.

CHROMOSOME DUPLICATION

During interphase, the cell duplicates each chromosome. During the mitotic phase, one copy of each chromosome is moved to each offspring cell. Here, we use just a single chromosome to represent the entire set of chromosomes found in the nucleus (46 in humans).

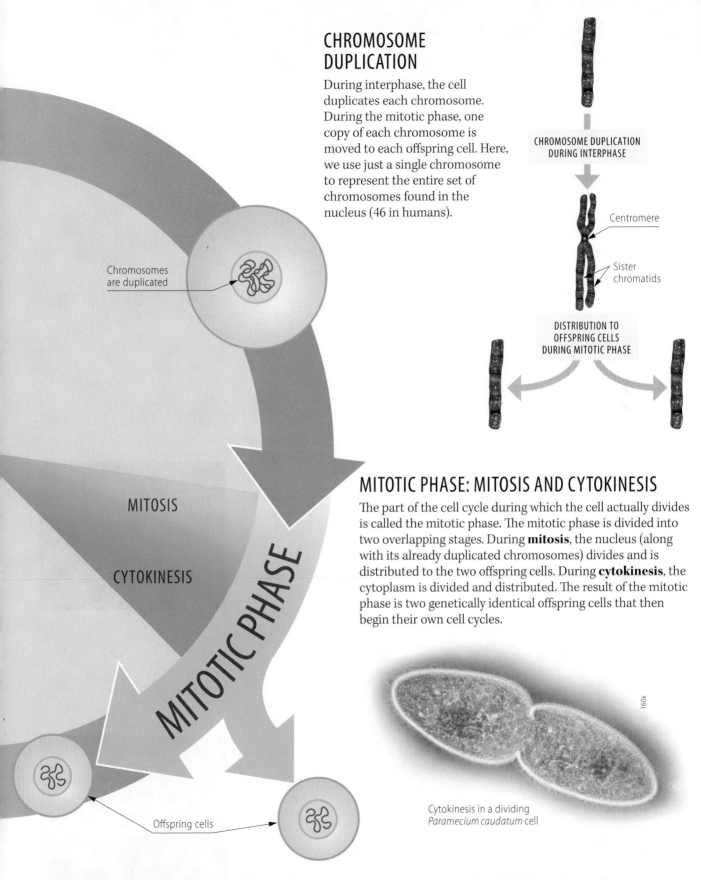

Chromosomes are duplicated

MITOSIS

CYTOKINESIS

MITOTIC PHASE

Offspring cells

CHROMOSOME DUPLICATION DURING INTERPHASE

Centromere

Sister chromatids

DISTRIBUTION TO OFFSPRING CELLS DURING MITOTIC PHASE

MITOTIC PHASE: MITOSIS AND CYTOKINESIS

The part of the cell cycle during which the cell actually divides is called the mitotic phase. The mitotic phase is divided into two overlapping stages. During **mitosis**, the nucleus (along with its already duplicated chromosomes) divides and is distributed to the two offspring cells. During **cytokinesis**, the cytoplasm is divided and distributed. The result of the mitotic phase is two genetically identical offspring cells that then begin their own cell cycles.

160x

Cytokinesis in a dividing *Paramecium caudatum* cell

CORE IDEA: The cell cycle is an ordered sequence of events that leads from the creation of a cell to the division of that cell. The cell cycle can be divided into two broad phases: interphase (during which the cell grows, duplicates its chromosomes, and prepares for division) and the mitotic phase (which includes mitosis and cytokinesis).

CORE QUESTION: Explain why the following statement is incorrect: "During mitosis, chromosomes are duplicated and distributed."

ANSWER: Because the chromosomes are already duplicated (during interphase) by the time the mitotic phase starts.

During mitosis, the nucleus is duplicated

An important outcome of the cell cycle—the ordered series of stages leading from the creation of a cell to the division of that cell—is the duplication and distribution of the nucleus, which contains the chromosomes. This is accomplished through a process called **mitosis**, the division of the nucleus into two offspring nuclei. During **interphase**—that is, before mitosis—the chromosomes are duplicated. During mitosis, the duplicated chromosomes are organized and separated. Here, you can see three representations of the events of mitosis: a simplified diagram, a more detailed computer-generated view, and on the right page, microscope images of a salamander cell undergoing mitosis.

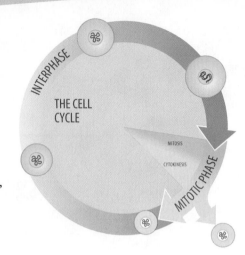

INTERPHASE

MITOSIS

CYTOKINESIS

MITOTIC PHASE

THE CELL CYCLE

A rapidly dividing cell can complete the steps of mitosis in less than a minute.

INTERPHASE

① EARLY INTERPHASE

During interphase, the chromosomes are uncondensed and loosely arranged in the nucleus. Each chromosome exists as a single very long, very thin piece of DNA with associated proteins. During this time, the cell carries out its normal activities.

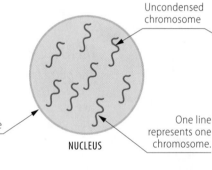

Uncondensed chromosome

Nuclear envelope

NUCLEUS

One line represents one chromosome.

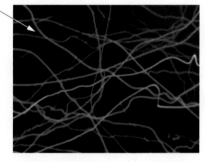

② CHROMOSOMES DUPLICATE

As the cell prepares to divide, each chromosome is duplicated. By the end of interphase, every chromosome in the nucleus has a duplicate, called a **sister chromatid**, attached to it at a structure called the **centromere**. The chromosomes are still uncondensed and so are not visible under a microscope.

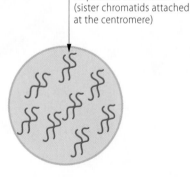

Duplicated chromosome (sister chromatids attached at the centromere)

Chromosome undergoing duplication

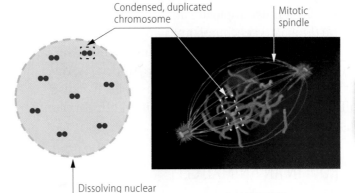

In steps 3–5, sister chromatids are represented as dots, as if looking at a top view of the chromosomes.

MITOSIS

3 CHROMOSOMES CONDENSE

At the start of mitosis, the nuclear membrane dissolves. The duplicated chromosomes become visible in the microscope as they condense (with the help of proteins) into tightly wound, thick, dense structures. The cell starts to lay down protein tracks (called the **mitotic spindle**) onto which the chromosomes can be organized.

Condensed, duplicated chromosome

Mitotic spindle

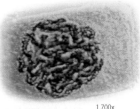

Dissolving nuclear membrane

4 CHROMOSOMES ALIGN

The duplicated chromosomes (pairs of sister chromatids attached at a centromere) line up in the center of the cell, with each chromosome attached to a separate track of the mitotic spindle.

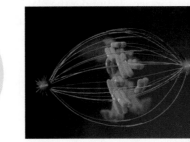

1,700x

5 CHROMOSOMES SPLIT

The sister chromatids split apart and the mitotic spindle retracts. As the tracks of the spindle retreat toward opposite ends of the cell, one chromatid from each pair of sister chromatids is dragged along with each track.

1,700x

6 NUCLEUS REFORMS

After the mitotic spindle is fully retracted, and each set of chromosomes is at one end of the cell, the nuclear membrane begins to re-form. The chromosomes uncondense. By the end, the nucleus has been duplicated into two nuclei, each with a full set of the original chromosomes.

Nuclear membrane re-forming

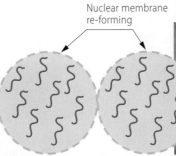

1,700x

CORE IDEA: Mitosis, the division of the nucleus, proceeds through a series of stages that organize and divide the duplicated chromosomes: The chromosomes condense, line up at the center of the cell, split into two groups, and then uncondense.

CORE QUESTION: What would happen if one pair of sister chromatids failed to split during mitosis?

ANSWER: One offspring cell would have one chromosome too many, and the other offspring cell would have one chromosome too few.

87

During cytokinesis, the cell is split in two

Before a cell can divide, it must duplicate all its chromosomes. During mitosis, the duplicated chromosomes are distributed into two nuclei, one for each of the new genetically identical offspring cells. The final step in the cell cycle is **cytokinesis**, the distribution of the cytoplasm to the two offspring cells. Because of structural differences, cytokinesis proceeds differently for plant and animal cells.

CLEAVAGE FURROW IN AN ANIMAL CELL

In animal cells, cytokinesis proceeds through a process called **cleavage**. The first sign of cleavage is the appearance of a **cleavage furrow**, an indentation around the equator of the cell. The cleavage furrow consists of a ring of protein filaments in the cytoplasm just under the plasma membrane. As the cells separate, the cleavage furrow contracts like the pulling of a drawstring, deepening the furrow. Eventually, the parent cell is pinched into two, leaving two independent offspring cells.

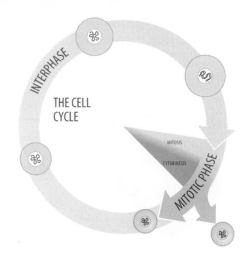

THE CELL
CYCLE

INTERPHASE

MITOSIS

CYTOKINESIS

MITOTIC PHASE

The proteins that form the cleavage furrow—called actin and myosin—are closely related to the proteins that allow muscles to contract.

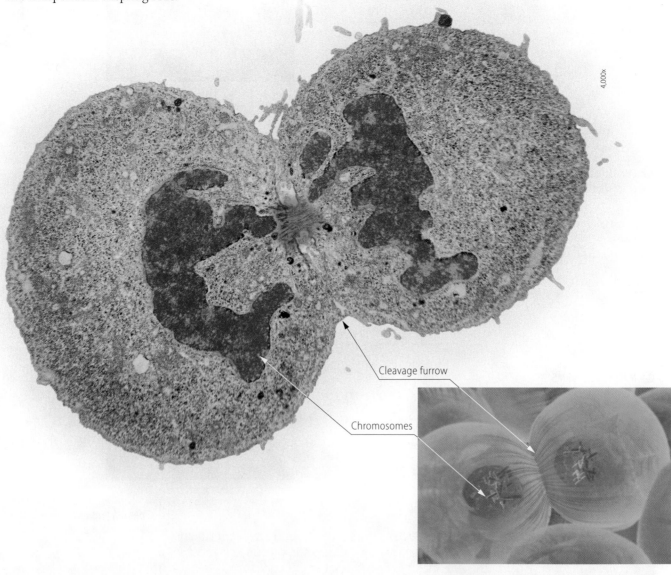

4,000x

Cleavage furrow

Chromosomes

CELL PLATE IN A PLANT CELL

Because plant cells have a stiff cell wall (which animal cells lack), cytokinesis proceeds differently in them than it does in animal cells. Plant cells divide their cytoplasm by forming a **cell plate**, a strip of membrane and cell wall materials that forms along the center line of the cell. The cell plate builds up and eventually fuses with the plasma membrane, separating the two offspring cells.

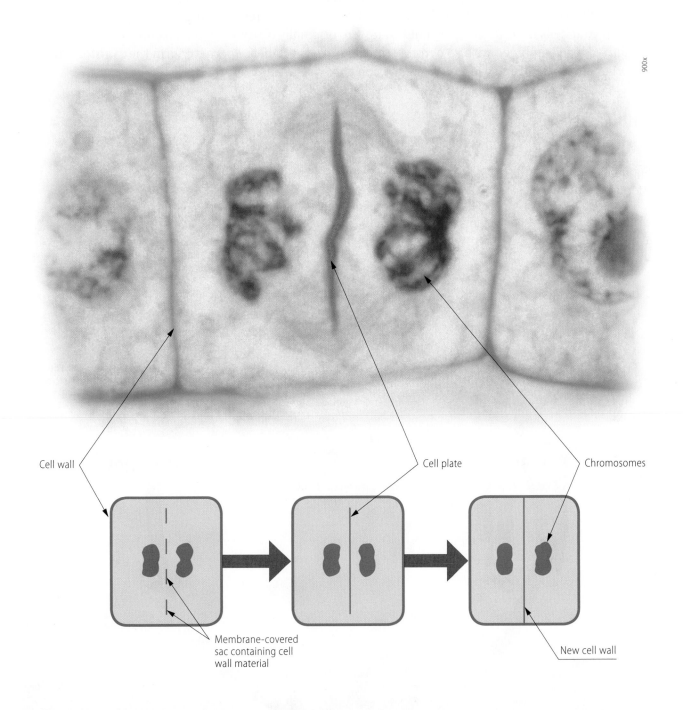

900x

Cell wall

Cell plate

Chromosomes

Membrane-covered sac containing cell wall material

New cell wall

CORE QUESTION: Why can't a plant cell pinch inward and form a cleavage furrow?

ANSWER: All plant cells are surrounded by a stiff cell wall that prevents pinching inward.

CORE IDEA: Cytokinesis, the division of the cytoplasm, is the last step in the cell cycle. In animal cells, cytokinesis proceeds via formation of a cleavage furrow. In plant cells, it proceeds via formation of a cell plate.

Gametes have half as many chromosomes as body cells

Even though all cells arise from previously existing cells, the cells of your body are genetically distinct from the cells of all other humans. Such variety is the result of **sexual reproduction**, the reproductive process that results from the fusion of **gametes** (sperm and egg). Such fusion is a key step in the **life cycle** of multicellular organisms—the stages leading from the adults of one generation to the adults of the next. In the human life cycle shown here (which is typical of many sexually reproducing animal species), you can see that all your cells have 46 chromosomes that were originally obtained when 23 chromosomes from your mother's egg combined with 23 chromosomes from your father's sperm to form the **zygote**, your single original cell. Repeated rounds of cell division gave rise to all your other cells.

THE HUMAN LIFE CYCLE

Every one of the trillions of cells in your body can trace its ancestry back through cell divisions to the zygote. Each gamete has a single set of chromosomes, a state called **haploid** (abbreviated as n). When gametes fuse, they form a **diploid** (abbreviated as $2n$) zygote, containing two matched sets of chromosomes. All of the **somatic cells** (body cells other than the reproductive gametes) of humans, most other animals, and many plants are diploid. Only gamete cells are haploid.

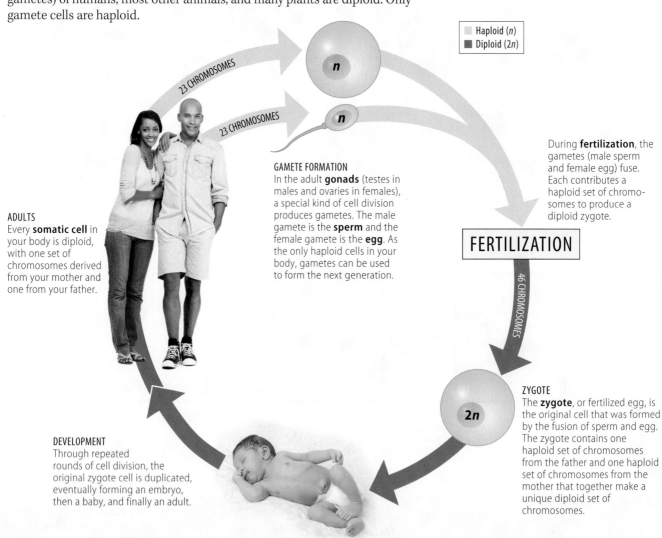

Haploid (n)
Diploid ($2n$)

23 CHROMOSOMES

23 CHROMOSOMES

n

n

During **fertilization**, the gametes (male sperm and female egg) fuse. Each contributes a haploid set of chromosomes to produce a diploid zygote.

FERTILIZATION

46 CHROMOSOMES

GAMETE FORMATION
In the adult **gonads** (testes in males and ovaries in females), a special kind of cell division produces gametes. The male gamete is the **sperm** and the female gamete is the **egg**. As the only haploid cells in your body, gametes can be used to form the next generation.

ADULTS
Every **somatic cell** in your body is diploid, with one set of chromosomes derived from your mother and one from your father.

ZYGOTE
The **zygote**, or fertilized egg, is the original cell that was formed by the fusion of sperm and egg. The zygote contains one haploid set of chromosomes from the father and one haploid set of chromosomes from the mother that together make a unique diploid set of chromosomes.

$2n$

DEVELOPMENT
Through repeated rounds of cell division, the original zygote cell is duplicated, eventually forming an embryo, then a baby, and finally an adult.

AN INVENTORY OF CHROMOSOMES

A **karyotype** is a photographic inventory of the chromosomes in one person's cells. A karyotype is made by breaking open a cell that is in the process of dividing, staining the chromosomes with dyes, taking a photograph with a microscope, and then arranging the chromosomes into matching pairs. If you look closely at the karyotype shown here, you can see that each chromosome is duplicated, consisting of two sister chromatids joined at the centromere. Notice also that every chromosome has a twin that resembles it in length and centromere position. Two chromosomes in a matching pair, called **homologous chromosomes**, carry genes controlling the same inherited characteristics. If a gene influencing eye color is located at a particular place on one chromosome, then the homologous chromosome has that same gene in the same location. However, the versions of the genes on the two homologous chromosomes may be the same or they may be different. For example, one chromosome may contain a gene that encodes for blue eyes, while the homologous chromosome contains a gene that encodes for eyes of a different color.

A person missing one sex chromosome can live a normal life, but missing an autosome is inevitably fatal.

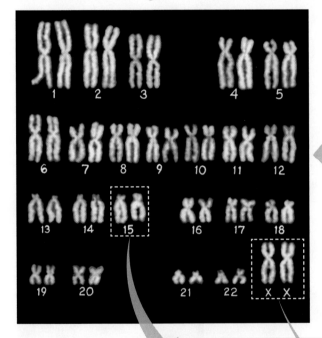

450x

CELL

ONE PAIR OF HOMOLOGOUS CHROMOSOMES

You have 46 chromosomes in your cells and they come in 23 sets of **homologous chromosomes**. The two chromosomes in a homologous pair are more than 99% identical to each other. They look identical under a microscope and contain the same genes in the same order (although the versions of the genes may vary). One member of each homologous pair was inherited from your mother, while the other member was inherited from your father.

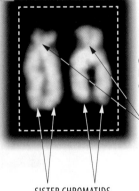

One pair of homologous chromosomes

Centromeres

SISTER CHROMATIDS
As a cell prepares to divide, each chromosome is duplicated. The two copies, called sister chromatids, are attached at the centromere.

SEX CHROMOSOMES
In humans, two of the 46 chromosomes are the **sex chromosomes** that determine sex (male vs. female). In human females, this pair of chromosomes is fully homologous (XX). But human males have an XY set of chromosomes that are homologous only in a small region. The other 44 (nonsex) chromosomes are called **autosomes**.

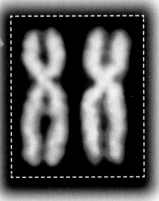

Two X chromosomes

CORE IDEA: The human life cycle involves fertilization (the fusion of haploid gametes to produce a diploid zygote), growth and development via cell division, and then production of gametes in the gonads. All human body cells have 46 chromosomes organized as 23 homologous pairs of chromosomes.

CORE QUESTION: If you examine a human cell and observe that it has 23 chromosomes including one Y chromosome, where must it be from?

ANSWER: It must be from a sperm cell (since it is haploid and has a Y chromosome).

Meiosis produces gametes

Sexual reproduction produces tremendous variety among offspring because each offspring inherits a unique combination of genes from its two parents. Sexual reproduction depends on **meiosis**, a type of cell division that produces gametes (sperm and egg) from cells in the gonads. The gametes produced via meiosis are haploid, with half the number of chromosomes as diploid body cells. For example, while cells in your liver, skin, or brain all have 46 chromosomes, your gametes have only 23. Thus, when two gametes fuse during fertilization, the resulting zygote will have 46 chromosomes.

MEIOSIS

Like mitosis, meiosis involves duplication of the chromosomes before division starts. But during meiosis, there are two rounds of cell division (whereas mitosis has only one). Thus, the chromosomes double, then are divided in half, then are divided in half again. The resulting cells have half the number of chromosomes as the starting cells.

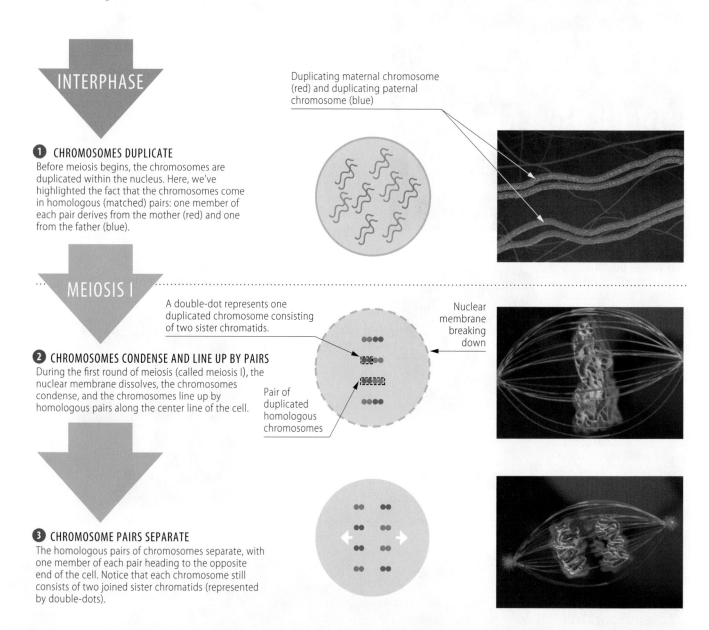

INTERPHASE

Duplicating maternal chromosome (red) and duplicating paternal chromosome (blue)

❶ CHROMOSOMES DUPLICATE
Before meiosis begins, the chromosomes are duplicated within the nucleus. Here, we've highlighted the fact that the chromosomes come in homologous (matched) pairs: one member of each pair derives from the mother (red) and one from the father (blue).

MEIOSIS I

A double-dot represents one duplicated chromosome consisting of two sister chromatids.

Nuclear membrane breaking down

❷ CHROMOSOMES CONDENSE AND LINE UP BY PAIRS
During the first round of meiosis (called meiosis I), the nuclear membrane dissolves, the chromosomes condense, and the chromosomes line up by homologous pairs along the center line of the cell.

Pair of duplicated homologous chromosomes

❸ CHROMOSOME PAIRS SEPARATE
The homologous pairs of chromosomes separate, with one member of each pair heading to the opposite end of the cell. Notice that each chromosome still consists of two joined sister chromatids (represented by double-dots).

In the testes of the human male, meiosis occurs hundreds of millions of times each day.

4 CYTOKINESIS

After the chromosome pairs separate, the cell divides into two cells. Inside each cell, the nucleus re-forms. There are now two offspring cells ready for meiosis II.

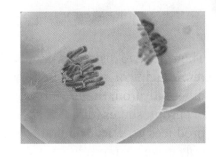

MEIOSIS II

5 CHROMOSOMES CONDENSE AND LINE UP

In each of the two cells produced by meiosis I, the chromosomes (still duplicated, with sister chromatids attached at a centromere) line up singly in the center of the cell.

6 CHROMOSOMES SEPARATE

The sister chromatids split apart, with one copy distributed to each side of the cell.

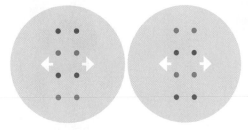
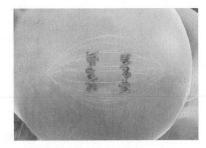

7 CYTOKINESIS II

At the end of meiosis II, each cell splits by cytokinesis, producing a total of four offspring cells, each with half the number of chromosomes as the starting cell. The nuclear envelopes re-form, and the chromosomes uncondense.

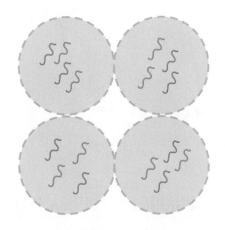
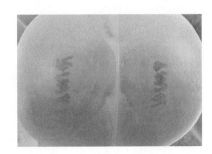

CORE IDEA: Meiosis is a type of cell division that produces haploid gametes from diploid body cells. One round of chromosome duplication is followed by two rounds of division. The result is four offspring cells, each with half the number of chromosomes as the starting cell.

CORE QUESTION: What is the key feature of meiosis I with respect to the arrangement of chromosomes?

ANSWER: The chromosomes line up by homologous pair (rather than singly, as they do in mitosis and meiosis II).

Mitosis and meiosis have important similarities and differences

There are two distinct types of cell division that play important roles during the life cycle of all sexually reproducing organisms. Both types are preceded by a duplication of chromosomes, but they proceed differently from there. Mitosis, which produces genetically identical offspring cells with a full complement of chromosomes (diploid cells), allows an organism to develop, grow, and repair itself. Meiosis, which produces genetically unique gametes with half the number of chromosomes (haploid cells), is involved only in sexual reproduction. As you can see from this summary, there are important differences in where, when, and how these two types of cell division occur.

Meiosis:

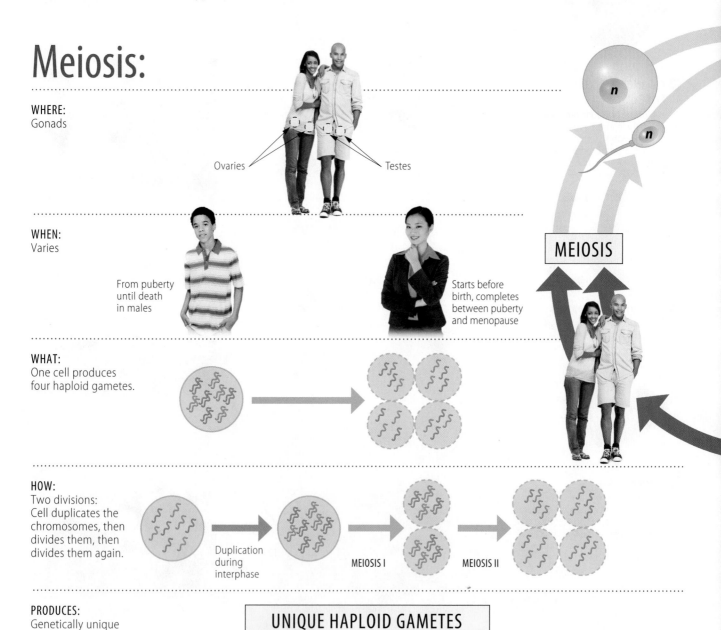

WHERE:
Gonads

Ovaries   Testes

WHEN:
Varies

From puberty until death in males

Starts before birth, completes between puberty and menopause

MEIOSIS

WHAT:
One cell produces four haploid gametes.

HOW:
Two divisions: Cell duplicates the chromosomes, then divides them, then divides them again.

Duplication during interphase

MEIOSIS I

MEIOSIS II

PRODUCES:
Genetically unique haploid gametes

UNIQUE HAPLOID GAMETES

Every living organism performs mitosis, but only sexually reproducing multicellular organisms undergo meiosis.

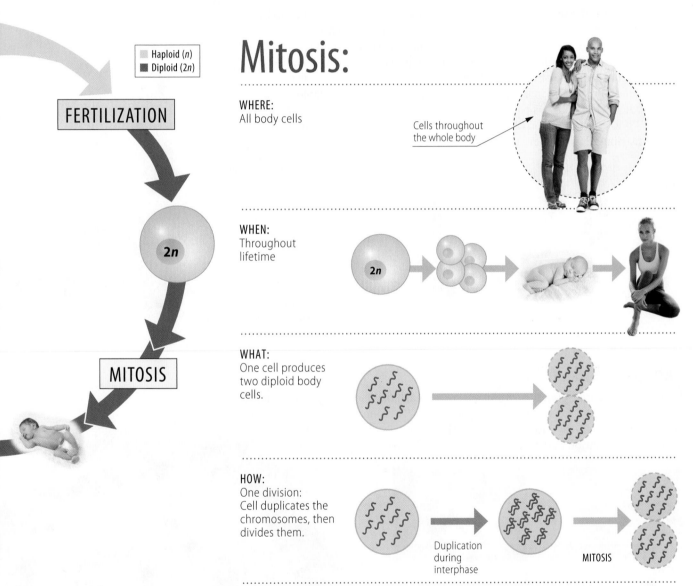

FERTILIZATION

Haploid (*n*)
Diploid (2*n*)

2*n*

MITOSIS

Mitosis:

WHERE:
All body cells

Cells throughout the whole body

WHEN:
Throughout lifetime

2*n*

WHAT:
One cell produces two diploid body cells.

HOW:
One division: Cell duplicates the chromosomes, then divides them.

Duplication during interphase

MITOSIS

PRODUCES:
Genetically identical diploid cells

IDENTICAL DIPLOID CELLS

CORE IDEA: Both mitosis and meiosis involve the duplication and distribution of chromosomes to offspring cells. However, mitosis involves one round of division that produces diploid cells, whereas meiosis involves two rounds that produce haploid cells.

CORE QUESTION: With respect to the chromosomes, is there any difference between a cell about to start mitosis and a cell about to start meiosis?

ANSWER: No; both have duplicated chromosomes.

Several processes produce genetic variation among sexually reproducing organisms

The process of sexual reproduction—involving **fertilization**, the union of sperm and egg—produces genetically unique offspring. Unless you happen to have an identical twin, you are genetically different from every other human. In fact, every organism produced through sexual reproduction is genetically distinct. How does such extreme variation arise? Here, you can see three ways in which sexual reproduction produces variations in organisms.

INDEPENDENT ASSORTMENT

During meiosis I (the first round of gamete formation), chromosomes line up by homologous pair. Each homologous pair contains one maternal chromosome (represented here in red) and one paternal chromosome (blue). The side-by-side orientation of each homologous pair of chromosomes is a matter of chance—it can be either red/blue or blue/red. Therefore, which member of any given homologous pair ends up in a particular gamete is random. For example, since each of your 23 homologous pairs of chromosomes could line up one of two ways, there are over 8 million (2^{23}) possible combinations of chromosomes that can be distributed to the gametes.

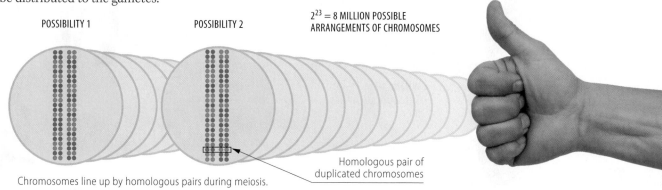

POSSIBILITY 1   POSSIBILITY 2

2^{23} = 8 MILLION POSSIBLE ARRANGEMENTS OF CHROMOSOMES

Chromosomes line up by homologous pairs during meiosis.

Homologous pair of duplicated chromosomes

RANDOM FERTILIZATION

Every human gamete contains 23 chromosomes, each of which is a random member from one of the 23 homologous pairs of the adult. When one gamete with random chromosomes fertilizes another gamete with random chromosomes, the result is tremendous variation in the 46 chromosomes that can wind up in the zygote.

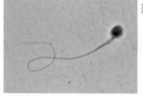

700x

SPERM
This sperm contains one of 8 million possible combinations of chromosomes.

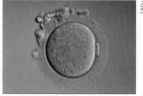

180x

EGG
This egg contains one of 8 million possible combinations of chromosomes.

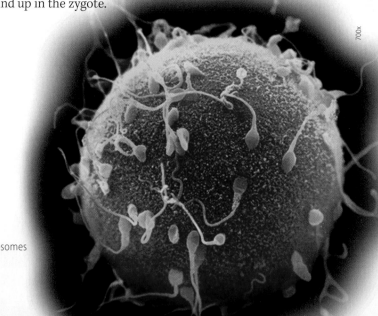

700x

ZYGOTE
8 million × 8 million = 64 trillion possible combinations of chromosomes in the zygote

The 23 pairs of homologous chromosomes in one of your cells can be arranged 8,388,608 different ways.

CROSSING OVER

So far, we have represented chromosomes as if they always remain whole. But, in fact, when you produce gametes, you do not actually pass on your chromosomes intact. When homologous chromosomes line up during meiosis, the maternal and paternal chromosomes swap pieces. This process, called **crossing over**, produces hybrid chromosomes that are partially paternal and partially maternal. As a result, new combinations of genes may be produced and passed on to offspring.

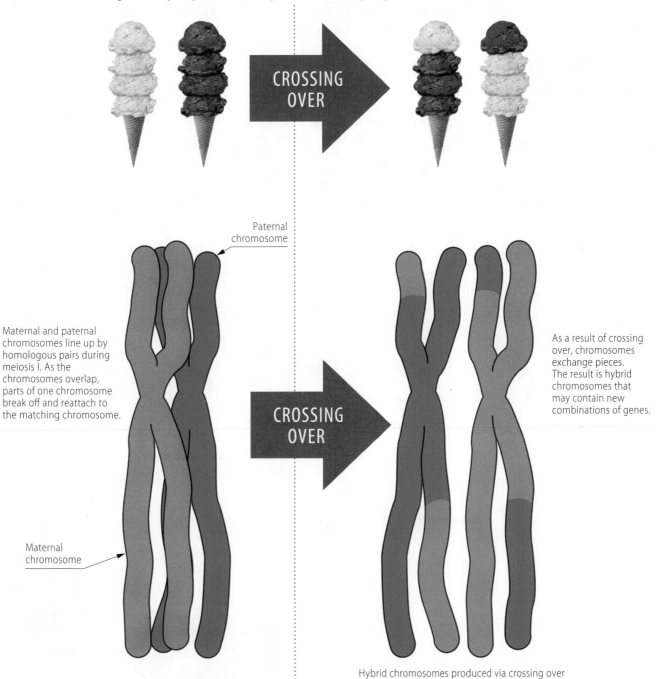

Paternal chromosome

Maternal and paternal chromosomes line up by homologous pairs during meiosis I. As the chromosomes overlap, parts of one chromosome break off and reattach to the matching chromosome.

CROSSING OVER

As a result of crossing over, chromosomes exchange pieces. The result is hybrid chromosomes that may contain new combinations of genes.

Maternal chromosome

Hybrid chromosomes produced via crossing over

CORE IDEA: Genetic variation is created during sexual reproduction through independent assortment of chromosomes, random fertilization, and crossing over. The first two of these processes shuffle chromosomes, while the third creates new hybrid chromosomes that may contain new combinations of genes.

CORE QUESTION: Is a chromosome in one of your gametes likely to be just like one that you received from your parents?

ANSWER: No. Due to crossing over, most chromosomes in your gametes are hybrids with parts from both your maternal and paternal chromosomes.

Mistakes during meiosis can produce gametes with abnormal numbers of chromosomes

When a sexually reproducing organism generates gametes through meiosis, these sex cells normally contain half the number of chromosomes as other body cells, a state called haploid, abbreviated n. But sometimes an error can occur during meiosis. The result may be gametes with an abnormal number of chromosomes. If this happens, and the abnormal gamete is involved in fertilization to produce a zygote, an organism with an unusual number of chromosomes may result. This is usually, but not always, fatal.

NONDISJUNCTION

Almost always, when chromosomes align in pairs and then separate during meiosis, exactly half of each set of chromosomes ends up in each offspring cell. But occasionally there is a mishap, called a **nondisjunction**, in which the chromosomes fail to separate properly. The result is gametes with unusual numbers of chromosomes (usually one too few or one too many). If an abnormal gamete is involved in fertilization, the resulting zygote will have an abnormal number of chromosomes. Most often, such a zygote cannot develop into a viable embryo. If it does develop, the resulting organism will usually have abnormalities.

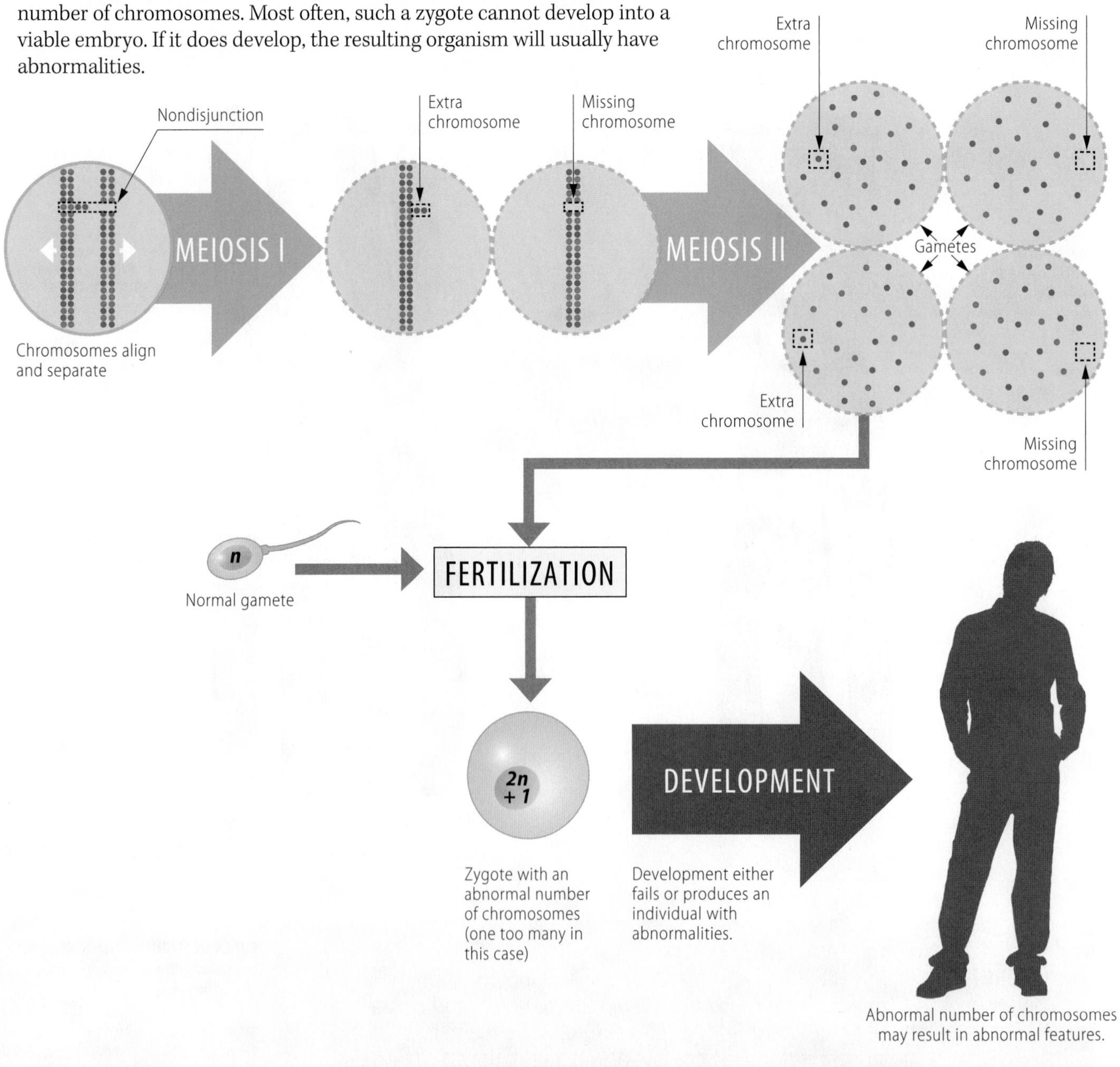

Nondisjunction

Chromosomes align and separate

MEIOSIS I

Extra chromosome

Missing chromosome

MEIOSIS II

Extra chromosome

Missing chromosome

Gametes

Extra chromosome

Missing chromosome

n

Normal gamete

FERTILIZATION

$2n$ $+ 1$

Zygote with an abnormal number of chromosomes (one too many in this case)

DEVELOPMENT

Development either fails or produces an individual with abnormalities.

Abnormal number of chromosomes may result in abnormal features.

TRISOMY 21

The vast majority of the time in humans, having too few or too many chromosomes is fatal, resulting in an early miscarriage. But sometimes such a zygote can survive, in which case the resulting person will have a condition caused by the abnormal number of chromosomes.

The most common chromosomal abnormality is **trisomy 21**, a condition in which a person has three copies of chromosome 21 (instead of the usual two) for a total of 47 chromosomes. Trisomy 21 produces **Down syndrome**. People with this condition will usually have a short stature, heart defects, an unusually shaped face, varying degrees of developmental and intellectual disabilities, and increased susceptibility to certain diseases.

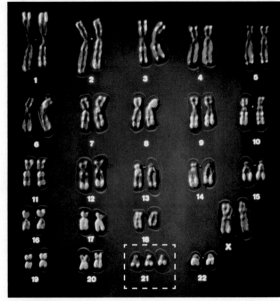

This image shows the complete set of chromosomes for an individual with Down syndrome. Notice the extra copy of chromosome 21.

SEX CHROMOSOME ABNORMALITIES

Nondisjunction can affect sex chromosomes as well as autosomes. If an embryo lacks an X chromosome, it cannot survive. But other abnormalities in the number of sex chromosomes—such as having a single X (and no Y), or extra Xs or Ys—have less dramatic effects. Sex chromosome abnormalities are usually not fatal, but each combination produces a characteristic syndrome whose effects can be mild (as in XXX) or profound (as in XO).

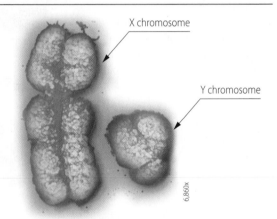

Human sex chromosomes

SEX CHROMOSOME ABNORMALITIES				
Sex chromosomes	Syndrome	Frequency in population	Sex	Effect
XXY	Klinefelter syndrome	$\frac{1}{2,000}$	Male	Normal intelligence but underdeveloped testes and some feminine features
XYY	Jacob's syndrome	$\frac{1}{2,000}$	Male	Often taller than average
XXX	Triple-X syndrome	$\frac{1}{1,000}$	Female	Normal
XO	Turner syndrome	$\frac{1}{5,000}$	Female	Short, web of skin connecting neck and shoulders, sterile, underdeveloped sex organs, normal intelligence

CORE IDEA: Nondisjunction is an error during meiosis that produces gametes with abnormal numbers of chromosomes. This is usually fatal but can occasionally produce an adult with a characteristic syndrome.

CORE QUESTION: What cellular process can result in a person with 47 chromosomes?

ANSWER: Nondisjunction during meiosis can produce gametes with abnormal numbers of chromosomes.

Mendel deduced the basic principles of genetics by breeding pea plants

Heredity is the transmission of traits from one generation to the next. **Genetics**, the scientific study of heredity, can trace its roots back to the 19th-century experiments of Gregor Mendel. Working with pea plants in the garden of his monastery, Mendel was able to deduce many of the basic principles of inheritance. Although at that time people had no knowledge of DNA or genes, Mendel determined that some discrete substances (which he called "heritable factors") were passed from parents to offspring during reproduction and that these factors retained their individual identities from generation to generation. From these simple ideas, Mendel deduced the underlying principles of genetics.

Gregor Mendel
(1822–1884) at age 34

CHARACTERS AND TRAITS

When discussing genetics, we usually focus on a **character**, an inherited feature that varies from one individual to another. Each character comes in two or more variations, called **traits**. Here, you can see traits of a pea plant character that Mendel studied as well as a more familiar example.

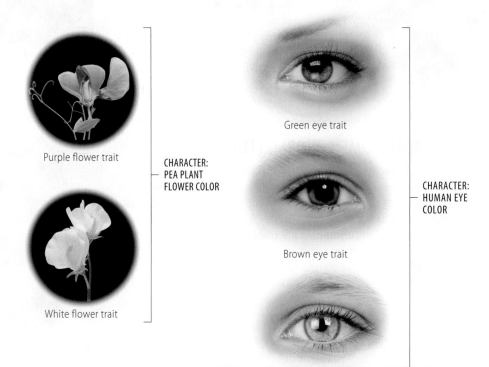

Purple flower trait

White flower trait

CHARACTER:
PEA PLANT
FLOWER COLOR

Green eye trait

Brown eye trait

Blue eye trait

CHARACTER:
HUMAN EYE
COLOR

ALLELES

Traits derive from genes, individual units of inheritance that comprise stretches of DNA on chromosomes. Many plants and animals (including peas and humans) are diploid, meaning that their chromosomes and genes come in matched sets. One member of each set comes from the organism's mother and one from the father. Alternate forms of a particular gene are called **alleles**. Two alleles for a gene may be identical (in which case the organism is said to be **homozygous** for that trait) or they may be different (in which case the organism is said to be **heterozygous** for that trait).

Matched set of chromosomes, one derived from the father (blue) and one derived from the mother (red)

Heterozygous (two different alleles) for this gene

Homozygous (two identical alleles) for this gene

B

D

b

D

5.11

In one set of experiments, Mendel bred and analyzed over 20,000 pea plants.

DOMINANT VS. RECESSIVE ALLELES

An individual who is heterozygous for a given character has two different alleles for a particular gene. Only one of these alleles will usually determine the organism's appearance for that character. This version of the gene is called the **dominant allele** (and is represented by an uppercase letter). The other allele, the **recessive allele** (represented by a lowercase letter), has no noticeable effect, although it may be passed on to offspring.

CHARACTER	PEA FLOWER COLOR	PEA SEED COLOR	LABRADOR COAT COLOR	HUMAN FRECKLES
Dominant trait	*P* (purple)	*Y* (yellow)	*B* (black)	*F* (freckled)
Recessive trait	*p* (white)	*y* (green)	*b* (chocolate)	*f* (not freckled)

GENOTYPE VS. PHENOTYPE

An organism's **phenotype** is its physical traits, while the organism's **genotype** is its underlying genetic makeup. One of the main principles of genetics is: *In combination with the environment, genotype causes phenotype through the action of genes.*

GENOTYPE

PHENOTYPE

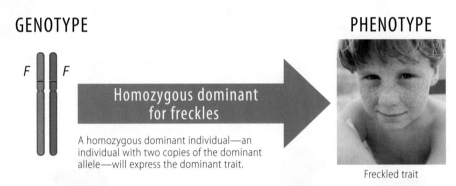

F *F*

Homozygous dominant for freckles

A homozygous dominant individual—an individual with two copies of the dominant allele—will express the dominant trait.

Freckled trait

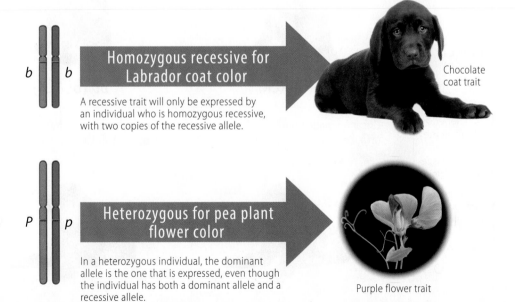

b *b*

Homozygous recessive for Labrador coat color

A recessive trait will only be expressed by an individual who is homozygous recessive, with two copies of the recessive allele.

Chocolate coat trait

P *p*

Heterozygous for pea plant flower color

In a heterozygous individual, the dominant allele is the one that is expressed, even though the individual has both a dominant allele and a recessive allele.

Purple flower trait

CORE IDEA: An organism's physical traits (phenotype) are caused by its genetic makeup (genotype). Most characters are controlled by two copies of a gene. The two copies are called alleles. The two alleles for a trait may be identical (homozygous) or they may be different (heterozygous).

CORE QUESTION: If a Labrador retriever has a black coat, can you tell for certain what coat color genes it has?

ANSWER: No. It may be homozygous dominant (*BB*) or heterozygous (*Bb*).

A Punnett square can be used to predict the results of a genetic cross

If you breed a black Labrador retriever with a chocolate Lab, what color can you expect the puppies in the litter to be? Such an experiment is called a **genetic cross**. In a genetic cross, two individuals of the **P generation** (parents) are crossed to produce the **F1 generation** (first-generation offspring). A genetic cross can be represented using a diagram called a **Punnett square**.

MONOHYBRID CROSS

A **monohybrid cross** is a genetic cross that follows just one character of the two parents. The Punnett square allows you to predict the genetic makeup (genotype) and appearance (phenotype) of the offspring by showing the probability of each possible outcome. The overall results are often expressed as a ratio, numbers that represent the relative number of individuals displaying each possible outcome.

B = black allele (dominant)
b = chocolate allele (recessive)

bb: Homozygous recessive chocolate-coated Labrador father

GAMETES (SPERM) PRODUCED

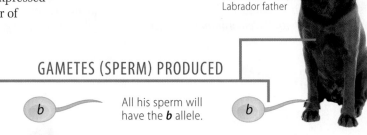

All his sperm will have the *b* allele.

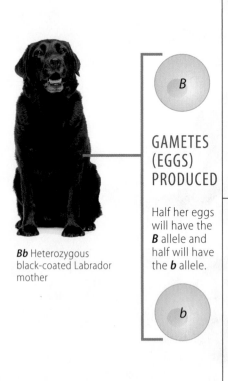

Bb Heterozygous black-coated Labrador mother

GAMETES (EGGS) PRODUCED

Half her eggs will have the *B* allele and half will have the *b* allele.

Offspring made from fertilization of *B* egg and *b* sperm

Bb (BLACK)

Offspring made from fertilization of *B* egg and *b* sperm

Bb (BLACK)

Offspring made from fertilization of *b* egg and *b* sperm

bb (CHOCOLATE)

Offspring made from fertilization of *b* egg and *b* sperm

bb (CHOCOLATE)

RESULTS: 2 *Bb* (BLACK) : 2 *bb* (CHOCOLATE)
Ratio of phenotypes = 1 black dog : 1 chocolate dog

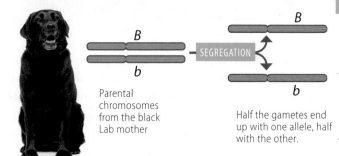

LAW OF SEGREGATION

Predicting the results of a genetic cross with a Punnett square requires following genes from one generation to the next. During the production of gametes, the pair of alleles for any given character separates, with one allele going into half the gametes and the other allele going into the other half of the gametes. This principle is called the **law of segregation** because the two members of the pair segregate (separate) during gamete formation. Since we cannot predict which allele will go into any given gamete, we need to consider all four possibilities (two possible gametes from the mother mixing with two possible gametes from the father), which is why the Punnett square has four boxes.

Parental chromosomes from the black Lab mother

Half the gametes end up with one allele, half with the other.

Most human characters—including eye color and hair color—are too complex to be predicted using a monohybrid cross.

TEST CROSS

Suppose you have a Labrador retriever with a chocolate coat. You can tell that its genotype must be *bb* because that is the only combination of alleles that produces the chocolate-coat trait. But what if you have a black Lab? It could have one of two possible genotypes—*BB* or *Bb*—and there is no way to tell which is correct by looking at the dog. To determine your dog's genotype, you could perform a **test cross**, a mating between an individual of dominant phenotype but unknown genotype (your black Lab) and a homozygous recessive individual—in this case, a *bb* chocolate Lab. As you can see here, the results of the test cross can tell you the genetic makeup of the original black dog.

Is the genotype of this black Lab **BB** or **Bb**? To find out, mate it with a chocolate Lab.

Chocolate Lab with genotype **bb**

If black dog has the *BB* genotype

If black dog has the *Bb* genotype

RESULTS: ALL BLACK
If all the offspring are black, the black parent most likely had genotype **BB**.

RESULTS: 2 BLACK : 2 CHOCOLATE
If any of the offspring are chocolate, the black parent must be heterozygous (**Bb**).

CORE IDEA: A Punnett square can be used to predict the offspring that will result from a genetic cross. A monohybrid cross is a genetic mating that follows a single trait. The law of segregation states that the two alleles for a character separate during gamete formation. Alleles from separate parents then join in the offspring.

CORE QUESTION: If you owned a chocolate Lab and wanted to know its genotype, would you learn anything from a test cross?

ANSWER: No. You already know that a chocolate Lab must be bb.

Mendel's law of independent assortment accounts for the inheritance of multiple traits

A **dihybrid cross** is one in which two separate characters are studied. For example, when Mendel bred his pea plants, he followed both seed color (yellow vs. green) and seed shape (smooth vs. wrinkled). During such a cross, Mendel deduced that each pair of alleles separates independently of the others during gamete formation. In other words, which allele is inherited for seed color has absolutely no effect on which allele is inherited for seed shape. This principle is called the **law of independent assortment**. This principle applies to all sexually reproducing species, including garden peas, Labrador retrievers, and humans. A Punnett square can be used to track the possibilities.

LAW OF INDEPENDENT ASSORTMENT

The law of independent assortment states that the inheritance of one character has no effect on the inheritance of another. The example shown here considers two characters for Labrador retrievers: coat color (where black is dominant to chocolate) and hearing (where the ability to hear is dominant to deafness). Each character is controlled by a pair of alleles located on homologous chromosomes. (All chromosomes in diploid individuals come in homologous pairs, with one member of each pair inherited from each parent.) When chromosomes separate during meiosis, each pair segregates independently; if a particular gamete receives a dominant allele for coat color, this has no effect on whether that gamete will receive a dominant allele for hearing. Independent assortment thus produces four kinds of gametes, each with a different combination of alleles.

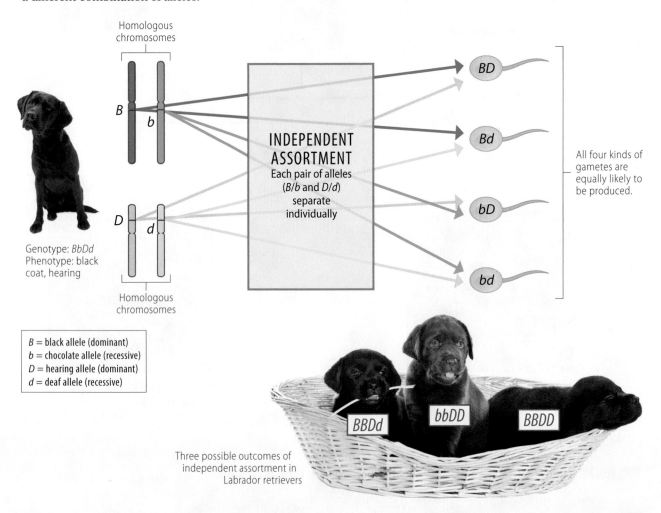

Homologous chromosomes

B
b

INDEPENDENT ASSORTMENT
Each pair of alleles (*B/b* and *D/d*) separate individually

D
d

Homologous chromosomes

Genotype: *BbDd*
Phenotype: black coat, hearing

BD

Bd

bD

bd

All four kinds of gametes are equally likely to be produced.

B = black allele (dominant)
b = chocolate allele (recessive)
D = hearing allele (dominant)
d = deaf allele (recessive)

BBDd

bbDD

BBDD

Three possible outcomes of independent assortment in Labrador retrievers

DIHYBRID CROSS

The law of independent assortment predicts that during a dihybrid cross—which tracks the genes and characters among offspring produced from parents who differ in two genes—each character should be inherited separately. As you can see here, offspring with all four possible combinations of traits are produced in a ratio that reflects the greater likelihood of dominant traits appearing.

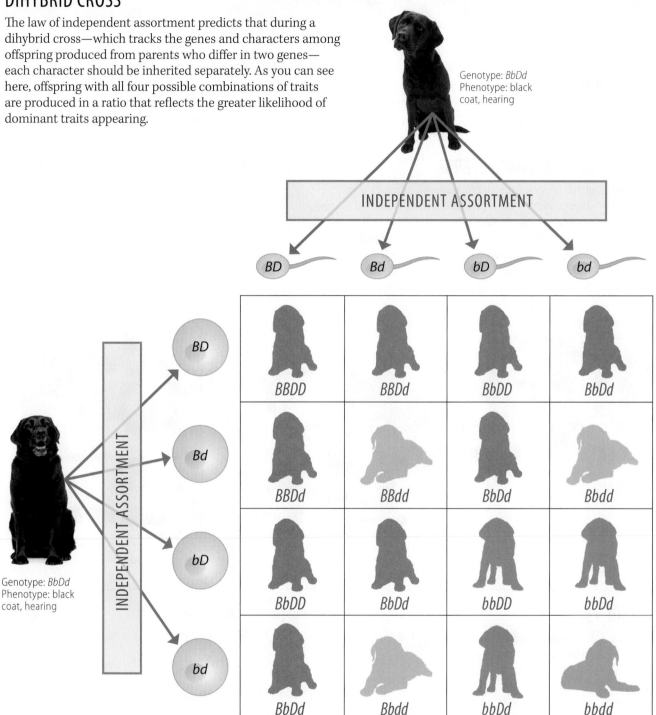

Genotype: *BbDd*
Phenotype: black coat, hearing

INDEPENDENT ASSORTMENT

BD   Bd   bD   bd

Genotype: *BbDd*
Phenotype: black coat, hearing

INDEPENDENT ASSORTMENT

BD   Bd   bD   bd

	BD	Bd	bD	bd
BD	BBDD	BBDd	BbDD	BbDd
Bd	BBDd	BBdd	BbDd	Bbdd
bD	BbDD	BbDd	bbDD	bbDd
bd	BbDd	Bbdd	bbDd	bbdd

RESULTS: **9** BLACK, HEARING : **3** BLACK, DEAF : **3** CHOCOLATE, HEARING : **1** CHOCOLATE, DEAF

Due to independent assortment, all four possible combinations of phenotypes are found among the offspring.

CORE IDEA: A dihybrid cross involves parents who differ in two characters. When forming gametes, the law of independent assortment states that alleles for each character will segregate (separate) independently of the alleles for other characters.

CORE QUESTION: You have a chocolate Lab that is heterozygous for the deafness allele. What gametes will this dog produce?

ANSWER: Since the parent is *bbDd*, half the gametes will be *bD* and half will be *bd*.

All modern Labrador retrievers are descended from two dogs bred in England in the 1880s.

Pedigrees can be used to trace traits in human families

The basic principles of Mendelian genetics apply to many human traits. However, human genetic analysis presents a unique challenge: We cannot control human matings! A Punnett square might thus be inadequate to analyze the genetic makeup of a real human family. There are, however, a number of other ways in which human genetics can be analyzed.

HUMAN GENETIC CHARACTERS

Here you can see several human characters—some benign, others that affect health—each believed to be controlled by a single gene. These characters therefore display simple dominant/recessive patterns. Notice that the dominant trait is not necessarily "normal" or more common than the recessive trait. The **wild-type trait** is the one most often seen in nature, while a **mutant trait** is a less common form.

DOMINANT	RECESSIVE
Freckles	No freckles
Normal pigmentation	Albinism
Healthy	Cystic fibrosis
Achondroplasia (dwarfism)	Normal height

☐ Mutant trait (less common)
☐ Wild-type trait (more common)

The freckled phenotype (left) is dominant to the non-freckled phenotype.

The albino phenotype (characterized by a lack of pigmentation, right) is recessive.

Dwarfs with achondroplasia are heterozygous, so two dwarf parents may have a child of normal height.

CARRIERS

Most human genetic disorders are recessive, so a person must have two copies of the disease allele to express the disease. A heterozygous individual—with one copy of the normal allele and one copy of the disease-causing allele—is said to be a **carrier** of the disease. Such a person does not have the disease but can pass the disease-causing allele along to their offspring. As you can see from this Punnett square, two carriers have a one-in-four chance of having a child with the condition and a two-in-four chance of having a child who is a carrier.

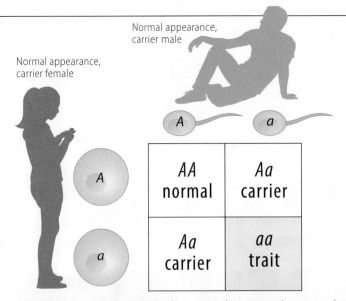

Normal appearance, carrier male

Normal appearance, carrier female

	A	a
A	AA normal	Aa carrier
a	Aa carrier	aa trait

RESULTS: 1 *AA* (NORMAL) : 2 *Aa* (CARRIER) : 1 *aa* (TRAIT)
Ratio of phenotypes: 3 normal : 1 trait

PEDIGREES

A genetic family history is called a **pedigree**. Although you may associate pedigrees with purebred animals, they can represent human matings just as well. To analyze a pedigree, you can apply Mendel's principles and some simple logic to figure out many of the genotypes.

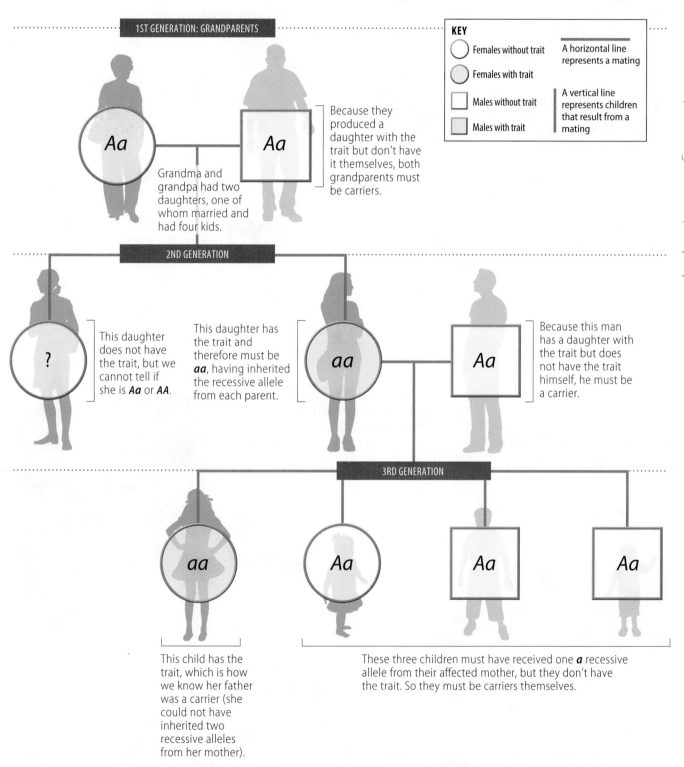

Cystic fibrosis is the most common lethal genetic disease in the United States, affecting about 30,000 people.

KEY
- Females without trait
- Females with trait
- Males without trait
- Males with trait
- A horizontal line represents a mating
- A vertical line represents children that result from a mating

1ST GENERATION: GRANDPARENTS

Aa *Aa*

Because they produced a daughter with the trait but don't have it themselves, both grandparents must be carriers.

Grandma and grandpa had two daughters, one of whom married and had four kids.

2ND GENERATION

? *aa* *Aa*

This daughter does not have the trait, but we cannot tell if she is *Aa* or *AA*.

This daughter has the trait and therefore must be *aa*, having inherited the recessive allele from each parent.

Because this man has a daughter with the trait but does not have the trait himself, he must be a carrier.

3RD GENERATION

aa *Aa* *Aa* *Aa*

This child has the trait, which is how we know her father was a carrier (she could not have inherited two recessive alleles from her mother).

These three children must have received one *a* recessive allele from their affected mother, but they don't have the trait. So they must be carriers themselves.

CORE IDEA: Mendel's principles can be applied to human characters. A carrier is a heterozygous individual who has a recessive disease-causing allele but does not display the disease. A pedigree can be used to track genetic traits in a family.

CORE QUESTION: Using the Punnett square shown earlier, what are the odds that the "?" daughter in the pedigree is in fact a carrier of the disease?

ANSWER: Two in three. We know she is not *aa*, but there is a one-third chance that she is *AA*.

The inheritance of many traits is more complex than Mendel's laws

Although many traits follow the simple patterns of Mendelian inheritance, others do not. In real life, you will often encounter inheritance patterns that are more complex than can be accounted for by the principles that Mendel deduced in the 19th century. Here, you can see examples of several such complexities.

INCOMPLETE DOMINANCE

In classic Mendelian genetics, a dominant allele produces the same appearance whether it is present in one or two copies. But for some characters, such as flower color in snapdragons, the heterozygous condition (with one copy each of the dominant and recessive allele) produces an intermediate appearance. This is called **incomplete dominance**.

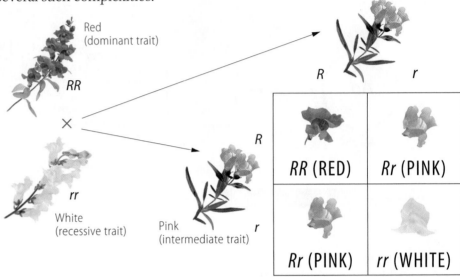

Red (dominant trait)
RR
×
White (recessive trait)
rr
Pink (intermediate trait)

	R	r
R	RR (RED)	Rr (PINK)
r	Rr (PINK)	rr (WHITE)

MULTIPLE ALLELES

Classic Mendelian genetics involves inheritance patterns with only two alleles per gene (R versus r, for example). But most genes occur in more than two forms, known as **multiple alleles**. Each individual carries, at most, two different alleles for a particular gene. But, because there are several alleles for the gene, multiple combinations of genes and therefore multiple phenotypes (physical traits) may result.

Human blood types are determined by a gene with three alleles: i, I^A, I^B. These three alleles can be combined in six ways:

GENOTYPE	PHENOTYPE (blood type)
ii	O
iI^A	A
iI^B	B
I^AI^A	A
I^AI^B	AB
I^BI^B	B

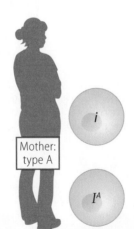

Father: type B

Mother: type A

i
I^A

i
I^B

	i	I^B
i	ii (type O)	iI^B (type B)
I^A	iI^A (type A)	I^AI^B (type AB)

RESULTS: 1 ii (TYPE O) : 1 iI^B (TYPE B) : 1 iI^A (TYPE A) : 1 I^AI^B (TYPE AB)

The rarest blood type in the United States is AB negative, found in only 1 person in 250.

PLEIOTROPY

In simple Mendelian genetics, one gene is responsible for one character. But in some cases, one gene influences many characters, a situation called **pleiotropy**. For example, a mutation in a single gene in humans can cause sickle-cell disease (also called sickle-cell anemia). As a result of the mutation, red blood cells deform to a sickle shape. Sickled cells may clog blood vessels and cause pain and damage to organs. Sickled cells are destroyed by the body, causing anemia and general weakness. Thus, a single mutation can cause many physical changes.

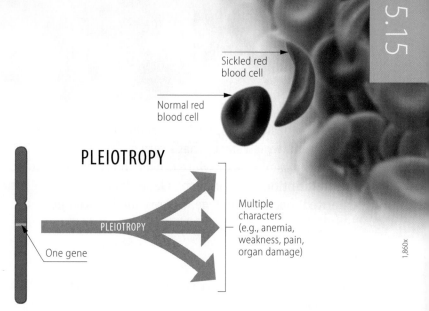

Sickled red blood cell

Normal red blood cell

PLEIOTROPY

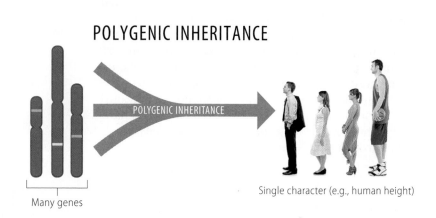

One gene → PLEIOTROPY → Multiple characters (e.g., anemia, weakness, pain, organ damage)

1,860x

POLYGENIC INHERITANCE

Traditional Mendelian genetics describes all characters as having just two traits; for example, pea plant flowers are either purple or white. However, many characters actually vary along a whole continuum of traits. This is often due to **polygenic inheritance**, the effect of many genes on a single character. For example, human height and skin color are each affected by several genes, so following a single one will never provide a reliable genetic description.

POLYGENIC INHERITANCE

Many genes → POLYGENIC INHERITANCE → Single character (e.g., human height)

ENVIRONMENT VS. GENETICS

Some characters, such as blood type, are completely genetically determined. Others, such as skin color, have both a genetic component and an environmental component. And some characters, such as body piercings, are entirely determined by the environment. There is an important difference between these two causes of appearance, however: Only genetic influences are inherited; any effects of the environment are not passed on to the next generation.

Your blood type trait is completely genetically determined.

Your skin color results from both genetic and environmental influences (e.g., sun exposure).

Many of your physical traits have no genetic basis.

CORE IDEA: Classical Mendelian genetics describes the inheritance of many characters, but others require extensions to the simple rules. Examples of complications include incomplete dominance, multiple alleles, pleiotropy, polygenic inheritance, and environmental influences.

CORE QUESTION: Which two extensions of Mendelian genetics discussed here are essentially the opposite of each other?

ANSWER: Pleiotropy is one gene affecting many characters, whereas polygenic inheritance occurs when many genes affect a single character.

Linked genes may not obey the law of independent assortment

One of the basic principles Mendel deduced during his experiments was the law of independent assortment, which states that every character is inherited independently. In other words, whether a pea plant has purple or white flowers has no effect on whether it has green or yellow seeds. However, this description is incomplete. Genes located close together on the same chromosome tend to be inherited together. Such genes are called **linked genes**, and display unique inheritance patterns.

LINKED GENES

Normally, the genes for each character sort into gametes independently of the genes for other characters. But if two genes are located near each other on a chromosome, then those genes will tend to be inherited together. As you can see in the example here, instead of producing the ratio expected from independently sorted chromosomes, the ratio that results from a cross involving linked genes is three double dominant to one double recessive. Notice that there are no offspring that are a mix of dominant and recessive traits. This result represents an exception to the law of independent assortment. With linked genes, whether or not an individual has a particular trait *is* affected by whether or not they have another trait because the genes are physically associated with each other on the same chromosome.

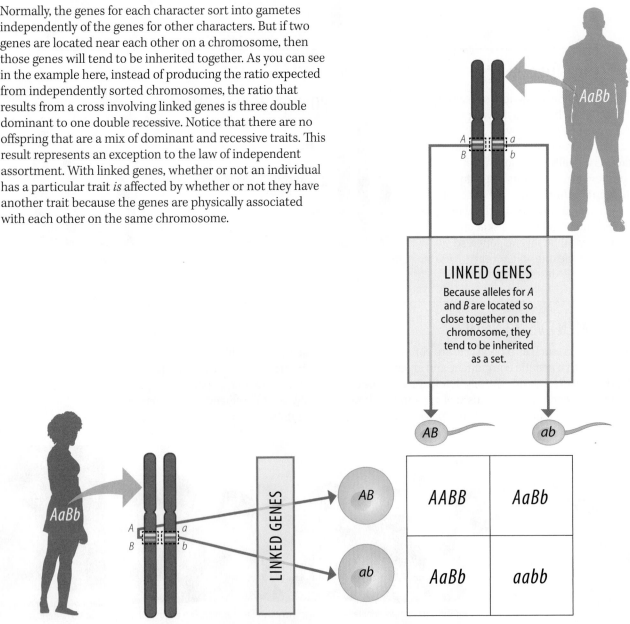

AaBb

LINKED GENES
Because alleles for *A* and *B* are located so close together on the chromosome, they tend to be inherited as a set.

AB   ab

AaBb

LINKED GENES

AB

ab

AABB	AaBb
AaBb	aabb

RESULTS: 1 *AABB* (DOUBLE DOMINANT) : 2 *AaBb* (DOUBLE DOMINANT) : 1 *aabb* (DOUBLE RECESSIVE)
These results differ from the ratio observed when the genes are unlinked and sort independently.

GENETIC RECOMBINATION

Genes are found on chromosomes, and the movement of chromosomes controls the inheritance of genetic traits. During gamete formation through the process of meiosis, homologous chromosomes pair up and stay physically attached to one another. At this time, the chromosomes may swap pieces in a process called crossing over. Crossing over produces hybrid chromosomes called **recombinant chromosomes** that contain parts of both original chromosomes. The process of crossing over may shuffle genes that are linked on the same chromosome. However, two genes that are located very near each other on a chromosome have little chance that a crossover will occur between them. The odds of a crossover event separating two genes is therefore related to how far apart those genes are on a chromosome.

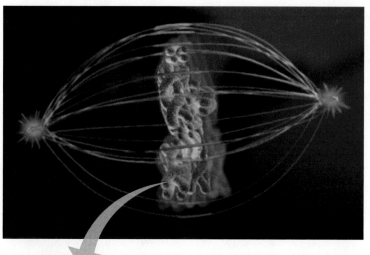

Chromosomes align by homologous pairs during meiosis.

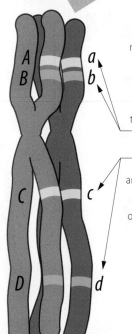

These two genes are located very near each other on the chromosome. The odds of a crossover separating gene *a* from gene *b* are therefore very low.

These two genes are located far from each other on the chromosome. The odds of a crossover separating gene *c* from gene *d* are therefore much higher than for *a* and *b*.

CROSSING OVER

Pair of homologous chromosomes overlapping

A crossover separated genes *C* and *D*, producing a hybrid chromosome, but genes *A* and *B* remained linked.

CORE IDEA: Genes are located on chromosomes, so the behavior of chromosomes accounts for the inheritance of genes. Genes located near each other on the same chromosome (called linked genes) are often inherited as a set, but crossing over can produce hybrid chromosomes with swapped genes.

CORE QUESTION: Can genes located on separate chromosomes be linked?

ANSWER: No. Only genes located near each other on the same chromosome can be linked and therefore inherited as a set.

Sex-linked genes display unusual inheritance patterns

Of the 46 human chromosomes, 44 are called **autosomes** and are found in all people. The other two chromosomes are the **sex chromosomes**, and these differ between males and females. As you can see here, the sex chromosomes have several special properties and produce unique patterns of inheritance.

SEX CHROMOSOMES

Like most mammals, humans have a pair of sex chromosomes—designated X and Y—that determines an individual's sex. Individuals with one X chromosome and one Y chromosome are males; XX individuals are females. Each sperm produced by a human male contains 22 autosomes and one sex chromosome; half will have an X chromosome and half will have a Y chromosome. Each egg produced by a human female will contain 22 autosomes and one X chromosome. An offspring's sex therefore depends on whether the sperm cell that fertilizes the egg bears an X or a Y chromosome.

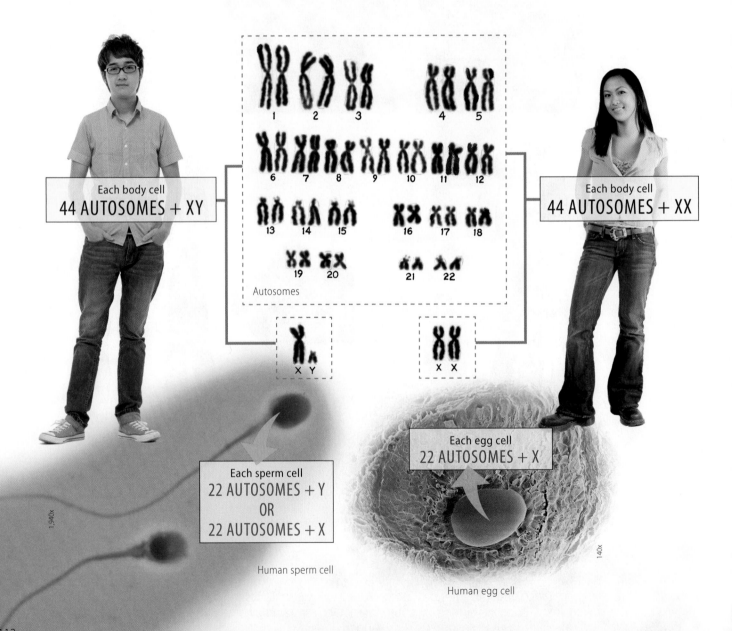

Each body cell
44 AUTOSOMES + XY

Each body cell
44 AUTOSOMES + XX

1 2 3 4 5
6 7 8 9 10 11 12
13 14 15 16 17 18
19 20 21 22

Autosomes

X Y   X X

Each egg cell
22 AUTOSOMES + X

Each sperm cell
22 AUTOSOMES + Y
OR
22 AUTOSOMES + X

1,940x

Human sperm cell

140x

Human egg cell

SEX-LINKED INHERITANCE

The Y chromosome is very small and carries virtually no genes other than the ones that confer maleness. However, several human characters are controlled by single genes located on the X chromosome. Such genes are called **sex-linked genes**. Characters controlled by sex-linked genes display unusual inheritance patterns because females have two X chromosomes while males have only one. Some types of color blindness are sex-linked, as is the recessive mutation that causes hemophilia, a disease characterized by excessive bleeding that was prevalent in the royal families of Europe. Because a single copy of the normal (dominant) gene will prevent the disease, hemophilia and other sex-linked recessive disorders are much more common among men (who have a single X chromosome) than among women (who have two).

The last royal family of Russia

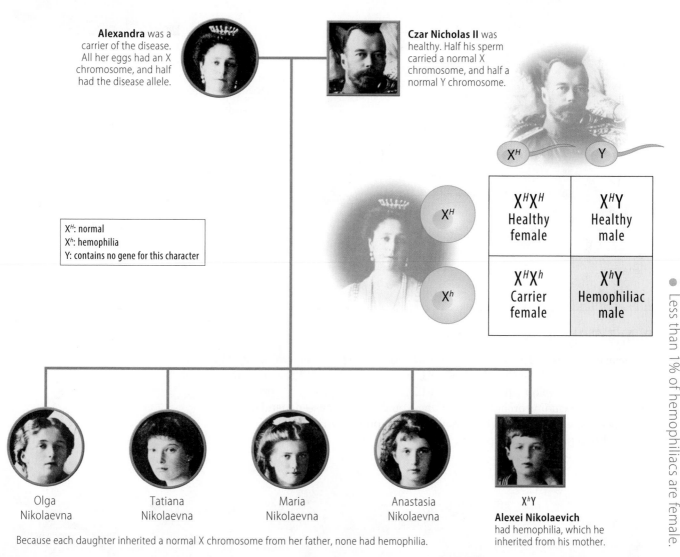

Alexandra was a carrier of the disease. All her eggs had an X chromosome, and half had the disease allele.

Czar Nicholas II was healthy. Half his sperm carried a normal X chromosome, and half a normal Y chromosome.

X^H: normal
X^h: hemophilia
Y: contains no gene for this character

$X^H X^H$ Healthy female	$X^H Y$ Healthy male
$X^H X^h$ Carrier female	$X^h Y$ Hemophiliac male

Olga Nikolaevna

Tatiana Nikolaevna

Maria Nikolaevna

Anastasia Nikolaevna

$X^h Y$
Alexei Nikolaevich had hemophilia, which he inherited from his mother.

Because each daughter inherited a normal X chromosome from her father, none had hemophilia.

Less than 1% of hemophiliacs are female.

CORE QUESTION: Is it possible for a female carrier to have a daughter who has hemophilia?

CORE IDEA: Human cells contain 44 autosomes and 2 sex chromosomes. If a person has two X sex chromosomes, she is female; males have one X and one Y. Because they are located on the X chromosome, sex-linked genes display unusual inheritance patterns.

ANSWER: Yes. If she mates with a male who has hemophilia, there is a one-quarter chance of having a daughter with hemophilia.

Nuclear transfer can be used to produce clones

The process of cell division produces new cells that are used by the body for growth, repair, and reproduction. Biologists have learned to artificially manipulate the natural process of cell division in order to produce **clones**, genetically identical individuals born of a single parent. Here you can see several applications of cloning in plants, animals, and humans.

NUCLEAR TRANSPLANTATION

Animals may be artificially cloned through the process of **nuclear transplantation**. In this procedure, a nucleus is removed from an adult donor cell, injected into a nucleus-free egg cell, and then induced to grow into an embryo that may give rise to a whole organism through cell division. The resulting individual will be a clone of the animal that provided the adult cell nucleus.

Researcher using a microscope to select an egg for cloning

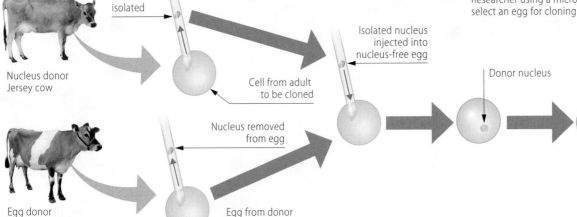

Nucleus donor Jersey cow

Nucleus isolated

Cell from adult to be cloned

Nucleus removed from egg

Egg from donor

Egg donor Jersey cow

Isolated nucleus injected into nucleus-free egg

Donor nucleus

Resulting embryo

PLANT CLONING

Artificial cloning is widely used in agriculture. Many crop plants are propagated by placing small samples, even individual cells, into a growth liquid. These individual cells divide and develop into new plants, which can then be grown as normal. Many crops—including orchids, navel oranges, and wine grapes—depend on this process. You may have taken advantage of this process yourself if you've ever grown a houseplant from a clipping.

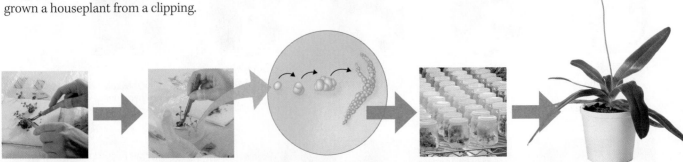

A plant is cut into small pieces.

Some of the plant tissue is placed in a growth medium.

Within the growth medium, some cells divide and form new plant tissue.

Tissue develops into new plants.

Plants grow into adults.

REPRODUCTIVE CLONING

An embryo produced through nuclear transplantation may develop into a new living individual, a process called **reproductive cloning**. If the animal to be cloned is a mammal, the embryo must be injected into the uterus of a surrogate mother. Scottish researchers used this method to produce Dolly the sheep in 1997, the first ever adult mammal clone. In subsequent years, biologists have cloned many mammal species such as dogs, cats, cows, and mice, as well as several rare and endangered species.

CLONED CREATURES

Prometea (right), the first cloned horse (*Equus caballus*)

Cloned rabbits

Embryo produced via nuclear transplantation injected into surrogate mother

Surrogate mother: Aberdeen Angus cow

Clone of the adult nucleus donor (Jersey cow)

STEM CELLS

An embryo (whether naturally occurring or produced via nuclear transplantation) is rich in cells called **stem cells**. In **therapeutic cloning**, stem cells are harvested from the cloned embryo. Embryonic stem cells have the potential to develop into every cell type in the body (such as muscle, bone, skin, or nerve cells). Researchers hope to someday use embryonic stem cells to treat a wide variety of diseases and injuries. Other sources of stem cells include adult bone marrow and umbilical cord blood. These types of stem cells cannot be grown into as wide a variety of cell types as embryonic stem cells, but they are easier to collect.

Adult stem cells removed from bone marrow can be used to generate blood cells.

Different growing conditions

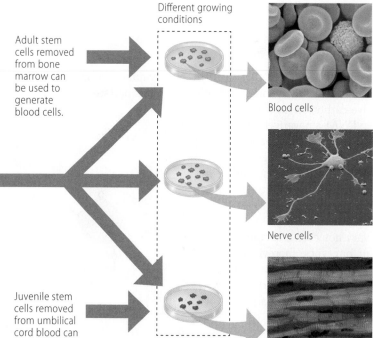

Blood cells
3,860x

Nerve cells
390x

Muscle cells
400x

Embryo produced via nuclear transplantation

Cells removed and grown in culture

Juvenile stem cells removed from umbilical cord blood can be used to generate several types of cells.

CORE IDEA: Cloning—the production of genetically identical offspring from a single parent—has several uses in agriculture and medicine. Nuclear transplantation can be used to produce a cloned embryo. This embryo can be used to produce a new individual (reproductive cloning) or it can be used to produce stem cells (therapeutic cloning).

CORE QUESTION: If a nucleus is removed from a brown mouse, injected into a cell from a white mouse, and the resulting embryo is implanted in a surrogate black mouse, what color will the babies be?

ANSWER: Brown (since that is the source of the nucleus).

DNA is a polymer of nucleotides

In the early 1900s, biologists hypothesized that some molecule acted as the chemical basis of inheritance, but no one knew what it was. By the 1950s, scientists had identified the hereditary material as **DNA** (which stands for **deoxyribonucleic acid**) and had figured out its three-dimensional structure. DNA is one type of **nucleic acid**, molecules that store information. Within the cell, DNA usually exists as one or more very long fibers called chromosomes, which contain genes, specific stretches of DNA that encode instructions for building proteins. The structure of DNA can be understood as a polymer, a large molecule made by repeating a smaller unit. In DNA, these individual units are called nucleotides.

THE DOUBLE HELIX

Every living cell contains one or more chromosomes (46 in the case of humans). Each chromosome consists of one very long molecule of DNA plus proteins that help organize and compact the DNA. The overall structure of DNA is a **double helix**, with two strands wrapped around each other like the twisting red and white stripes of a candy cane. In the center of the double helix, hydrogen bonds between bases hold the two strands together. The precise chemical structure of DNA is common to all life on Earth.

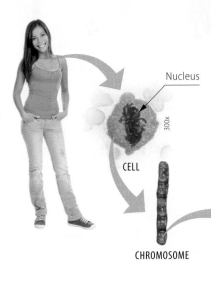

Nucleus

CELL

CHROMOSOME

300x

DOUBLE HELIX OF DNA

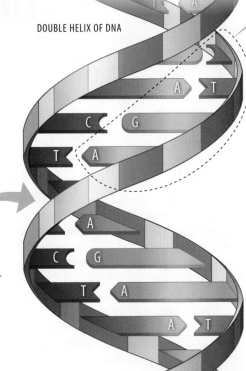

NUCLEOTIDES

Each molecule of DNA is made from individual subunits called **nucleotides**. Each nucleotide contains three parts: a central five-carbon **sugar** (called deoxyribose in DNA), a negatively charged **phosphate**, and a **base** made from one or two rings of nitrogen and carbon. The sugar and phosphate are identical among all DNA nucleotides. The DNA bases vary, however, and come in four types, abbreviated as A, G, T, and C. All genetic information is encoded using this four-letter alphabet.

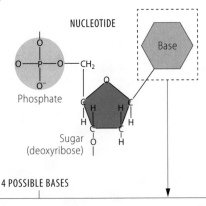

NUCLEOTIDE

Base

Phosphate

Sugar (deoxyribose)

4 POSSIBLE BASES

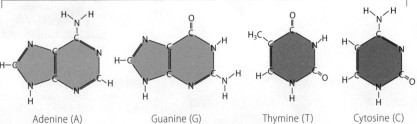

Adenine (A)   Guanine (G)   Thymine (T)   Cytosine (C)

POLYNUCLEOTIDE

One molecule of DNA contains two **polynucleotides** wrapped around each other, forming a double helix—one polynucleotide is the red stripe of the candy cane and the other polynucleotide is the white stripe. Each polynucleotide is a long strand of individual nucleotides. Hence the name: "poly" means "many," so a "polynucleotide" means "many nucleotides." A polynucleotide can contain any combination of the four bases along its length.

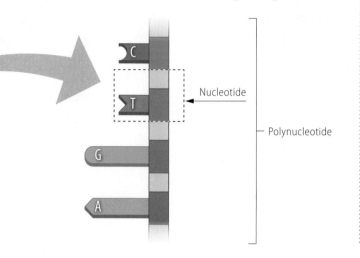

SUGAR–PHOSPHATE BACKBONE

A polynucleotide consists of bases attached to a **sugar–phosphate backbone**. The backbone has alternating sugar and phosphate groups that form a long chain. The backbone of the polynucleotide is identical among all DNA molecules; it is the bases that change from one molecule to the next.

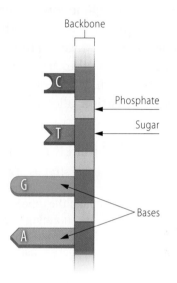

The DNA backbone from one of your cells has the same chemical structure as the DNA backbone from any other cell on Earth.

BASE PAIRS

Each nucleotide contains one of four different bases: adenine (abbreviated A), guanine (G), cytosine (C), or thymine (T). Because of their chemical makeup, each kind of base can form hydrogen bonds with only one other kind of base: A with T, and C with G. The hydrogen bonds within **base pairs** hold the two strands of the double helix together. Due to the base-pairing rules, the sequence of nucleotides along one polynucleotide within a double helix dictates the identity of nucleotides along the other polynucleotide.

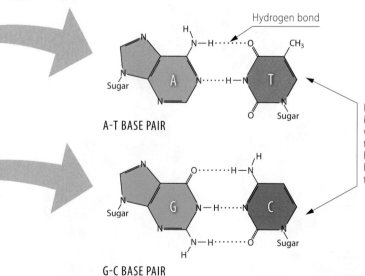

In a molecule of DNA, an A-T base pair is held together by 2 hydrogen bonds, whereas a C-G base pair is held together by 3 hydrogen bonds. These hydrogen bonds between the bases hold the two polynucleotide strands together in a DNA double helix.

CORE IDEA: A molecule of DNA is a double helix made of two intertwined polynucleotide strands. Each strand is a long string of nucleotides. Each nucleotide consists of the same sugar and phosphate, and one of four possible bases.

CORE QUESTION: If you were to compare a nucleotide from your DNA to a random one extracted from a bacterium, which components would be exactly the same and which might differ?

ANSWER: The backbone (sugar and phosphate) will always be identical, but the base might differ.

During DNA replication, a cell duplicates its chromosomes

For the continuity of life to be maintained, a complete set of genetic instructions in the form of DNA must pass from one generation to the next. For this to occur, there must be a means of precisely copying DNA molecules. The structure of DNA allows it to perform this function because each DNA strand can serve as a mold, or template, to guide production of the other strand. In other words, if you know the sequence of bases in one strand of the double helix, you can easily determine the sequence of bases in the other strand by applying the base-pairing rules: A pairs with T (and T with A), and G pairs with C (and C with G). This simple scheme underlies **DNA replication**, allowing molecules of DNA to be precisely copied.

SEMI-CONSERVATIVE REPLICATION

During DNA replication, the two strands of an original DNA molecule separate from each other. Due to the base-pairing rules, each separated strand can serve as a template to precisely rebuild the other strand. In the end, two new DNA molecules are produced, each containing one newly created strand and one strand from the original molecule. DNA replication is said to be **semi-conservative** because each new molecule conserves half of the original molecule.

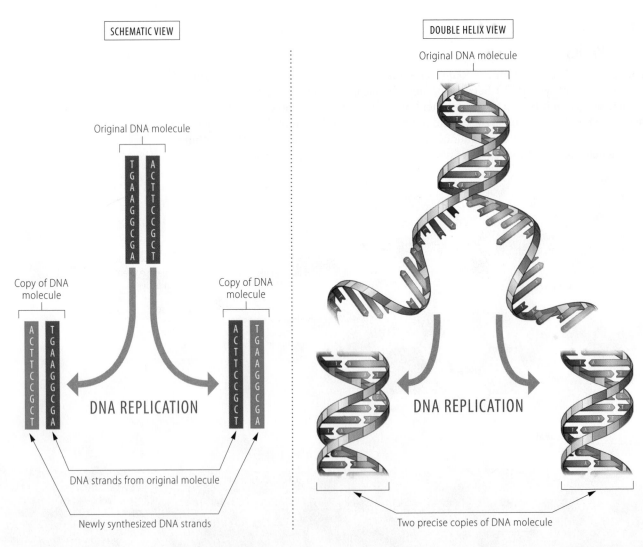

SCHEMATIC VIEW

DOUBLE HELIX VIEW

Original DNA molecule

Original DNA molecule

Copy of DNA molecule

Copy of DNA molecule

DNA REPLICATION

DNA REPLICATION

DNA strands from original molecule

Newly synthesized DNA strands

Two precise copies of DNA molecule

THE PROCESS OF DNA REPLICATION

Within a cell, the process of DNA replication is accomplished by a series of enzymes that work together to duplicate a DNA molecule.

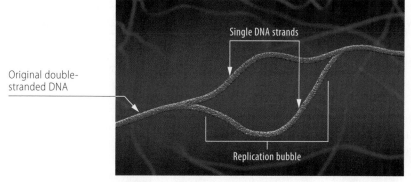

Original double-stranded DNA

Single DNA strands

Replication bubble

① DOUBLE HELIX IS PEELED APART

DNA replication begins when an enzyme called helicase attaches to specific DNA sequences called origins of replication. Starting at that point, the enzyme peels apart the two DNA strands of the double helix from each other, forming a replication bubble. Because the double helix is peeled apart, the bases in each separated DNA strand within the replication bubble are now exposed.

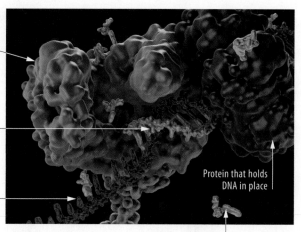

DNA polymerase (orange)

New DNA strand in the process of being synthesized (light blue)

DNA strand from original molecule (dark blue)

Protein that holds DNA in place

Free DNA nucleotide

② NEW STRANDS ARE SYNTHESIZED

Each of the two separated DNA strands is bound by an enzyme called DNA polymerase. This enzyme builds a new DNA molecule that is complementary to the existing strand. DNA polymerase matches up a correct free nucleotide on the separated strand, A with T, C with G, etc. This happens to both peeled-apart DNA strands simultaneously, with each original DNA strand serving as a template to build a new complementary DNA strand.

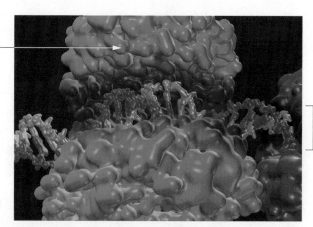

DNA ligase (green)

New DNA molecule

③ DNA FRAGMENTS ARE FUSED TOGETHER

DNA polymerase creates the new DNA molecule in fragments. An enzyme called DNA ligase joins the individual fragments, fusing them together into the final DNA molecule.

CORE IDEA: Genetic instructions are passed down via DNA replication: the double helix of a DNA molecule is peeled apart, and each separated strand serves as a template to build a new strand following the base-pairing rules.

CORE QUESTION: If a DNA molecule is chemically modified to glow and then replicated, will either, neither, or both of the two new molecules glow?

ANSWER: Because each new DNA molecule receives half of the original, both will glow (but half as much).

DNA directs the production of proteins via RNA

Nucleic acids are information storage molecules that contain directions for the production of proteins. Two types of nucleic acids can be found within all living cells: DNA (deoxyribonucleic acid) and **RNA (ribonucleic acid)**. These two types of nucleic acid have several structural similarities and a few important differences. Under the direction of DNA, RNA carries instructions that result in the production of proteins. These proteins then perform the cellular work that is essential to life.

NUCLEIC ACIDS: DNA AND RNA

A molecule of DNA shares many structural similarities with a molecule of RNA. Both are nucleic acids. Both are polymers of nucleotides, each of which consists of a sugar, a phosphate, and a base. However, DNA and RNA also have three important structural differences.

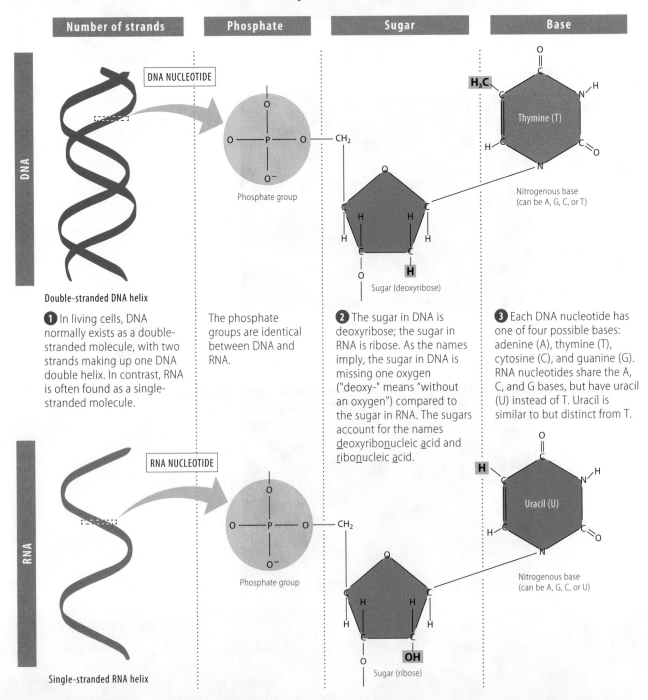

| Number of strands | Phosphate | Sugar | Base |

Double-stranded DNA helix

DNA NUCLEOTIDE

Phosphate group

Sugar (deoxyribose)

Thymine (T)

Nitrogenous base (can be A, G, C, or T)

1 In living cells, DNA normally exists as a double-stranded molecule, with two strands making up one DNA double helix. In contrast, RNA is often found as a single-stranded molecule.

The phosphate groups are identical between DNA and RNA.

2 The sugar in DNA is deoxyribose; the sugar in RNA is ribose. As the names imply, the sugar in DNA is missing one oxygen ("deoxy-" means "without an oxygen") compared to the sugar in RNA. The sugars account for the names deoxyribonucleic acid and ribonucleic acid.

3 Each DNA nucleotide has one of four possible bases: adenine (A), thymine (T), cytosine (C), and guanine (G). RNA nucleotides share the A, C, and G bases, but have uracil (U) instead of T. Uracil is similar to but distinct from T.

RNA NUCLEOTIDE

Phosphate group

Sugar (ribose)

Uracil (U)

Nitrogenous base (can be A, G, C, or U)

Single-stranded RNA helix

THE FLOW OF GENETIC INFORMATION

DNA is able to act as the molecule of heredity because it can direct the production of proteins. To do this, DNA first directs the production of RNA, which in turn controls the manufacture of proteins. Proteins then perform the majority of cellular functions. This molecular "chain of command" within the cell is often described as DNA makes RNA, which makes proteins. It is through these processes that DNA is able to control your physical traits.

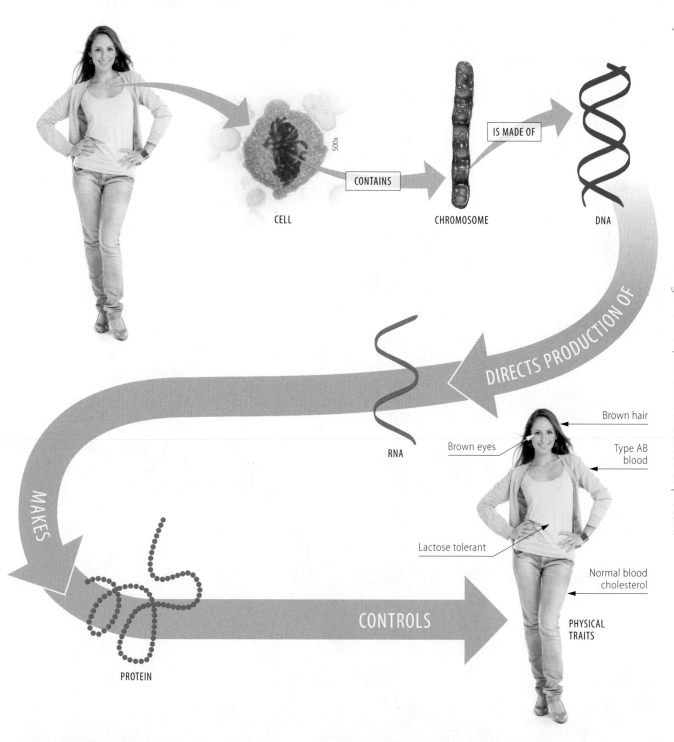

Each one of your cells contains instructions for building over 100,000 different kinds of protein.

CELL

CONTAINS

CHROMOSOME

IS MADE OF

DNA

500x

DIRECTS PRODUCTION OF

RNA

MAKES

PROTEIN

CONTROLS

Brown hair

Brown eyes

Type AB blood

Lactose tolerant

Normal blood cholesterol

PHYSICAL TRAITS

CORE IDEA: DNA and RNA are both nucleic acids with some common features and some differences. Within a cell, DNA directs the production of RNA, which in turn directs the production of proteins.

CORE QUESTION: Do your chromosomes contain any uracil (U) bases?

ANSWER: No. Your chromosomes are made of DNA, and U is found only in RNA.

Genetic information flows from DNA to RNA to protein

Every trait that you were born with is encoded by the DNA of your genes. But DNA itself does not produce your appearance. Rather, DNA directs the production of proteins, and it is these proteins that are responsible for your physical traits. The overall flow of information in a cell starts with DNA in the nucleus. The information is encoded into RNA that leaves the nucleus. Finally, this RNA is used to produce proteins by ribosomes in the cytoplasm. The genetic information that is passed along undergoes two transformations in the journey from DNA to RNA to protein—transcription and translation.

FLOW OF GENETIC INFORMATION THROUGH THE CELL

The information in a DNA gene is converted to an RNA molecule through the process of **transcription** in the nucleus. The RNA leaves the nucleus, exiting through an opening in the nuclear envelope called a nuclear pore. The RNA molecule makes its way to the ribosomes, where the message it contains is converted to protein through the process of **translation**.

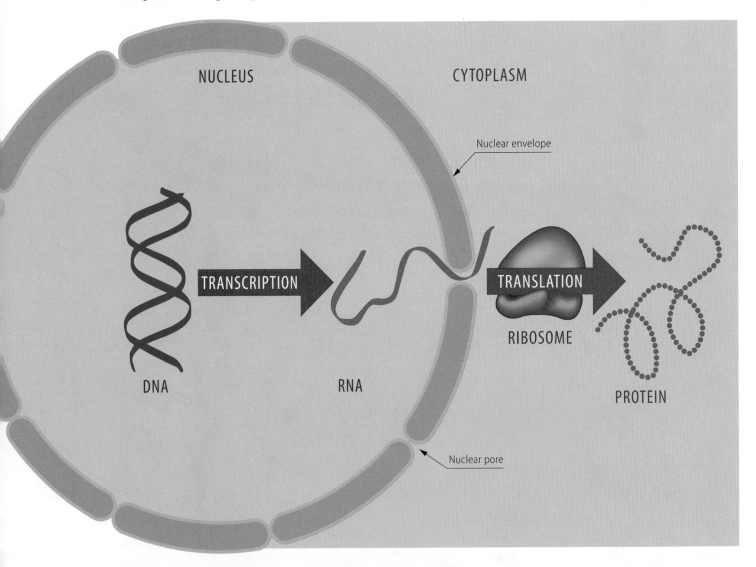

TRANSCRIPTION OVERVIEW

In a eukaryotic cell, the flow of genetic information always starts in the nucleus (that's where the DNA is!). DNA directs the production of RNA through the process of transcription. The molecule that results is called **messenger RNA (mRNA)**. Transcription follows the base-pairing rules (C is transcribed to G, T to A, etc.) with one important exception: uracil (U) is used in RNA instead of thymine (T), so an A nucleotide in DNA is transcribed to a U nucleotide in RNA. After it's made, mRNA exits the nucleus and travels to the ribosomes for translation.

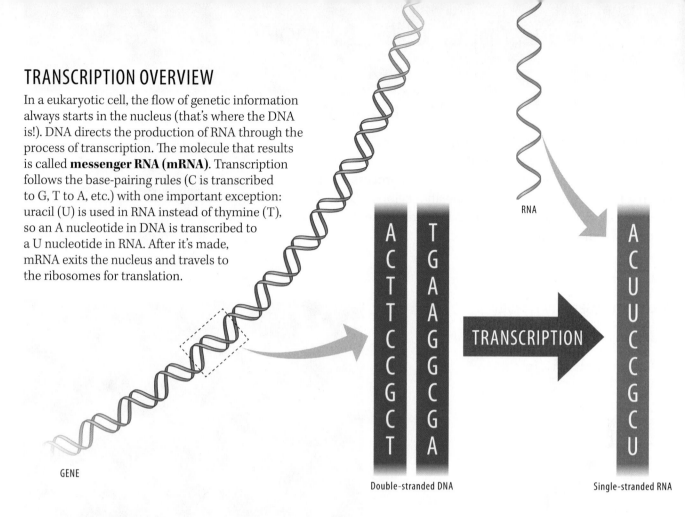

RNA

GENE

| A C T T C C G C T | T G A A G G C G A | TRANSCRIPTION | A C U U C C G C U |

Double-stranded DNA

Single-stranded RNA

TRANSLATION OVERVIEW

During the process of translation, the sequence of RNA nucleotides on a molecule of mRNA serves as instructions to build a sequence of amino acids. A molecule of mRNA contains successive **codons**, sequences of three nucleotides, each of which specifies one amino acid. Within the cytoplasm, the ribosomes read each successive codon and attach the proper amino acid to a growing chain. The molecule that results, consisting of a string of amino acids, is called a polypeptide. After further modifications, the polypeptide can serve as a functioning protein.

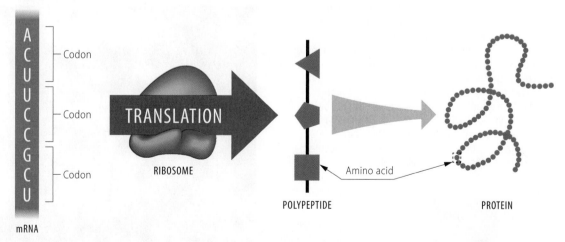

A C U U C C G C U — Codon

mRNA

TRANSLATION

RIBOSOME

Amino acid

POLYPEPTIDE

PROTEIN

CORE IDEA: Genetic information flows from DNA to messenger RNA in the nucleus through the process of transcription. At the ribosomes in the cytoplasm, each mRNA codon is translated into an amino acid of a protein.

CORE QUESTION: In one of your cells, could transcription and translation of one mRNA take place simultaneously?

ANSWER: No. Transcription takes place in the nucleus (where the DNA is), but translation occurs in the cytoplasm (where the ribosomes are).

Transcription creates a molecule of RNA from a molecule of DNA

Transcription is the transfer of information from DNA to a molecule of **messenger RNA (mRNA)**. In eukaryotic cells, transcription takes place in the nucleus. The process begins when RNA polymerase binds to a specific region of DNA. After a molecule of RNA is synthesized, it is altered in several ways before it departs the nucleus.

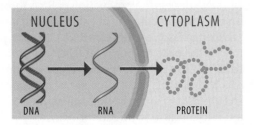

NUCLEUS   CYTOPLASM

DNA   RNA   PROTEIN

1 RNA POLYMERASE BINDS A PROMOTER

Transcription begins when an enzyme called **RNA polymerase** binds to a DNA sequence called a **promoter**. The promoter acts as a "start here" signal, marking the beginning of a gene. Promoters play a key role in regulating genes: By controlling whether or not RNA polymerase can bind the promoter, a cell can turn a gene on or off.

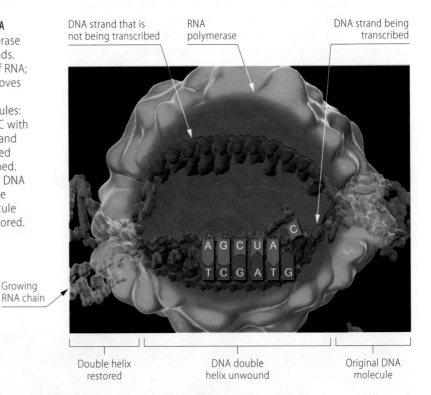

Promoter

RNA polymerase

DNA

2 RNA POLYMERASE SYNTHESIZES A MOLECULE OF RNA

After it binds the DNA at the promoter, RNA polymerase peels open the double helix, exposing the two strands. One strand serves as a template for the formation of RNA; the other DNA strand is unused. RNA polymerase moves down the DNA, creating a new RNA molecule one nucleotide at a time by following the base-pairing rules: A matches with T, U matches with A, G with C, and C with G. As RNA synthesis continues, the growing RNA strand peels away from the DNA, allowing the two separated DNA strands to rejoin in the region already transcribed. Transcription ends when RNA polymerase reaches a DNA "stop" sequence called a **terminator**. At this point, the polymerase molecule detaches from the RNA molecule and the gene, and the original DNA molecule is restored.

DNA strand that is not being transcribed

RNA polymerase

DNA strand being transcribed

A G C U A C
T C G A T G

Growing RNA chain

Double helix restored

DNA double helix unwound

Original DNA molecule

❸ RNA SPLICING

In a eukaryotic cell, the molecule of RNA that is produced during transcription is processed before it leaves the nucleus. For example, stretches of the RNA that do not actually encode for amino acids—called **introns**—are removed. The regions that do code for amino acids—called **exons**—are then spliced together. This is called **RNA splicing**. In some cases, one RNA molecule may be spliced in multiple ways, resulting in several proteins from one gene. In addition to splicing, extra nucleotides are added as a "cap" and "tail" to the RNA.

❹ THE mRNA LEAVES THE NUCLEUS

After processing, the finalized RNA molecule is now called messenger RNA. It leaves the nucleus through an opening in the nuclear membrane called a nuclear pore. In the cytoplasm, the molecule of mRNA will be used as instructions for generating a protein.

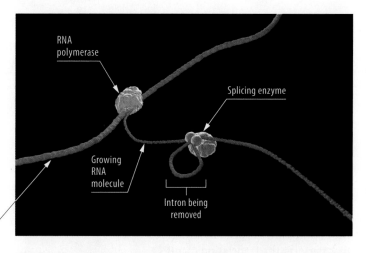

RNA polymerase

Splicing enzyme

Growing RNA molecule

DNA strand

Intron being removed

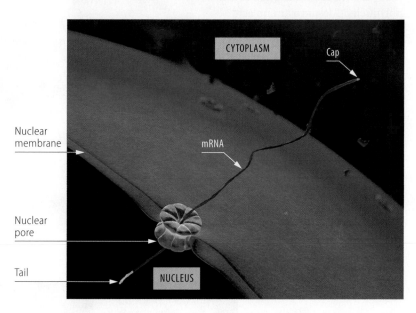

CYTOPLASM

Cap

Nuclear membrane

mRNA

Nuclear pore

Tail

NUCLEUS

One study in mice found that a molecule of mRNA lasts an average of 9 hours within a cell.

WHAT IS A GENE?

Biologists first defined a gene as a stretch of DNA responsible for the production of one enzyme, a protein that promotes a chemical reaction. Soon, this definition was expanded to include any protein, since many gene-encoded proteins (such as the protein keratin found in your hair and nails) are not enzymes. Then biologists realized that some proteins are made of multiple polypeptides—the protein hemoglobin in your red blood cells, for example, is made by combining polypeptides produced by two different genes—which required another tweak to the definition of a gene. Paradoxically, the more biologists learn about the makeup of the human genome, the more difficult it becomes to find a fully correct definition of a gene. Today, there is no simple, agreed-upon definition that accurately describes all known genes. For now, we can define a gene as a discrete unit of hereditary information consisting of a specific nucleotide sequence in DNA. This seemingly simple topic remains a subject of heated discussion and debate among biologists.

CORE IDEA: During transcription, the DNA double helix separates, and one strand is used to generate a molecule of RNA. The RNA is processed to become messenger RNA which then exits the nucleus via a nuclear pore.

CORE QUESTION: Why might a molecule of mRNA be shorter than the gene from which it is derived?

ANSWER: If introns are removed, the mRNA would be shorter.

Translation involves the coordination of three kinds of RNA

Translation is a process through which a molecule of messenger RNA (mRNA) is used to produce a molecule of protein. Translation takes place within the cytoplasm in cellular structures called ribosomes. There, molecules of transfer RNA (tRNA) translate triplets of RNA nucleotides into amino acids using a specific genetic code.

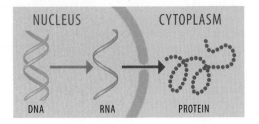

RIBOSOMES

Ribosomes are the cellular structures that perform the translation of mRNA polynucleotides into amino acid polypeptides. Ribosomes themselves are made from proteins combined with a type of RNA called **ribosomal RNA (rRNA)**. Within each ribosome are binding sites for the mRNA (which carries the message) and molecules of transfer RNA (tRNA). The ribosome acts like a vice holding the various components together. The ribosome can read the mRNA three nucleotides at a time. Each set of three, called a **codon**, encodes one amino acid.

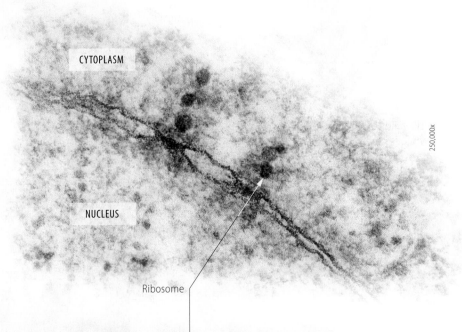

250,000x

Ribosome

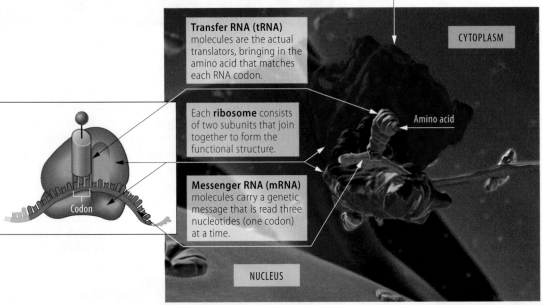

Transfer RNA (tRNA) molecules are the actual translators, bringing in the amino acid that matches each RNA codon.

Each **ribosome** consists of two subunits that join together to form the functional structure.

Messenger RNA (mRNA) molecules carry a genetic message that is read three nucleotides (one codon) at a time.

Codon

Amino acid

CYTOPLASM

NUCLEUS

The amino acid glutamic acid, when combined with a single atom of sodium, forms the flavoring agent MSG.

TRANSFER RNAs

Translation of one language into another requires an interpreter, someone or something that can recognize the words of the first language and convert them to the other. Cellular translation depends on molecules called **tRNAs**, or **transfer RNAs**. Transfer RNAs can speak the "RNA language" through a structure at one end called the **anticodon**. The anticodon binds to the codon of a messenger RNA via the RNA base-pairing rules (A with U, C with G, etc.). The other end of the tRNA molecule holds the amino acid that corresponds to that codon. The transfer RNA thus comes pre-loaded with the amino acid that matches the RNA sequence in the codon.

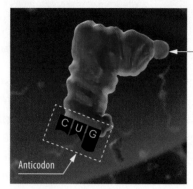

Attached amino acid (leucine)

Anticodon

TRANSFER RNA

THE GENETIC CODE

A molecule of mRNA is a string of codons. Each codon consists of three consecutive RNA bases that together encode for one amino acid. The correspondence between an RNA codon and its amino acid is called the **triplet code**. Notice that every RNA codon encodes just one possible amino acid (CUG always encodes for leucine, for example), but a given amino acid may be encoded by multiple codons (AAA and AAG both encode for lysine, for example). There is one **start codon** (AUG) that always signals the start of the genetic message; this codon also codes for the amino acid methionine. There are three **stop codons**, any one of which can signal the end of the message. Each codon between the start and stop codons encodes for one amino acid.

Second base of RNA codon

		U	C	A	G	
U		UUU UUC — Phenylalanine (Phe) UUA UUG — Leucine (Leu)	UCU UCC UCA UCG — Serine (Ser)	UAU UAC — Tyrosine (Tyr) UAA Stop UAG Stop	UGU UGC — Cysteine (Cys) UGA Stop UGG Tryptophan (Trp)	U C A G
C		CUU CUC CUA CUG — Leucine (Leu)	CCU CCC CCA CCG — Proline (Pro)	CAU CAC — Histidine (His) CAA CAG — Glutamine (Gln)	CGU CGC CGA CGG — Arginine (Arg)	U C A G
A		AUU AUC AUA — Isoleucine (Ile) AUG — Methionine (Met) or start	ACU ACC ACA ACG — Threonine (Thr)	AAU AAC — Asparagine (Asn) AAA AAG — Lysine (Lys)	AGU AGC — Serine (Ser) AGA AGG — Arginine (Arg)	U C A G
G		GUU GUC GUA GUG — Valine (Val)	GCU GCC GCA GCG — Alanine (Ala)	GAU GAC — Aspartic acid (Asp) GAA GAG — Glutamic acid (Glu)	GGU GGC GGA GGG — Glycine (Gly)	U C A G

First base of RNA codon

Third base of RNA codon

CORE IDEA: Translation is accomplished by ribosomes, made from rRNA and protein. Ribosomes use two kinds of RNA to produce a string of amino acids. The genetic code dictates the correspondence between RNA triplets and amino acids.

CORE QUESTION: What amino acids would be produced by an mRNA with the sequence AUG-ACU-AAU-AGU-UGA?

ANSWER: Methionine-threonine-asparagine-serine (notice the start and stop codons at the ends).

Translation creates a molecule of protein via the genetic code

During the process of translation, ribosomes use information coded in messenger RNA (mRNA) polynucleotides to create corresponding amino acid polypeptides. Each ribosome contains binding sites for the mRNA created via transcription in the nucleus as well as for molecules of transfer RNA (tRNA). Translation begins when ribosome parts assemble around a molecule of mRNA. Translation then proceeds in stepwise fashion, building a polypeptide one amino acid at a time. At the end of the mRNA, the newly formed polypeptide is released, and the ribosome disassembles into its parts.

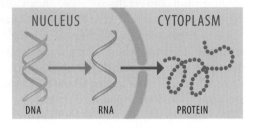

NUCLEUS   CYTOPLASM

DNA   RNA   PROTEIN

INITIATION

❶ RIBOSOME ASSEMBLES

At first, the ribosome exists as two separate components: the large and small subunits. Translation begins when an mRNA molecule binds to a small ribosomal subunit. A molecule of transfer RNA (tRNA) then binds to the start codon of the mRNA, where translation is to begin. This first tRNA carries the amino acid methionine (Met); its anticodon, UAC, binds to the start codon, AUG. Next, a large ribosomal subunit binds to the small one, creating a complete ribosome.

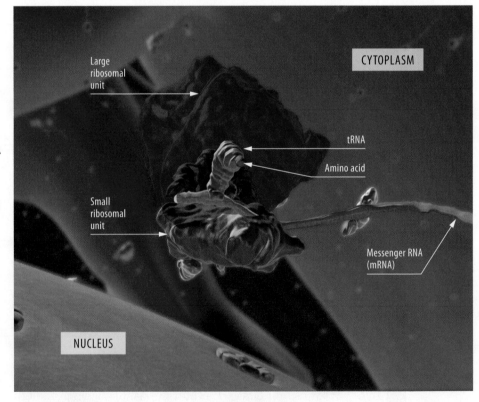

Large ribosomal unit

CYTOPLASM

tRNA

Amino acid

Small ribosomal unit

Messenger RNA (mRNA)

NUCLEUS

The 2009 Nobel Prize in Chemistry was awarded to scientists who determined the precise structure of a bacterial ribosome.

ELONGATION

② POLYPEPTIDE GROWS LONGER

Once initiation is complete, additional amino acids are added one by one. The anticodon of an incoming tRNA molecule, carrying its amino acid, pairs with the mRNA codon via the RNA base pairing rules. The new amino acid is added to the end of the growing polypeptide chain. The old, empty tRNA (the one that brought the amino acid during the previous step) exits the ribosome, and the ribosome moves the remaining tRNA to the spot just vacated, creating space for the next tRNA to enter. This process continues, with each entering tRNA bringing a new amino acid to be added.

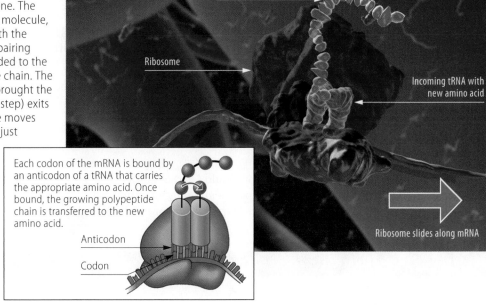

Growing polypeptide

Ribosome

Incoming tRNA with new amino acid

Ribosome slides along mRNA

Each codon of the mRNA is bound by an anticodon of a tRNA that carries the appropriate amino acid. Once bound, the growing polypeptide chain is transferred to the new amino acid.

Anticodon

Codon

TERMINATION

③ RIBOSOME DISASSEMBLES

Elongation continues until the ribosome reaches a stop codon on the mRNA. Stop codons—UAA, UAG, and UGA—do not code for amino acids but instead indicate that translation should cease. The completed polypeptide, typically several hundred amino acids long, is freed. The ribosome splits back into its subunits, releasing the mRNA and tRNAs.

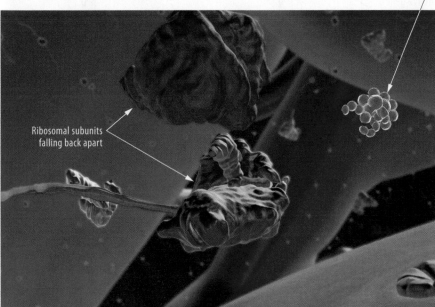

Completed polypeptide

Ribosomal subunits falling back apart

Amino acid

CORE IDEA: Translation begins when a ribosome assembles from its subunits at the start codon of an mRNA. Elongation then proceeds, adding one amino acid at a time. When a stop codon is reached, the ribosome machinery disassembles.

CORE QUESTION: If an mRNA contains 300 nucleotides (not including the start and stop codons), how many amino acids does it encode?

ANSWER: 100 (3 mRNA nucleotides = 1 amino acid).

Gene expression is regulated in several ways

Every cell in your body contains every one of your genes. And yet each type of cell looks and acts differently than other types because each type of cell produces different proteins. This is due to **gene regulation**, mechanisms that turn on certain genes while keeping other genes turned off in a particular cell. Genes contain information that is used to produce proteins. So turning a gene off means that **gene expression**—the flow of genetic information from DNA to RNA to protein—is not being completed. The pathway from gene to functioning protein is a long one, providing a number of points where the process can be regulated—turned on or off, sped up, or slowed down.

GENE REGULATION IN ACTION

If you examined the genes from a set of cells, you would see that certain genes are turned on in some cells and not in others. In the examples below, only intestinal cells produce the digestive enzyme lactase, and none of the cell types shown produce hemoglobin.

GENE TYPE	INTESTINAL CELL	NERVE CELL	WHITE BLOOD CELL
	400x	300x	660x
Gene for a glucose-digesting enzyme	√ (Active gene)	√	√
Antibody gene			√
Lactase gene	√		
Hemoglobin gene			

X CHROMOSOME INACTIVATION

If a chromosome is condensed (compacted by proteins), then the many genes that it contains are inaccessible and will not be used to produce proteins. One intriguing case is seen in female mammals, where one X chromosome in each body cell is highly compacted and almost entirely inactive. This **X chromosome inactivation** ensures that males (who have only one X chromosome) and females (who have two) have the same number of active genes. The inactive X in each cell condenses into a compact object called a **Barr body**, which is visible under a microscope.

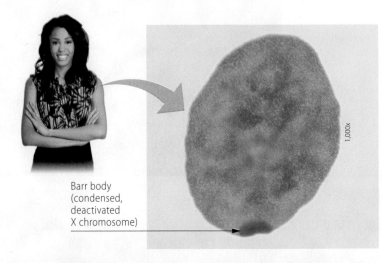

Barr body (condensed, deactivated X chromosome)

1,000x

CONTROL OF GENE EXPRESSION

There are many points along the path from gene to protein where gene expression may be regulated.

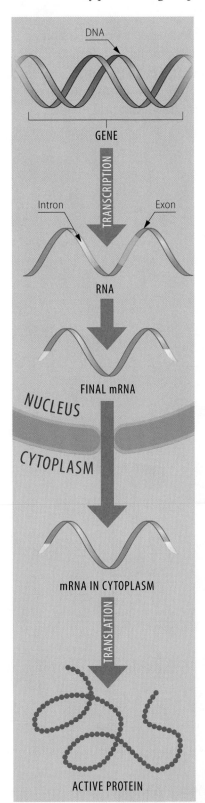

DNA

GENE

TRANSCRIPTION

Intron   Exon

RNA

FINAL mRNA

NUCLEUS

CYTOPLASM

mRNA IN CYTOPLASM

TRANSLATION

ACTIVE PROTEIN

❶ TRANSCRIPTION FACTORS

In a typical eukaryotic cell, the majority of genes are in the "off" state, meaning they are not being used to produce proteins. A series of proteins called **transcription factors** must bind to DNA before transcription can begin.

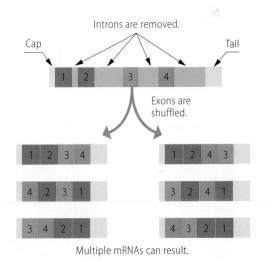

DNA regions where transcription factor proteins can bind

DNA of gene

Transcription factor proteins

RNA polymerase

Transcription produces messenger RNA.

❷ CONTROL OF RNA

Before leaving the nucleus, the RNA transcribed from a gene may be altered in several ways. For example, a cap and a tail are added. Additionally, noncoding regions (called **introns**) may be spliced out, and the remaining protein-coding regions (called **exons**) may be rearranged. Such exon shuffling can produce several messenger RNA molecules from a single gene. Another type of regulation occurs when small RNA molecules called **microRNAs** bind to mRNA molecules, preventing them from producing protein.

Introns are removed.

Cap   Tail

1 2   3   4

Exons are shuffled.

1 2 3 4   1 2 4 3

4 2 3 1   3 2 4 1

3 4 2 1   4 3 2 1

Multiple mRNAs can result.

❸ PROTEIN CONTROL: INITIATION, ACTIVATION, BREAKDOWN

The process of translation (when a molecule of messenger RNA is used to produce a protein) offers further opportunities for gene regulation. For example, the cell can control whether translation proceeds, how proteins are modified after translation, and when proteins are broken down.

Researchers have found a gene that, through RNA processing, codes for seven different proteins.

CORE IDEA: Through gene regulation (the turning on and off of genes), cells control gene expression, the production of proteins. There are several points along the path from DNA to RNA to protein that can be regulated.

CORE QUESTION: How can one gene direct the production of several different kinds of proteins?

ANSWER: One gene may have several exons, which can be spliced together in many different ways to produce different proteins.

Signal transduction pathways can control gene expression

Communication within a cell is important. For example, genetic information is carried from the nucleus to the cytoplasm by molecules of messenger RNA (mRNA). But communication among cells is also important. In a multicellular organism, cells work together within tissues and organs to perform various functions. Multicellular life therefore depends on cell-to-cell signaling. Communication from one cell to another plays a variety of important roles in your body and the bodies of all multicellular organisms.

SIGNAL TRANSDUCTION

Within a multicellular organism, cells typically communicate with each other by producing molecules that exit one cell and bind to a receptor protein on the outside of another cell. This binding triggers a **signal transduction pathway**, a series of relay molecules that convey the message from the outside of the cell into the cell's interior cytoplasm and eventually to the nucleus. The signal usually results in the turning on or off of one or more genes.

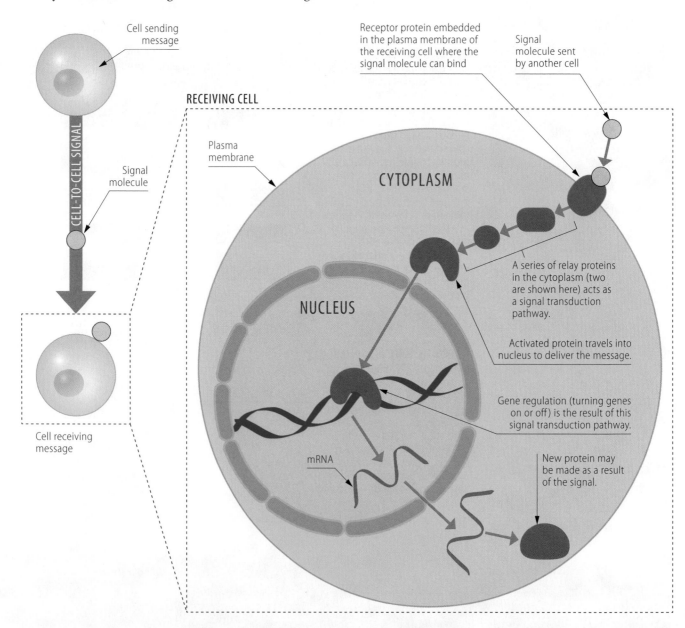

Cell sending message

Receptor protein embedded in the plasma membrane of the receiving cell where the signal molecule can bind

Signal molecule sent by another cell

CELL-TO-CELL SIGNAL

Signal molecule

RECEIVING CELL

Plasma membrane

CYTOPLASM

NUCLEUS

A series of relay proteins in the cytoplasm (two are shown here) acts as a signal transduction pathway.

Activated protein travels into nucleus to deliver the message.

Gene regulation (turning genes on or off) is the result of this signal transduction pathway.

New protein may be made as a result of the signal.

Cell receiving message

mRNA

CELL-TO-CELL SIGNALING IN A DEVELOPING EMBRYO

One example of the importance of cell-to-cell signaling and the control of gene expression is the growth of an animal from an embryo into an adult. This process, called **development**, involves frequent cell division (to increase body size) that must be carefully coordinated. Chemical signals that are passed between neighboring cells help the developing organism to properly control the formation of organs as well as the overall organization of the body.

Proper development depends on chemical signals passed between neighboring cells, telling embryonic cells precisely what to do and when.

A human embryo

DEVELOPMENT

A human baby

The mechanism of **induction** occurs when one group of cells influences the development of an adjacent group of cells. Its effect is to switch a set of genes on or off in the target cells. For example, inductive signals can cause cells to change shape, migrate, or even to destroy other cells, as in the removal of webbing in the developing hand.

Developing hand

INDUCTION

Developed hand

Homeotic genes are master control genes. In a developing embryo, homeotic genes produce protein signals that turn groups of other genes on and off. Homeotic genes help establish the overall structure of an organism. For example, specific homeotic genes direct which end of the embryo will be the head, while others direct which body parts will develop in which locations.

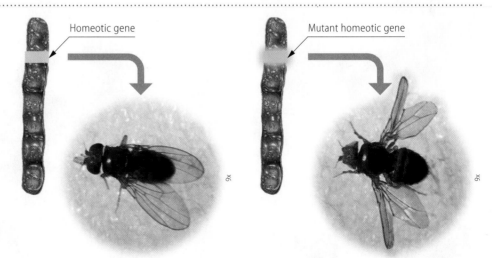

Homeotic gene

Mutant homeotic gene

Normal fruit fly

Fruit fly with extra set of wings

A mutation in a homeotic gene can cause a fly to grow an extra set of legs where its antennae should be.

CORE IDEA: Cells can communicate with each other by secreting proteins that trigger signal transduction pathways. The result is gene regulation. Such communication is particularly important in the developing embryo.

CORE QUESTION: How can one gene be responsible for a very large effect, such as establishing which end of the body is the head?

ANSWER: One homeotic gene can produce a protein that turns groups of other genes on or off.

Mutations can have a wide range of effects

A **mutation** is any change in the nucleotide sequence of DNA. Occasionally, mutations lead to improvements in an organism; that is, the change makes the organism better suited to its environment. Such changes are the raw material of evolution by natural selection. Much more often, however, mutations are harmful and will not be favored by natural selection. There are a wide variety of mutations. Some involve just a single DNA nucleotide, others affect a group of nucleotides, and some can even change a large region of a gene.

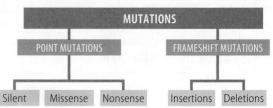

MUTATIONS		
POINT MUTATIONS		FRAMESHIFT MUTATIONS
Silent Missense Nonsense		Insertions Deletions

MUTAGENS

Why do mutations happen? Some mutations occur spontaneously due to errors that occasionally (but rarely) happen during DNA replication. Other mutations are caused by **mutagens**, physical or chemical factors in the environment that can damage DNA. Many mutagens are **carcinogens**, cancer-causing agents. For example, exposure to UV radiation in sunlight can cause mutations that lead to melanoma (skin cancer). Avoiding mutagens—for example, by using sunscreen, not smoking, and avoiding a high-fat diet—can help reduce your risk of developing cancer.

CANCER-CAUSING SUBSTANCES

Ultraviolet radiation

X-rays

Tobacco products

High-fat foods

CANCER-PREVENTING SUBSTANCES

Vitamins C and E

Sunscreen

Fresh vegetables

Fresh fruits

Sickle-cell disease is caused by the replacement of an A with a T at the 17th nucleotide of one of the genes for hemoglobin.

POINT MUTATIONS

The simplest type of mutation is a **point mutation**, the substitution of one DNA nucleotide for another. As you can see here, there are several types of point mutations with varying effects.

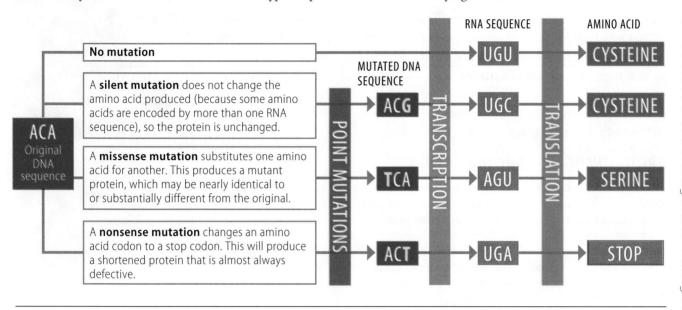

No mutation

A **silent mutation** does not change the amino acid produced (because some amino acids are encoded by more than one RNA sequence), so the protein is unchanged.

A **missense mutation** substitutes one amino acid for another. This produces a mutant protein, which may be nearly identical to or substantially different from the original.

A **nonsense mutation** changes an amino acid codon to a stop codon. This will produce a shortened protein that is almost always defective.

FRAMESHIFT MUTATIONS

After a DNA gene is transcribed into a molecule of messenger RNA, that molecule is read as a series of triplet codons. A codon is three consecutive nucleotides. Mutations that add nucleotides (called insertions) or remove nucleotides (called deletions) may alter the triplet reading frame. If an insertion or deletion does throw off the reading frame, it is called a frameshift mutation. Frameshift mutations often result in the production of a completely different (and, presumably, defective) protein.

PART OF THE GENETIC CODE
Use this table to translate DNA to RNA to protein in the figure below.

DNA	TAC	TCA	ACG	GAT	GCC	AAC	GGA	TGC
RNA	AUG	AGU	UGC	CUA	CGG	UUG	CCU	ACG
AMINO ACID	Met	Ser	Cys	Leu	Arg	Leu	Pro	Thr

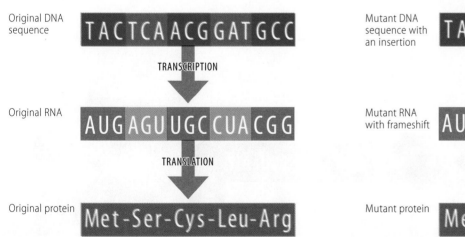

Original DNA sequence **TACTCAACGGATGCC**

TRANSCRIPTION

Original RNA **AUGAGUUGCCUACGG**

TRANSLATION

Original protein **Met-Ser-Cys-Leu-Arg**

Mutant DNA sequence with an insertion **TACT A CAACGGATGCC**

TRANSCRIPTION

Mutant RNA with frameshift **AUGAUGUUGCCUACGG**

TRANSLATION

Mutant protein **Met-Met-Leu-Pro-Thr**

CORE IDEA: Mutations, changes to the nucleotide sequence of DNA, can occur spontaneously or be caused by mutagens. Point mutations occur at a single nucleotide, whereas insertions and deletions can affect many codons.

CORE QUESTION: Why do nonsense mutations often have a strong effect?

ANSWER: Because an amino acid codon is changed to a stop codon, resulting in a shortened (and therefore probably defective) protein.

135

Loss of gene expression control can result in cancer

Cell division is a normal, necessary process that produces new cells from preexisting cells. Cells control the rate and timing of cell division via a mechanism called the cell cycle control system. However, a cell occasionally loses the ability to control its cell cycle. The result is a **tumor**, a mass of body cells that is growing out of control. If a tumor can spread to other tissues, the person is said to have **cancer**. What causes a tumor to form? Tumors often result from mutations in genes that normally regulate the cell cycle.

Mammogram showing a breast tumor

PROTO-ONCOGENES AND ONCOGENES

Within each cell, a **cell cycle control system** regulates the timing of cell duplication. The cell cycle control system consists of proteins that integrate information from the environment and communicate "start" or "stop" signals to the nucleus. Since each protein of the cell cycle control system is encoded by a gene, a mutation in one of those genes may produce a faulty protein and therefore a cell that fails to respond to the normal control signals. Such a cell runs through the cycle again and again, producing new cells even when it should not.

NORMAL

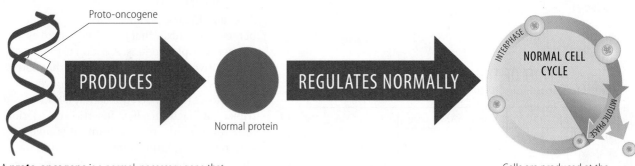

A **proto-oncogene** is a normal, necessary gene that produces a protein that properly regulates the cell cycle.

Cells are produced at the proper place and time.

MUTATION

An **oncogene** is a mutated proto-oncogene. It produces an abnormal protein that fails to regulate the cell cycle. The result is out-of-control growth.

Cells are produced continuously, not when and where they should be.

GROWTH FACTORS

An oncogene, a gene that leads to uncontrolled cell growth, is a mutated version of a proto-oncogene, a gene that produces a protein that normally regulates the cell cycle. One such type of normal protein is a **growth factor**, a protein that stimulates cell division. Normally, a growth factor only stimulates growth when it is appropriate. But a mutation in a growth factor gene can produce a hyperactive protein that promotes cell division even when it should not. A tumor may result.

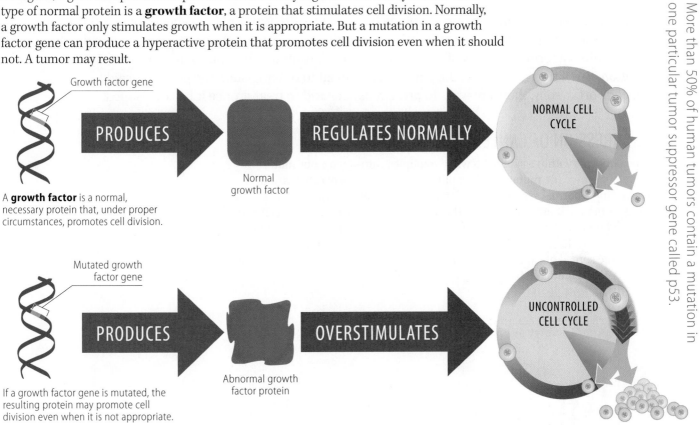

A **growth factor** is a normal, necessary protein that, under proper circumstances, promotes cell division.

If a growth factor gene is mutated, the resulting protein may promote cell division even when it is not appropriate.

TUMOR SUPPRESSOR GENES

Tumor suppressor genes normally code for proteins that inhibit cell division. A mutation that deactivates a tumor suppressor gene may result in uncontrolled growth.

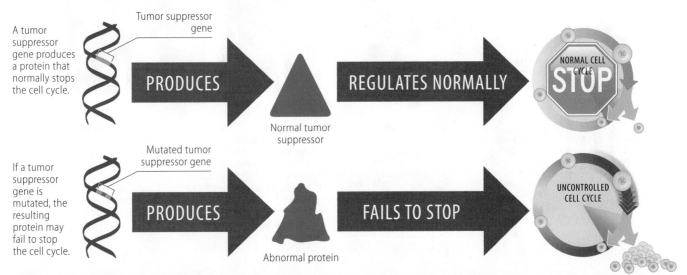

A tumor suppressor gene produces a protein that normally stops the cell cycle.

If a tumor suppressor gene is mutated, the resulting protein may fail to stop the cell cycle.

CORE IDEA: Cancer is caused by out-of-control cell growth due to a breakdown of the cell cycle control system. This can occur when proto-oncogenes (such as genes that code for growth factors and tumor suppressors) are mutated to oncogenes.

More than 50% of human tumors contain a mutation in one particular tumor suppressor gene called p53.

CORE QUESTION: How could the duplication of a gene result in cancer?

ANSWER: If that gene produces a growth factor, duplication of the gene may result in overstimulation of growth.

Cancer is caused by out-of-control cell growth

Cancer is caused by abnormal growth of the body's own cells. Cancer always begins when the DNA within a single cell undergoes a series of mutations. These mutations cause the cell to divide uncontrollably, forming a mass of bodily tissue called a tumor. The tumor may stay in its original location. But if the tumor gains the potential to spread to other tissues, the person is said to have cancer. Fortunately, there are ways to prevent cancer and to treat it once it has developed.

PROGRESSION OF CANCER

Cancer begins within a single cell when multiple proto-oncogenes mutate into oncogenes. The mutations disrupt the normal control of cell division, converting a normal cell into a cell that multiplies continuously, a process called transformation. If not destroyed by the immune system, the mutated cell and its descendants will form a tumor. Different types of cancer are named for where they originate; for example, lung cancer starts in the lung and may spread from there. The spread of cancer cells from their site of origin to sites distant in the body is called **metastasis**.

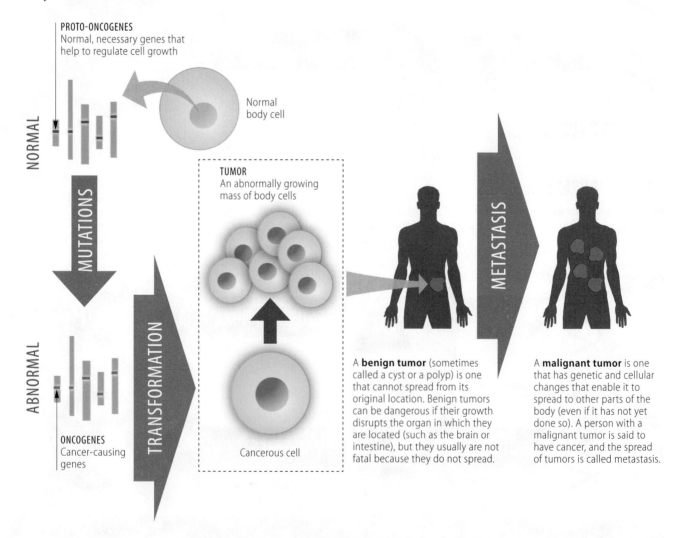

PROTO-ONCOGENES
Normal, necessary genes that help to regulate cell growth

Normal body cell

NORMAL

MUTATIONS

ABNORMAL

ONCOGENES
Cancer-causing genes

TRANSFORMATION

TUMOR
An abnormally growing mass of body cells

Cancerous cell

METASTASIS

A **benign tumor** (sometimes called a cyst or a polyp) is one that cannot spread from its original location. Benign tumors can be dangerous if their growth disrupts the organ in which they are located (such as the brain or intestine), but they usually are not fatal because they do not spread.

A **malignant tumor** is one that has genetic and cellular changes that enable it to spread to other parts of the body (even if it has not yet done so). A person with a malignant tumor is said to have cancer, and the spread of tumors is called metastasis.

CANCER TREATMENT

Most people diagnosed with cancer have three treatment options, sometimes referred to as "slash, burn, and poison." If a tumor is benign, surgery is often sufficient to completely remove it. If a tumor is malignant, surgery to remove as much as possible is usually combined with radiation therapy and/or chemotherapy. Radiation therapy is often localized, while chemotherapy can help destroy cancer cells throughout the whole body.

Surgery can remove a tumor. This may be sufficient if the tumor is benign. Here, a surgeon prepares a laser to be used in removing a tumor.

Radiation therapy exposes specific areas of the body with tumors to high-energy radiation. Such radiation disrupts cell division, which may slow down or kill cancer cells, but can also damage normal body cells.

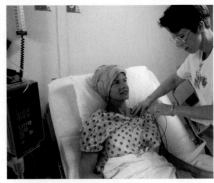

Chemotherapy uses drugs that disrupt cell division. Because the drugs are administered via the bloodstream, they affect cells throughout the body. Chemotherapy may have negative side effects, such as fatigue, hair loss, and nausea.

CANCER PREVENTION

Some mutations that can lead to cancer are inherited or occur spontaneously and so cannot be prevented. Most cases of cancer, however, are caused by carcinogens, agents in the environment that promote the development of cancer. Although even the healthiest person may develop cancer, there are ways that you can reduce your cancer risk and increase your odds of survival if you do develop it.

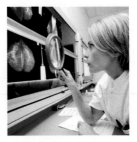

REGULAR SCREENINGS
Early detection—particularly of skin, breast, cervical, testicular, prostate, colon, and oral cancers—can greatly increase your chances of survival.

NOT SMOKING
Avoiding tobacco products is the single best lifestyle decision you can make to reduce your risk of cancer.

HEALTHY DIET
Eating a high-fiber, low-fat diet can reduce your risk of colon and other cancers.

SUN PROTECTION
UV radiation in sunlight can cause skin cancer. Cover up, avoid sunburns, and use sunscreen.

EXERCISE
Regular physical activity (30 minutes at a time, 3 times per week) has been shown to reduce the risk of several cancers.

CORE IDEA: All tumors start in a single cell. If a tumor cannot move beyond its original location, it is benign; if it is capable of spreading, it is malignant and causes cancer. There are several ways to prevent and treat cancer.

CORE QUESTION: If a person has breast cancer, how can it spread to the lungs?

ANSWER: Breast cancer starts in breast tissue, but if it metastasizes, it can spread to other tissues.

Genetic engineering involves manipulating DNA for practical purposes

The history of science demonstrates that after researchers understand a natural phenomenon, they often use this knowledge to manipulate the natural world. So it is with our understanding of DNA. Gaining knowledge of the structure and function of DNA led to the development of several important technologies that are used to manipulate DNA for practical purposes. Here, you can see some of the basic methods for manipulating DNA and some of the products that result.

GENETIC ENGINEERING

Genetic engineering is only the latest field in a long history of humans manipulating nature.

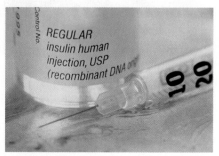

Biotechnology is the manipulation of organisms or their components to make useful products. Biotechnology is as old as human civilization. For example, prehistorical humans used yeast to brew beer and make bread, and selectively bred livestock.

DNA technology is a set of methods for studying and manipulating genetic material. DNA technology has become increasingly important to our society since it was first developed in the 1970s.

Genetic engineering is the direct manipulation of genes for practical purposes, such as the production of pharmaceutical products. This often involves creating recombinant DNA, a segment of DNA containing sequences from two different sources, which may even be different species.

HOW TO CLONE A GENE

A typical genetic engineering challenge is to produce large quantities of a desired human protein. This objective can be achieved through **gene cloning**, where the gene that synthesizes the protein is isolated and inserted into a piece of bacterial DNA called a plasmid. As the bacteria multiply, large amounts of the gene, and thus the protein, are produced. Here, you can see such a gene cloning experiment, using human insulin (a protein hormone) as an example.

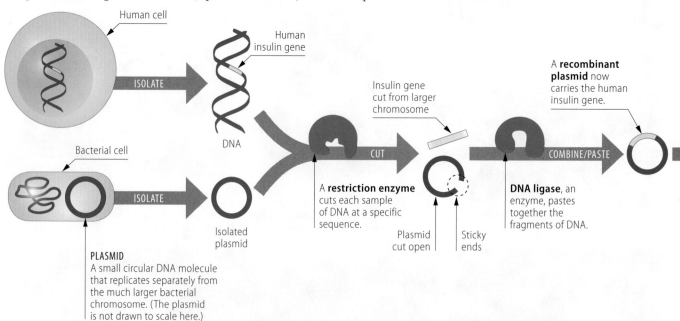

RESTRICTION ENZYMES

One of the most important steps in any genetic engineering project is to cut and paste specific pieces of DNA. This is achieved using **restriction enzymes**, proteins that cut DNA only at specific nucleotide sequences. Each kind of restriction enzyme recognizes just one short sequence of DNA (usually four to eight nucleotides long) and cuts the DNA at a specific location within that sequence. When exposed to the same restriction enzyme, every matching DNA sequence will be cut exactly the same way, and no other sequences will be cut.

In 1982, the Food and Drug Administration approved Humulin (human insulin produced by bacteria), the world's first genetically engineered pharmaceutical product.

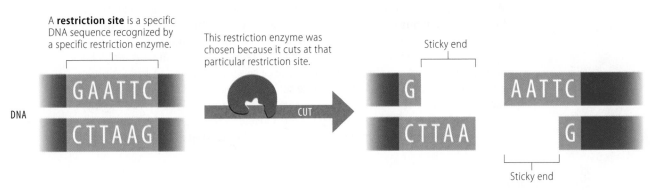

A **restriction site** is a specific DNA sequence recognized by a specific restriction enzyme.

This restriction enzyme was chosen because it cuts at that particular restriction site.

Sticky end

DNA

GAATTC
CTTAAG

CUT

G
CTTAA

AATTC
G

Sticky end

The resulting fragments are called **restriction fragments**. Many restriction enzymes produce fragments with single-stranded regions called **sticky ends**, which can join to other pieces of DNA with complementary DNA sequences.

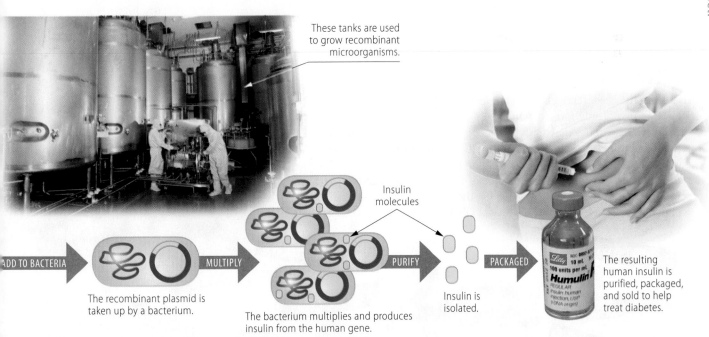

These tanks are used to grow recombinant microorganisms.

Insulin molecules

ADD TO BACTERIA — The recombinant plasmid is taken up by a bacterium.

MULTIPLY — The bacterium multiplies and produces insulin from the human gene.

PURIFY — Insulin is isolated.

PACKAGED — The resulting human insulin is purified, packaged, and sold to help treat diabetes.

CORE IDEA: Genetic engineering, the manipulation of genes for practical purposes, is an application of biotechnology. A typical genetic engineering challenge is to clone a gene in order to produce large quantities of a desired protein.

CORE QUESTION: In the procedure shown, why were the plasmid and the human DNA cut with the same restriction enzyme?

ANSWER: Using the same restriction enzyme ensures that both pieces of DNA will have matching sticky ends that allow them to bind together.

DNA may be manipulated many ways within the laboratory

In recent decades, biologists have learned to manipulate DNA in the laboratory in many different ways. By combining these techniques, scientists have created genetically modified organisms, identified disease-causing genes, and even created artificial life-forms. Here you can see a survey of some of these important genetic engineering methods.

GENOMIC LIBRARIES

When preparing **recombinant DNA**—a DNA molecule containing nucleotides from more than one source—it is often useful to start with a whole **genome**, an organism's entire set of DNA (about 3 billion nucleotides in humans). The entire genome is cut up with restriction enzymes to create a large set of fragments. Each of these fragments can then be inserted into a separate plasmid (a small circular bit of bacterial DNA), producing a large set of plasmids, each of which contains one segment of DNA. This is a **genomic library**, a collection of cloned DNA fragments that includes an organism's entire genome. Once created, a genomic library can be used to hunt for and manipulate any gene from the starting organism.

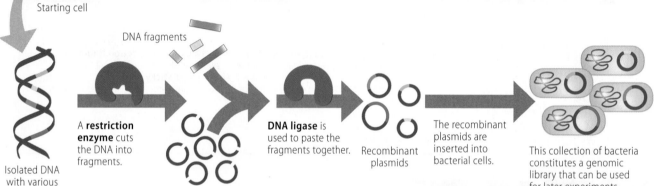

Starting cell

DNA fragments

A **restriction enzyme** cuts the DNA into fragments.

DNA ligase is used to paste the fragments together.

Recombinant plasmids

The recombinant plasmids are inserted into bacterial cells.

This collection of bacteria constitutes a genomic library that can be used for later experiments.

Isolated DNA with various genes

Plasmids

NUCLEIC ACID PROBES

It is often the case that a genetic engineering experiment produces many DNA fragments, only some of which contain the gene of interest. For example, a researcher may wish to select just one DNA fragment from among an entire genomic library. To find the right gene, a researcher can use a **nucleic acid probe**, a complementary molecule made using radioactive or fluorescent building blocks. Because it has a sequence that matches the desired target (by following the base-pairing rules), the probe will bind to and help visualize the target DNA.

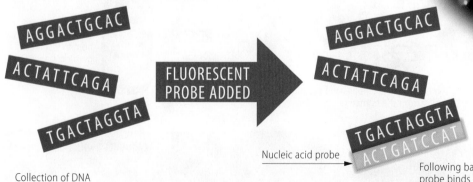

AGGACTGCAC

ACTATTCAGA

TGACTAGGTA

FLUORESCENT PROBE ADDED

AGGACTGCAC

ACTATTCAGA

TGACTAGGTA
ACTGATCCAT

Nucleic acid probe

Collection of DNA fragments to be searched

Following base pairing rules, the probe binds only to the desired fragment, making it visible.

These chromosomes (red) were mixed with a fluorescent probe (yellow) complementary to DNA sequences found only at a few locations within each chromosome.

1,400x

DNA SYNTHESIS

DNA can be created from scratch in the laboratory using an automated DNA synthesizer. These machines can quickly and accurately produce customized DNA molecules of any sequence up to lengths of a few hundred nucleotides. Many such synthesized fragments can be joined to produce DNA molecules of almost any length.

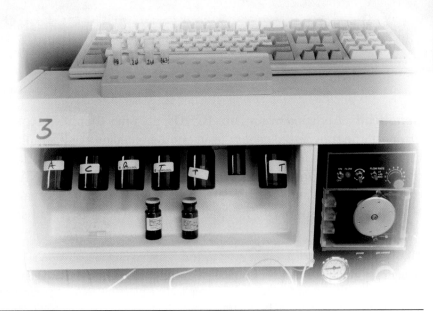

In 2010, researchers used DNA synthesis machines to produce the first bacterium with a completely synthetic genome.

COMPLEMENTARY DNA

Another tool in the genetic engineering toolbox starts with the messenger RNA (mRNA) molecules produced in the nucleus of eukaryotic cells. Within the cells, these mRNAs are made from active genes and are used to produce proteins. An enzyme called **reverse transcriptase** can synthesize DNA molecules from the collection of mRNAs within the cell. The result is **complementary DNA (cDNA)** representing just the genes that were being transcribed in the cell at the time.

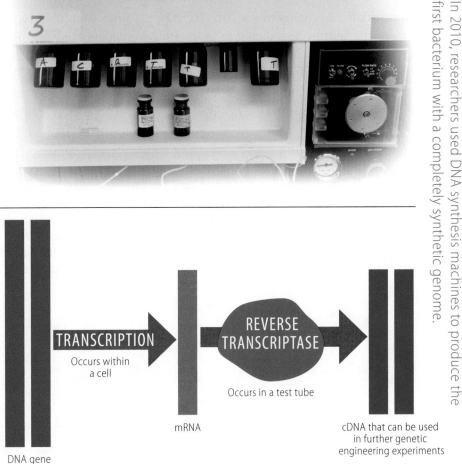

TRANSCRIPTION
Occurs within a cell

REVERSE TRANSCRIPTASE
Occurs in a test tube

DNA gene

mRNA

cDNA that can be used in further genetic engineering experiments

A fluorescence microscope can be used to visualize fluorescently labeled nucleic acids.

CORE IDEA: Scientists can manipulate DNA in various ways: it can be isolated from a cell and put into a genomic library, visualized using nucleic acid probes, synthesized directly, or produced from a cell's messenger RNA.

CORE QUESTION: If you produce a genomic library with all the genes in a human cell, how can you find the few bacteria that contain a particular gene?

ANSWER: You can use a probe that is complementary to the desired gene.

Plants and animals can be genetically modified

Genetically modified organisms (GMOs) are ones that have acquired one or more genes by artificial means. If the transferred gene is from another species (for example, a goat carrying a gene from a human), then the organism is called a **transgenic organism**. The production of transgenic GMOs is one of the most widespread applications of genetic engineering. In fact, DNA technology has largely replaced traditional animal and crop breeding programs that aim to increase the productivity of many food sources. However, as the uses of GMOs have spread, some people have raised safety concerns.

TRANSGENIC CROPS

Genetically modified food crops can be produced by inserting a desired gene into a plasmid, a small, circular, independently replicating piece of DNA originally isolated from a bacterium. The plasmid acts as a temporary DNA carrier, allowing a gene of interest to be inserted into the genome of a plant. The result is a genetically modified plant that expresses the trait from the newly inserted gene. Genetically engineered food crops are widely consumed in the United States. Some agricultural scientists hope that GM crops can increase food production, pest resistance, and the nutritional value of crops.

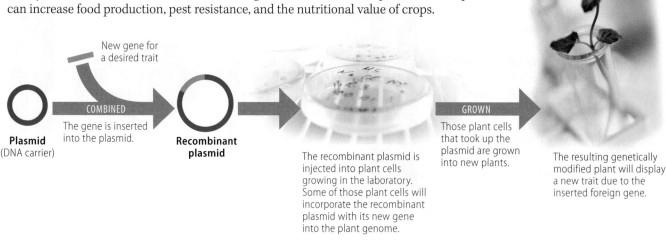

New gene for a desired trait

Plasmid (DNA carrier)

COMBINED
The gene is inserted into the plasmid.

Recombinant plasmid

The recombinant plasmid is injected into plant cells growing in the laboratory. Some of those plant cells will incorporate the recombinant plasmid with its new gene into the plant genome.

GROWN
Those plant cells that took up the plasmid are grown into new plants.

The resulting genetically modified plant will display a new trait due to the inserted foreign gene.

Examples of genetically modified food crops

Bt corn has been genetically modified to express a protein from the bacterium *Bacillus thuringiensis*. This protein acts as an insecticide, selectively killing caterpillars that attack the corn plant.

Golden Rice is a transgenic variety, created in 2000, with a few daffodil genes that produce beta-carotene, which our body uses to make vitamin A. This rice could help prevent vitamin A deficiency—and the resulting blindness—among the billions of people who depend on rice as their staple food.

In Hawaii, the ring spot virus seemed poised to devastate the papaya industry until a GM papaya variety resistant to the virus was introduced in 1992.

The majority of corn, soybean, and cotton grown in the United States is genetically modified.

TRANSGENIC ANIMALS

Continuing the millennia-old tradition of selectively breeding farm animals, agricultural scientists are working on various genetic modifications to food animals to make them healthier or more productive. To date, however, none of these animals are in our food supply. (As of mid-2013, the Food and Drug Administration was still considering whether to grant approval to a variety of fast-growing transgenic salmon.) Additionally, pharmaceutical companies have produced various GMOs that secrete medically useful human proteins using the procedure shown here.

The new protein can be harvested, purified, and used to treat medical conditions.

Erythropoietin
(treatment for anemia)

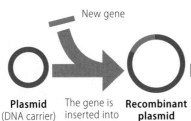

New gene

Plasmid
(DNA carrier)

The gene is inserted into the plasmid.

Recombinant plasmid

The recombinant plasmid can be injected into mammalian cells growing in the laboratory. Some of these cells will take up the new gene and produce its protein.

The recombinant plasmid can be injected into mammalian embryos.

The recombinant embryos are then implanted in surrogate mothers.

GIVE BIRTH

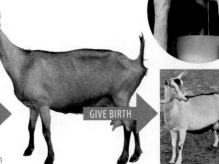

The new protein can be harvested from the milk of the transgenic offspring, purified, and used to treat medical conditions. For example, the goat shown to the right carries a gene for a human blood protein.

SAFETY AND ETHICAL CONCERNS

The use of genetically modified organisms in agriculture has rapidly replaced traditional plant and animal breeding programs. As with every new technology, genetic engineering carries the potential for harm as well as benefit. Regulations are still being drafted, so it is important for all citizens to be educated about the possible risks and benefits of this new technology.

Pros	Cons
No scientific evidence has been found that GMOs pose any special risks to humans or the environment.	GMOs may be hazardous to human health in ways that we cannot yet understand.
Voluntary and mandatory safety measures are in place, with a ban on certain dangerous lines of research.	GMOs might escape into the environment and cause unforeseen problems by passing genes to wild organisms.
GMOs can be more nutritious than standard crops and can be grown in a wider variety of habitats.	Widespread use of GMOs may promote a loss of biodiversity.
Strains of microorganisms used in recombinant DNA experiments are genetically crippled so they cannot survive in the wild.	There is no standard for labeling GMOs in the United States, so consumers don't know what they are getting.

CORE IDEA: Genetic engineers can use plasmids to create transgenic plants and animals. While GM plants currently make up a significant part of our food supply, GM animals do not. The use of such organisms carries risks and benefits.

CORE QUESTION: What is a major similarity in the production of GM plants and mammals?

ANSWER: Both use a plasmid to introduce foreign DNA.

PCR can be used to multiply samples of DNA

The **polymerase chain reaction (PCR)** is a laboratory technique by which a specific segment of DNA can be targeted and copied quickly and precisely. Through PCR, a scientist can obtain enough DNA from even a trace amount of blood or other tissue to allow further experimentation. Starting from just a single copy, automated PCR can generate billions of copies of a DNA segment in just a few hours, and it can do so with an extremely high rate of accuracy. Biologists use PCR to determine DNA profiles, to study relationships among organisms, and to hunt for disease-causing genes.

THE PCR TECHNIQUE

In PCR, a sample of DNA is subjected to rounds of heating and cooling in a machine called a **thermal cycler**. During each round, the quantity of a desired region of DNA is doubled. The key to PCR is **DNA polymerase**, a naturally occurring enzyme that synthesizes a new DNA strand that is complementary to a single-stranded DNA sample. The DNA polymerase used in PCR was originally isolated from a prokaryote that lives in natural hot springs, and so can tolerate the high temperatures that would otherwise inactivate a standard DNA polymerase.

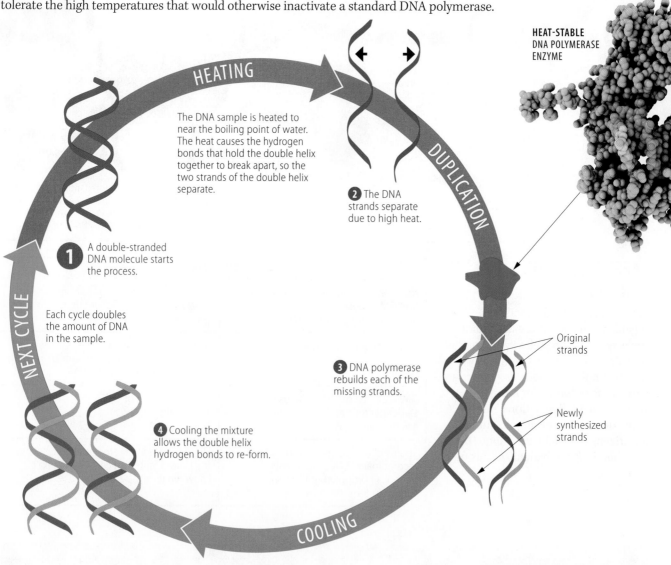

HEAT-STABLE DNA POLYMERASE ENZYME

HEATING

The DNA sample is heated to near the boiling point of water. The heat causes the hydrogen bonds that hold the double helix together to break apart, so the two strands of the double helix separate.

DUPLICATION

2 The DNA strands separate due to high heat.

1 A double-stranded DNA molecule starts the process.

Each cycle doubles the amount of DNA in the sample.

NEXT CYCLE

3 DNA polymerase rebuilds each of the missing strands.

Original strands

Newly synthesized strands

4 Cooling the mixture allows the double helix hydrogen bonds to re-form.

COOLING

PCR has been used to study DNA extracted from a 40,000-year-old frozen wooly mammoth.

HOW DNA SEGMENTS ARE TARGETED AND COPIED

After a sample of DNA has been collected (for example, blood evidence from a crime scene), an investigator does not copy the entire genetic content of the sample. Instead, PCR is typically used to copy one specific region within the larger mass of DNA. To do this, a researcher uses **primers**, short (usually 15 to 20 nucleotides long) chemically synthesized single-stranded DNA molecules with sequences that are complementary to sequences at each end of the target sequence. DNA polymerase can then bind the primers and synthesize new DNA molecules using free nucleotides included in the mixture. By adding the correct primers along with the other needed ingredients, a large sample of the DNA within that specific region can be obtained. The DNA sample produced via PCR can then be used for further experiments.

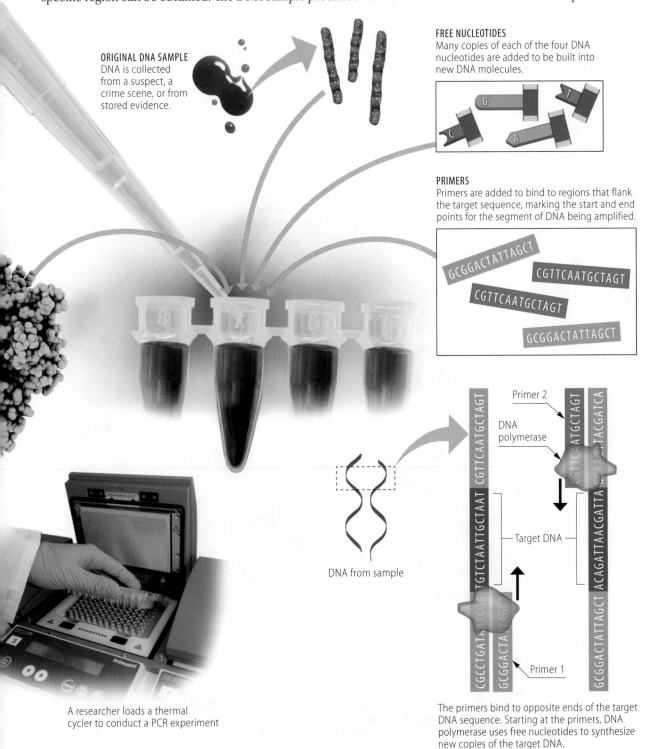

ORIGINAL DNA SAMPLE DNA is collected from a suspect, a crime scene, or from stored evidence.

FREE NUCLEOTIDES Many copies of each of the four DNA nucleotides are added to be built into new DNA molecules.

PRIMERS Primers are added to bind to regions that flank the target sequence, marking the start and end points for the segment of DNA being amplified.

GCGGACTATTAGCT
CGTTCAATGCTAGT
CGTTCAATGCTAGT
GCGGACTATTAGCT

Primer 2
DNA polymerase
Target DNA
Primer 1
DNA from sample

A researcher loads a thermal cycler to conduct a PCR experiment

The primers bind to opposite ends of the target DNA sequence. Starting at the primers, DNA polymerase uses free nucleotides to synthesize new copies of the target DNA.

CORE IDEA: Each round of the polymerase chain reaction (PCR) uses ingredients—heat-stable DNA polymerase enzyme, DNA nucleotides, a sample to be studied, and primers that target a specific sequence—to precisely double the amount of DNA.

CORE QUESTION: If you wish to compare DNA from a suspect and a crime scene, would you use PCR to duplicate the entire genome?

ANSWER: No. PCR duplicates only a small part of the DNA that is marked by the specific primers chosen.

DNA profiles are based on STR analysis

DNA profiling is a set of laboratory techniques that lets you determine with certainty whether two samples of DNA came from the same individual. Imagine that you have a sample of DNA from a crime scene, and a second sample from a suspect. How can you prove they match? Theoretically, you could compare the sequences of the entire set of DNA in each sample. But this is very impractical. Instead, modern DNA profiling techniques rely on comparing small regions of DNA within the genome that are known to vary among the population.

SHORT TANDEM REPEATS

DNA profilers focus on specific sites within the genome that are known to vary considerably from person to person. Scattered throughout the genome are **short tandem repeats (STRs)**, sites where a short nucleotide sequence is repeated many times in a row. The locations of these sites within chromosomes and the sequences that are repeated are identical from person to person. But the number of repeats varies widely within the human population.

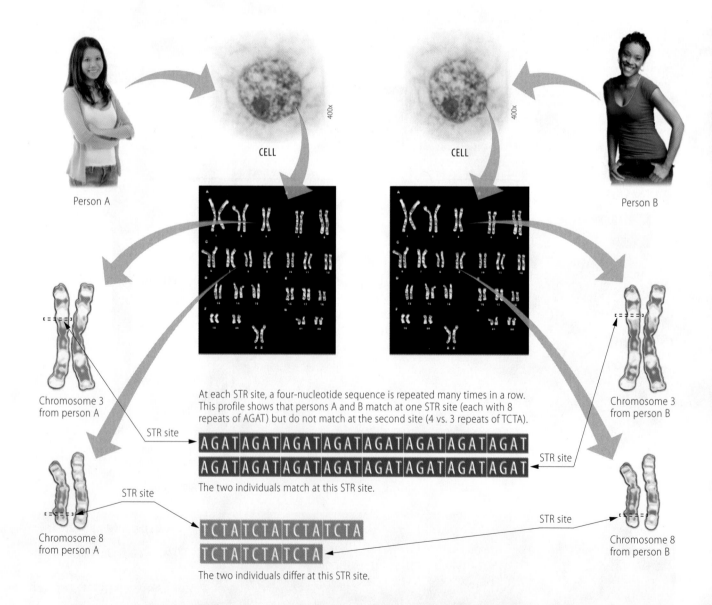

Person A

CELL

CELL

Person B

Chromosome 3 from person A

Chromosome 3 from person B

At each STR site, a four-nucleotide sequence is repeated many times in a row. This profile shows that persons A and B match at one STR site (each with 8 repeats of AGAT) but do not match at the second site (4 vs. 3 repeats of TCTA).

STR site

STR site

AGAT AGAT AGAT AGAT AGAT AGAT AGAT AGAT
AGAT AGAT AGAT AGAT AGAT AGAT AGAT AGAT

The two individuals match at this STR site.

STR site

TCTA TCTA TCTA TCTA
TCTA TCTA TCTA

The two individuals differ at this STR site.

Chromosome 8 from person A

STR site

Chromosome 8 from person B

STR ANALYSIS

The current method for generating a DNA profile relies on **STR analysis**, a comparison of the lengths of short tandem repeat sequences at 13 predefined sites within the human genome. At each site, a four-nucleotide sequence is repeated between 3 and 50 times in a row. These sites vary so widely that no two humans have ever had the same number of repeats at all 13 sites (except identical twins). To perform an STR analysis, a forensic scientist obtains DNA from two or more samples and then uses the polymerase chain reaction (PCR) to produce large quantities for comparison. Here, the process is simplified to include just 2 STR sites.

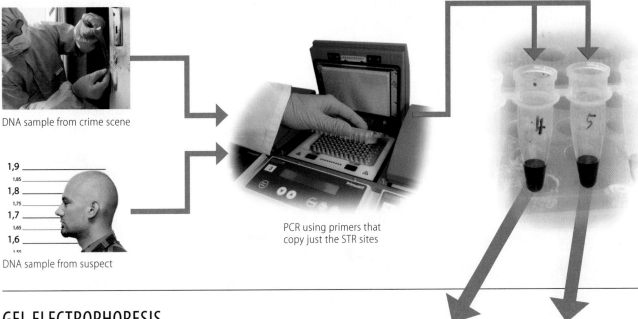

DNA sample from crime scene

DNA sample from suspect

PCR using primers that copy just the STR sites

Output from PCR (large amounts of the DNA at the STR sites)

STR analysis proved that a 9,000-year-old skeleton found in a cave in England was a direct ancestor of a schoolteacher who lived nearby.

GEL ELECTROPHORESIS

Once an STR analysis sample has been prepared by PCR, it needs to be analyzed. The process of **gel electrophoresis** allows visualization of DNA samples based on their length. A researcher loads samples of DNA into the top of the gel and then applies electrical current. All DNA molecules are negatively charged, so they migrate through the gel toward the positive pole. A dense thicket of fibers within the gel slows the migration. As a result, smaller pieces of DNA migrate toward the bottom of the gel faster than larger pieces. The large pieces travel slower, staying nearer to the top of the gel. After the power is turned off and the DNA stops migrating, staining reveals the locations of the bands. In this case, you can see that one STR site (the blue bands) exactly matches between the two samples, but the other one (purple bands) is different. This indicates that the two DNA samples came from different individuals—that is, that the suspect's DNA was not recovered at the crime scene.

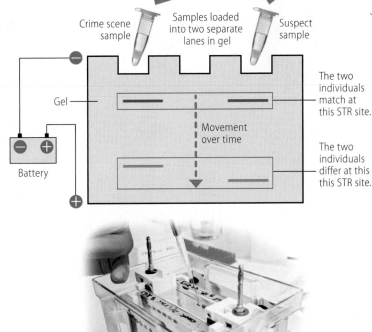

Crime scene sample

Samples loaded into two separate lanes in gel

Suspect sample

Gel

Battery

Movement over time

The two individuals match at this STR site.

The two individuals differ at this this STR site.

A gel electrophoresis chamber

CORE IDEA: DNA profiling can prove that two DNA samples came from the same individual. PCR creates a sample of DNA containing specific short tandem repeats that vary widely among people. Gel electrophoresis visualizes these sites.

CORE QUESTION: Looking at the gel, which DNA segment is longer: the matching STR site (in blue) or the non-matching STR site (purple)?

ANSWER: The matching site (blue) is longer since it migrated a shorter distance.

Whole genomes can be sequenced and mapped

In 1995, a team of biologists announced that they had determined the DNA sequence of the entire genome of *Haemophilus influenzae*, a disease-causing bacterium. This marked the first successful experiment in **genomics**, the science of studying the complete sets of genes (genomes) and their interactions. In recent years, scientists have developed many detailed genome maps from a variety of species, including our own. The methods of genomics are relatively simple, but the results teach us much about our own species and the evolutionary relationships among life on Earth.

THE HUMAN GENOME PROJECT

In 2003, researchers announced they had sequenced virtually all of the genes from a human. The **Human Genome Project** found that our chromosomes (22 that we all share, plus the sex chromosomes X and Y) contain about 21,000 genes within about 3 billion DNA nucleotides. Recently, the genomes of several other humans have been sequenced, allowing comparisons within our species. Additionally, the genomes of many other organisms have been sequenced, allowing cross-species comparisons. The data below relate some of the interesting data that genomics studies have revealed.

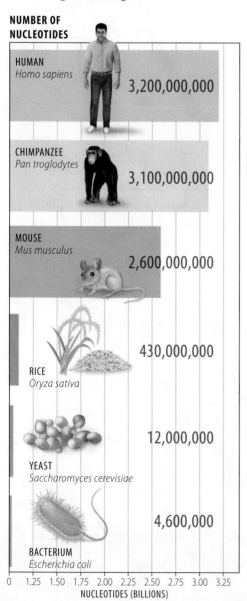

NUMBER OF NUCLEOTIDES

HUMAN
Homo sapiens
3,200,000,000

CHIMPANZEE
Pan troglodytes
3,100,000,000

MOUSE
Mus musculus
2,600,000,000

RICE
Oryza sativa
430,000,000

YEAST
Saccharomyces cerevisiae
12,000,000

BACTERIUM
Escherichia coli
4,600,000

0 1.25 1.50 1.75 2.00 2.25 2.50 2.75 3.00 3.25
NUCLEOTIDES (BILLIONS)

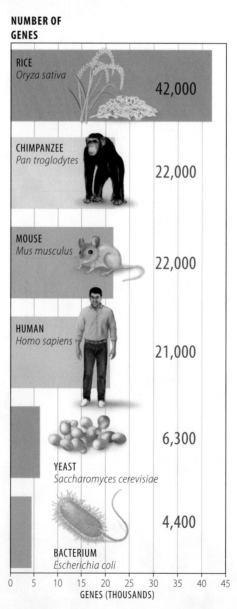

NUMBER OF GENES

RICE
Oryza sativa
42,000

CHIMPANZEE
Pan troglodytes
22,000

MOUSE
Mus musculus
22,000

HUMAN
Homo sapiens
21,000

YEAST
Saccharomyces cerevisiae
6,300

BACTERIUM
Escherichia coli
4,400

0 5 10 15 20 25 30 35 40 45
GENES (THOUSANDS)

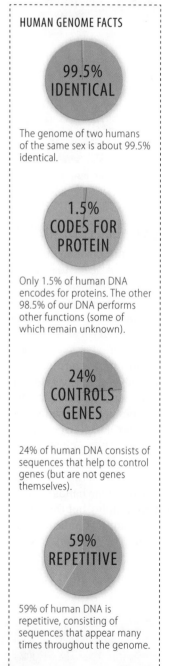

HUMAN GENOME FACTS

99.5% IDENTICAL

The genome of two humans of the same sex is about 99.5% identical.

1.5% CODES FOR PROTEIN

Only 1.5% of human DNA encodes for proteins. The other 98.5% of our DNA performs other functions (some of which remain unknown).

24% CONTROLS GENES

24% of human DNA consists of sequences that help to control genes (but are not genes themselves).

59% REPETITIVE

59% of human DNA is repetitive, consisting of sequences that appear many times throughout the genome.

WHOLE-GENOME SHOTGUN METHOD

The set of techniques used to sequence an entire genome from an organism is called the **whole-genome shotgun method**. As you can see here, the method combines several separate techniques that together can produce a wealth of data very quickly.

GENOME(S)

DNA is obtained from one or more individuals.

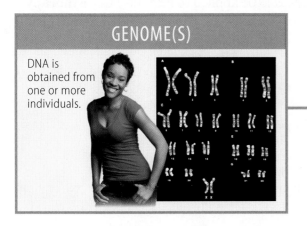

DNA FRAGMENTS

The DNA is digested with a variety of restriction enzymes, which chop up the DNA into small segments.

SEQUENCES OF DNA

The sequence of each DNA fragment is determined by an automated sequencing machine.

PROTEOMICS

Although DNA is the molecule that ultimately controls cells, proteins actually perform the tasks that keep all cells functioning. Scientists are thus interested in examining the complete set of proteins encoded by a genome, a field called **proteomics**. The number of proteins in humans (about 100,000) vastly exceeds the number of genes (about 21,000), so an understanding of what all these proteins are and what tasks they perform is at the forefront of biological research.

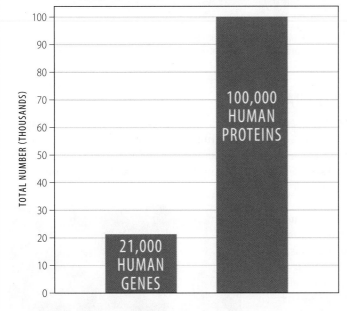

COMPARING THE NUMBER OF GENES AND PROTEINS IN HUMANS

(Y-axis: TOTAL NUMBER (THOUSANDS), 0 to 100)

21,000 HUMAN GENES

100,000 HUMAN PROTEINS

OVERLAPPING SEQUENCES

Computer programs use overlapping regions from each fragment to determine the original order of the sequences.

FINAL SEQUENCE

The complete genome is uploaded to a database.

```
CAATTCAGGGTGGTGAATGTGAAAGCAGGCATGACAGGAGTA
ACCCACTAAGGTATTTTCATGGCGACCAAGAGGATTAAGTAT
ATACACCCAGGGGGCGGAATGAAAGCGTAACAATGGCGACAT
GGTAACGAGGTAACAACCATGCGAGTGCATAACGGAGTGATC
CGCGTAAGGAAATCCATTATGTACTATTTCACACAGGAAACA
ACCCACTAAGGTATTTTCATGGCGACCAAGAGGATTAAGTAT
```

CORE IDEA: By digesting a cell's DNA with enzymes and sequencing the fragments, the entire genome of an organism can be determined. The human genome contains 21,000 genes that encode for 100,000 different proteins.

CORE QUESTION: Looking at the two bar graphs on the left page: Does the number of genes in a genome always correlate with the total size of the genome?

ANSWER: No. Rice has far less DNA but far more genes than the human genome.

Gene therapy aims to cure genetic diseases

Many human diseases are caused by a mutation in a single gene. Biologists have learned to genetically engineer bacteria, plants, and animals—can these techniques also be applied to humans? **Gene therapy** is the alteration of a person's genes in order to treat a disease. Medical researchers have proven that gene therapy can work in theory, and they have even cured some people. However, the promise of gene therapy has—so far—outpaced the actual successes.

OBTAINING A HEALTHY GENE

Gene therapy begins with isolation of the normal gene from a healthy person. Enzymes are used to produce an RNA version of the target DNA gene. This RNA gene is then combined with an infectious, but harmless, retrovirus (a virus with an RNA genome). The retrovirus most commonly used in human gene therapy is a crippled version of a cold virus, capable of infecting human cells but unable to cause harm. This recombinant retrovirus is ready for therapy.

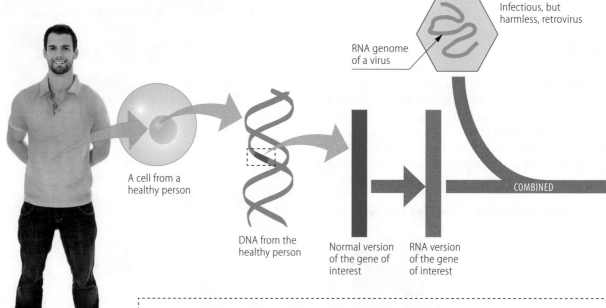

Infectious, but harmless, retrovirus

RNA genome of a virus

A cell from a healthy person

DNA from the healthy person

Normal version of the gene of interest

RNA version of the gene of interest

COMBINED

HEALTHY PERSON

In a 2009 gene therapy trial, several children were cured of a genetic form of progressive blindness through injection of a recombinant virus.

A CASE STUDY IN GENE THERAPY

Gene therapy has been used to treat severe combined immunodeficiency (SCID), a disease caused by the absence of an enzyme required by the immune system. SCID patients, who must remain isolated within protective "bubbles," succumb to infections that would be easily fought off by a normal immune system. Since 2000, gene therapy has cured 22 children with SCID. The celebrations of this medical breakthrough were short lived, however; four of the patients developed leukemia, and one died, because the retrovirus caused some blood cells to become cancerous. Thus, although gene therapy remains promising, there is little evidence of safe and effective application. Active research continues, with new, tougher safety guidelines.

INJECTING THE HEALTHY GENE

Once a recombinant retrovirus has been engineered to carry the healthy gene, that gene needs to get inside a diseased person's cells. Bone marrow cells are ideally suited for this purpose because they multiply—continuously producing the normal protein—for a long time. The goal of gene therapy is to cure the patient, perhaps permanently, with just a single injection.

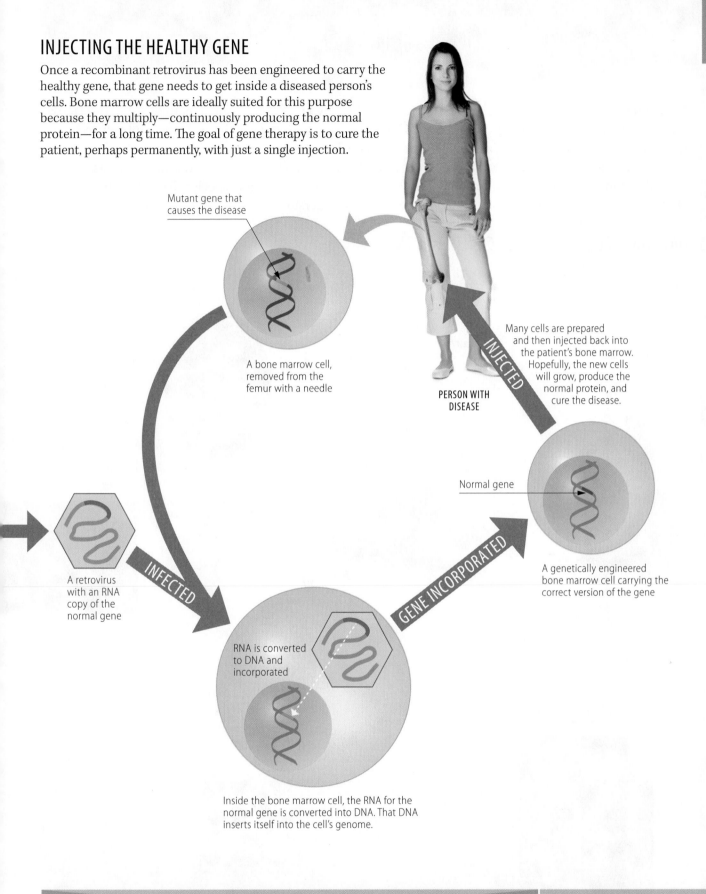

Mutant gene that causes the disease

A bone marrow cell, removed from the femur with a needle

A retrovirus with an RNA copy of the normal gene

INFECTED

RNA is converted to DNA and incorporated

Inside the bone marrow cell, the RNA for the normal gene is converted into DNA. That DNA inserts itself into the cell's genome.

GENE INCORPORATED

Normal gene

A genetically engineered bone marrow cell carrying the correct version of the gene

INJECTED

Many cells are prepared and then injected back into the patient's bone marrow. Hopefully, the new cells will grow, produce the normal protein, and cure the disease.

PERSON WITH DISEASE

CORE IDEA: Human gene therapy involves the production of a recombinant retrovirus carrying an RNA version of a normal human gene. The virus can infect bone marrow cells, transferring the proper gene to a diseased individual.

CORE QUESTION: Why are viruses particularly good carriers of healthy human genes for therapy?

ANSWER: Viruses naturally infect human cells, making them well suited to insert the normal DNA into a patient with a disease.

153

Darwin's influences and experiences led him to publish his theory of evolution

November 24, 1859, is a landmark date in the history of biology. On that day, British naturalist Charles Darwin published *On the Origin of Species by Means of Natural Selection*. In this classic work, Darwin introduced the concepts of natural selection and evolution (which he described as "descent with modification"). The thinking behind his concepts was quite different from the prevailing intellectual view that Earth was only a few thousand years old and occupied by entirely unrelated and unchanging species. To understand how Darwin arrived at his revolutionary ideas, you need to first understand his scientific and cultural context.

DARWIN'S CULTURAL AND SCIENTIFIC CONTEXT

1700s

1800s

ANCIENT GREEKS
Around 350 B.C.

The ancient Greeks, led by the philosopher Aristotle, believed that species were permanent and unchanging.

FOSSILS
1700s

In the 1700s, the discovery of fossils suggested that the Earth might be much older than previously thought. It also suggested that, in previous times, organisms related to, but distinct from, today's creatures lived on Earth. Scientists began to suspect that fossils might represent extinct, ancient forms of modern creatures.

DARWIN'S BIRTH
1809

Charles Darwin was born at home in Shrewsbury, England, in 1809. Even as a young boy, he had ambitions as a naturalist.

LYELL
1830

In 1830, Scottish geologist Charles Lyell published *Principles of Geology*. In this classic text, Lyell suggested that Earth was very old and gradually changed through slow, accumulating processes such as erosion and earthquakes.

THE MIDDLE AGES
400–1400

For over a millenium, Judeo-Christian scholars followed a literal interpretation of the Bible, teaching that the Earth was created around 6,000 years ago, and that all life remains as it was originally created.

LAMARCK
1800s

French naturalist Jean-Baptiste de Lamarck was one of the first to suggest that life evolves through physical changes that enable organisms to succeed in their environment.

The A&E program *Biography* ranked Charles Darwin as the fourth most influential person of the second millennium.

DARWIN'S EDUCATION
1825–1831

Although he originally entered medical school, Darwin ultimately enrolled in Cambridge University to become a minister. He graduated with a B.A. degree.

DARWIN'S ADULT LIFE
1830s–1850s

By the end of his voyage, Darwin was convinced that Earth and its life were very old and constantly changing. He spent decades reading, analyzing his specimens (such as his beetle collection), and discussing ideas with colleagues.

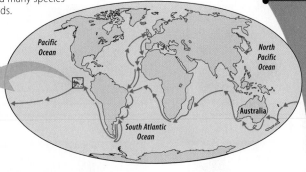

Sample of beetles from Darwin's collection

WALLACE
1850s

By the mid-1850s, Darwin's colleague Alfred Russel Wallace developed and was preparing to publish his own ideas of natural selection. This prompted Darwin, after decades of carefully collecting evidence and formulating logical arguments, to present his papers publicly.

ALFRED RUSSEL WALLACE (1823–1913)

Wallace's notes and sketch for a hornbill bird (*Buceros cassidix*), 1856

1830s

THE VOYAGE OF THE BEAGLE
1831–1836

Just after graduation, 22-year-old Darwin accepted a post as the shipboard naturalist aboard the HMS *Beagle* for its round-the-world surveying voyage. Whenever he could, Darwin ventured ashore to collect fossils and specimens of modern plants and animals. Darwin was struck by the diversity of life on the Galápagos Islands, off the western coast of South America, which included many species found nowhere but on one of the islands.

1850s

Map of Galápagos Islands

Giant tortoises (*Geochelone nigra*) on Isabela Island, Galápagos

Marine iguanas (*Amblyrhynchus cristatus*) from the Galápagos Islands

THE ORIGIN OF SPECIES
1859

In 1859, Darwin published *The Origin of Species*. In this book, Darwin presented a strong logical argument for evolution by natural selection as well as a mountain of supporting evidence. His book became an instant best seller and remains, even 150 years later, the most influential book in biology.

CORE IDEA: Influenced by the scientific and cultural changes of the day—and his travels on the HMS *Beagle* and subsequent research—Darwin's *The Origin of Species* presented ideas that were at odds with prevailing notions about life on Earth.

CORE QUESTION: How many years passed between Darwin setting off on his *Beagle* voyage and the publication of his book?

ANSWER: 28 (1831 to 1859).

155

Unequal reproductive success leads to natural selection

In *The Origin of Species*, Darwin made two important points. First, he presented the concept of evolution (which he termed "descent with modification"): that all living species have descended from a succession of ancestral species, each of which accumulated modifications that helped them survive in their habitat. Second, Darwin addressed the question of how evolution occurs: through **natural selection**, the process by which organisms with certain traits are more likely to survive and reproduce than other organisms. Darwin's two concepts are supported by logic and by numerous examples from the natural world.

OBSERVATION: OVERPRODUCTION

Every population has the potential to greatly increase its numbers very quickly. Rabbits, like most populations, can multiply rapidly.

OBSERVATION: LIMITED RESOURCES

The amount of resources in the environment—living space, water, sunlight, etc.—stays relatively constant. For example, new watering holes are not created each year.

CONCLUSION: COMPETITION

Because more individuals are born than can possibly be supported by the environment, there is a constant competition among organisms for the limited resources available. Not every gazelle that is born can possibly survive to adulthood.

OBSERVATION: VARIABILITY

Individuals in a population vary in many inherited traits. You know that no two people are the same; careful observation reveals that no two individuals of any population (including ladybugs and gazelles) are exactly alike.

CONCLUSION: NATURAL SELECTION

Those individuals with variations that make them best suited to their environment will, on average, be more likely to survive and reproduce. Darwin named this unequal reproductive success "natural selection." For example, faster gazelles will more often escape lions.

OBSERVATION: HERITABILITY

2,100x

The traits of an organism are likely to be passed to the next generation. For example, fast gazelles are likely to have, on average, fast offspring.

CONCLUSION: EVOLUTION

Because traits are passed from one generation to the next, and because certain members are more likely to survive and reproduce, a population will change over time, becoming better suited to its environment. Darwin called this descent with modification. We call it evolution by natural selection.

THE ARGUMENT FOR NATURAL SELECTION

In *The Origin of Species*, Darwin presented the theory of natural selection based on simple observations and obvious conclusions. Above, you can follow Darwin's original logical argument.

EVOLUTION VIA NATURAL SELECTION IN ACTION

Evolution by natural selection has shaped life on Earth for over 3 billion years. But that doesn't mean that these mechanisms operated only in the distant past. Indeed, evolution continues to shape our world in ways that are directly relevant to your life, as you can see in this example.

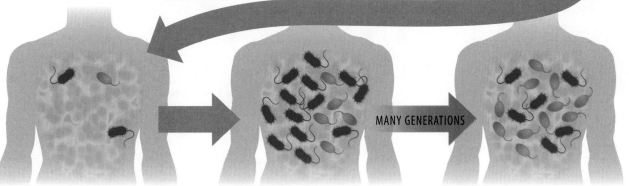

Bacterium

Bacterium with random mutation that confers antibiotic resistance

Antibiotic added

1 POPULATION WITH VARIATION
The members of this population of bacteria differ in their resistance to an antibiotic. Some bacteria have a mutation that makes them more likely to survive the antibiotic.

2 CHANGE IN THE ENVIRONMENT
The environment in which the bacteria are living is changed by the application of an antibiotic.

MANY GENERATIONS

3 ELIMINATION OF INDIVIDUALS WITH UNDESIRABLE TRAITS
Most of the bacteria that lack the resistance gene die off. The few with the resistance gene are more likely to survive.

4 REPRODUCTION OF SURVIVORS
Due to natural selection, certain individuals are more likely to reproduce than others. The heritability of the resistance gene ensures that it is passed on to offspring.

5 CHANGE TO THE POPULATION OVER TIME
Applied over many generations, the frequency of antibiotic-resistant bacteria within the population increases, rendering the antibiotic less effective.

Penicillin, once the most widely used antibiotic, is now nearly useless due to the evolution of penicillin-resistant bacteria.

IMPORTANT POINTS ABOUT EVOLUTION

Although simple in concept, many people have mistaken ideas about evolution. Here, you can read the correct way to interpret three important points about evolution.

INDIVIDUALS DON'T EVOLVE

A common misconception is that individuals evolve. However, although natural selection acts on individuals, evolution is a generation-to-generation change in populations. An organism does not evolve, but a population of organisms may.

NATURAL SELECTION WORKS WITH HERITABLE TRAITS

Another misconception is that all aspects of an individual's physical characteristics—such as muscle mass, hair style, and so forth—are passed on to the offspring. However, only traits that are encoded within genes and passed on to the next generation can possibly affect the reproductive success of that generation.

EVOLUTION DOES NOT HAVE A GOAL

Finally, there is a misconception that evolution results in steady progress through each generation. However, natural selection cannot anticipate what an organism will need in the future. Evolution always occurs in response to the local environmental conditions at that time.

CORE IDEA: Evolution by natural selection is a logical consequence of some easily observed natural phenomena. You can see evidence of this mechanism in the world around you.

CORE QUESTION: Would you expect evolution to proceed more or less quickly in a rapidly changing environment?

ANSWER: Because natural selection favors organisms best suited to the local environment, rapid change will promote faster evolution.

The fossil record provides important evidence for evolution

Darwin's original formulation of the theory of natural selection was based on observations of the world around him. In particular, Darwin (and other naturalists of his day) was struck by the appearance of fossil forms that were similar to, but distinct from, modern forms. The fossil record remains one of the best and most easily observed lines of evidence in support of evolutionary theory.

THE FOSSIL RECORD

As water runs over land, eroded sand and silt are carried into the sea. Dead organisms that settle into the sediment leave impressions. Over millions of years, these sediments pile up and are compressed. Layers of rock build up, with each layer containing fossils of the organisms that died at a particular time. This is the **fossil record**, the ordered sequence of fossils found in layers of rock, with older fossils buried further down, and newer fossils near the surface.

Fossilized fish called *Lepidotes* from 180 million years ago

Newer layer

Older layer

Layers of sediment

Sediment carried by rivers empties into the oceans and buries dead organisms in successive layers over time.

Excavating sediment reveals an ordered series of fossils, with the oldest buried the deepest down.

Trilobite (*Huntonia linguifer*) from about 600 million years ago

RADIOMETRIC DATING

How do scientists know the age of fossils? The most common dating method is **radiometric dating**, which is based on the breakdown of the radioactive isotopes found in all living matter.

This graph shows the half-life of carbon-14. Every 5,600 years, half the C-14 present in any sample will break down into C-12. By measuring the percentage of C-14 in a sample, the age of that sample is revealed.

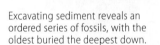

RADIOACTIVE CARBON-14 (as % of living organism's C-14 to C-12 ratio)

100
75
50
25
0

0.0 5.6 11.2 16.8 22.4 28.0 33.6 39.2 44.8 50.4
TIME (thousands of years)

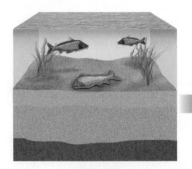

Carbon-14 is a radioactive isotope of carbon that living organisms incorporate into their bodies from the environment. When an organism dies, its body contains a known percentage of C-14.

Over time, the C-14 in the organism's body breaks down into normal carbon (C-12). No new C-14 enters since the organism is dead and not metabolizing, so the amount of C-14 goes down steadily over time.

When a fossil is found, the amount of C-14 remaining can be measured to give a reliable estimate of the age of the fossil.

A GALLERY OF FOSSILS

Although many fossils form through the accumulation of sedimentary rock, fossils may form in various other ways.

Sedimentary fossils form when minerals from surrounding sediment seep into a dead organism and replace the organic matter. Left, petrified wood; right, fossil of duck-billed dinosaur (*Gryposaurus notabilis*) that lived around 70 million years ago.

Organisms that died relatively recently can sometimes be found frozen in ice. This 10,000-year-old mammoth was found in the permafrost of northern Russia in 1989.

Amber, fossilized tree sap, can sometimes contain organisms within it, preserving them in great detail.

Sometimes, the original organism disappears, but it leaves behind a mold that is later filled in with minerals.

Trace fossils hint at an organism's behavior through footprints or tracks. These five parallel *Apatosaurus* tracks from about 150 million years ago are preserved in the Morrison Formation in Colorado.

EVIDENCE FOR EVOLUTION WITHIN THE FOSSIL RECORD

How does the fossil record provide evidence of evolution? The fossil record reveals an ordered appearance of life on Earth, from simple prokaryotes over 3 billion years ago through progressively more complex and modern-looking species. Such evidence is clearly at odds with the view of life as static and unchanging. Even more significantly, fossil hunters have found many transitional forms, such as whales that still had their rear legs. Here, you can see a drawing of a *Basilosaurus*, an ancient whale species that died out about 34 million years ago. These whales, although entirely aquatic, had not yet lost their hind limbs.

Basilosaurus

CORE IDEA: Fossils form when organisms die, fall into accumulating sediment, and the sediment is compressed into rock. Fossils provide evidence for evolution through their appearance in layers and by providing evidence of transitional forms.

CORE QUESTION: If you found a fossil in your backyard that had 25% the level of atmospheric C-14, how old would you estimate it is?

ANSWER: About 11,200 years; half of the radioactivity would disappear in 5,600 years, and half of that remaining in the next 5,600 years.

Much evidence for evolution is found in the natural world

Evolution is a testable theory with a great deal of supporting evidence gathered over hundreds of years. The fossil record provides some of the best evidence in favor of the gradual adaptation of populations to their local environment. Several other lines of evidence that support evolution by natural selection are observable in our natural world.

BIOGEOGRAPHY

Biogeography, the study of the geographic distribution of species, provides much evidence for the evolution of life on Earth. Consider, for example, the dominance of marsupial mammals (animals in whom embryonic development is completed in a pouch) in Australia, compared to the dominance of placental mammals (in whom embryonic development is completed within the placenta) elsewhere. How can evolution account for this? The prevailing view is presented here.

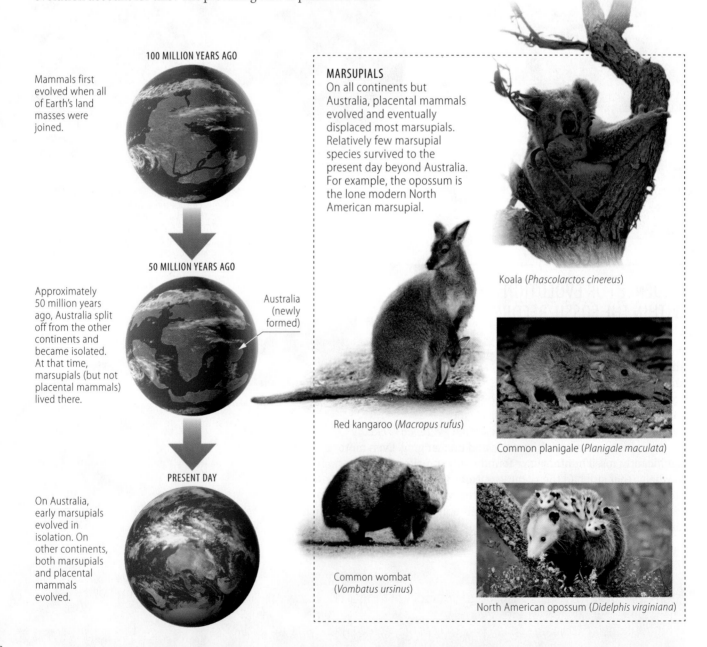

100 MILLION YEARS AGO

Mammals first evolved when all of Earth's land masses were joined.

50 MILLION YEARS AGO

Approximately 50 million years ago, Australia split off from the other continents and became isolated. At that time, marsupials (but not placental mammals) lived there.

Australia (newly formed)

PRESENT DAY

On Australia, early marsupials evolved in isolation. On other continents, both marsupials and placental mammals evolved.

MARSUPIALS
On all continents but Australia, placental mammals evolved and eventually displaced most marsupials. Relatively few marsupial species survived to the present day beyond Australia. For example, the opossum is the lone modern North American marsupial.

Koala (*Phascolarctos cinereus*)

Red kangaroo (*Macropus rufus*)

Common planigale (*Planigale maculata*)

Common wombat (*Vombatus ursinus*)

North American opossum (*Didelphis virginiana*)

COMPARATIVE ANATOMY

Comparisons of the body structures of modern organisms, a discipline called comparative anatomy, can provide insight into evolutionary history.

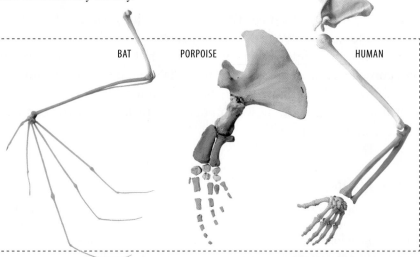

Examination of animal forelimbs—here, a bat wing, a porpoise flipper, and a human arm—show that they are all constructed from similar bones. If each limb had been designed for its current use, then each would have a unique structure optimized for its unique environment. Instead, it is more logical to conclude that the similarities among these limbs reflect their shared evolution from a common ancestor. Starting from an ancient mammal with this skeleton structure, millions of years of evolution resulted in the variety of mammals we see today.

BAT   PORPOISE   HUMAN

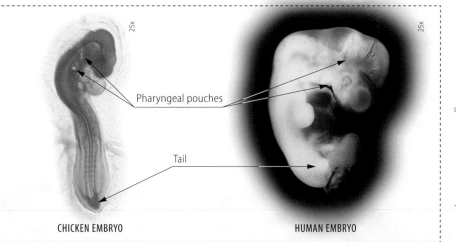

Comparing animal embryos reveals common structures that may not be apparent in the adult organism. For example, both human and chicken embryos have a stage with tails and pharyngeal pouches (which develop into gills in fishes, but become incorporated into features of the head in chickens and humans). Comparison of embryos can allow us to infer a shared evolutionary past.

Pharyngeal pouches

Tail

CHICKEN EMBRYO   HUMAN EMBRYO

On average, more than 98% of your genes are identical to the genes of a chimp.

BIOINFORMATICS

All organisms use genes encoded in their hereditary DNA to produce proteins. By comparing the sequences of DNA and proteins from different organisms, scientists can gain insight into their shared evolutionary history. If the sequences of genes and proteins in two species are closely matched, the species have a recent common ancestor. The greater the similarities, the more closely related the two species. This approach, in which computational tools are used to make biological comparisons, is called **bioinformatics**. A great wealth of sequence data has been gathered in recent years, allowing many evolutionary relationships to be uncovered. For example, the data in the graph show that the selected DNA sequences from a chimpanzee most closely match those of a human, and then a gorilla. Such data provide insights into the evolutionary history of primates.

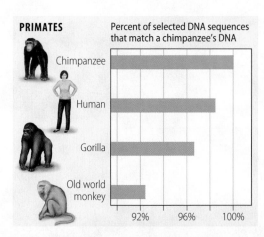

PRIMATES

Percent of selected DNA sequences that match a chimpanzee's DNA

Chimpanzee
Human
Gorilla
Old world monkey

92%   96%   100%

CORE IDEA: The natural world abounds with observable evidence of evolution, such as the distribution of species (biogeography), the comparison of anatomical structures, and the comparison of DNA and protein sequences (bioinformatics).

CORE QUESTION: Your leg is used for running, but a kangaroo's is used for hopping. Why then do they have such similar structures?

ANSWER: Both legs evolved from a common mammal ancestor.

Populations are the units of evolution

When Darwin published *The Origin of Species* in 1859, scientists did not yet understand the nature of DNA, genes, or the molecular basis of heredity. Darwin knew that traits tended to be passed down, but the mechanism of this inheritance was completely unknown. By the mid-1900s, biologists had begun to meld what was known about genetics with what was known about evolution. Through this synthesis, evolution can now be understood by following the changes to DNA and genes over generations.

POPULATIONS

One of the most common misconceptions about evolution is that individuals evolve. Although natural selection does act on individuals—tending to favor the survival and reproduction of the most fit individuals in an environment—evolution is defined only in terms of changes in a population over time. A **population** is a group of individuals of the same species living in the same place at the same time. Members of a population are capable of meeting and mating, and tend to be more closely related to each other than they are to members of other populations. A population is the smallest unit that can evolve.

Bacteria multiply so quickly that even relatively rare mutations can rapidly spread through a population.

BIRDS: SAME POPULATION
The birds shown here would be considered to be the same population because they are capable of meeting and mating.

SQUIRRELS: DIFFERENT POPULATION
If two groups of squirrels cannot cross a geographic barrier, they would represent two isolated populations. Each population will evolve independently.

FISH: SAME POPULATION
The fish shown here would be considered to be the same population because they are capable of meeting and mating.

Bird A

Squirrel A

Bird B

Squirrel B

Fish A

Fish B

GENE POOLS

Every member of a species (such as the tulips shown here) has a unique set of traits. How does such genetic variation arise? Mutations (random changes to DNA) and sexual reproduction (which randomly mixes and matches genes) both change the **gene pool**, all the forms of all the genes in a population at any one time. Natural selection acts on the gene pool; those traits that enhance survival are more likely to be passed on to the next generation and therefore will be represented with increasing frequency in the gene pool.

MICROEVOLUTION

The gene pool consists of all versions of all the genes carried by all the individuals in a population. The gene pool of a population changes over time. For example, if one version of a gene confers an advantage to survival and reproduction (such as the red gene in the hypothetical example involving rabbits shown below), that version of the gene will be favored by natural selection. Over time, we expect that version of the gene to become more frequent within the gene pool. Other versions of the gene that do not confer such an advantage (the blue gene) will become less prevalent. This generation-to-generation change in the gene pool is called **microevolution**, which is evolution occurring on its smallest scale.

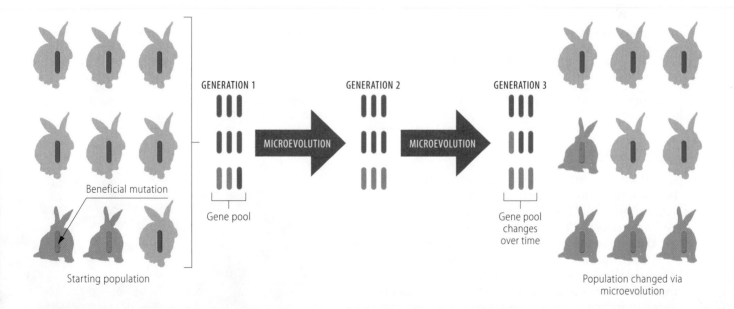

Beneficial mutation

Starting population

GENERATION 1
MICROEVOLUTION
GENERATION 2
MICROEVOLUTION
GENERATION 3

Gene pool

Gene pool changes over time

Population changed via microevolution

CORE IDEA: Evolution acts on populations—members of the same species capable of interbreeding—by altering the gene pool (all versions of all genes in the population). Microevolution is a generation-to-generation change in the gene pool.

CORE QUESTION: In the example of microevolution, how did the red version of the gene first arise in the population?

ANSWER: Through mutation.

Evolution proceeds through several mechanisms

Evolution ("descent with modification") on its smallest scale is a generation-to-generation change in a population's gene pool, the total of all versions of all genes. Taken over many generations, such microevolution can result in the gradual adaptation of species to the local environment. But how do changes in genes occur? By what mechanisms does evolution proceed? Several different mechanisms can contribute to microevolution.

GENETIC VARIATION

Changes to the genetic makeup of a population can arise via two mechanisms: mutations (which create new genes) and sexual recombination (which shuffles existing genes). Both of these processes are random. Once created, the environment favors the survival and spread of those new combinations of genes that enhance survival and reproductive success.

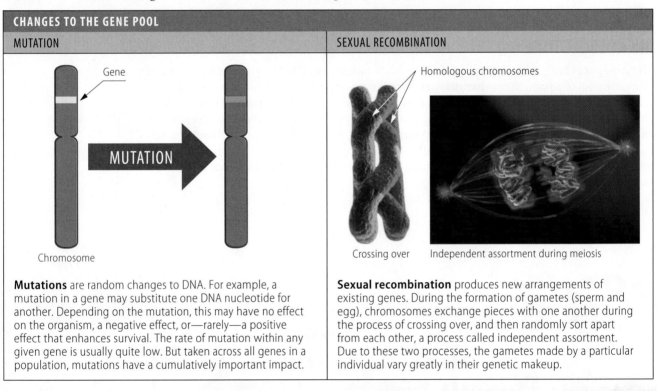

CHANGES TO THE GENE POOL

MUTATION	SEXUAL RECOMBINATION

Gene
MUTATION
Chromosome

Homologous chromosomes
Crossing over   Independent assortment during meiosis

Mutations are random changes to DNA. For example, a mutation in a gene may substitute one DNA nucleotide for another. Depending on the mutation, this may have no effect on the organism, a negative effect, or—rarely—a positive effect that enhances survival. The rate of mutation within any given gene is usually quite low. But taken across all genes in a population, mutations have a cumulatively important impact.

Sexual recombination produces new arrangements of existing genes. During the formation of gametes (sperm and egg), chromosomes exchange pieces with one another during the process of crossing over, and then randomly sort apart from each other, a process called independent assortment. Due to these two processes, the gametes made by a particular individual vary greatly in their genetic makeup.

DARWINIAN FITNESS

Natural selection favors those organisms that are most fit in their present environment. But what does it mean to say that an organism is "fit"? Does that mean that this organism is the strongest, or the fastest, or the most able to find food? In fact, **Darwinian fitness** is defined as the contribution that an individual makes to the gene pool of the next generation in comparison to the contributions from other individuals. In other words, fitness is measured only in terms of the number of healthy offspring produced. This is the only way that an organism can contribute to the evolutionary path of a population. There are many sorts of adaptations that can improve fitness. For example, well-camouflaged organisms (such as the two stone curlew chicks shown here) do not have to be fast or strong to be fit.

MECHANISMS OF EVOLUTION

What mechanisms are responsible for the changes in a gene pool over successive generations? Here, we examine four means by which evolution may occur. Some combination of these mechanisms operates on all evolving populations.

MECHANISMS OF EVOLUTION

GENETIC DRIFT

Although natural selection tends to favor the best-adapted individuals, there is also random chance. **Genetic drift** is a change in a gene pool due to chance. In a large population, the death or migration of a few individuals will have a minimal impact. But if the population is small or isolated, such random changes can have a significant impact on the gene pool.

GENE FLOW

Most populations are not isolated. Individuals may migrate into or emigrate out of a population. This is called **gene flow** and it tends to make separate populations more genetically similar as they exchange genes. For example, as we modernize, previously isolated populations of humans have begun to meet, mate, and exchange genes, rendering the human population more genetically uniform. Among plants, drifting pollen may transfer genes between otherwise separate populations, causing them to become more genetically similar over time.

THE BOTTLENECK AND FOUNDER EFFECTS

Sometimes, a population is drastically reduced in numbers. Disaster (earthquake, flood, fire, etc.) may kill off most of a population. Or perhaps a few individuals make it to a new and isolated habitat, such as an unoccupied island. The small surviving population is unlikely to have the same gene pool as the larger population from which it was derived. By chance, some versions of genes will be more or less common in the few individuals than in the larger group. A change in the gene pool due to such a reduction in population size is called the **bottleneck effect**; if this change is due to the establishment of a small population it is called the **founder effect**. For example, cheetahs were hunted to near extinction in the 1800s. Due to this bottleneck, today's surviving population has a much more uniform gene pool than past populations.

SEXUAL SELECTION

Sexual selection is a form of natural selection whereby certain individuals are more likely to attract mates and, therefore, have more offspring. Many male animals have adornments that can increase their chances of being chosen by a female, such as the colorful plumage of the bird of paradise. In some species (such as elephant seals), males compete in combat or ritualized displays to win the right to mate with the herd's females. Every time the adornment results in a mating, the genes responsible for that feature are more likely to be passed on to the next generation and make a larger contribution to the gene pool of the population.

Elephant seal
(*Mirounga angustirostris*)

Bird of paradise
(*Paradisaea apoda*)

CORE IDEA: Gene pools change on a generation-by-generation basis via various means, such as mutations, sexual recombination, genetic drift, gene flow, the bottleneck and founder effects, and sexual selection.

CORE QUESTION: On the Galápagos Islands, Darwin observed bird populations that appeared to have originated with a few birds that migrated from South America. Which evolutionary mechanism does this demonstrate?

ANSWER: The founder effect.

Macroevolution encompasses large-scale changes

Microevolution is the generation-to-generation change in the total genetic makeup of a population. Although microevolution accounts for the selection of small-scale adaptations that make organisms better suited for their environment, it cannot account for larger-scale changes such as the appearance of new species. The process of **macroevolution** encompasses major changes in the history of life such as the origin of new species, mass extinctions, and the diversification of new forms of life.

SPECIATION

Speciation, the evolutionary formation of new species, increases the diversity of life on Earth. Throughout Earth's history, ancestral species gave rise to one or more new species, which in turn evolved again and again, giving rise to the incredible diversity we see today. Speciation may occur through two different mechanisms—nonbranching and branching evolution—only one of which increases the number of species.

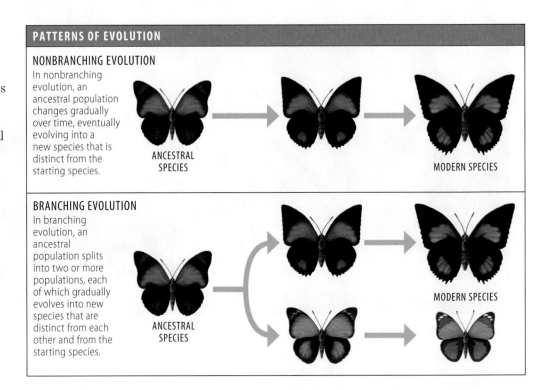

PATTERNS OF EVOLUTION

NONBRANCHING EVOLUTION
In nonbranching evolution, an ancestral population changes gradually over time, eventually evolving into a new species that is distinct from the starting species.

ANCESTRAL SPECIES

MODERN SPECIES

BRANCHING EVOLUTION
In branching evolution, an ancestral population splits into two or more populations, each of which gradually evolves into new species that are distinct from each other and from the starting species.

ANCESTRAL SPECIES

MODERN SPECIES

NOVEL FEATURES

Throughout the history of life on Earth, novel features have evolved. Ancient birds evolved from flightless reptiles, humans evolved bipedalism (an upright, two-legged posture) and flowering plants evolved from nonflowering predecessors. How does such large-scale innovation occur? The fossil record shows that structures that serve one role can gradually change to serve another. New features may arise gradually; each species is adapted for its environment, but the lineage of species shows the evolution of a new structure. For example, birds evolved from one lineage of dinosaurs. These dinosaur ancestors had a very lightweight skeleton that may have helped them be agile and elude predators. Over time, this lineage evolved the ability to take long hops, then short glides, and eventually full-fledged flight. While each species was adapted for its own environment, viewed over a long time, the evolution of flight is a good example of a large-scale macroevolutionary change.

Model of a 125-million-year-old *Caudipteryx*, a dinosaur that had feathers but likely did not fly

MASS EXTINCTIONS

One phenomenon that promotes macroevolution can be clearly seen in the fossil record: periodic mass extinctions. The evolutionary history of life on Earth has been punctuated by brief, catastrophic periods. There have been five such upheavals during which the majority (between 50% and 90%) of living species suddenly died out.

mya = million years ago
1,000 mya = 1 billion years ago

About 250 million years ago, "the Great Dying" wiped out nearly all life on Earth.

4,600 mya: Formation of Earth | 3,500 mya: Oldest known fossil | 2,100 mya: Oldest eurkaryotic fossil | Present day
4,000 mya | 3,000 mya | 2,000 mya | 1,000 mya | 0

500 mya | 400 mya | 300 mya | 200 mya | 100 mya | 0

● 450 mya: About 80% of all species became extinct during the second largest mass extinction.

● 370 mya: About 70% of all species died out.

● 251 mya: About 90% of all species died out in the largest mass extinction in the Earth's history.

● 205 mya: About 50% of species became extinct, including many amphibians, leaving the dinosaurs as the dominant land animals.

● 65 mya: Approximately 50% of species became extinct, including the dinosaurs. Afterwards, mammals became the dominant land animals.

65 MILLION YEARS AGO

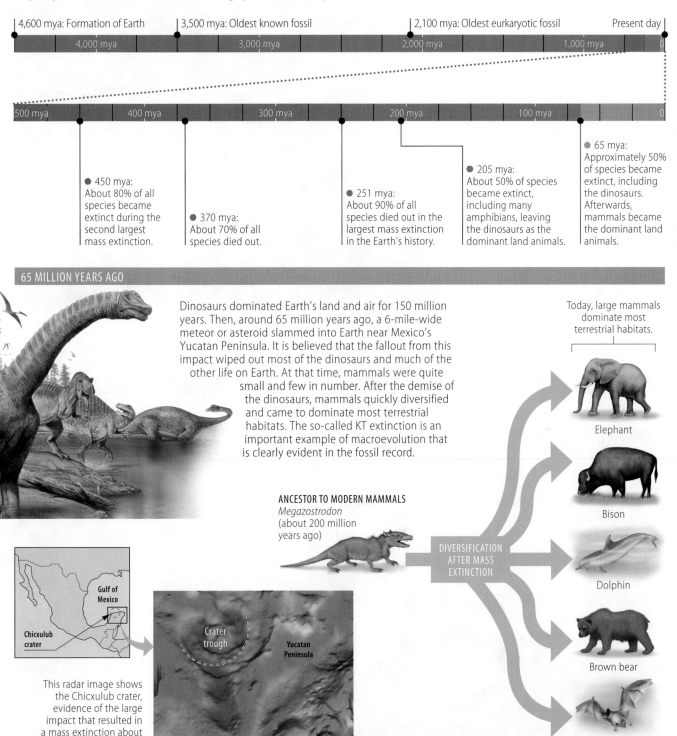

Dinosaurs dominated Earth's land and air for 150 million years. Then, around 65 million years ago, a 6-mile-wide meteor or asteroid slammed into Earth near Mexico's Yucatan Peninsula. It is believed that the fallout from this impact wiped out most of the dinosaurs and much of the other life on Earth. At that time, mammals were quite small and few in number. After the demise of the dinosaurs, mammals quickly diversified and came to dominate most terrestrial habitats. The so-called KT extinction is an important example of macroevolution that is clearly evident in the fossil record.

Today, large mammals dominate most terrestrial habitats.

Elephant

Bison

ANCESTOR TO MODERN MAMMALS
Megazostrodon
(about 200 million years ago)

DIVERSIFICATION AFTER MASS EXTINCTION

Dolphin

Gulf of Mexico

Chicxulub crater

Crater trough

Yucatan Peninsula

This radar image shows the Chicxulub crater, evidence of the large impact that resulted in a mass extinction about 65 million years ago.

Brown bear

Parti-coloured bat

CORE IDEA: Macroevolution, major changes in the history of life, includes the formation of new species (speciation), the evolution of biological novelty, and mass extinctions followed by diversification of the survivors.

CORE QUESTION: What would happen if nonbranching evolution was the only form of macroevolution?

ANSWER: There would only be one species of life on Earth, since only branching evolution increases the number of species.

The geological record ties together the history of Earth and its life

The history of life on Earth is intertwined with the history of the Earth itself. Like the life that inhabits it, our planet is dynamic, changing over time in ways both gradual and abrupt. Studying the history of Earth informs our study of the evolution of life.

THE GEOLOGIC RECORD

Changes in Earth's geology are always accompanied by changes in Earth's biodiversity. The more than four-billion-year history of Earth is divided into four broad divisions called eras. The names and precise dates of the eras are not as important as the realization that each represents a distinct period in the history of life. The three most recent eras are further subdivided into periods.

mya = million years ago
1,000 mya = 1 billion years ago

Computer image of meteoric bombardment during the formation of Earth around 4.5 billion years ago

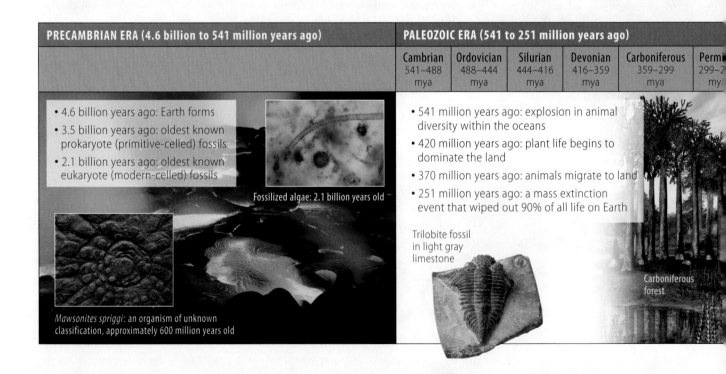

4,600 mya: Formation of Earth	3,500 mya: Oldest known fossil		2,000 mya: First multicelled organisms	Present day
4,000 mya	3,000 mya	2,000 mya	1,000 mya	0

PRECAMBRIAN ERA (4.6 billion to 541 million years ago)		PALEOZOIC ERA (541 to 251 million years ago)					
		Cambrian 541–488 mya	Ordovician 488–444 mya	Silurian 444–416 mya	Devonian 416–359 mya	Carboniferous 359–299 mya	Permi 299–2 my

- 4.6 billion years ago: Earth forms
- 3.5 billion years ago: oldest known prokaryote (primitive-celled) fossils
- 2.1 billion years ago: oldest known eukaryote (modern-celled) fossils

Fossilized algae: 2.1 billion years old

Mawsonites spriggi: an organism of unknown classification, approximately 600 million years old

- 541 million years ago: explosion in animal diversity within the oceans
- 420 million years ago: plant life begins to dominate the land
- 370 million years ago: animals migrate to land
- 251 million years ago: a mass extinction event that wiped out 90% of all life on Earth

Trilobite fossil in light gray limestone

Carboniferous forest

PLATE TECTONICS

Although the land beneath your feet seems quite steady, the surface of Earth is actually in constant motion. Huge, irregularly shaped masses of the Earth's crust called **tectonic plates** float atop a very hot layer of rock called the **mantle**. Plates meet at fault lines, where slow, steady contact between plates creates geological upheavals such as earthquakes, mountain building, and volcanoes. The movement of tectonic plates continuously rearranges the geography of the continents. Such upheavals can be catastrophic in the short term, and can alter the evolution of life on Earth in the long term.

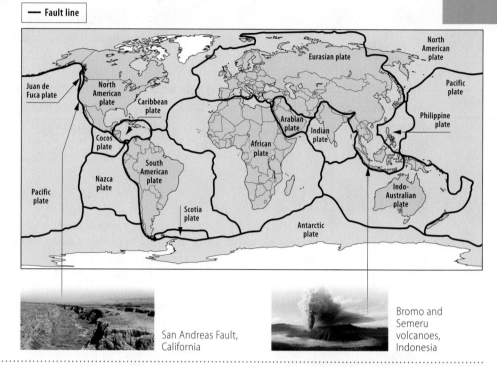

— Fault line

North American plate
Eurasian plate
Pacific plate
Philippine plate
Juan de Fuca plate
North American plate
Caribbean plate
Arabian plate
Indian plate
Cocos plate
African plate
South American plate
Nazca plate
Pacific plate
Indo-Australian plate
Scotia plate
Antarctic plate

San Andreas Fault, California

Bromo and Semeru volcanoes, Indonesia

The movement of the Earth's plates significantly affected the evolution of life by bringing species into and out of contact on changing continents.

Pangea

200 MILLION YEARS AGO
All the continents were combined into the single land mass Pangea.

Atlantic Ocean

100 MILLION YEARS AGO
Diverging plates begin to open up the Atlantic Ocean and separate the continents.

Because they are on different tectonic plates, North America and Europe are drifting apart at a rate of about one inch per year.

MESOZOIC ERA (251 to 65 million years ago)			CENOZOIC ERA (65 million years ago to today)	
Triassic 251–200 mya	**Jurassic** 200–146 mya	**Cretaceous** 146–65 mya	**Tertiary** 65–2.6 mya	**Quaternary** 2.6 mya–today

- 230 million years ago: first dinosaurs
- 100 million years ago: flowering plants begin to dominate the land

- 65 million years ago: extinction of dinosaurs and diversification of mammals
- 200,000 years ago: appearance of anatomically modern humans

Jurassic forest

Stegoceras, bipedal dinosaurs

Megazostrodon, early mammal

Darwinius masillae fossil, early primate from around 47 mya

Paranthropus boisei, a hominid that lived in Africa around 1.75 mya

Modern Africa

CORE IDEA: Geologists recognize four broad eras in the history of Earth, each marked by the appearance of distinctive life. The continuous movement of the tectonic plates that make up Earth's surface has a profound impact on the history of life.

CORE QUESTION: What event marks the start of the era in which humans evolved?

ANSWER: The extinction of the dinosaurs.

Species are maintained by reproductive barriers

Even a young child can recognize that all dogs—even those from very different breeds—belong to one group, but no cats belong to that group. We all have an intuitive sense of species. But how do biologists recognize members of the same species and distinguish them from members of other species? Species are recognized by their ability to naturally and successfully breed.

WHAT IS A SPECIES?

The word "species" is derived from a Latin word meaning "appearance," and we often use looks to distinguish different species. However, as the birds in the example below illustrate, this is not a good way to tell species apart. Appearance alone cannot be used to tell one species from another. The most commonly used definition of **species** is a population that is capable of interbreeding to produce healthy offspring that can themselves reproduce. Thus, even though two dogs of different breeds look very different, they can (for the most part) breed with each other to produce healthy offspring. Similarly, all humans are members of the same species.

However, it is not possible to apply the above definition of species in all cases. For example, bacteria reproduce asexually, preventing the given definition from being useful. Also, the definition cannot be applied to extinct species since there is no way to tell from fossils if life forms were capable of mating with each other. In such cases, physical appearance is usually the best way to distinguish species.

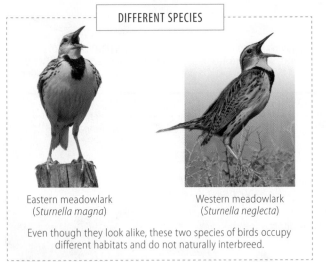

DIFFERENT SPECIES

Eastern meadowlark
(*Sturnella magna*)

Western meadowlark
(*Sturnella neglecta*)

Even though they look alike, these two species of birds occupy different habitats and do not naturally interbreed.

SAME SPECIES

All dogs are members of the species *Canis lupus familiaris*.

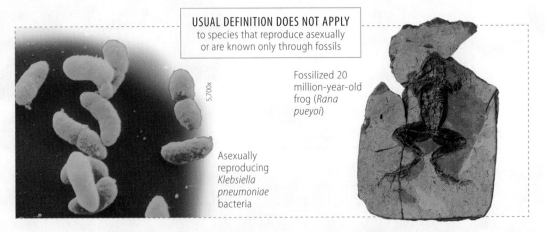

USUAL DEFINITION DOES NOT APPLY
to species that reproduce asexually
or are known only through fossils

5,700x

Asexually reproducing *Klebsiella pneumoniae* bacteria

Fossilized 20 million-year-old frog (*Rana pueyoi*)

Some snails are prevented from mating with members of other species because their shells spiral in the opposite direction, preventing genital contact.

REPRODUCTIVE BARRIERS

If we define a species as a group of individuals capable of successfully interbreeding, then what prevents members of different species from reproducing with each other? The separation of species occurs because one or more **reproductive barriers** prevent members of different species from interbreeding. There are a variety of reproductive barriers that affect different steps in the creation of healthy offspring.

REPRODUCTIVE BARRIERS

BEHAVIORAL ISOLATION

Members of a species often identify each other through specific rituals, markings, or smells. A peahen, for example, will only mate with a peacock who displays his large and colorful feathers.

MATING TIME DIFFERENCES

Many species are able to reproduce only at specific times. Any attempts by members of another species to mate outside of this time frame will not be successful. For example, different species of toad will only breed at different times of year (early summer versus late summer, for example).

HABITAT ISOLATION

If species live in slightly different habitats, they may never meet. For example, there are three different species of bristlecone pine trees. Each one grows only in areas that are well separated from the other species by mountains.

MECHANICAL INCOMPATIBILITY

Members of different species often cannot mate because their anatomies are incompatible. For example, in orchids, the pollen-bearing organs of many species have a specific structure that allows only one specific pollinator species to land and collect pollen.

GAMETIC INCOMPATIBILITY

The gametes (sperm and egg) of different species usually cannot fertilize each other. This is particularly important during spawning season in aquatic habitats, when individuals of many different species release gametes into the same body of water. Multiple species of sea urchins, for example, may live alongside each other, but the sperm of one species cannot fertilize the egg of another.

HYBRID WEAKNESS

If two members of different species do manage to mate, a hybrid organism may result. Such hybrids may be unfit, or they may be sterile and unable to reproduce themselves, or they may produce unfit offspring. For example, a mule, a hybrid produced from a female horse and a male donkey, is always sterile.

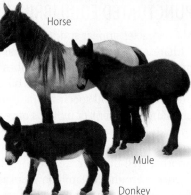

Horse

Mule

Donkey

CORE IDEA: Biologists usually define species as a group of individuals capable of successfully interbreeding. One or more reproductive barriers prevent members of different species from breeding.

CORE QUESTION: Of the six various reproductive barriers presented, which one(s) come into play after a mating has actually occurred?

ANSWER: Only gametic incompatibility and hybrid weakness.

Speciation can occur through various mechanisms

When considering macroevolution, major changes in the evolutionary history of life, we often focus on speciation, the formation of new species. Speciation occurs when one ancestral species evolves into one or more new species. A key event in this process occurs when a single population is separated into two or more populations that are isolated from each other. Once separated, each population is subject to unique selection pressures—natural selection acting within each environment—and can therefore follow its own evolutionary path. Eventually, this may lead to populations that are no longer capable of meeting and mating; in other words, they become separate species.

GRADUATED MODEL

The fossil record reveals examples of species that changed slowly and gradually over long periods of time. In such cases, a species acquires small adaptations to its environment that, taken together over millions of years, transform an ancestral species to a modern one. Such a model for the evolution of new species is called the graduated model. The graduated model does not mean that evolution proceeds in a fully linear, stepwise fashion; successive species often coexist and overlap.

55–45 MILLION YEARS AGO
Eohippus

40–30 MILLION YEARS AGO
Mesohippus

32–25 MILLION YEARS AGO
Miohippus

26–2 MILLION YEARS AGO
Pliohippus

5 MILLION YEARS AGO–TODAY
Equus ferus caballus
(Modern horse)

GRADUAL CHANGE OVER A LONG PERIOD OF TIME

PUNCTUATED EQUILIBRIUM

Many species appear suddenly in the fossil record, persist with little change, then suddenly disappear. The term **punctuated equilibrium** describes such periods of stasis interrupted by occasional bursts of speciation. For example, ancestral species of mammals were relatively unchanged for tens of millions of years until a rapid series of speciations following the extinction of the dinosaurs (rapid in an evolutionary sense, as it took a few million years). Another well-documented case of punctuated equilibrium occurred around 530 million years ago. This period, called the Cambrian explosion due to its rapid diversification of species, saw the relatively rapid evolution of a great many new animal species (shown in an artist's rendition). Although still requiring millions of years to be completed, the rate of evolution during this time was an order of magnitude higher than the normal rate.

ALLOPATRIC SPECIATION

Speciation often begins when a population is separated into two or more populations. But how does such separation occur? Sometimes, a physical barrier—such as the formation of a valley, mountain range, or body of water—arises that physically isolates populations from each other. This is called **allopatric speciation**.

White-tailed antelope squirrel
(*Ammospermophilus leucurus*)

Harris's antelope squirrel
(*Ammospermophilus harrisii*)

6 MILLION YEARS

Before formation of the Grand Canyon (around 6 million years ago), a single population of squirrels inhabited the area.

Formation of the Grand Canyon produced two isolated habitats. One species of squirrel is now found exclusively on each side of the canyon. Birds, whose movement is not affected by the canyon, have not diverged into separate species.

ALLOPATRIC SPECIATION: SPECIES DIVERGE DUE TO A PHYSICAL BARRIER

Archaeological evidence suggests that wheat was first cultivated near the modern country of Jordan around 10,000 years ago.

SYMPATRIC SPECIATION

Speciation may occur quite suddenly due to large-scale genetic changes. This is an example of **sympatric speciation**, the emergence of a new species right in the midst of its parent population without geographic isolation. In the plant kingdom, accidents during cell division can result in organisms with multiple sets of chromosomes. If the new organism self-fertilizes, a new species may arise in just one generation. This mechanism of speciation produced many of our food crops, including apples, coffee, and wheat. For example, around 8,000 years ago, two ancestral species of wheat produced a sterile hybrid. An error in cell division caused this hybrid to double its chromosomes. Self-fertilization then produced today's species of modern bread wheat. Since the new species arose amidst the ancestral population, this is an example of sympatric speciation.

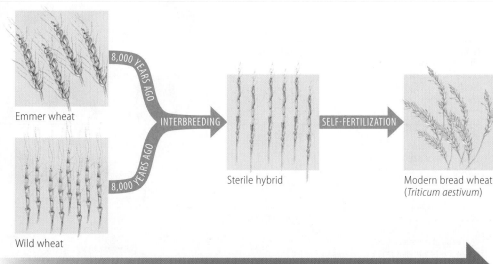

Emmer wheat

8,000 YEARS AGO

Wild wheat

8,000 YEARS AGO

INTERBREEDING

Sterile hybrid

SELF-FERTILIZATION

Modern bread wheat
(*Triticum aestivum*)

SYMPATRIC SPECIATION: NEW SPECIES ARISES IN MIDST OF OLD SPECIES

CORE IDEA: New species may form over long periods of time (graduated model) or relatively rapidly (punctuated equilibrium), and after geographic isolation (allopatric speciation) or within a parent species (sympatric speciation).

CORE QUESTION: If a lake rises and isolates a series of islands, what kind of speciation is most likely to occur among the land-dwelling organisms?

ANSWER: Since the species become physically isolated from one another, allopatric speciation is likely to occur.

Taxonomy is the classification of life

Taxonomy is the identification, naming, and classification of species. A formalized naming system eases communication among scientists and helps researchers unambiguously identify an organism or recognize a new species when discovered. All life is classified into one of three large groups called **domains**. Starting with a domain, every organism can be placed into the **taxonomic hierarchy**, an ordered series of progressively smaller categories. The hierarchy ends with the species name.

DOMAIN
The broadest category of classification; of the three domains currently recognized, Bacteria and Archaea are prokaryotes, while Eukarya contains organisms with eukaryotic cells.

EUKARYA
Eukaryotes

KINGDOM
Within the domain Eukarya, there are three kingdoms that are generally agreed upon—Plantae, Fungi, and Animalia—and many others that are under debate.

ANIMALIA
Animals

PHYLUM
The different phyla of animals are distinguished by certain fundamental characteristics of their body forms.

CHORDATA
Chordates

CLASS
The phylum of chordates includes such classes as reptiles, amphibians, mammals, and several classes of fish.

MAMMALIA
Mammals

ORDER
Within the class of mammals, the order of carnivores is recognized by dental anatomy.

CARNIVORA
Carnivores

FAMILY
The order of carnivores contains at least a dozen families including the canines and the felines.

FELIDAE
Felines

GENUS
A genus contains one or more closely related species. The four members of the genus *Panthera* are the only cats that can roar.

Panthera
Big cats

THE TAXONOMIC HIERARCHY
Biologists classify life using an ordered hierarchy that runs from domains to species. This example shows the classification of the tiger.

SPECIES
The most narrow category of classification; species are recognized by their ability to interbreed with each other.

Panthera tigris
Tiger

THE THREE-DOMAIN SYSTEM

Although other taxonomic systems have been used in the past, biologists today use a **three-domain system**. In the current classification scheme, all life is classified into one of three basic groups: two domains of **prokaryotes**—Bacteria and Archaea—and one domain of **eukaryotes**, called Eukarya. The domain Eukarya is divided into **kingdoms**. Biologists generally agree on three kingdoms: Plantae, Fungi, and Animalia. Biologists have not yet agreed on the classification of the remaining eukaryotes into kingdoms. Instead, all eukaryotes that do not fit the definition of plant, fungus, or animal are grouped together as the **protists**, a highly diverse taxonomic grab bag. It is important to understand that classifying Earth's diverse species is a work in progress; the exact classifications of many organisms will inevitably change as we learn more about organisms and their evolution.

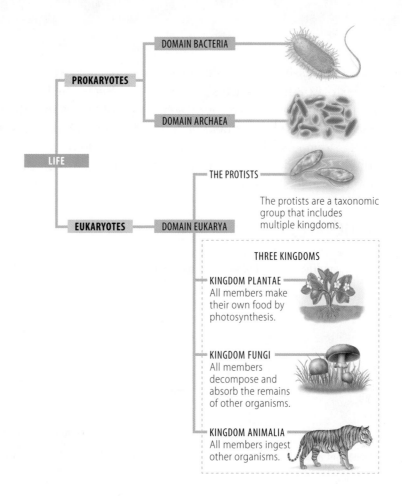

The protists are a taxonomic group that includes multiple kingdoms.

THREE KINGDOMS

KINGDOM PLANTAE
All members make their own food by photosynthesis.

KINGDOM FUNGI
All members decompose and absorb the remains of other organisms.

KINGDOM ANIMALIA
All members ingest other organisms.

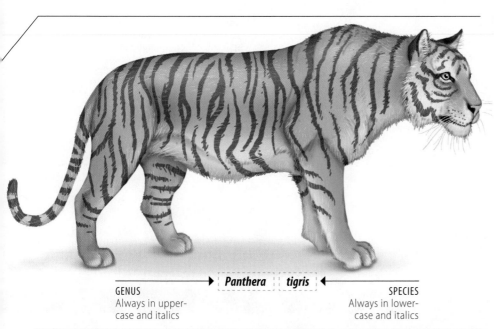

GENUS
Always in upper-case and italics

Panthera *tigris*

SPECIES
Always in lower-case and italics

BINOMIAL NOMENCLATURE
The last two groups in the taxonomic hierarchy (genus and species) are used to form a binomial (two-part) Latin name, sometimes called a "scientific name."

CORE IDEA: Taxonomy is the classification of life into a hierarchical system that runs from domain (most broad) to species (most narrow). Species are identified using a two-name convention: *Genus species*.

CORE QUESTION: The scientific name for the wolf is *Canis lupus*. What level of taxonomic classification does each word represent?

ANSWER: *Canis* is the genus name, *lupus* is the species name.

Phylogenetic trees represent evolutionary history

Systematics is the classification of organisms. One goal of systematics is to reflect the evolutionary history of the organisms presented. **Phylogenetic trees** are one way to accomplish this. A phylogenetic tree represents a hypothesis about the evolutionary history of related species. The relative placement of different species on phylogenetic trees is based on several lines of evidence, such as their appearance in the fossil record, and (when available) an analysis of their DNA and proteins (called bioinformatics). As you can see here, phylogenetic trees can help you understand and investigate the interrelationships of life on Earth.

PHYLOGENETIC TREES

A phylogenetic tree presents a hypothesis about the evolutionary relatedness of a group of species. Knowing how to properly read a phylogenetic tree can offer great insights into the origin of species.

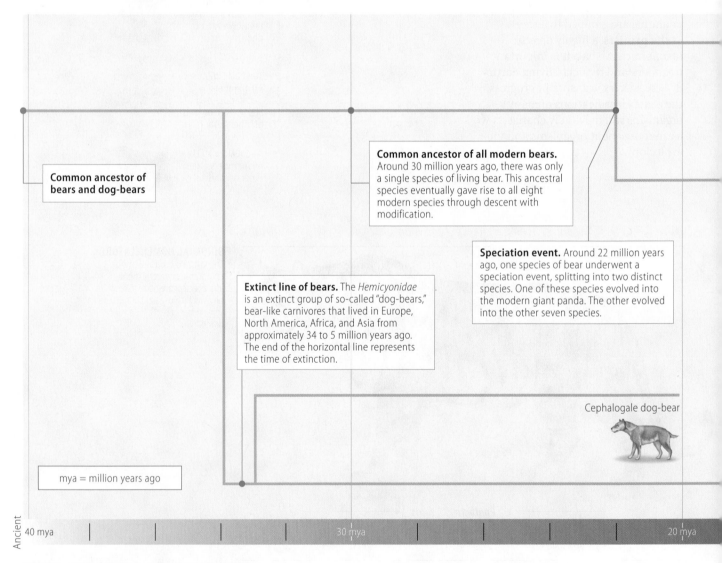

Common ancestor of bears and dog-bears

Common ancestor of all modern bears. Around 30 million years ago, there was only a single species of living bear. This ancestral species eventually gave rise to all eight modern species through descent with modification.

Speciation event. Around 22 million years ago, one species of bear underwent a speciation event, splitting into two distinct species. One of these species evolved into the modern giant panda. The other evolved into the other seven species.

Extinct line of bears. The *Hemicyonidae* is an extinct group of so-called "dog-bears," bear-like carnivores that lived in Europe, North America, Africa, and Asia from approximately 34 to 5 million years ago. The end of the horizontal line represents the time of extinction.

Cephalogale dog-bear

mya = million years ago

Ancient   40 mya   30 mya   20 mya

TIME: MILLIONS OF YEARS AGO

CLADISTICS

Imagine a family tree that starts with your great grandfather and contains all of his descendants. This is analogous to a **clade**, a group that consists of an ancestral species and all its descendants—a distinct branch in the tree of life. The analysis of clades is called **cladistics**. In this example, you can see that all reptiles form an ingroup, one cluster that is descended from a common ancestor. The frog represents an outgroup, a species that diverged from the others before their common ancestor. An ingroup will have distinct characteristics that separate it from the outgroup. In this case, all reptiles have a hard-shelled amniotic egg that can survive on dry land, whereas amphibians require water to keep their soft eggs viable. Cladistic analyses help clarify evolutionary relationships.

COMMON ANCESTOR

AMNIOTIC EGG (ancestor to all reptiles)

FROG — OUTGROUP Amphibian

SNAKE

LIZARD — INGROUP Reptile clade

BIRD

What about koala bears? They are marsupials and not members of the bear family.

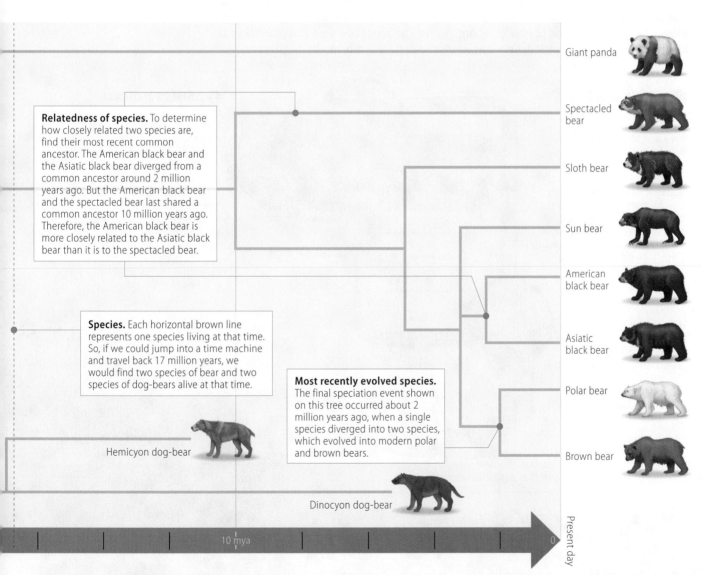

Giant panda

Spectacled bear

Sloth bear

Sun bear

American black bear

Asiatic black bear

Polar bear

Brown bear

Relatedness of species. To determine how closely related two species are, find their most recent common ancestor. The American black bear and the Asiatic black bear diverged from a common ancestor around 2 million years ago. But the American black bear and the spectacled bear last shared a common ancestor 10 million years ago. Therefore, the American black bear is more closely related to the Asiatic black bear than it is to the spectacled bear.

Species. Each horizontal brown line represents one species living at that time. So, if we could jump into a time machine and travel back 17 million years, we would find two species of bear and two species of dog-bears alive at that time.

Most recently evolved species. The final speciation event shown on this tree occurred about 2 million years ago, when a single species diverged into two species, which evolved into modern polar and brown bears.

Hemicyon dog-bear

Dinocyon dog-bear

10 mya

0

Present day

CORE IDEA: Phylogenetic trees present hypotheses about the shared evolutionary history of a group of organisms. Reading phylogenetic trees can provide insights into clades, any group of species descended from a common ancestor.

CORE QUESTION: After the polar and brown bears, which two bear species are most closely related to each other?

ANSWER: The American black bear and the Asiatic black bear shared a common ancestor until about 2 million years ago, making them the next most closely related.

Biologists hypothesize that life originated in a series of stages

Although Earth teems with life today, it was not always so. Our planet first formed around 4.6 billion years ago when gravitational attraction caused clouds of gas, dust, and rocky debris to coalesce into the planets of our solar system. Under constant bombardment by meteorites, young Earth was so hot that it remained molten. As our planet cooled, the outermost layer solidified into a thin crust. By about 3.9 billion years ago, Earth had solidified enough to support life. Although we will never know for certain how life on Earth originated billions of years ago, scientists do hypothesize about what may have occurred.

BIOGENESIS

All cells arise from the reproduction of preexisting cells. But this principle raises an obvious question: How could the first living cells arise? One hypothesis is that chemical and physical processes within the unique conditions of the primordial Earth resulted in **biogenesis**, the formation of new living organisms. The series of stages discussed here is highly speculative and much debate surrounds the details of this model.

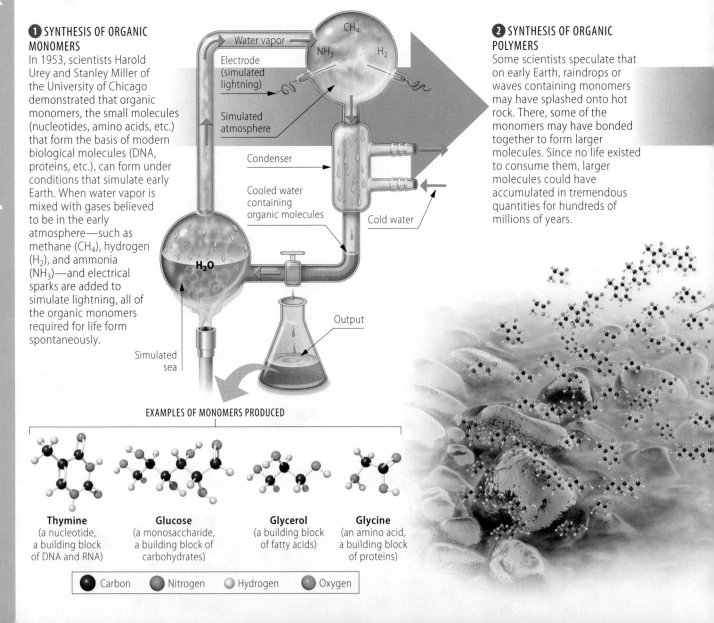

❶ SYNTHESIS OF ORGANIC MONOMERS

In 1953, scientists Harold Urey and Stanley Miller of the University of Chicago demonstrated that organic monomers, the small molecules (nucleotides, amino acids, etc.) that form the basis of modern biological molecules (DNA, proteins, etc.), can form under conditions that simulate early Earth. When water vapor is mixed with gases believed to be in the early atmosphere—such as methane (CH_4), hydrogen (H_2), and ammonia (NH_3)—and electrical sparks are added to simulate lightning, all of the organic monomers required for life form spontaneously.

❷ SYNTHESIS OF ORGANIC POLYMERS

Some scientists speculate that on early Earth, raindrops or waves containing monomers may have splashed onto hot rock. There, some of the monomers may have bonded together to form larger molecules. Since no life existed to consume them, larger molecules could have accumulated in tremendous quantities for hundreds of millions of years.

Water vapor

CH_4

NH_3 H_2

Electrode (simulated lightning)

Simulated atmosphere

Condenser

Cooled water containing organic molecules

Cold water

H_2O

Output

Simulated sea

EXAMPLES OF MONOMERS PRODUCED

Thymine
(a nucleotide, a building block of DNA and RNA)

Glucose
(a monosaccharide, a building block of carbohydrates)

Glycerol
(a building block of fatty acids)

Glycine
(an amino acid, a building block of proteins)

● Carbon ● Nitrogen ○ Hydrogen ● Oxygen

CONDITIONS ON EARLY EARTH

Earth's early atmosphere was very different than it is now. Scientists speculate that frequent volcanic eruptions produced a thick atmosphere containing water vapor, nitrogen-containing gases, carbon dioxide, methane, ammonia, hydrogen gas, and hydrogen sulfide. As Earth cooled, water vapor condensed and fell as rain, creating vast oceans. The stage was set for the formation of the first life. Notice that there was no oxygen gas in the early atmosphere. This came later as a result of the metabolism of living organisms.

❸ ORIGIN OF SELF-REPLICATING MOLECULES

The continued existence of life depends on the passing down of information from one generation to the next. This requires a molecule that can copy itself. Scientists hypothesize that a key stage in the origin of life was the spontaneous formation of RNA molecules that were capable of duplicating themselves. Unlike DNA, which requires proteins to replicate, RNA can self-replicate, leading to speculation that RNA was the original genetic material. Once formed randomly, natural selection would occur: those RNA varieties that could replicate fastest and best would produce more copies of themselves.

❹ FORMATION OF THE FIRST CELLS

Lipids created in the primordial oceans would spontaneously come together, forming bubbles in the same manner that oil comes together when mixed with water. These tiny fat bubbles would trap water inside. If the internal water contained self-replicating molecules, the result could be very primitive cells: cytoplasm (organic molecules dissolved in water) surrounded by a membrane.

EVOLUTION BY NATURAL SELECTION

Until around 2.5 billion years ago, there was very little oxygen in Earth's atmosphere because there were few photosynthetic organisms.

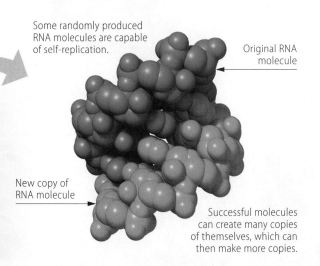

Some randomly produced RNA molecules are capable of self-replication.

Original RNA molecule

New copy of RNA molecule

Successful molecules can create many copies of themselves, which can then make more copies.

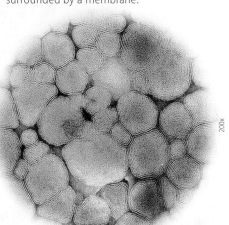

200x

Primitive cells would be subject to natural selection, and the most successful would duplicate and thrive.

CORE IDEA: Biogenesis, the formation of new life on Earth, may have proceeded through a series of steps that produced organic monomers, polymers, and finally, primitive cells containing self-replicating molecules.

CORE QUESTION: Why is it likely that RNA preceded DNA as the original genetic material?

ANSWER: DNA requires proteins to replicate, whereas RNA can copy itself.

179

Prokaryotes have unique cellular structures

The first life to evolve on Earth—and the only kinds of life to exist for over 1 billion years— were **prokaryotes**, organisms with prokaryotic cells. Today, prokaryotes continue to have an immense impact on our world. They are found everywhere there is life, including many places where no other life can survive. Close examination reveals some unique structures and lifestyle features of prokaryotes.

DOMAINS

Biologists classify all life on Earth into three domains. The domains Bacteria and Archaea both contain organisms with prokaryotic cells, but the members of these two domains differ in several molecular details, such as the precise organization of the chromosomes. All other life—including plants, fungi, animals, and protists—consists of organisms with eukaryotic cells and so is placed in the domain Eukarya.

LIFE		
PROKARYOTES Cells that are relatively small and more primitive, and lack membrane-bound organelles		**EUKARYOTES** Cells that are relatively large and more complex, and contain membrane-bound organelles
Domain Archaea	Domain Bacteria	Domain Eukarya

STRUCTURAL FEATURES

All prokaryotes are single celled, but these individual cells come in a wide variety of shapes and sizes. Some features that are unique to prokaryotic cells are shown here.

Many prokaryotic species are mobile, having one or more **flagella** that propel their cells through their environment.

Nearly all prokaryotes have a **cell wall** located outside the plasma membrane. The cell wall provides protection and allows prokaryotes to thrive in a wide range of environments.

17,000x

The cell wall of many prokaryotes is covered by a sticky **capsule** that helps the cell stick to surfaces and provides protection.

Helicobacter pylori, the bacterium that causes stomach ulcers

Some prokaryotes can form an **endospore**, a thick-shelled container that protects the cell when conditions become harsh (too dry, hot or cold, toxic, etc.). When the environment becomes hospitable again, the endospore can resume its normal state.

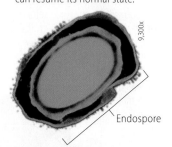

9,300x

Endospore

Bacillus anthracis, the bacterium that causes anthrax

REPRODUCTION BY BINARY FISSION

Many prokaryotes reproduce by a process called **binary fission**, which literally means "splitting in half." A single cell can become two genetically identical cells, which can become four, and then eight, and so forth, doubling with each generation. Under ideal conditions, binary fission can result in a huge population of prokaryotes in a short period of time.

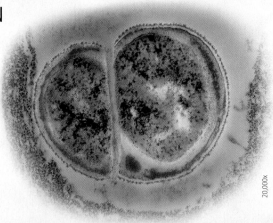

Alloiococcus otitidis bacterium, which causes middle ear infections, undergoing cell division via binary fission

20,000x

PROKARYOTIC SHAPES

COCCI	BACILLI	SPIRAL
Cocci are spherical cells that may be found alone, in chains, or in clusters.	**Bacilli** are rod-shaped cells that may be found singly or in chains.	Most prokaryotes with a curved or spiral shape occur singly.

1,700x

4,000x

7,000x

Streptococcus mutans bacteria are commonly found in the human mouth, where they contribute to tooth decay. This species usually lives as chains of individual spherical cells.

This strain of *Escherichia coli* bacteria, called O157:H7, can cause potentially deadly outbreaks of food poisoning.

Borrelia burgdorferi bacteria, the cause of Lyme disease, are passed to humans through the bites of infected ticks.

PROKARYOTE NUTRITION

Prokaryotes display much greater nutritional diversity than eukaryotes. Nearly all eukaryotes survive by photosynthesis or by consuming other organisms. In contrast, many prokaryotes can produce their own food using energy and molecular building blocks obtained directly from chemicals in the environment.

7,300x

Prokaryotes (including *Archaeoglobus fuligidus*, shown above) form the base of a food chain surrounding a hydrothermal vent several kilometers beneath the surface of the Atlantic Ocean. Far from any sunlight, this species of prokaryote can obtain nutrition from the sulfur-rich superheated water emitted by the vent.

BIOFILMS

Prokaryotes often form **biofilms**, organized colonies of one or several species (perhaps in conjunction with some microscopic eukaryotes) attached to a surface, such as rocks or living tissue. Sticky molecular "glue" may hold the biofilm together and make it difficult to remove.

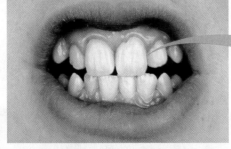

A teenager's teeth after chewing a plaque discloser tablet. Plaque shows up as a blue stain.

3,300x

Bacterial plaque on a human tooth

There are more prokaryotes living in your mouth right now than the total number of humans who have ever lived.

CORE IDEA: Two domains of life on Earth—Archaea and Bacteria—consist of prokaryotes, with relatively small, simple cells. Prokaryotes exhibit a wide range of shapes, structures, features, and nutritional habits.

CORE QUESTION: In terms of size and complexity, how do fungal cells compare to bacterial cells?

ANSWER: Fungal cells are eukaryotic, so they are relatively larger and more complex.

Archaea are found in extreme habitats

The prokaryotes (organisms consisting of a single, relatively small and simple cell that lacks membrane-bound organelles) are divided into two domains. One of these is the domain **Archaea**, a name that invokes their ancient (archaic) roots. Some archaea are called extremophiles ("lovers of the extreme") because they can thrive in habitats where no other types of organisms can survive. The archaea that live in extreme ecological niches have unusual adaptations that enable them to metabolize and reproduce under such conditions.

LIFE		
Prokaryotes		Eukaryotes
Domain Archaea	Domain Bacteria	Domain Eukarya

METHANOGENS

Methanogens live in anaerobic (oxygen-free) environments, where they emit methane gas as a waste product of their metabolism. Methanogens are frequently found in the thick mud at the bottom of a swamp or bog; their emitted methane is called "swamp gas." Methanogens also thrive in the oxygen-free conditions within landfills; this is why landfills often require a pipe to release the gas (lest it build up and explode!). Great numbers of methanogens can also be found in the digestive tracts of grazing animals such as cows and deer.

A well collects methane gas from decaying garbage within a landfill. The methane, emitted by prokaryotes, can be burned to generate electricity.

Swamp gas rises to the surface of the Suwannee River in Okefenokee National Wildlife Refuge in southern Georgia.

Methanococcoides burtonii are cold-loving methanogenic archaea discovered in Antarctica. This species can survive in temperatures as low as 27° Fahrenheit (F), or −2.5° Celsius (C).

HALOPHILES

Halophiles are archaea that thrive in very salty environments, such as Utah's Great Salt Lake, the Dead Sea in the Middle East, and saltwater evaporating ponds used to produce sea salt. Many species of halophiles can tolerate salt concentrations 5 to 10 times higher than that of seawater. Extremely salty environments often turn colors as a result of the dense growth and colorful pigments of halophiles.

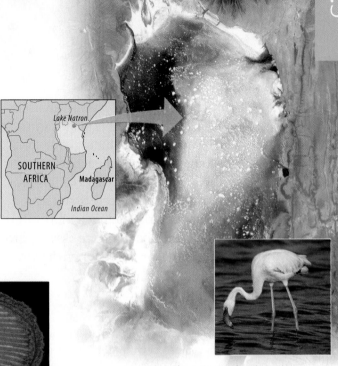

Archaea are among the few forms of life that can tolerate the high salt concentrations found in the Dead Sea, Israel.

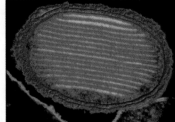

16,000x

Ectothiorhodospira lives in salty pools, and prefers high temperatures (around 176°F, or 80°C) and an alkaline pH (between 7–9).

Located in Africa's Great Rift Valley, Lake Natron (seen in a satellite photo) is considered to be the world's most caustic lake. The orange color of the salt crusts is due to red pigments from salt-loving archaea, which provide coloring and food for the flamingoes that feed here.

THERMOPHILES

The **thermophiles** ("heat lovers") are archaea that live and grow in high-temperature environments. Some thermophiles even thrive in deep-sea hydrothermal vents, where emitted gases can raise temperatures well above boiling. Others can be found in hot springs or geysers. Still others can grow in environments that are both hot and highly acidic. Some biologists speculate that such environments may be similar to those faced by the first organisms billions of years ago.

The Grand Prismatic Spring displays the colors of its thermophilic inhabitants at Yellowstone National Park, Wyoming.

The archaea within the digestive system of a typical cow produces over 100 pounds of methane gas each year.

CORE IDEA: The archaea are prokaryotes often found growing in extreme environments. They include methanogens (which produce methane gas), halophiles (salt lovers), and thermophiles (heat lovers).

CORE QUESTION: Would you expect to find methanogens growing in an aquarium?

ANSWER: No, because methanogens thrive in an oxygen-free environment.

Bacteria are very numerous and common

Of the two domains of prokaryotes, you are most likely to interact with **bacteria** because they thrive in most of Earth's habitats—including in and on your body! Although each individual bacterium is microscopic, their collective impact is enormous. We most often hear about bacteria that harm human health, but bacteria are much more often helpful to people. In fact, bacteria are essential to our own health and the health of our environment.

LIFE		
Prokaryotes		Eukaryotes
Domain Archaea	Domain Bacteria	Domain Eukarya

HELPFUL BACTERIA

Bacteria are indispensable components of our environment. Bacteria thrived on the Earth for a billion years without eukaryotes; yet without bacteria, eukaryotes would be doomed! Here, you can see several ways that bacteria have a positive impact on human society.

SEWAGE TREATMENT
In a treatment plant, solid material from raw sewage is mixed with prokaryotes (both bacteria and archaea). The microbes decompose the sludge, helping to recycle the nutrients contained within it.

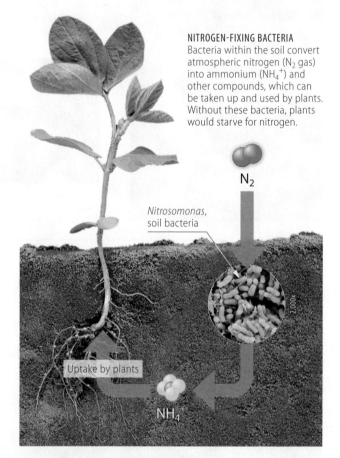

NITROGEN-FIXING BACTERIA
Bacteria within the soil convert atmospheric nitrogen (N_2 gas) into ammonium (NH_4^+) and other compounds, which can be taken up and used by plants. Without these bacteria, plants would starve for nitrogen.

N_2

Nitrosomonas, soil bacteria

Uptake by plants

NH_4^+

DECOMPOSERS
Bacteria are important decomposers, breaking down dead organisms and returning their nutrients to the environment. Without such chemical recycling, plants and animals could not survive because many needed nutrients would be bound up in dead matter.

BIOREMEDIATION
Bacteria can be used for bioremediation, the use of organisms to remove pollutants from the environment. In this photo, workers in Orange County, California, are spraying fertilizer that stimulates naturally occurring oil-eating bacteria.

HARMFUL BACTERIA

When you think of bacteria, you probably think about **pathogens**, the relatively few species that can cause serious illness. Bacterial infections do, in fact, account for about half of all human diseases. Although you are constantly bombarded by pathogens, your body's defenses keep most of them in check. Occasionally, however, a pathogen successfully invades the body, and illness results. Here, you can see a variety of diseases caused by pathogenic bacteria.

Yersinia pestis
18,000x

BUBONIC PLAGUE

Bubonic plague is caused by *Yersinia pestis*, a bacteria carried by rodents and their fleas. During the 1300s, the Black Death swept through Europe, killing about one-third of the population. During the last worldwide plague pandemic in the 1890s, houses in China were burned to prevent spreading of the disease.

STAPH INFECTIONS

Staphylococcus aureus is a bacteria normally found on human skin. If internalized, *S. aureus* can release a toxin that causes a number of diseases, including flesh-eating disease and toxic shock syndrome. Recently, *S. aureus* has become an even bigger health threat as it evolves resistance to antibiotics. MRSA (methicillin-resistant *Staphylococcus aureus*) can be spread in hospitals and locker rooms.

ANTHRAX

Anthrax is a disease, normally found in cattle, that is caused by the bacterium *Bacillus anthracis*. In 2001, anthrax spores were mailed to members of the media and the U.S. Senate. Five people died after inhaling the spores into their lungs, where the spores germinated and caused anthrax infection.

1,300x
Staphylococcus aureus

2,000x
Bacillus anthracis

LYME DISEASE

Lyme disease is caused by *Borrelia burgdorferi*, a type of spiral bacteria that may be passed from deer to humans via ticks. Lyme disease can be identified early by a red bull's-eye rash. Antibiotics can cure Lyme disease if administered soon after the rash appears.

SALMONELLA

One of the most common causes of food poisoning is a genus of bacteria called *Salmonella*. Unsanitary practices can lead to infection of food. If not killed by cooking, *Salmonella* eaten with food may cause a potentially serious gastrointestinal infection.

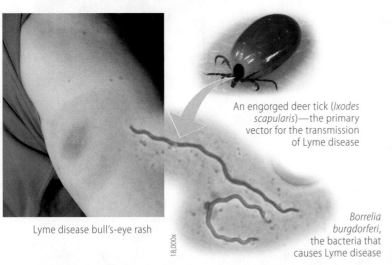
An engorged deer tick (*Ixodes scapularis*)—the primary vector for the transmission of Lyme disease

Lyme disease bull's-eye rash

18,000x

Borrelia burgdorferi, the bacteria that causes Lyme disease

11,000x
Salmonella

CORE IDEA: Nearly all life on Earth (including humans) depends on bacteria due to their ability to fix nitrogen and decompose wastes. Yet bacteria also cause a number of serious human illnesses.

CORE QUESTION: Why is *Staphylococcus* infection becoming an increasingly serious health threat?

ANSWER: Because more and more *Staphylococcus* bacteria are evolving resistance to antibiotics.

Bacteria can transfer DNA

Each individual bacterium has one chromosome, a closed loop of DNA containing genes. Bacterial reproduction usually involves binary fission: duplication of the chromosome followed by splitting of the cell in two, producing two genetically identical copies of the original parent cell. Because binary fission is an asexual process, involving only a single parent, the bacteria in a colony are genetically identical to each other. But this does not mean that bacteria lack ways to produce new combinations of genes. In fact, bacteria have several mechanisms by which genes can be transferred from one bacterium to another. Once transferred, new DNA can be integrated into the bacterial chromosome.

TRANSFORMATION

Transformation is the uptake of naked DNA from the surrounding environment into a bacterial cell. A dead bacterium may release DNA into the environment. Such pieces of DNA may then be taken up by other bacteria. If so, these bacteria are said to be "transformed." For example, DNA left from the dead cells of a drug-resistant strain of bacteria could be taken up by nonresistant bacteria, transforming them such that they become resistant to the drugs as well.

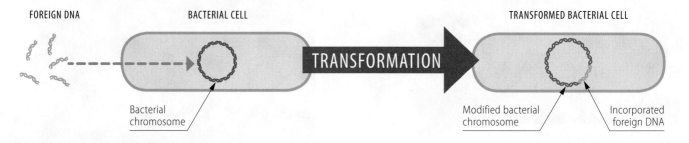

FOREIGN DNA — BACTERIAL CELL — TRANSFORMATION — TRANSFORMED BACTERIAL CELL

Bacterial chromosome

Modified bacterial chromosome — Incorporated foreign DNA

TRANSDUCTION

Transduction is the transfer of bacterial genes by a **bacteriophage** (also called a **phage**), a virus that infects bacteria. When a virus infects a bacterial cell, the viral DNA directs production of new viruses using the bacterium's cellular machinery. The host cell's chromosome becomes fragmented, and some bits of host DNA may end up as part of the viral chromosome. When the host cell eventually bursts, releasing new viruses, a bacteriophage with the bacterial DNA incorporated may later infect a new bacterial cell and inject its DNA, including the DNA stowaway from the former host cell. The virus may thus act as a DNA shuttle, transferring genes from one bacterium to another.

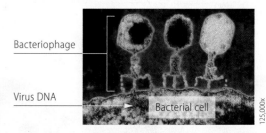

Bacteriophage

Virus DNA

Bacterial cell

125,000×

This micrograph shows a bacteriophage called T4 infecting an *E. coli* bacterium. After the "legs" of the phage touch the cell surface, the virus punctures the cell membrane and injects its DNA.

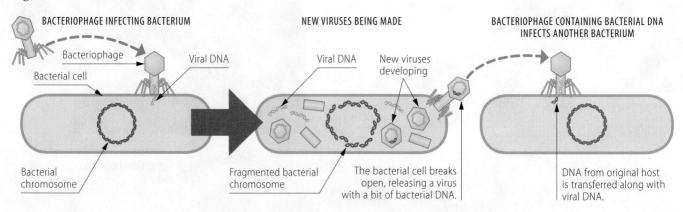

BACTERIOPHAGE INFECTING BACTERIUM

Bacteriophage

Bacterial cell

Viral DNA

Bacterial chromosome

NEW VIRUSES BEING MADE

Viral DNA

New viruses developing

Fragmented bacterial chromosome

The bacterial cell breaks open, releasing a virus with a bit of bacterial DNA.

BACTERIOPHAGE CONTAINING BACTERIAL DNA INFECTS ANOTHER BACTERIUM

DNA from original host is transferred along with viral DNA.

CONJUGATION

Conjugation is the transfer of DNA between two bacterial cells through a physical bridge. The donor cell uses a hollow extension called a **sex pilus** (plural, *pili*) to form a physical connection to a recipient cell. The donor cell transfers a copy of its chromosome through this mating bridge.

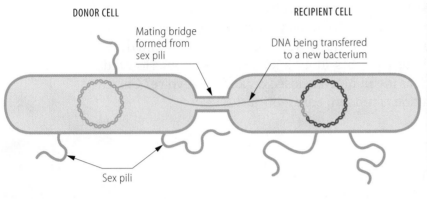

DONOR CELL — Mating bridge formed from sex pili

RECIPIENT CELL — DNA being transferred to a new bacterium

Sex pili

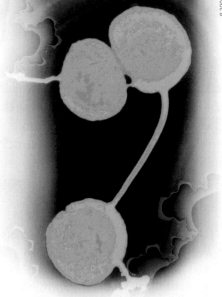

Neisseria gonorrhoeae bacteria undergoing conjugation

PLASMIDS

A **plasmid** is a small, circular DNA molecule—a "mini chromosome" typically containing just a few genes—that resides in the cytoplasm of a bacterium and can replicate independently of the bacterial chromosome. Like the chromosome, genes on the plasmid are used to produce proteins that can perform functions within the cell. A plasmid may be duplicated and passed from one cell to another, carrying along whatever genes it contains. For example, some plasmids contain genes that produce proteins that deactivate antibiotics. These antibiotic-resistance genes can be transferred from one bacterium to another, rendering a whole population of bacteria resistant to a drug.

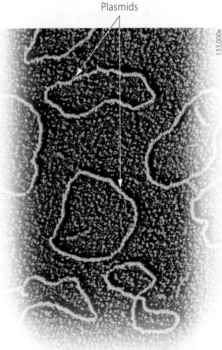

Plasmids

Plasmids from the bacterium *Escherichia coli*

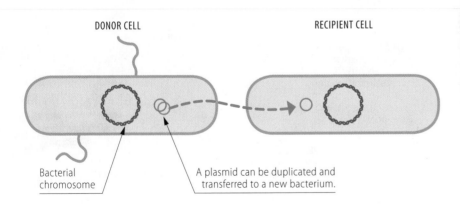

DONOR CELL

RECIPIENT CELL

Bacterial chromosome

A plasmid can be duplicated and transferred to a new bacterium.

CORE IDEA: Although they reproduce asexually, bacteria have several mechanisms for transferring DNA between cells, including transformation, transduction, conjugation, and the transfer of plasmids.

CORE QUESTION: Which form of bacterial DNA transfer could be called "bacterial sex"?

ANSWER: Conjugation, since it involves the direct transfer of DNA from one individual bacteria to another through special anatomical structures.

Eukaryotic cells evolved from prokaryotic cells

One of the most significant events in the history of life on Earth was the evolution of complex **eukaryotic cells** from simpler prokaryotic ones. The first prokaryotes appeared approximately 3.5 billion years ago, and they were the sole form of life on Earth for around 1.4 billion years. Then, around 2.1 billion years ago, the first eukaryotic cells appeared. These cells are the ancestors of modern eukaryotes, including all plants, fungi, animals, and protists. How did the uniquely eukaryotic features—such as membrane-enclosed organelles—evolve? Current hypotheses propose a multistage evolution of modern eukaryotic cells.

ENDOMEMBRANE SYSTEM

One prominent feature that can be found in all eukaryotic cells is the **endomembrane system**, a series of internal membranes and membrane-enclosed organelles that are largely interconnected. The endomembrane system includes the plasma membrane, endoplasmic reticulum, nuclear envelope, and the Golgi apparatus. Prokaryotic cells lack internal membranes. It is reasonable to imagine that the internal membranes of eukaryotes originated through inward folding of the external plasma membrane of an ancestral cell.

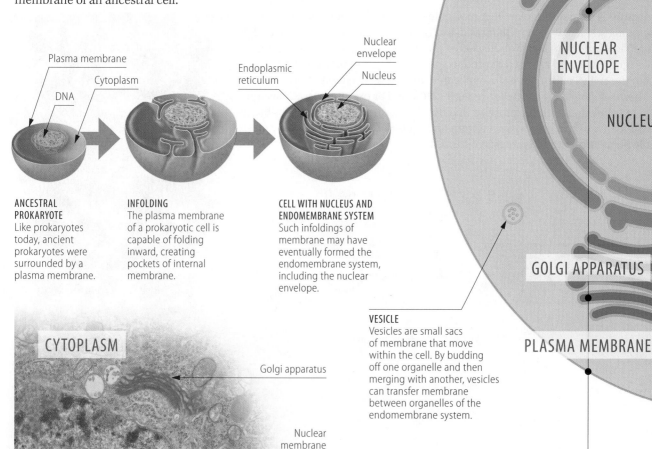

ANCESTRAL PROKARYOTE
Like prokaryotes today, ancient prokaryotes were surrounded by a plasma membrane.

Plasma membrane
Cytoplasm
DNA

INFOLDING
The plasma membrane of a prokaryotic cell is capable of folding inward, creating pockets of internal membrane.

CELL WITH NUCLEUS AND ENDOMEMBRANE SYSTEM
Such infoldings of membrane may have eventually formed the endomembrane system, including the nuclear envelope.

Endoplasmic reticulum
Nuclear envelope
Nucleus

ENDOPLASMIC RETICULUM

NUCLEAR ENVELOPE

NUCLEUS

VESICLE
Vesicles are small sacs of membrane that move within the cell. By budding off one organelle and then merging with another, vesicles can transfer membrane between organelles of the endomembrane system.

GOLGI APPARATUS

PLASMA MEMBRANE

CYTOPLASM

Golgi apparatus

Nuclear membrane

NUCLEUS

Micrograph of a human brain cell with visible organelles of the endomembrane system

20,000x

ENDOMEMBRANE SYSTEM
A series of cellular organelles connected by internal membranes

The name "endoplasmic reticulum" derives from Latin meaning "little net in the cytoplasm."

ENDOSYMBIOSIS

Evidence suggests that a key stage in the evolution of eukaryotic cells was **endosymbiosis**, which occurs when one species lives inside another host species. Two eukaryotic organelles—mitochondria and chloroplasts—appear to have evolved from small, free-living prokaryotes that were engulfed. Once established inside, endosymbiosis benefited both organisms, offering an evolutionary advantage that eventually led to complete interdependence. Today, modern eukaryotic cells act as one working unit with inseparable organelles. A key piece of evidence in favor of the endosymbiotic hypothesis is that modern-day mitochondria and chloroplasts retain their own DNA, RNA, and proteins that resemble prokaryotic versions more closely than they resemble eukaryotic ones.

Ribosomes

CHLOROPLAST

MITOCHONDRIA

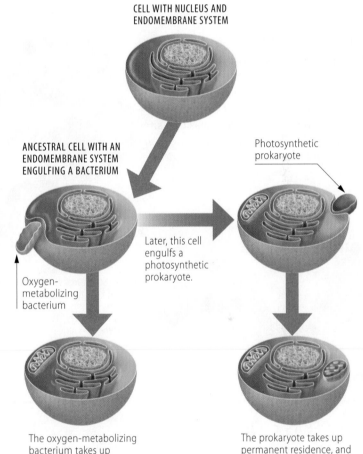

CELL WITH NUCLEUS AND ENDOMEMBRANE SYSTEM

ANCESTRAL CELL WITH AN ENDOMEMBRANE SYSTEM ENGULFING A BACTERIUM

Photosynthetic prokaryote

Later, this cell engulfs a photosynthetic prokaryote.

Oxygen-metabolizing bacterium

The oxygen-metabolizing bacterium takes up permanent residence, and its descendants eventually evolve into mitochondria.

The prokaryote takes up permanent residence, and its descendants eventually evolve into chloroplasts. The ancestral eukaryote now resembles a modern photosynthetic eukaryote in several important ways: it has an endomembrane system, mitochondria, and chloroplasts.

ENDOSYMBIOSIS
The internalization of previously free-living prokaryotes, which eventually evolved into eukaryotic organelles

CORE IDEA: Around 2.1 billion years ago, ancestral prokaryotic cells evolved into the first eukaryotic cells through an inward folding of the plasma membrane and the engulfment and incorporation of other free-living prokaryotic cells.

CORE QUESTION: How might a bit of membrane start as part of the plasma membrane but later become part of the nuclear membrane?

ANSWER: A vesicle may bud off one part of the endomembrane system and join to another.

Protists are very diverse

The term **protist** is used to describe all eukaryotes that do not belong to the plant, animal, or fungus kingdoms. Since they first appeared around 2.1 billion years ago, a great variety of protists have evolved. Most are unicellular and microscopic, but some are multicellular and large. Because all protists have eukaryotic cells (which include a nucleus containing chromosomes, and other membrane-bound organelles), even the simplest protist is much more complex than any prokaryote. But beyond the nature of their cells, it is hard to make generalizations about protists because they come in such wide variety, live in a great variety of habitats, and display a wide variety of lifestyles.

PROTOZOANS

Protozoans are protists that obtain nutrients primarily by eating. Some ingest bacteria and other protists, some absorb organic nutrients from their surroundings, and some live as parasites in animals. They are found in almost all aquatic and moist habitats.

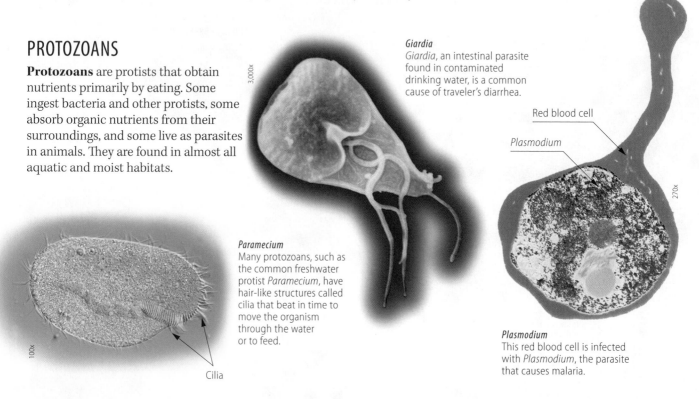

Giardia
Giardia, an intestinal parasite found in contaminated drinking water, is a common cause of traveler's diarrhea.

Red blood cell

Plasmodium

3,000x

270x

100x

Paramecium
Many protozoans, such as the common freshwater protist *Paramecium*, have hair-like structures called cilia that beat in time to move the organism through the water or to feed.

Cilia

Plasmodium
This red blood cell is infected with *Plasmodium*, the parasite that causes malaria.

AMOEBAS

Amoebas are single-celled protists with great flexibility in their body form; they can assume almost any shape as they crawl over surfaces or engulf food. Most move and feed using pseudopodia ("fake feet"), temporary extensions of the cell body.

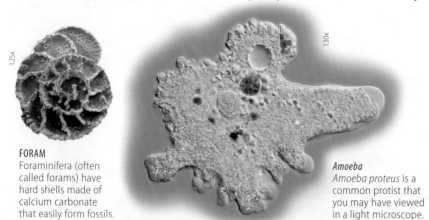

125x

130x

FORAM
Foraminifera (often called forams) have hard shells made of calcium carbonate that easily form fossils.

Amoeba
Amoeba proteus is a common protist that you may have viewed in a light microscope.

SLIME MOLDS

Slime molds are protists that resemble fungi in appearance and lifestyle. The web-like body of a slime mold provides a large surface area for contact with the environment. Slime molds are decomposers, using pseudopodia to engulf and break down the bodies of dead organisms.

SLIME MOLD
You can find plasmodial slime molds growing among leaf litter on the forest floor. Even though they can grow to several inches in size, the entire body consists of a single cell with multiple nuclei.

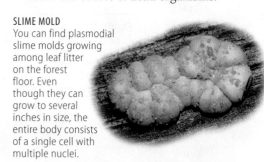

ALGAE

Algae (singular, *alga*) are photosynthetic protists able to produce their own food from sunlight. Algae may be unicellular, they may live in colonies, or they may be multicellular. Since they are producers, they form the base of many aquatic food chains.

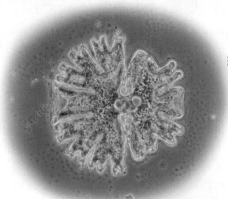
800x

GREEN ALGAE
Unicellular green algae flourish in lakes, ponds, and improperly tended pools and aquariums.

(Right margin, rotated text:) Some dinoflagellates are bioluminescent, producing their own internal light that can be seen in shallow ocean waters.

DINOFLAGELLATES

Dinoflagellates are abundant in plankton. An overabundance of dinoflagellates, sometimes caused by fertilizer runoff, can produce population explosions called red tide. Red tides can be dangerous to people because some dinoflagellates release metabolic toxins.

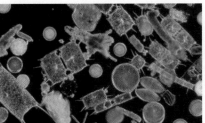
80x

Symbiodinium
Coral reefs depend on *Symbiodinium*, single-celled algae that inhabit the bodies of reef-building coral animals, providing the coral with food.

250x

DIATOMS
Diatom bodies are surrounded by a shell made from silica, the mineral found in glass.

SEAWEEDS

Seaweeds are large, multicellular marine algae that grow on rocky shores and in shallow marine habitats. Seaweeds are classified into three groups by the pigment present in their chloroplasts. Many coastal people harvest seaweeds and eat them in soups or as sushi wrappers.

RED ALGAE
Red algae are most common in relatively deep, warm, tropical waters.

GREEN ALGAE
Many green algae form edible seaweeds that grow in shallow coastal waters.

BROWN ALGAE
Also known as kelp, brown algae grow in large deep-water marine forests.

CORE IDEA: Protists display a wide variety of lifestyles and body forms. Common protists include protozoans, amoebas, slime molds, algae, and seaweeds.

CORE QUESTION: Are all protists microscopic?

ANSWER: No. Most are, but some (such as slime molds and seaweed pluralize seaweed to seaweeds) are big enough to see with the naked eye.

The origin of multicellular life was a major milestone in the evolution of life on Earth

One of the most significant events in the evolution of life on Earth was the emergence of multicellular organisms around 1.2 billion years ago. Actually, it is incorrect to refer to this as if it were a singular event; evidence indicates that multicellularity evolved multiple times among ancient protists, each time leading to many new and diverse species. Although it is impossible to say for certain what occurred over 1 billion years ago, biologists hypothesize a series of steps that may have occurred in the evolution of multicellularity.

THE EVOLUTION OF MULTICELLULAR LIFE

Colonies are loose physical associations of individual free-living cells. Colonial organisms probably represent an evolutionary intermediate between independent unicellular protists and multicellular organisms. Natural selection would favor those colonies that increased efficiency by evolving a division of labor. Over time, cells within the colonies became more and more specialized. Eventually, they were no longer capable of living on their own. The result is a multicellular eukaryote.

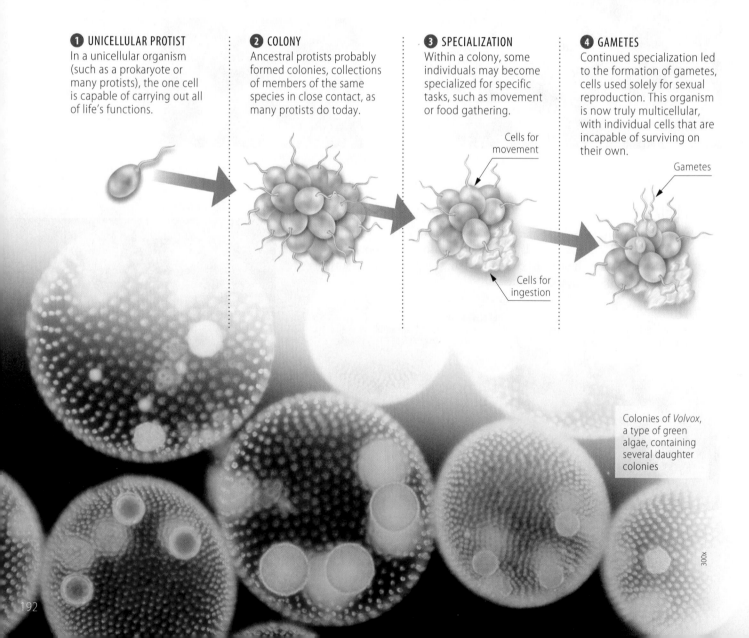

1 UNICELLULAR PROTIST
In a unicellular organism (such as a prokaryote or many protists), the one cell is capable of carrying out all of life's functions.

2 COLONY
Ancestral protists probably formed colonies, collections of members of the same species in close contact, as many protists do today.

3 SPECIALIZATION
Within a colony, some individuals may become specialized for specific tasks, such as movement or food gathering.

Cells for movement

Cells for ingestion

4 GAMETES
Continued specialization led to the formation of gametes, cells used solely for sexual reproduction. This organism is now truly multicellular, with individual cells that are incapable of surviving on their own.

Gametes

Colonies of *Volvox*, a type of green algae, containing several daughter colonies

300x

A BRIEF HISTORY OF THE EVOLUTION OF LIFE ON EARTH

As summarized on this time line, the evolution of multicellular organisms was just one of many important stops along the continuous path of the evolution of life on Earth. Keep in mind that all of these dates are subject to revision as more evidence accumulates.

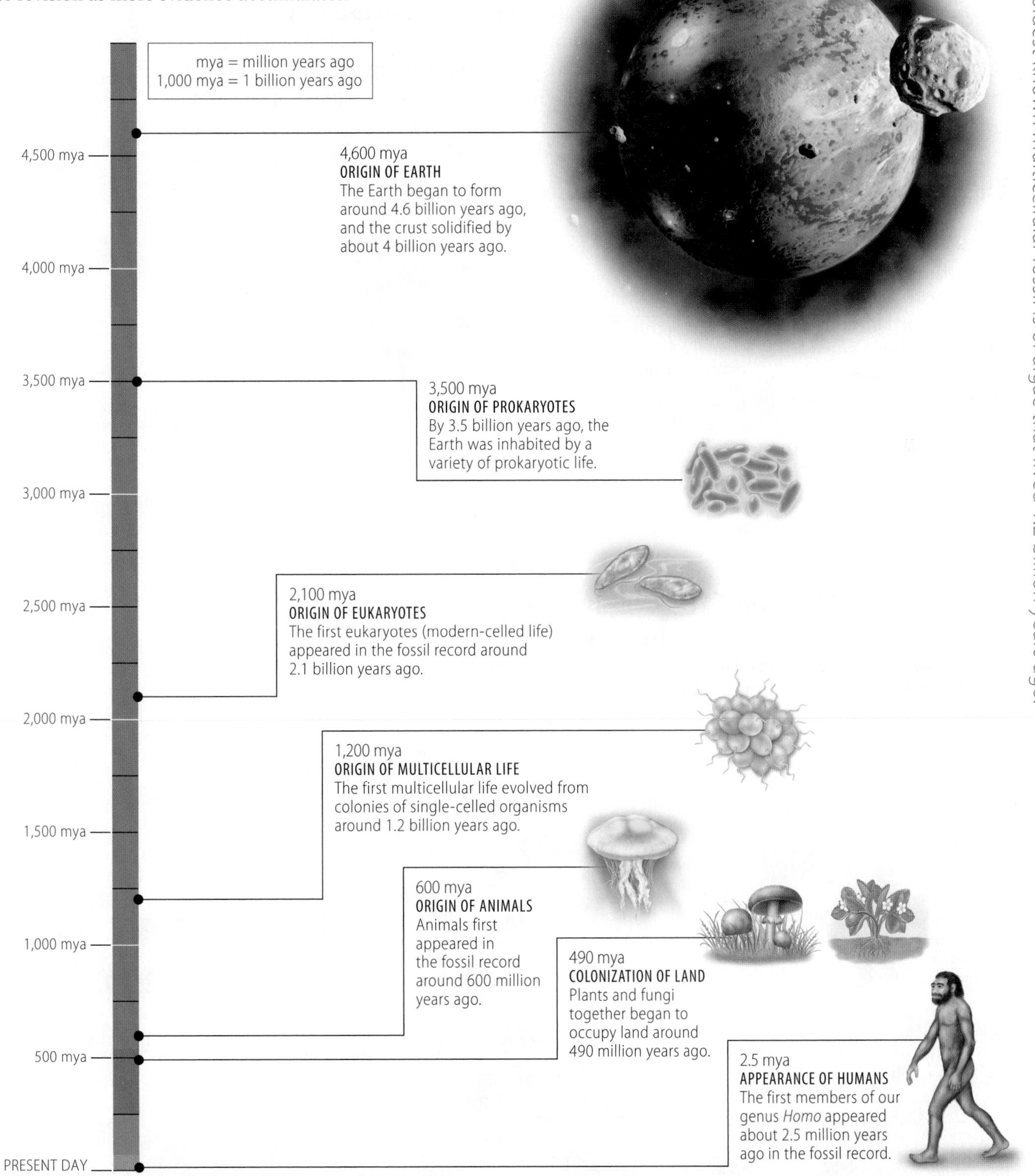

The oldest known multicellular fossil is of algae that lived 1.2 billion years ago.

mya = million years ago
1,000 mya = 1 billion years ago

4,600 mya
ORIGIN OF EARTH
The Earth began to form around 4.6 billion years ago, and the crust solidified by about 4 billion years ago.

3,500 mya
ORIGIN OF PROKARYOTES
By 3.5 billion years ago, the Earth was inhabited by a variety of prokaryotic life.

2,100 mya
ORIGIN OF EUKARYOTES
The first eukaryotes (modern-celled life) appeared in the fossil record around 2.1 billion years ago.

1,200 mya
ORIGIN OF MULTICELLULAR LIFE
The first multicellular life evolved from colonies of single-celled organisms around 1.2 billion years ago.

600 mya
ORIGIN OF ANIMALS
Animals first appeared in the fossil record around 600 million years ago.

490 mya
COLONIZATION OF LAND
Plants and fungi together began to occupy land around 490 million years ago.

2.5 mya
APPEARANCE OF HUMANS
The first members of our genus *Homo* appeared about 2.5 million years ago in the fossil record.

4,500 mya — 4,000 mya — 3,500 mya — 3,000 mya — 2,500 mya — 2,000 mya — 1,500 mya — 1,000 mya — 500 mya — PRESENT DAY

CORE IDEA: The evolution of multicellular eukaryotes from colonial protists was one of the major episodes in the evolution of life on Earth.

CORE QUESTION: What is the difference between the cells of colonies versus the cells of truly multicellular organisms?

ANSWER: The cells of colonies are all identical, while the cells of multicellular organisms are specialized and incapable of living independently.

193

Viruses are nonliving parasites

Although a **virus** shares some of the characteristics of living organisms (such as having a highly ordered structure), most biologists do not consider viruses to be alive. For example, a virus has no cells and cannot reproduce on its own—two of the defining characteristics of life. A virus has only one chance for survival: It must infect a living cell and direct that cell's internal machinery to make more viruses, a parasitic existence that contributes nothing to the survival of the host.

VIRUS STRUCTURE

Viruses have simple structures. In a sense, a virus is nothing more than "genes in a box": nucleic acid (either DNA or RNA) wrapped in a protein container called a **capsid**. Some viruses contain other structures (such as a membrane envelope), but nearly all viruses contain the three parts shown here.

NUCLEIC ACID
All viruses have genes made of nucleic acid. While living cells always use double-stranded DNA, viruses may use DNA or RNA, and it may be single- or double-stranded.

RECOGNITION SPIKE
Each virus has one or more proteins that protrudes into the environment and is capable of binding to proteins on the outside of its host cells. This protein-protein recognition is very specific, causing each type of virus to infect a specific type of cell.

CAPSID
The outer coat of a virus, called the capsid, is made of one or a few different protein molecules repeated in a regular pattern. Viral genes code for the capsid protein(s). Once produced inside a cell, the parts spontaneously self-assemble into a new capsid.

THE LYTIC AND LYSOGENIC CYCLES OF BACTERIOPHAGE

Viruses that infect bacteria are called **bacteriophage** ("bacteria eaters," also called **phages** for short). Once a bacterial cell is infected, the virus can enter one of two life cycles: the lytic cycle (left) or the lysogenic cycle (right).

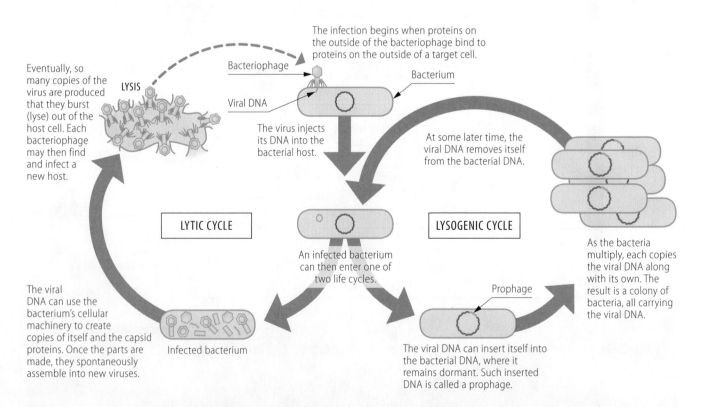

Eventually, so many copies of the virus are produced that they burst (lyse) out of the host cell. Each bacteriophage may then find and infect a new host.

LYSIS

The infection begins when proteins on the outside of the bacteriophage bind to proteins on the outside of a target cell.

Bacteriophage

Viral DNA

Bacterium

The virus injects its DNA into the bacterial host.

At some later time, the viral DNA removes itself from the bacterial DNA.

LYTIC CYCLE

LYSOGENIC CYCLE

An infected bacterium can then enter one of two life cycles.

As the bacteria multiply, each copies the viral DNA along with its own. The result is a colony of bacteria, all carrying the viral DNA.

Prophage

The viral DNA can use the bacterium's cellular machinery to create copies of itself and the capsid proteins. Once the parts are made, they spontaneously assemble into new viruses.

Infected bacterium

The viral DNA can insert itself into the bacterial DNA, where it remains dormant. Such inserted DNA is called a prophage.

A GALLERY OF VIRUSES

You can probably name several viruses that cause disease in humans. In fact, viruses are capable of infecting a broad variety of animals and plants.

INFLUENZA VIRUS

No virus poses a greater threat to human health than the influenza virus. Like many animal viruses, this one has an outer envelope of membrane (similar to the plasma membrane of cells) surrounding a protein coat with protruding spikes. During the 1918 influenza pandemic, 20–50 million people died worldwide. New strains occasionally arise that present novel threats to human health.

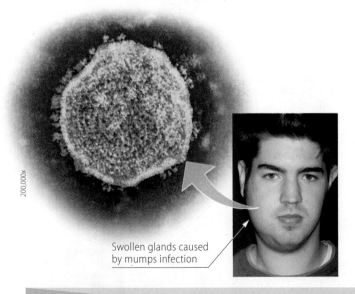

120,000x

HERPESVIRUS

Herpesvirus is a family of DNA viruses that cause many human diseases, including chicken pox, shingles, cold sores (caused by a virus called herpes simplex 1), and genital herpes (herpes simplex 2). Even if no symptoms are present, herpesvirus DNA may remain dormant inside nerve cells. Stress can trigger new virus production and a reappearance of symptoms. Herpesvirus infections are therefore permanent.

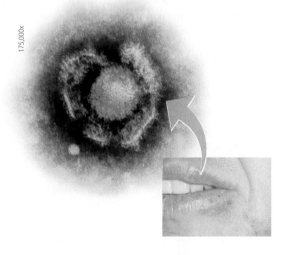

175,000x

MUMPS VIRUS

The virus that causes mumps—like the viruses that cause colds, measles, and polio—has RNA as its genetic material. Once inside the host cell, the viral RNA can direct reproduction of itself and produce capsid proteins. The disease mumps has become quite rare in industrialized nations due to widespread vaccination.

200,000x

Swollen glands caused by mumps infection

TOBACCO MOSAIC VIRUS

The tobacco mosaic virus was the first virus to be discovered (in the late 1800s). This virus—with proteins arranged in a spiral rod surrounding an RNA genome—damages tobacco leaf cells, causing significant agricultural damage.

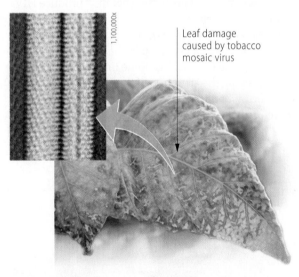

1,100,000x

Leaf damage caused by tobacco mosaic virus

More people died during the 18-month influenza pandemic of 1918–1919 than have ever died of AIDS.

CORE IDEA: Viruses are nonliving parasites that infect a host and then use the host's cellular machinery to produce more copies of themselves. Viruses can infect humans, other animals, and plants.

CORE QUESTION: What determines which host cells a virus can infect?

ANSWER: Proteins on the outside of the virus must match proteins on the outside of the host cell.

The deadly disease **AIDS** (acquired immunodeficiency syndrome) is caused by **HIV** (human immunodeficiency virus). AIDS is a global health scourge, particularly in southern Africa and southeastern Asia, with millions of people newly infected each year. So far, AIDS has claimed over 30 million lives. As biologists learn more about this virus, they are discovering that HIV has some unique features that contribute to its deadly nature.

STRUCTURE OF HIV

In some ways, HIV is a typical virus. In other ways, it is quite unique. HIV is spherical, and is about 1/60th the size of the blood cells it attacks, making it fairly large for a virus. Soon after its discovery, biologists realized that understanding the structure of HIV is key to understanding its deadly attack.

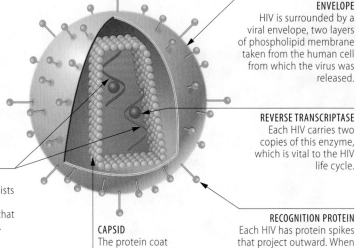

ENVELOPE
HIV is surrounded by a viral envelope, two layers of phospholipid membrane taken from the human cell from which the virus was released.

REVERSE TRANSCRIPTASE
Each HIV carries two copies of this enzyme, which is vital to the HIV life cycle.

RNA GENOME
The HIV genome consists of two copies of single-stranded RNA that encodes 9 viral genes.

CAPSID
The protein coat surrounding the HIV genome consists of about 2,000 copies of a single protein.

RECOGNITION PROTEIN
Each HIV has protein spikes that project outward. When these spikes contact a susceptible cell, infection begins.

ANTI-HIV DRUGS

There is no cure for AIDS, but progression of the disease can be slowed by anti-HIV drugs. One type of medicine—called protease inhibitors—interferes with viral proteins called proteases that help produce HIV proteins. Another type of drug, called AZT, inhibits the action of HIV reverse transcriptase (RT). The RT enzyme performs a vital step in the HIV life cycle: it converts the HIV RNA genome into a DNA copy. By mimicking the structure of the DNA/RNA base thymine (abbreviated T), AZT shuts down RT and therefore halts HIV reproduction.

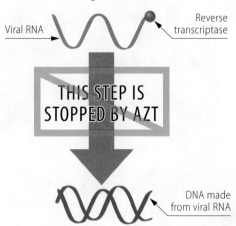

Viral RNA

Reverse transcriptase

THIS STEP IS STOPPED BY AZT

DNA made from viral RNA

Notice that the structure of the drug AZT closely mimics the structure of a T nucleotide. The structures are similar enough that reverse transcriptase uses AZT instead of T, but the small difference (highlighted in pink) prevents AZT from being incorporated into a growing viral DNA molecule. This effectively halts HIV reproduction.

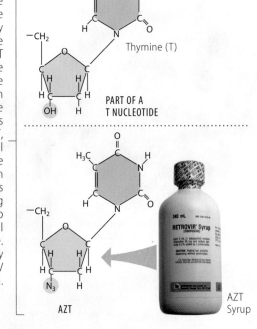

Thymine (T)

PART OF A T NUCLEOTIDE

AZT

AZT Syrup

HIV LIFE CYCLE

HIV is a **retrovirus**, a virus with an RNA genome that is converted to DNA prior to viral reproduction. In other words, retroviruses reverse the normal flow of genetic information: The information flows from RNA to DNA, rather than from DNA to RNA as it does in all living cells. Once it enters the host cell, HIV is able to direct the production of many copies of itself.

HIV is deadly because it destroys helper T cells, a key component of the human immune system. Once immunity is destroyed, the body becomes susceptible to opportunistic infections that are easily fought off by a healthy immune system. It is these infections that actually kill the patient.

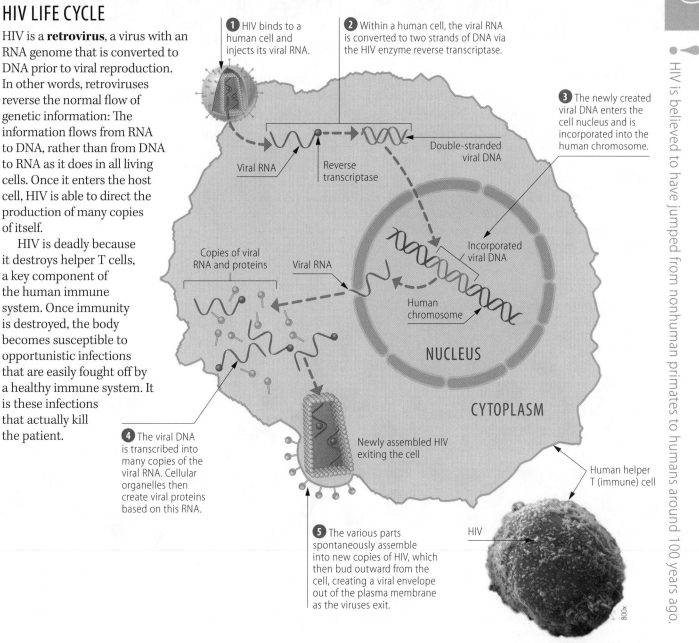

1 HIV binds to a human cell and injects its viral RNA.

2 Within a human cell, the viral RNA is converted to two strands of DNA via the HIV enzyme reverse transcriptase.

3 The newly created viral DNA enters the cell nucleus and is incorporated into the human chromosome.

Viral RNA

Reverse transcriptase

Double-stranded viral DNA

Copies of viral RNA and proteins

Viral RNA

Incorporated viral DNA

Human chromosome

NUCLEUS

CYTOPLASM

4 The viral DNA is transcribed into many copies of the viral RNA. Cellular organelles then create viral proteins based on this RNA.

Newly assembled HIV exiting the cell

Human helper T (immune) cell

HIV

5 The various parts spontaneously assemble into new copies of HIV, which then bud outward from the cell, creating a viral envelope out of the plasma membrane as the viruses exit.

800x

HIV is believed to have jumped from nonhuman primates to humans around 100 years ago.

EMERGING VIRUSES

HIV is an example of an **emerging virus**, one that has rapidly come to the attention of scientists. Why do new viruses suddenly appear? Some arise through mutations of older viruses, while some expand their range quickly after being confined to an isolated area. Recently emerged viruses include the H1N1 influenza virus, West Nile virus, SARS, and the Ebola virus. Because few effective antiviral drugs are available, preventing the spread of a virus is often the best solution to stopping a widespread epidemic.

Red Cross workers spray disinfectant around the intensive care room at Kelle Hospital in northwestern Congo.

CORE IDEA: HIV is a retrovirus, with an RNA genome that is converted to DNA during the process of infecting human helper T immune cells. HIV relies on the action of the enzyme reverse transcriptase for reproduction.

CORE QUESTION: Does infection by HIV itself normally cause death?

ANSWER: No. HIV cripples the immune system, leaving a person susceptible to other deadly infections.

Prions and viroids are nonliving parasites even smaller than viruses

A **pathogen** is an agent that causes disease. Many pathogens are living organisms, such as bacteria, fungi, and protists. But some pathogens are not alive. Viruses, for example, do not display all the characteristics that define life. In addition, there are two other classes of pathogens that are even smaller and simpler than viruses: prions and viroids.

PRIONS

A **prion** is an infectious protein. Prions are actually misshapen versions of normal brain proteins. By interfering with the proper assembly of brain tissue, prions cause a number of diseases in a variety of animals. Prion diseases are incurable and inevitably fatal.

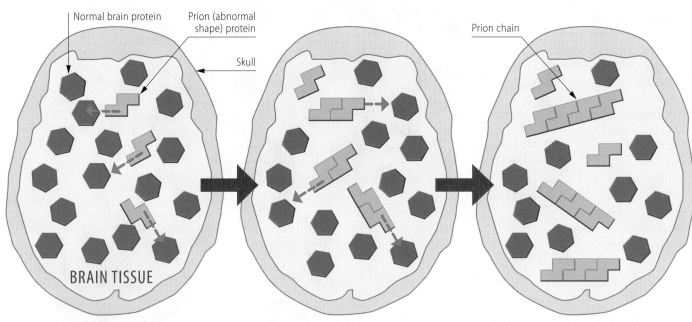

When a prion protein enters brain tissue, it converts normal brain proteins into misshapen prion proteins.

As more copies of the misfolded protein are made, they start to assemble into clusters.

Eventually, the clusters accumulate, disrupting normal brain tissue, and causing symptoms of the prion disease.

CREUTZFELDT-JAKOB DISEASE
Creutzfeldt-Jakob disease (CJD) is an incurable degenerative disease of the human brain that is contracted from consumption of mad-cow-infected beef. It is inevitably fatal. Normal sterilization procedures do not destroy prion proteins, so prevention of exposure is the only current medical approach. After infection, holes develop in brain tissue, giving it a sponge-like texture.

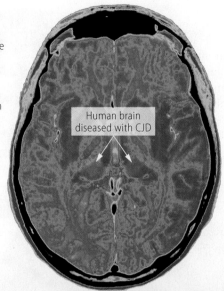

Human brain diseased with CJD

A computer rendering of a prion protein

PRION DISEASES IN ANIMALS

MAD COW DISEASE
Prions in cattle cause mad cow disease, formally called bovine spongiform encephalopathy (BSE). An outbreak of BSE in England, which caused a few hundred people to become sick after eating infected beef, was controlled by a mass culling (via incineration) of millions of cows.

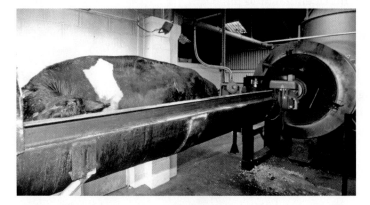

CHRONIC WASTING DISEASE
Prions cause chronic wasting disease in several animals, including deer, elk, and moose. Found in a dozen U.S. states and two Canadian provinces, the condition is named for the fact that it leads to chronic, fatal weight loss.

KURU
First described among one tribe in Papua New Guinea in the 1950s, kuru is a degenerative human brain disease. Kuru is transmitted by cannibalistic ingestion of an infected person's brain, which transfers the prion to a new host.

SCRAPIE
In sheep, prions cause scrapie, a disease named for the fact that continuous itching makes an animal scratch its fur off. First identified in the 1700s, scrapie does not appear to be transmissible to humans.

By eliminating ritualistic cannibalism (thought to pass on a "life force"), the spread of kuru has ceased among indigenous Indonesian tribes.

VIROID DISEASES IN PLANTS

Viroids are small circular single-stranded RNA molecules that are capable of infecting and causing disease in plants. Typically only a few hundred nucleotides long, viroids do not encode proteins but can replicate in host plant cells by using the host's own cellular enzymes. Viroids disrupt normal plant growth, causing abnormal development and stunted growth.

APPLE SCAR SKIN VIROID
Widespread in both Asia and North America, this viroid causes cracking of the fruit skin, devastating a crop.

POTATO SPINDLE TUBER VIROID
This viroid, the first to be discovered, can infect all species of potatoes and tomatoes. Infection causes cracking and distortion of tubers and fruit.

PEACH LATENT MOSAIC VIROID
The peach latent mosaic viroid, first discovered in the 1980s in Spain, can infect all known peach and nectarine species throughout the world. It is so named because it typically lies dormant for 5–7 years before symptoms appear.

CORE IDEA: Nonliving infectious pathogens include prions, misfolded proteins capable of clustering together and disrupting brain function, and viroids, small circular RNA molecules that can infect plants.

CORE QUESTION: What is the primary difference between a viroid and a prion?

ANSWER: A viroid is made of RNA, whereas a prion is made of protein.

Fungi are a diverse group of eukaryotes

Fungi are a large and diverse group with over 100,000 species identified so far. All fungi are **eukaryotes** (that is, composed of cells with nuclei and other membrane-surrounded organelles), most are multicellular, and all acquire nutrients by absorption. DNA studies indicate that the lineages that gave rise to fungi (and animals) diverged from a protist ancestor more than 1 billion years ago. The oldest fungi fossils date from only 460 million years ago, perhaps because fungi often fossilize poorly. The classification of fungi is not settled; one widely accepted phylogenetic tree divides the kingdom Fungi into five groups. Because they affect our lives in many ways, you are probably already familiar with many different fungi.

DECOMPOSERS

Fungi play a vital role in most ecosystems because they decompose dead organisms and other organic materials, recycling their essential chemical elements back to the environment in forms that other organisms can use. Without decomposers, needed nutrients would accumulate in nonliving matter, unable to be returned to the food chain. Fungi secrete powerful enzymes into their environment that digest large biological molecules—breaking down proteins into amino acids, for example. Once broken down, the small nutrient molecules are absorbed directly into the cells of the fungus.

These honey mushrooms (*Armillaria mellea*) are digesting the remains of this fallen log. Through such decomposition, fungi recover nutrients from dead matter.

A GALLERY OF FUNGI

There are a wide variety of fungi. Many are helpful to humans, but some are harmful.

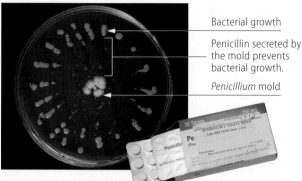

Bacterial growth

Penicillin secreted by the mold prevents bacterial growth.

Penicillium mold

PENICILLIN
Some fungi produce antibiotics that are used to treat bacterial diseases. The mold *Penicillium* produces the antibiotic penicillin, which is effective at preventing bacterial growth.

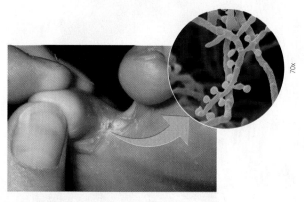

PARASITIC FUNGI
About 50 species of fungi are known to infect humans, causing such diseases as vaginal yeast infections, lung infections, and a skin disease called ringworm. A fungal infection in the lower extremities causes athlete's foot, which is highly contagious. Antifungal medications can quickly cure most fungal infections.

MOLDS
Molds are rapidly growing fungi that reproduce asexually by producing spores. Molds grow quickly over their food sources, which are often our food sources as well. Enzymes secreted by fungi growing on a loaf of bread digest the bread's starch into glucose molecules, which the fungal cells absorb.

Yeast
Bud

YEAST
Yeast are unicellular fungi that reproduce asexually by a process called budding. Humans have used yeast for thousands of years to make bread and to ferment grains and fruit.

Orange fungus cells

Green alga cells

LICHEN
Although it appears to be a single organism, a lichen is actually a symbiotic association of a unicellular, photosynthetic alga (or, sometimes, a bacterium) and a fungus. The fungus receives food produced by its photosynthetic partner while simultaneously providing a safe habitat, water, and minerals.

EDIBLE FUNGI
Fungi are a common food source for people. Many varieties of mushrooms are edible. Some people forage for mushrooms, but only experts can distinguish edible varieties from poisonous ones. Blue cheese gets its distinctive flavor from the fungi used to produce it. A fungus called corn smut can damage crops, but many people consider the infected ears a gourmet delicacy. Truffles, which grow underground in European forests, are highly prized by gourmets and fetch high prices.

Button mushrooms

Corn smut

Blue cheese

Black truffles

Despite its appearance, a mushroom is more closely related to you than it is to any plant.

CORE IDEA: Fungi are decomposers that break down large molecules in their environment and absorb the resulting small molecules. Such decomposition helps recycle nutrients. There is a wide variety of fungi, both helpful and harmful.

CORE QUESTION: Why is the action of fungi as decomposers so important?

ANSWER: Without decomposers, nutrients would accumulate in dead matter and never be recycled.

Fungi have specialized structures and means of reproduction

The kingdom Fungi is a large and diverse collection of species. But whether mold or mushroom, all fungi share certain structural and reproductive features: All fungi have a relatively simple body plan, and most can reproduce in more than one way.

ANATOMY OF A MUSHROOM

Mushrooms are probably the fungi that are most familiar to you. The body of a mushroom consists of an aboveground reproductive structure (what we call a "mushroom") that extends from a belowground structure called a **mycelium**. Both parts are made of fibers called **hyphae**.

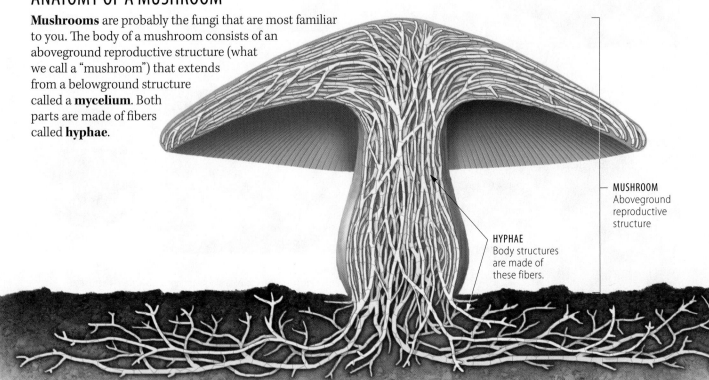

MUSHROOM
Aboveground reproductive structure

HYPHAE
Body structures are made of these fibers.

MYCELIUM
Belowground supporting structure

MUSHROOM
A mushroom consists of tightly packed hyphae that extend upward from a larger underground mycelium. A mushroom forms as a mycelium absorbs water, causing it to swell upward—explaining why mushrooms often appear after a rainstorm. The mushroom is a reproductive structure that produces tiny reproductive cells called spores. Spores are dispersed through the air.

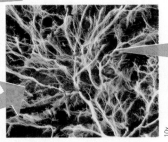

MYCELIUM
Although a fungus cannot move, its underground mycelium (plural, *mycelia*) grows rapidly, adding hyphae that mingle with organic matter. As a mycelium extends its hyphae through a food source, the fungus feeds by breaking down and absorbing organic matter.

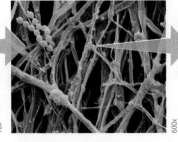

HYPHAE
Fungal bodies (both the aboveground mushroom and the belowground mycelium) are made of threadlike hyphae (singular, *hypha*). Each hypha is a chain of interconnected cells; pores within the cross-walls allow materials and even cellular organelles to flow between cells.

CHITIN
Fungal cell walls are made of chitin, a strong but flexible polysaccharide. Chitin also makes up the hard exoskeleton of many insects.

FUNGAL REPRODUCTION

Fungi can reproduce either asexually or sexually. In asexual reproduction, specialized cells on the underside of the mushroom cap produce **spores**, specialized sex cells that contain a single (haploid) set of chromosomes. Once produced and released (usually in tremendous numbers), the spores are dispersed on air currents or in water. If a spore settles in a moist place with food, it can germinate to produce a new haploid mycelium. Fungi reproduce sexually when the hyphae of two parents join and the haploid cells of each fuse, resulting in a zygote with diploid cells, ones that have two sets of chromosomes. The diploid cells then divide, producing genetically distinct haploid spores.

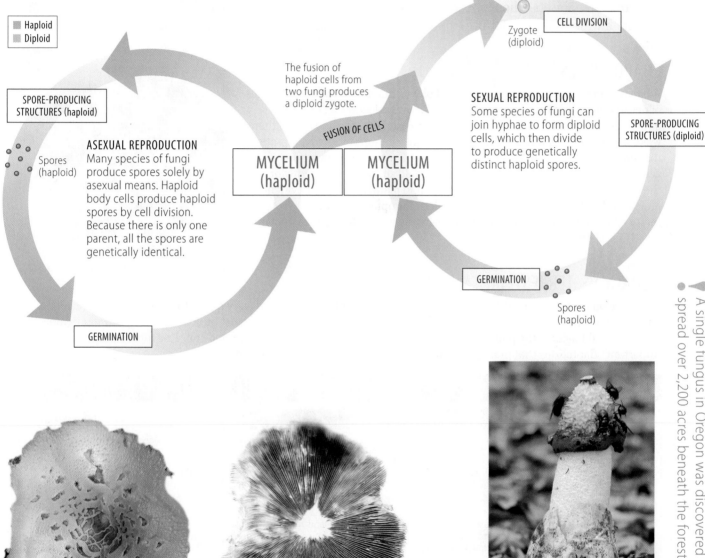

- Haploid
- Diploid

SPORE-PRODUCING STRUCTURES (haploid)

Spores (haploid)

ASEXUAL REPRODUCTION
Many species of fungi produce spores solely by asexual means. Haploid body cells produce haploid spores by cell division. Because there is only one parent, all the spores are genetically identical.

GERMINATION

The fusion of haploid cells from two fungi produces a diploid zygote.

FUSION OF CELLS

MYCELIUM (haploid)   MYCELIUM (haploid)

Zygote (diploid)

CELL DIVISION

SEXUAL REPRODUCTION
Some species of fungi can join hyphae to form diploid cells, which then divide to produce genetically distinct haploid spores.

SPORE-PRODUCING STRUCTURES (diploid)

GERMINATION

Spores (haploid)

A single fungus in Oregon was discovered to have a mycelium that spread over 2,200 acres beneath the forest.

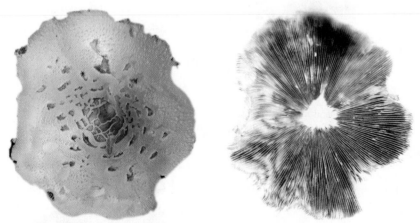

SPORE PRINT
If you place a mushroom cap (left) on a piece of paper, the spores will fall off to form a spore print (right). The color of the spores can aid in identification—as with this parasol mushroom (*Chlorophyllum molybdites*).

STINKHORN
Several species of the family *Phallaceae* (named for its phallic shape) are called stinkhorns. Their spores emit a slime with the scent of rotting meat. Flies are attracted to the smell, become coated with the slime, and spread the spores.

CORE QUESTION: When you eat a mushroom, what are you actually eating?

CORE IDEA: Fungal bodies consist of hyphae organized into belowground mycelia (which grow into and digest food) and aboveground mushrooms (which produce spores for reproduction). Many fungi reproduce both asexually and sexually.

ANSWER: The reproductive organ of a fungus.

Plants have unique adaptations that allow them to survive on land

Around 500 million years ago, a sea-dwelling algae—the ancestor to all land plants—began to evolve adaptations that allowed for survival on dry land. Such features distinguish a plant from algae and help plants thrive in terrestrial habitats. The pioneering plants opened a vast new terrestrial frontier and set the stage for the explosive diversification of plant life that followed.

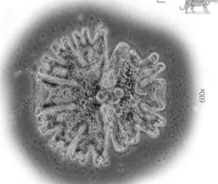

CHAROPHYTES

A half billion years ago, there were no land plants, but algae covered the edges of oceans, lakes, and other bodies of water. Natural selection would favor individuals that could survive further and further up the tide line where there was no competition for resources. Over time, populations evolved adaptations that enabled them to survive occasional drying. Eventually, populations accumulated enough adaptations that they could survive permanently beyond the edge of the water, making them the first land plants. Molecular evidence indicates that a modern group of algae called charophytes—relatively complex multicellular photosynthetic eukaryotes—are the closest living ancestors of these first plants.

600x

Micrasterias furcata, a type of charophyte algae

AQUATIC VS. TERRESTRIAL ENVIRONMENTS

The move on to land required several evolutionary adaptations that allowed early plants to thrive in terrestrial environments. Life on land poses many challenges: Land-dwelling organisms must be able to obtain and retain moisture, remain upright against the pull of gravity, and obtain resources from both the air and soil. This table summarizes some important adaptations that make plants well suited to their terrestrial environment.

COMPARING ALGAE TO PLANTS		
An alga in water	**ALGAE** / **PLANTS**	A plant on land
	Moisture	
	There is plenty of moisture available in the water surrounding the alga. / The plant body must obtain moisture from the soil and retain it within the plant's roots, stems, and leaves.	
	Support	
	The alga body is buoyed by the surrounding water; minimal stiffness is required. / Plants must have stiff structures that support the plant against gravity.	
	Reproduction	
	Gametes are dispersed in water and are in no danger of drying out. / Gametes and embryos must be protected from drying out.	
	Anchorage	
	There is minimal anchorage; a structure called a holdfast grips the underlying surface. / Plants require roots to stay firmly in place.	
	Nutrients	
	Nutrients can be obtained from the surrounding water via the entire body surface of the alga. / Nutrients must be absorbed via roots and distributed to the rest of the plant.	
	Photosynthesis	
	Photosynthesis is performed by the entire body. / Photosynthesis is performed mainly in leaves.	

TERRESTRIAL ADAPTATIONS

A **plant** is a terrestrial multicellular eukaryote that conducts photosynthesis, producing organic molecules from inorganic molecules using the energy of sunlight. But there are protists—such as some seaweeds—that are also multicellular, eukaryotic, and photosynthetic. What distinguishes plants from algae and other photosynthetic organisms is a set of structural adaptations that allow them to thrive on land. These structures include ones that allow plants to obtain resources from two very different sources: air and soil.

PLANT ORGAN SYSTEMS

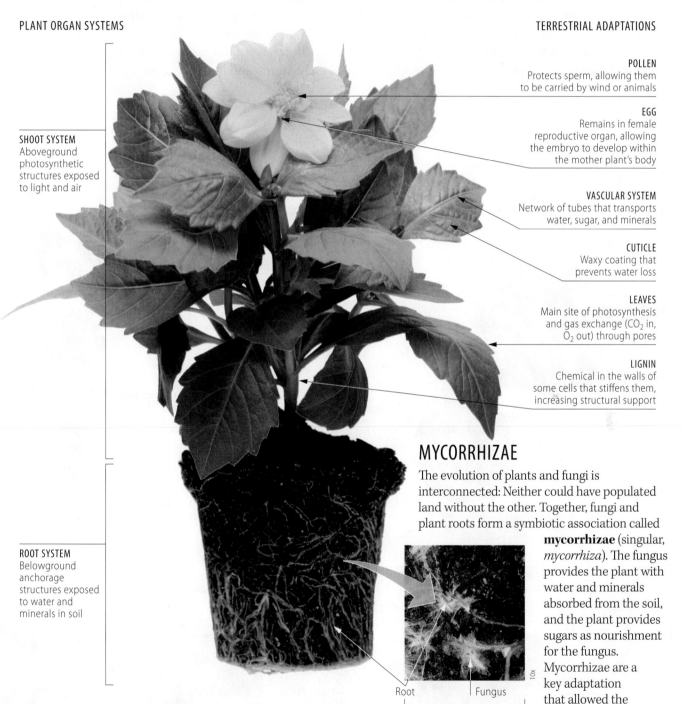

SHOOT SYSTEM
Aboveground photosynthetic structures exposed to light and air

ROOT SYSTEM
Belowground anchorage structures exposed to water and minerals in soil

Root | Fungus

10x

MYCORRHIZA

TERRESTRIAL ADAPTATIONS

POLLEN
Protects sperm, allowing them to be carried by wind or animals

EGG
Remains in female reproductive organ, allowing the embryo to develop within the mother plant's body

VASCULAR SYSTEM
Network of tubes that transports water, sugar, and minerals

CUTICLE
Waxy coating that prevents water loss

LEAVES
Main site of photosynthesis and gas exchange (CO_2 in, O_2 out) through pores

LIGNIN
Chemical in the walls of some cells that stiffens them, increasing structural support

MYCORRHIZAE

The evolution of plants and fungi is interconnected: Neither could have populated land without the other. Together, fungi and plant roots form a symbiotic association called **mycorrhizae** (singular, *mycorrhiza*). The fungus provides the plant with water and minerals absorbed from the soil, and the plant provides sugars as nourishment for the fungus. Mycorrhizae are a key adaptation that allowed the colonization of land.

CORE IDEA: Plants evolved from algae that accumulated adaptations, such as ones for obtaining nutrients and preventing gametes from drying out. Such adaptations enable plants to survive the unique challenges of living on land.

CORE QUESTION: What problem associated with living on land do mycorrhizae help a plant overcome?

ANSWER: The need to absorb water and minerals from the soil.

Plant bodies consist of roots, stems, and leaves

Living on land poses several challenges for a plant. Chief among them is the need to obtain resources from two very different places. Water and mineral nutrients are found mainly in the soil, whereas light and carbon dioxide are primarily available in the air. Modern land plants interact with each of these environments using specific structures: **roots**, the belowground structures, and **shoots**, the aboveground structures, including stems, leaves, and flowers. Neither roots nor shoots can survive without the other. Living in the dark, roots would starve without sugar and other organic nutrients produced via photosynthesis in stems and leaves. Conversely, stems and leaves depend on the water and minerals absorbed by roots.

ROOT SYSTEM

The roots of a plant anchor it in the soil, absorb water and minerals, transport nutrients to other parts of the plant, and store food. In addition, as you can see in the table below, many plant roots also form symbioses, relationships between species living in direct contact with each other.

Grasses have broad, fibrous, shallow roots that can hold on to soil, making grasses good ground cover.

Dandelions and many other plants have a deep, vertical taproot that extends far down into the soil.

Near the tips of all roots are tiny projections called root hairs. These cellular extensions greatly increase the surface area for absorption.

20x

Large taproots, such as those found in carrots, turnips, sugar beets, and sweet potatoes, store food in the form of starch and sugars. If not harvested by humans, the plants can use this stored food during times of growth.

PLANT ROOT SYMBIOSIS	
Root Nodules	**Mycorrhizae**
Some plant families, including legumes— peas, beans, peanuts, and many other plants that produce pods— have root nodules that house bacteria. The bacteria provide nitrogen-containing nutrients to the plant.  15x — Root nodule	Many plant roots form mutualistic (mutually beneficial) associations with fungi. Such mycorrhizae (singular, *mycorrhiza*) benefit the plant by increasing water and mineral absorption, and benefit the fungi by providing nourishment. 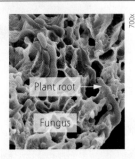 700x — Plant root — Fungus

ROOTS

SHOOT SYSTEM

The shoots of a plant are the aboveground structures, including stems, leaves, and structures for reproduction, such as cones or flowers.

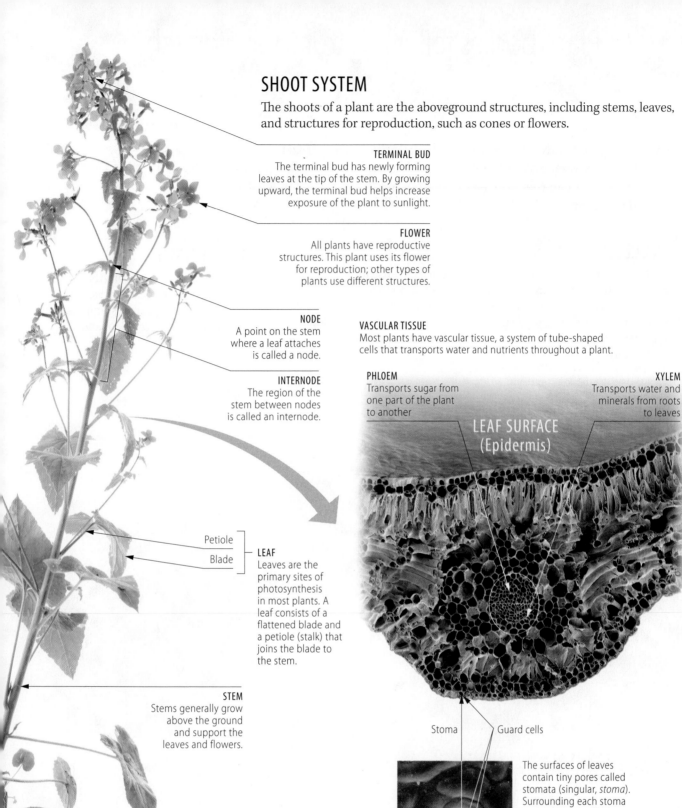

TERMINAL BUD
The terminal bud has newly forming leaves at the tip of the stem. By growing upward, the terminal bud helps increase exposure of the plant to sunlight.

FLOWER
All plants have reproductive structures. This plant uses its flower for reproduction; other types of plants use different structures.

NODE
A point on the stem where a leaf attaches is called a node.

INTERNODE
The region of the stem between nodes is called an internode.

VASCULAR TISSUE
Most plants have vascular tissue, a system of tube-shaped cells that transports water and nutrients throughout a plant.

PHLOEM
Transports sugar from one part of the plant to another

XYLEM
Transports water and minerals from roots to leaves

**LEAF SURFACE
(Epidermis)**

210x

Petiole

Blade

LEAF
Leaves are the primary sites of photosynthesis in most plants. A leaf consists of a flattened blade and a petiole (stalk) that joins the blade to the stem.

STEM
Stems generally grow above the ground and support the leaves and flowers.

Stoma   Guard cells

400x

The surfaces of leaves contain tiny pores called stomata (singular, *stoma*). Surrounding each stoma are two guard cells that regulate the opening of the pore. Gas exchange (CO_2 in and O_2 out) occurs through the stomata.

The root hairs of a single sunflower, if laid end to end, would stretch for miles.

CORE IDEA: Plants thrive in terrestrial environments because they have evolved roots and shoots. A variety of plant structures absorb, transport, and retain water and nutrients, collect light for photosynthesis, and exchange gases.

CORE QUESTION: Which prominent reproductive structure shown on this page is not found in all plants?

ANSWER: Flowers (other plants have other structures for reproduction).

Plant bodies follow a structural hierarchy

Like your body, the body of a plant contains cells that are organized into tissues, which are organized into organs, which are organized into systems. The anatomy of a plant can therefore be understood by examining each level in this structural hierarchy.

PLANT CELLS

The **cell** is the fundamental unit of all life, so it makes sense for an exploration of plant anatomy to begin there. In addition to features shared with other eukaryotic cells (a nucleus, mitochondria, ribosomes, etc.), most plant cells have some or all of the unique structures shown here, including chloroplasts, a central vacuole, a cell wall, and plasmodesmata.

PLANT TISSUES

A **tissue** is a group of cells that work together to perform a specific function. Two of the most important tissues in plants are xylem and phloem.

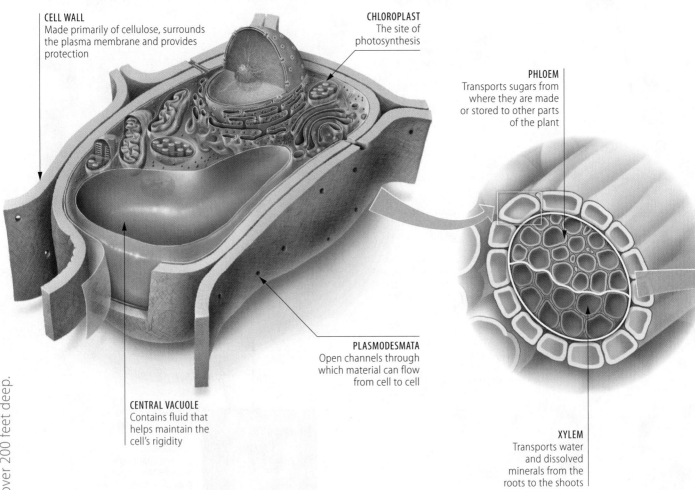

CELL WALL
Made primarily of cellulose, surrounds the plasma membrane and provides protection

CHLOROPLAST
The site of photosynthesis

PHLOEM
Transports sugars from where they are made or stored to other parts of the plant

PLASMODESMATA
Open channels through which material can flow from cell to cell

CENTRAL VACUOLE
Contains fluid that helps maintain the cell's rigidity

XYLEM
Transports water and dissolved minerals from the roots to the shoots

A tree growing near Tucson, Arizona, was found to have roots over 200 feet deep.

PLANT ORGANS

An **organ** consists of several types of tissue that work together to carry out a particular function. For example, a leaf—an organ that specializes in photosynthesis—contains xylem and phloem tissues, as you can see in the leaf cross section shown here. Other plant organs include roots and stems.

PLANT TISSUE SYSTEMS

In your body, organs (such as your heart) are organized into organ systems (such as your circulatory system). Similarly, plant tissues are organized into three **tissue systems**: the dermal tissue system, ground tissue system, and vascular tissue system. Each plant organ—root, stem, or leaf—contains tissues from all three tissue systems.

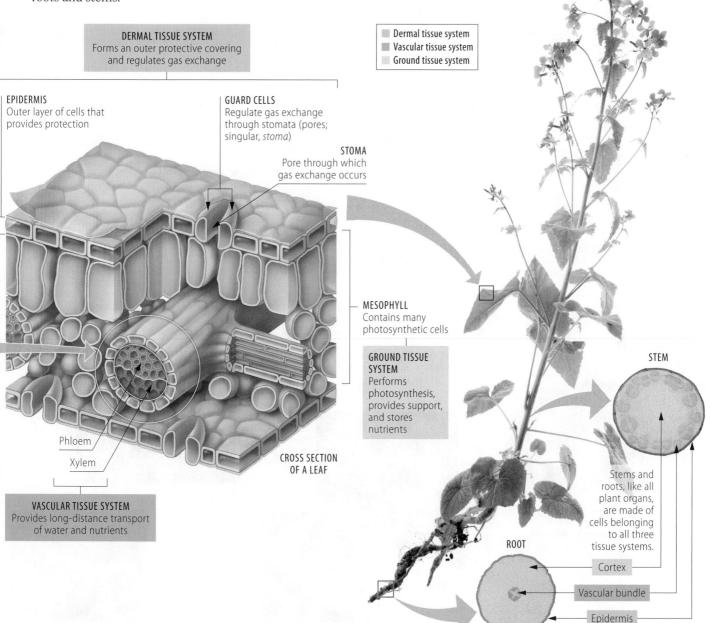

Dermal tissue system
Vascular tissue system
Ground tissue system

DERMAL TISSUE SYSTEM
Forms an outer protective covering and regulates gas exchange

EPIDERMIS
Outer layer of cells that provides protection

GUARD CELLS
Regulate gas exchange through stomata (pores; singular, *stoma*)

STOMA
Pore through which gas exchange occurs

MESOPHYLL
Contains many photosynthetic cells

GROUND TISSUE SYSTEM
Performs photosynthesis, provides support, and stores nutrients

Phloem
Xylem

CROSS SECTION OF A LEAF

VASCULAR TISSUE SYSTEM
Provides long-distance transport of water and nutrients

STEM

Stems and roots, like all plant organs, are made of cells belonging to all three tissue systems.

ROOT

Cortex
Vascular bundle
Epidermis

CORE IDEA: Plant anatomy can be understood on several levels: Cells work together to form tissues, which can be organized into tissue systems. Each plant organ includes tissues from all three systems.

CORE QUESTION: Which tissue system would you expect contains cells with the most chloroplasts?

ANSWER: The ground tissue system, where photosynthesis occurs.

Four major groups of plants have evolved

The fossil record chronicles four major branch points in the evolution of land plants. Each of these branches represents the appearance of novel adaptations that opened new opportunities on land and resulted in the evolution of a new group of modern plants. Today, there are four major groups of land plants, one that began at each branch point.

MAJOR EVENTS IN THE EVOLUTION OF PLANTS

This timeline and phylogenetic tree summarize the evolution of the four major groups of modern plants.

> mya = million years ago
> 1,000 mya = 1 billion years ago

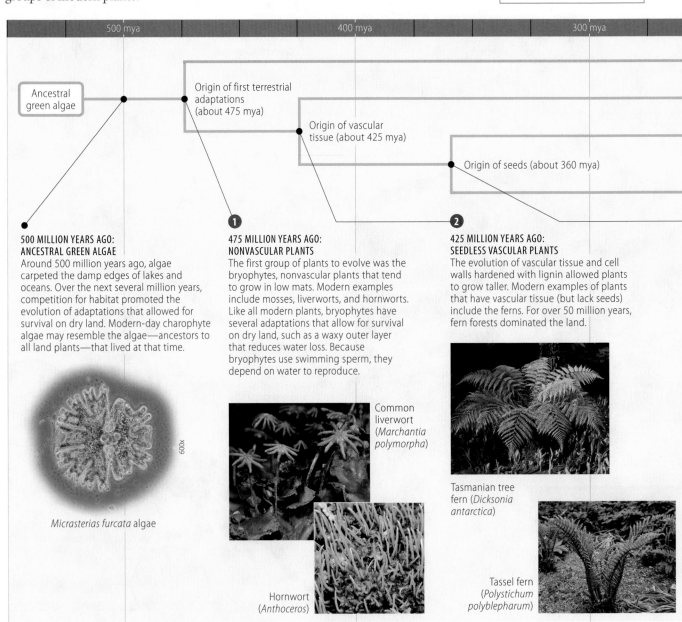

500 MILLION YEARS AGO: ANCESTRAL GREEN ALGAE
Around 500 million years ago, algae carpeted the damp edges of lakes and oceans. Over the next several million years, competition for habitat promoted the evolution of adaptations that allowed for survival on dry land. Modern-day charophyte algae may resemble the algae—ancestors to all land plants—that lived at that time.

Micrasterias furcata algae

600x

475 MILLION YEARS AGO: NONVASCULAR PLANTS
The first group of plants to evolve was the bryophytes, nonvascular plants that tend to grow in low mats. Modern examples include mosses, liverworts, and hornworts. Like all modern plants, bryophytes have several adaptations that allow for survival on dry land, such as a waxy outer layer that reduces water loss. Because bryophytes use swimming sperm, they depend on water to reproduce.

Common liverwort (*Marchantia polymorpha*)

Hornwort (*Anthoceros*)

425 MILLION YEARS AGO: SEEDLESS VASCULAR PLANTS
The evolution of vascular tissue and cell walls hardened with lignin allowed plants to grow taller. Modern examples of plants that have vascular tissue (but lack seeds) include the ferns. For over 50 million years, fern forests dominated the land.

Tasmanian tree fern (*Dicksonia antarctica*)

Tassel fern (*Polystichum polyblepharum*)

Ancestral green algae

Origin of first terrestrial adaptations (about 475 mya)

Origin of vascular tissue (about 425 mya)

Origin of seeds (about 360 mya)

500 mya 400 mya 300 mya

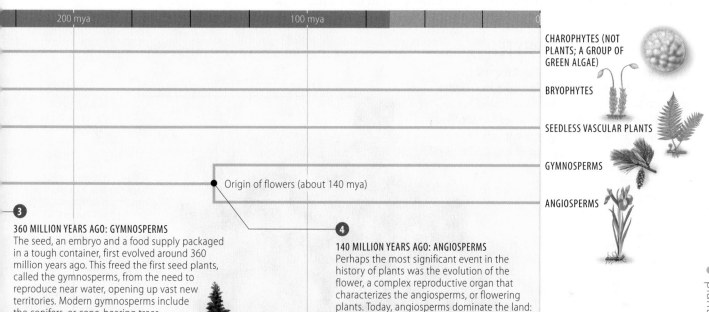

200 mya   100 mya   0

CHAROPHYTES (NOT PLANTS; A GROUP OF GREEN ALGAE)

BRYOPHYTES

SEEDLESS VASCULAR PLANTS

GYMNOSPERMS

ANGIOSPERMS

Origin of flowers (about 140 mya)

3

360 MILLION YEARS AGO: GYMNOSPERMS
The seed, an embryo and a food supply packaged in a tough container, first evolved around 360 million years ago. This freed the first seed plants, called the gymnosperms, from the need to reproduce near water, opening up vast new territories. Modern gymnosperms include the conifers, or cone-bearing trees.

4

140 MILLION YEARS AGO: ANGIOSPERMS
Perhaps the most significant event in the history of plants was the evolution of the flower, a complex reproductive organ that characterizes the angiosperms, or flowering plants. Today, angiosperms dominate the land: The vast majority of living plants—including nearly all of our important food crops—are angiosperms.

Mediterranean cypress (*Cupressus sempervirens*)

Douglas fir (*Pseudotsuga menziesii*)

Columbine (*Aquilegia*)

Sunflowers (*Helianthus annuus*)

Botanists have named 250,000 species of flowering plants on Earth today.

CORE IDEA: The evolutionary history of plants records the appearance of successive adaptations for living on land, resulting in four major groups of modern plants: bryophytes, seedless vascular plants, gymnosperms, and angiosperms.

CORE QUESTION: Products harvested from which group of plants are most likely to be found in your refrigerator?

ANSWER: Angiosperms.

Bryophytes are seedless, nonvascular plants

The first plants evolved from algae approximately 475 million years ago. These plants had two important adaptations for living on land: a waxy outer layer that helps to retain moisture, and structures that provide internal protection for gametes and embryos. But because this first group lacked other structures that are common today, such as seeds and vascular (transport) tissue, they had to live near water and could survive only in damp environments. The modern plants most closely related to these first plants are called **bryophytes**. Like their ancestors, modern bryophytes grow only in moist habitats.

CHAROPHYTES (NOT PLANTS; A GROUP OF GREEN ALGAE)

BRYOPHYTES

SEEDLESS VASCULAR PLANTS

GYMNOSPERMS

ANGIOSPERMS

ANATOMY OF A BRYOPHYTE

Bryophytes come in two distinct forms. The green, spongelike plant that forms a mat (what you probably think of as feathery "moss") is called the **gametophyte**. The other form, called a **sporophyte**, has a stalk with a capsule at its tip; you can find these poking out of a mat of moss if you look carefully.

SPORE CAPSULE
Bryophytes use sperm-containing **spores** stored within a capsule to reproduce. Once released, sperm will dry out unless they stay within water. Thus bryophytes require a moist environment to reproduce. When conditions permit, spores can develop into new gametophyte plants.

SPOROPHYTE
Cells of the sporophyte are diploid, containing two sets of chromosomes in matched pairs. The sporophyte produces spores, tough haploid cells that can survive harsh conditions.

CUTICLE
Like all land plants, bryophytes have a waxy layer, called the **cuticle**, that helps retain moisture within the plant body.

Common hair moss (*Polytrichum commune*)

GAMETOPHYTE
The gametophyte is the dominant form of most bryophytes. Its cells are haploid, containing a single, unmatched set of chromosomes. As its name suggests, the gametophyte produces haploid gametes (sperm and egg).

MAT
Because bryophytes lack lignin, a chemical that hardens the cell walls of most plants, they cannot grow tall. Instead, they grow in low dense mats, with many individuals packed together, each supporting the other.

A GALLERY OF BRYOPHYTES

MOSS
The most common bryophytes are mosses. Here, you can see Sphagnum moss (*Sphagnum palustre*) growing in a bog. Centuries of accumulated dead moss can be dug up and sold as "peat moss" to gardeners, valued because of its ability to absorb and retain moisture.

LIVERWORT
Leafy liverwort (*Lophocolea heterophylla*) is one of about 9,000 species of liverwort that have been identified. It grows on damp bark.

HORNWORT
Field hornwort (*Anthoceros agrestis*) has horn-like sporophytes. There are about 100 species of hornwort that have been identified.

ALTERNATION OF GENERATIONS

Plants have life cycles very different from ours: The gametophyte and sporophyte take turns producing each other. The gametophyte is haploid, with a single set of chromosomes, whereas the sporophyte is diploid, with a double set of chromosomes. This type of life cycle, called alternation of generations, is unique to plants and multicellular green algae.

GAMETOPHYTE
Haploid gametophytes produce haploid gametes: sperm and eggs. Bryophytes are the only plants where the gametophyte is the larger, more obvious plant.

Spores (haploid)

Mitosis

Mitosis

GAMETES
Sperm and eggs (haploid)

GERMINATION
After being released and settling into a hospitable habitat, haploid spores germinate and develop into new haploid gametophytes via mitosis.

FERTILIZATION
After swimming through water to an egg, a haploid sperm unites with a haploid egg to form a diploid zygote.

Meiosis

SPORE PRODUCTION
Within a spore capsule at the tip of a sporophyte stalk, cell division by meiosis produces haploid spores.

Spore capsule

Zygote (diploid)

Mitosis

■ Haploid
■ Diploid

SPOROPHYTE
The diploid zygote develops into a diploid sporophyte.

CORE QUESTION: When you look at a mat of moss, what structure are you seeing? Are the cells within it haploid or diploid?

CORE IDEA: Mosses are the most common bryophytes, seedless nonvascular plants. The bryophyte life cycle involves alternation of generations, where separate haploid (gametophyte) and diploid (sporophyte) generations take turns producing each other.

ANSWER: The gametophyte; haploid.

Vascular tissue transports water and nutrients

One of the primary challenges for a plant living on land is the distribution of vital materials (such as water, minerals, and sugar) between distant parts of the body. One of the most important evolutionary adaptations of land plants is **vascular tissue**, a system of tubes that acts as a transport system within the plant. Nearly all plants—except for bryophytes, such as moss—have vascular tissue. Within the vascular tissue, a variety of mechanisms are used to move materials between plant organs.

CHAROPHYTES (NOT PLANTS; A GROUP OF GREEN ALGAE)

BRYOPHYTES

SEEDLESS VASCULAR PLANTS

GYMNOSPERMS

ANGIOSPERMS

PHLOEM AND XYLEM

The vascular tissue system contains two types of specialized tissue: xylem and phloem. **Xylem** contains dead, hollow cells lined up as tiny pipes that convey water and minerals up from the roots. **Phloem** contains living cells that transport sugars from where they are made or stored to other parts of the plants where they are needed.

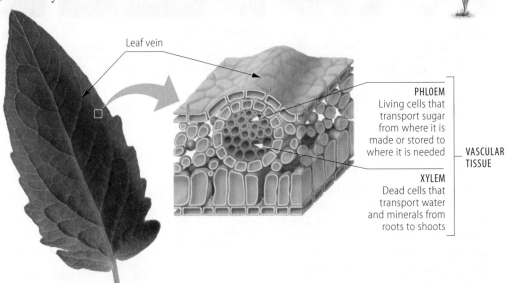

Leaf vein

PHLOEM
Living cells that transport sugar from where it is made or stored to where it is needed

VASCULAR TISSUE

XYLEM
Dead cells that transport water and minerals from roots to shoots

PHLOEM: SUGAR TRANSPORT

Phloem consists of living sugar-conducting cells arranged end to end into long tubes. Phloem sap—a solution of sugar (primarily sucrose) and other nutrients dissolved in water—flows through the tubes. Phloem sap always flows from a **sugar source**—cells where sugar is available, such as photosynthetic leaves where sugar is made, or organs such as tubers that store sugar—to a **sugar sink**, a site in the plant where the sugar will be used, such as a growing tip. Within phloem, sugar always flows from a source to a sink. Some structures, such as tubers, can be a sugar sink in the summer (as excess sugar is being stored) and a source in the spring (when the stored sugar is used to restart growth).

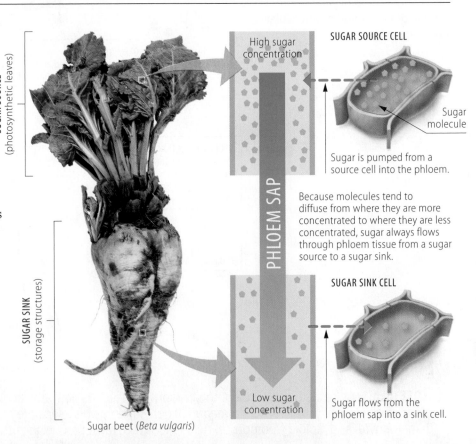

SUGAR SOURCE (photosynthetic leaves)

SUGAR SINK (storage structures)

Sugar beet (*Beta vulgaris*)

High sugar concentration

PHLOEM SAP

SUGAR SOURCE CELL

Sugar molecule

Sugar is pumped from a source cell into the phloem.

Because molecules tend to diffuse from where they are more concentrated to where they are less concentrated, sugar always flows through phloem tissue from a sugar source to a sugar sink.

SUGAR SINK CELL

Sugar flows from the phloem sap into a sink cell.

Low sugar concentration

During springtime, xylem sap contains dissolved sugars; obtained from sugar maples and boiled down about 40-fold, this sap is sold as maple syrup.

XYLEM: TRANSPIRATION

To thrive, a plant must transport large quantities of water absorbed through the roots to the rest of the plant body, even the very tips of the shoots. The combination of water and minerals that is transported within xylem is called xylem sap. Xylem sap is moved through the plant body via **transpiration**, a highly efficient process that uses physical forces to move a long string of water from the roots to the tips of leaves.

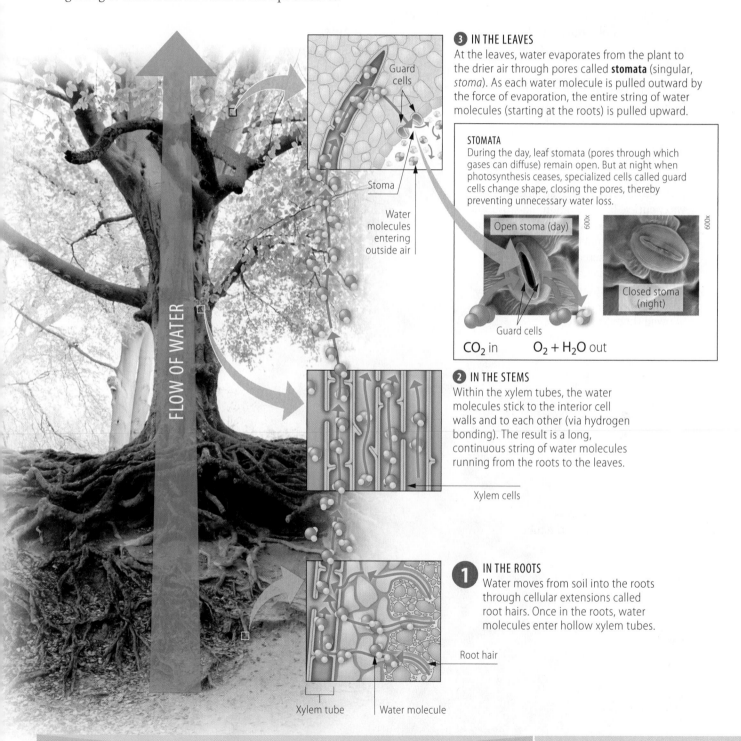

3 IN THE LEAVES

At the leaves, water evaporates from the plant to the drier air through pores called **stomata** (singular, *stoma*). As each water molecule is pulled outward by the force of evaporation, the entire string of water molecules (starting at the roots) is pulled upward.

STOMATA

During the day, leaf stomata (pores through which gases can diffuse) remain open. But at night when photosynthesis ceases, specialized cells called guard cells change shape, closing the pores, thereby preventing unnecessary water loss.

Open stoma (day) 600x

Closed stoma (night) 600x

Guard cells

CO_2 in $O_2 + H_2O$ out

2 IN THE STEMS

Within the xylem tubes, the water molecules stick to the interior cell walls and to each other (via hydrogen bonding). The result is a long, continuous string of water molecules running from the roots to the leaves.

Xylem cells

1 IN THE ROOTS

Water moves from soil into the roots through cellular extensions called root hairs. Once in the roots, water molecules enter hollow xylem tubes.

Guard cells

Stoma

Water molecules entering outside air

FLOW OF WATER

Xylem tube   Water molecule   Root hair

CORE IDEA: Vascular tissue allows most land plants to transport water, nutrients, and other substances through the plant body. Xylem primarily transports water and minerals from the roots. Phloem primarily transports sugars.

CORE QUESTION: Does a plant have to expend energy to transport water from the roots to the leaves?

ANSWER: No; physical properties (such as evaporation and the sticking together of water molecules) provide all the forces.

Ferns are seedless vascular plants

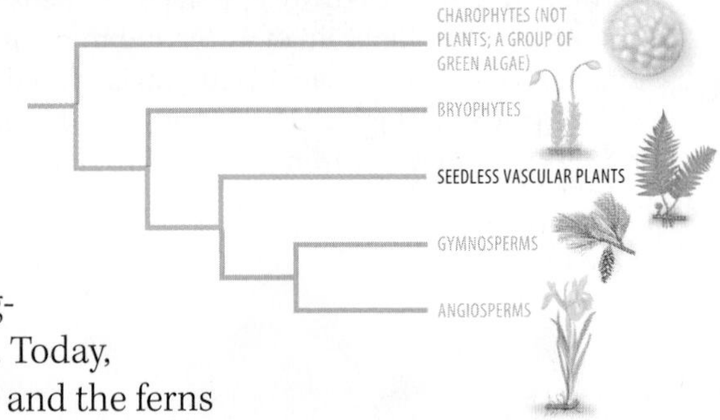

About 50 million years after the appearance of the first land plants, a major leap in plant evolution occurred with the origin of the **seedless vascular plants**. These plants have two significant adaptations not found in earlier plants: supporting cells hardened by lignin that allows these plants to stand up straight and tall, and **vascular tissue** that allows long-distance transport of water and nutrients through the plant body. Today, there are two groups of living seedless vascular plants: lycophytes and the ferns (which are far more common).

CHAROPHYTES (NOT PLANTS; A GROUP OF GREEN ALGAE)

BRYOPHYTES

SEEDLESS VASCULAR PLANTS

GYMNOSPERMS

ANGIOSPERMS

ANATOMY OF A FERN

Ferns display the features that first allowed plants to survive in terrestrial habitats: a waxy cuticle layer to retain moisture, and structures that allow gametes and embryos to develop internal to the plant body. The seedless vascular plants also display novel adaptations—lignin and vascular tissue—that allowed them to venture further into the landscape.

VASCULAR TISSUE
Tubes of vascular tissue transport water and nutrients through the plant body.

SPORE CAPSULE
Structures on the underside of fern leaves contain huge numbers of spores (haploid reproductive cells).

SPOROPHYTE
The fern plant that you typically see growing in the woods is the sporophyte generation.

FROND
Large leaves have many leaflets radiating from a central vein.

LIGNIN
The cells of ferns are stiffened with a complex organic polymer called **lignin**, which allows the plant to stand up straight and tall.

ROOTS
Water and minerals absorbed by a well-developed root system are transported to the rest of the plant body via vascular system tissue.

GAMETOPHYTE
A fern gametophyte is a tiny plant that grows on or just below the surface of the soil.

A GALLERY OF SEEDLESS VASCULAR PLANTS

LYCOPHYTE
Lycophytes, such as this *Huperzia selago*, first appeared around 410 million years ago, making them one of the longest-surviving groups of vascular plants.

CARBONIFEROUS FOREST
During the Carboniferous period, from about 360 to 300 million years ago, ancient ferns and other seedless vascular plants formed vast tropical forests (as shown in this artist's rendition). The fossilized remains of these plants gradually hardened into coal.

FERN
Modern ferns, such as these sword ferns (*Polystichum munitum*), can be found in the tropics and in temperate woodlands.

FERN REPRODUCTION

Although they have several adaptations that allow them to survive farther inland, seedless vascular plants—like the bryophytes that originated before them— are still tied to water for reproduction. Ferns require moist conditions for fertilization to occur because their sperm have whip-like flagella and must swim through water to reach eggs. Like the bryophytes, ferns alternate generations, with gametophyte and sporophyte each producing the other.

GAMETOPHYTE
The haploid gametophyte of a fern produces haploid gametes: sperm and eggs.

Mitosis

Mitosis

Spores (haploid)

GERMINATION
After being released and settling into a hospitable habitat, haploid spores develop into new haploid gametophytes via mitosis.

Meiosis

GAMETES
Sperm and eggs (haploid)

FERTILIZATION
After swimming through water to an egg, a haploid sperm unites with a haploid egg to form a diploid zygote.

SPORE PRODUCTION
Through meiosis, cells in the diploid sporophyte produce haploid spores that are dispersed via the wind.

New diploid sporophyte

Spore capsule

Zygote (diploid)

Mitosis

Mitosis

■ Haploid
■ Diploid

SPOROPHYTE
Through rounds of mitosis, a young sporophyte (right) develops into a mature diploid sporophyte (left; the fern that you typically see) with spore capsules on the underside of the leaves.

CORE IDEA: Seedless vascular plants have some important adaptations for living on land, such as vascular tissue and lignin-hardened cell walls. Fern reproduction involves an alternation of generations and requires a moist environment.

CORE QUESTION: Why can't ferns thrive in the desert?

ANSWER: Seedless vascular plants release sperm that must swim through water to reach an egg.

The first plants to evolve seeds were gymnosperms

The first plants with seeds evolved around 360 million years ago. A **seed**, consisting of an embryo packaged with a food supply inside a protective coating, is one of the key adaptations that allowed plants to spread across the land into a variety of new habitats. Because a developing embryo is packaged inside a seed, seed plants do not need to reproduce near water. The most successful of the first seed plants were the **gymnosperms**, plants that contain seeds within cones. Gymnosperms include many familiar varieties of evergreens, plants that retain their leaves all year.

CHAROPHYTES (NOT PLANTS; A GROUP OF GREEN ALGAE)

BRYOPHYTES

SEEDLESS VASCULAR PLANTS

GYMNOSPERMS

ANGIOSPERMS

ANATOMY OF A GYMNOSPERM

Notice that this gymnosperm has two types of cones: ovule-producing and pollen-producing.

NEEDLES
The needle-shaped leaves of gymnosperms have a thick cuticle and pores located in recessed pits. Both of these adaptations help a gymnosperm survive dry spells. Even in winter, the needles can perform photosynthesis.

OVULE-PRODUCING CONE
Scales contain the cells that, after pollination, develop into seeds.

Scale

Seed

SEED
After the arrival of sperm, ovules—located on the scales of female cones—develop into seeds. The seed contains an embryo, a food supply, and a tough exterior capable of surviving until favorable conditions promote germination.

POLLEN-PRODUCING CONE
These cones release millions of pollen grains at a time.

750x

POLLEN
A pollen grain houses the cells that develop into sperm. Windborne pollen carries the sperm to the egg. This adaptation allows gymnosperms to reproduce even in the absence of water.

WOOD
Gymnosperms often have thick trunks that consist of vascular tissue hardened with lignin. This wood gives the tree firm support and can be harvested as lumber.

A GALLERY OF GYMNOSPERMS

Although found in small numbers in the Southern Hemisphere, gigantic forests of gymnosperms cover much of northern America, Europe, and Asia. You probably rely on products harvested from gymnosperms every day, such as building lumber and the wood pulp used to make paper.

Ginkgo leaves

CONIFERS
Most gymnosperms are conifers, cone-bearing trees such as pine, spruce, cedar, redwood, and fir.

BRISTLECONE PINE
Bristlecone pine trees (three species of the genus *Pinus*) are native to California and can live over 4,500 years, making them among the longest-lived organisms on Earth.

REDWOOD
Coastal redwoods (*Sequoia sempervirens*) are native to northern California and are the world's tallest trees, rising over 300 feet in the air.

GINKGO TREE
Ginkgo trees (*Ginkgo biloba*) are easily recognizable by their distinctive fan-shaped leaves.

GYMNOSPERM REPRODUCTION

In most gymnosperms, cones house all reproductive stages: spores, eggs, sperm, zygotes, and embryos. A typical gymnosperm bears two types of cones that develop into male and female gametophytes.

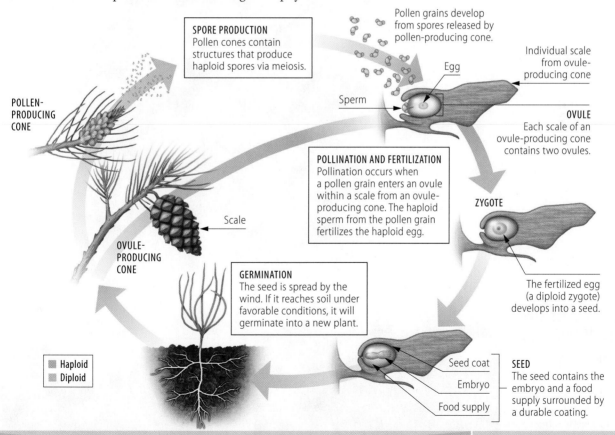

SPORE PRODUCTION
Pollen cones contain structures that produce haploid spores via meiosis.

Pollen grains develop from spores released by pollen-producing cone.

Egg

Sperm

Individual scale from ovule-producing cone

OVULE
Each scale of an ovule-producing cone contains two ovules.

POLLEN-PRODUCING CONE

POLLINATION AND FERTILIZATION
Pollination occurs when a pollen grain enters an ovule within a scale from an ovule-producing cone. The haploid sperm from the pollen grain fertilizes the haploid egg.

ZYGOTE

Scale

OVULE-PRODUCING CONE

GERMINATION
The seed is spread by the wind. If it reaches soil under favorable conditions, it will germinate into a new plant.

The fertilized egg (a diploid zygote) develops into a seed.

Haploid
Diploid

Seed coat
Embryo
Food supply

SEED
The seed contains the embryo and a food supply surrounded by a durable coating.

CORE IDEA: Gymnosperms (cone-bearing plants) bear their embryos in seeds, which include a food supply and a protective coating. Fertilization occurs after pollen released into the air contacts ovules in the scale of a cone.

CORE QUESTION: A person who is allergic to evergreen pollen is actually allergic to what?

ANSWER: The sperm-bearing reproductive structures of gymnosperms.

Angiosperms dominate the modern landscape

The vast majority of modern plants (over 250,000 species and counting) are **angiosperms**, the flowering plants. Since the evolution of the flower around 140 million years ago, angiosperms have dominated the landscape. Most of our food, as well as countless other products on which our modern society depends, comes from angiosperms. The most important adaptation of angiosperms is the **flower**, a complex reproductive structure that houses the ovary and develops into fruits and seeds.

CHAROPHYTES (NOT PLANTS; A GROUP OF GREEN ALGAE)

BRYOPHYTES

SEEDLESS VASCULAR PLANTS

GYMNOSPERMS

ANGIOSPERMS

ANATOMY OF AN ANGIOSPERM

Angiosperms have many familiar structures, but it is the reproductive flower that distinguishes this lineage of plants.

TERMINAL BUD
Tip of the shoot that grows upward

FLOWER
Reproductive organ

LEAF
Main photosynthetic organ

BLADE
Flattened region

PETIOLE
Stalk that joins leaf to stem

AXILLARY BUD
Growing shoot where a stem meets a leaf

STEM
Provides support to the aboveground structures

ROOTS
Absorb water and minerals

A GALLERY OF ANGIOSPERMS

Angiosperms include most of our food crops, such as fruits, vegetables, nuts, berries, beans, and grains.

Melon cactus (*Melocactus intortus*)

Bread wheat (*Triticum aestivum*)

MONOCOTS VS. DICOTS

Botanists classify angiosperms into two groups: monocots and dicots. A **monocot** embryo has one **cotyledon**, the first leaf (or leaves) to emerge from a sprouted seed. A **dicot** embryo has two cotyledons. There are several other important structural differences between monocots and dicots.

MONOCOT

One cotyledon

Fibrous root system

Leaf veins usually parallel

Floral parts usually in multiples of three

DICOT

Two cotyledons

Taproot usually present

Leaf veins usually branched

Floral parts usually in multiples of four or five

Worldwide, corn provides more calories to humans than any other plant.

Calamagrostis acutiflora, an ornamental grass

Scarlet runner bean (*Phaseolus coccineus*)

Coconut palms (*Cocos nucifera*)

CORE IDEA: Angiosperms dominate the modern landscape. The key adaptation of angiosperms is the flower, a reproductive organ that protects the developing seed. Two groups of angiosperms, monocots and dicots, differ in their anatomy.

CORE QUESTION: What important reproductive feature do all angiosperms—from grasses to trees—share?

ANSWER: All angiosperms use a flower as their reproductive organ.

221

Flowers, fruit, and seeds aid angiosperm reproduction

If you give your loved one a bouquet of flowers, or a basket of fruit, what is the true nature of your gift? Flowers and fruit—their appealing smells, colors, and tastes—are all about one thing: reproduction. From roses to ragweed, flowers display and make available a plant's sex organs. The development of tasty fruit helps to spread seeds (plant embryos) around the landscape. In fact, plants have evolved many mechanisms that use other forces—wind, water, and animals—to reproduce their own kind.

ANATOMY OF A FLOWER

Several parts of a flower—sepals, petals, stamens, and carpels—are actually modified leaves.

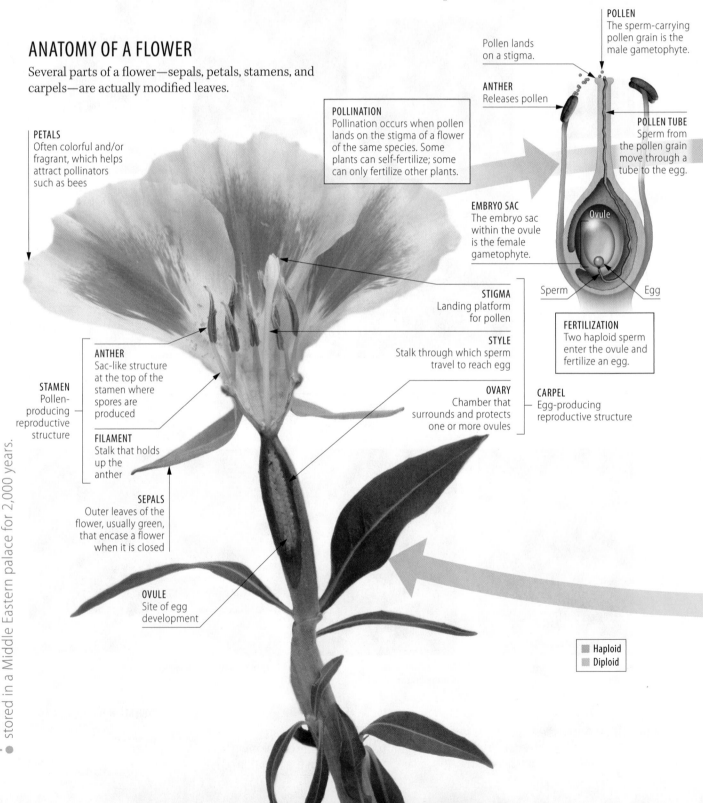

POLLEN
The sperm-carrying pollen grain is the male gametophyte.

Pollen lands on a stigma.

ANTHER
Releases pollen

POLLEN TUBE
Sperm from the pollen grain move through a tube to the egg.

POLLINATION
Pollination occurs when pollen lands on the stigma of a flower of the same species. Some plants can self-fertilize; some can only fertilize other plants.

EMBRYO SAC
The embryo sac within the ovule is the female gametophyte.

Ovule

Sperm   Egg

FERTILIZATION
Two haploid sperm enter the ovule and fertilize an egg.

PETALS
Often colorful and/or fragrant, which helps attract pollinators such as bees

STIGMA
Landing platform for pollen

STYLE
Stalk through which sperm travel to reach egg

OVARY
Chamber that surrounds and protects one or more ovules

CARPEL
Egg-producing reproductive structure

STAMEN
Pollen-producing reproductive structure

ANTHER
Sac-like structure at the top of the stamen where spores are produced

FILAMENT
Stalk that holds up the anther

SEPALS
Outer leaves of the flower, usually green, that encase a flower when it is closed

OVULE
Site of egg development

Haploid
Diploid

In 2005, botanists successfully germinated a date palm seed that had been stored in a Middle Eastern palace for 2,000 years.

FRUIT

After fertilization, the ovary expands and thickens to form a protective container around the seeds. The mature ovary, now called a **fruit**, protects seeds and aids in their dispersal by attracting animals. After eating the fruit, an animal will usually deposit the indigestible seed (along with a supply of fertilizer) some distance away from the parent plant. Fruits, like flowers, are found only in angiosperms.

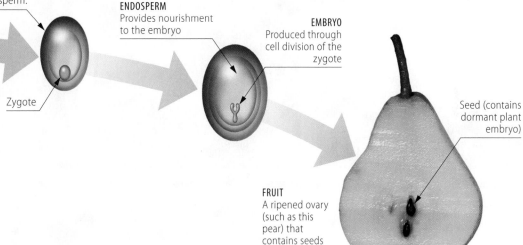

Fruits can be hard and dry (such as beans, nuts, and grains) or soft and fleshy (such as grapes, bananas, and apples).

One haploid sperm fertilizes the haploid egg to create a diploid zygote. The other sperm fuses with another cell to form the endosperm.

Zygote

ENDOSPERM
Provides nourishment to the embryo

EMBRYO
Produced through cell division of the zygote

FRUIT
A ripened ovary (such as this pear) that contains seeds

Seed (contains dormant plant embryo)

Cotyledon (seed leaf)

GERMINATION
When conditions are suitable, seed germination begins: The seed takes up water, expands, and the embryo within resumes its growth.

SEED
An embryonic plant, packed with food, inside a protective container

Cotyledon(s)

Endosperm (food supply, absorbed during germination)

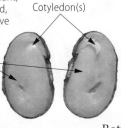

SEEDS

Both gymnosperms (such as pine trees and other conifers) and angiosperms (flowering plants) have seeds. But only the seeds of angiosperms are protected within the ovary, which provides nourishment. After dispersal, seeds can remain dormant for long periods until favorable conditions exist. At that point, the seed germinates (sprouts) and the embryonic plant emerges as a seedling.

CORE IDEA: The flower is the reproductive organ of angiosperms. Structures within the flower produce pollen, receive pollen in fertilization, and house and protect the developing embryo, which is eventually dispersed as a seed.

CORE QUESTION: Write one sentence that describes the relationship between flower, fruit, and seed.

ANSWER: After fertilization, the ovule of a flower develops into a fruit, which encloses the seed.

Angiosperms grow in length and in thickness

Most animals grow to a certain size and then stop. Many plants, on the other hand, grow as long as they live. Plants can thereby achieve great sizes and cover large areas. Angiosperms display three different life spans; they can be annuals, biennials, or perennials. During their lifetime, angiosperms grow in two different directions: in length (called primary growth) and in thickness (called secondary growth).

ANGIOSPERM LIFE SPANS

Flowering plants can be grouped into three categories, depending on their life spans.

ANNUALS

Annuals such as corn germinate from seeds, grow, mature, reproduce, and die in a single growing season.

Corn seedling

Flowering

Fruiting

YEAR ONE

BIENNIALS

Biennials like carrots live for two years, with flower and seed production occurring during the second year.

Carrot seedlings

Fruiting

Flowering

YEAR ONE

YEAR TWO

PERENNIALS

Perennials like blueberries live and reproduce via flower and fruit for many years.

Flowering

Fruiting

EVERY SPRING

EVERY SUMMER

9.13

Analysis of tree rings provides important evidence about global climate change.

LENGTHENING BY PRIMARY GROWTH

In all angiosperms, tissues called **meristems** are responsible for growth. A meristem contains unspecialized cells that, when conditions permit, can generate new cells via cell division (mitosis). Meristem tissues at the tips of roots and stems enable a plant to grow in length. This is called **primary growth**.

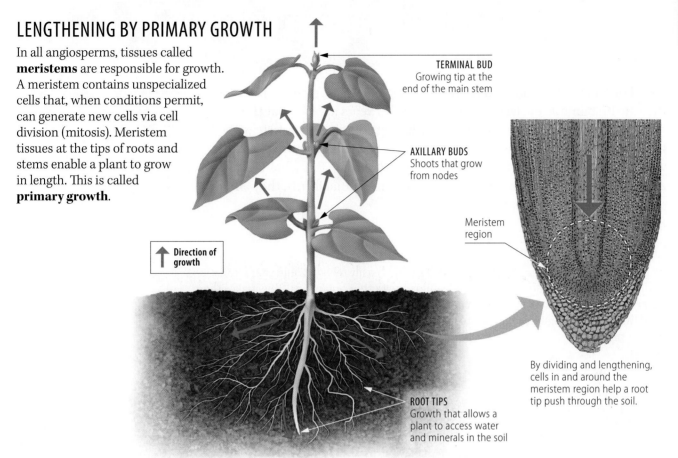

TERMINAL BUD
Growing tip at the end of the main stem

AXILLARY BUDS
Shoots that grow from nodes

Direction of growth

Meristem region

ROOT TIPS
Growth that allows a plant to access water and minerals in the soil

By dividing and lengthening, cells in and around the meristem region help a root tip push through the soil.

THICKENING BY SECONDARY GROWTH

Secondary growth is the thickening of stems and roots. Each year, meristem tissues produce new layers of cells, and old cells die. Over the years, the layers accumulate to form wood, which consists of mature, mostly dead xylem tissue that thickens the plant. Secondary growth is responsible for the accumulation of growth rings in a tree trunk. Not only can tree rings help pinpoint the age of a tree, but comparisons of the thickness of the rings can give insight into the climate that existed at the time.

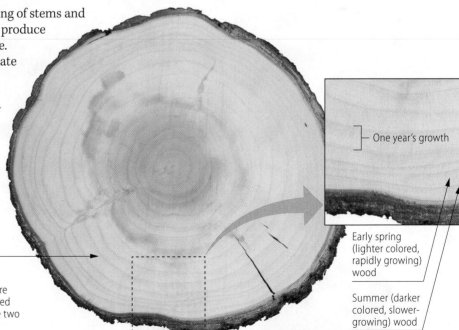

One year's growth

Early spring (lighter colored, rapidly growing) wood

Summer (darker colored, slower-growing) wood

GROWTH RINGS
Growth rings occur because the wood that grows in early spring consists of cells that are larger with thinner walls than those produced during summer. The boundary between the two types of cells is visible as a ring.

CORE IDEA: Flowering plants can be grouped according to whether they grow for just one season (annuals), two seasons (biennials), or for many seasons (perennials). Plants can grow in length (primary growth) and in thickness (secondary growth).

CORE QUESTION: Which type of growth (primary and/or secondary) are meristems responsible for?

ANSWER: Meristems are responsible for both primary (lengthening) and secondary (thickening) growth.

Animals are consumers that evolved from colonial protists

Based on the fossil record and DNA evidence, biologists hypothesize that the first animal evolved around 600 million years ago. At that time, the ancient oceans had abundant life, including colonies of protists with flagella, whiplike appendages used to propel the organism through the water. Such a colony likely evolved into the first animal. But what is an animal? Although there is tremendous variety in animal forms, they all share certain defining characteristics.

DOMAIN BACTERIA

DOMAIN ARCHAEA

The protists (Multiple Kingdoms)

Kingdom Plantae

DOMAIN EUKARYA

Kingdom Fungi

Kingdom Animalia

CHARACTERISTICS OF ANIMALS

All animals share a set of common features. The most easily recognizable characteristic of all animals is the mode of nutrition: All animals eat other organisms. This makes them **heterotrophs**, organisms that obtain nutrients and materials for building body structures from their environment (as opposed to autotrophs, organisms that make their own food, such as plants).

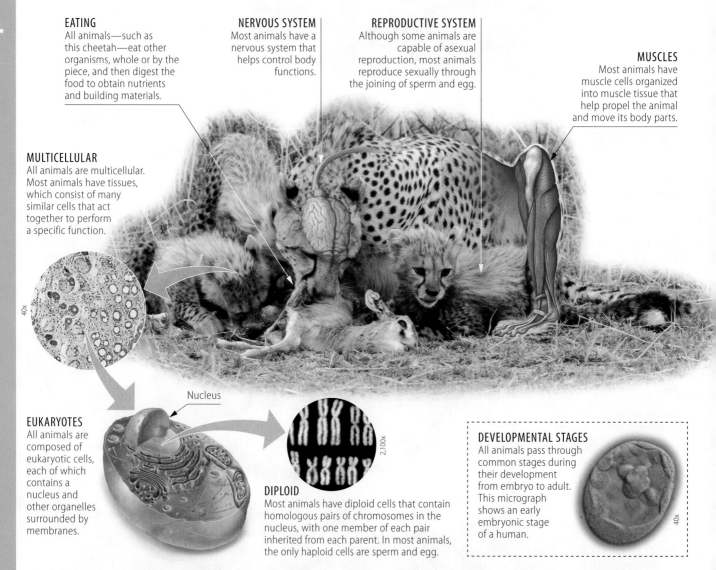

EATING
All animals—such as this cheetah—eat other organisms, whole or by the piece, and then digest the food to obtain nutrients and building materials.

NERVOUS SYSTEM
Most animals have a nervous system that helps control body functions.

REPRODUCTIVE SYSTEM
Although some animals are capable of asexual reproduction, most animals reproduce sexually through the joining of sperm and egg.

MUSCLES
Most animals have muscle cells organized into muscle tissue that help propel the animal and move its body parts.

MULTICELLULAR
All animals are multicellular. Most animals have tissues, which consist of many similar cells that act together to perform a specific function.

40x

Nucleus

EUKARYOTES
All animals are composed of eukaryotic cells, each of which contains a nucleus and other organelles surrounded by membranes.

2,100x

DIPLOID
Most animals have diploid cells that contain homologous pairs of chromosomes in the nucleus, with one member of each pair inherited from each parent. In most animals, the only haploid cells are sperm and egg.

DEVELOPMENTAL STAGES
All animals pass through common stages during their development from embryo to adult. This micrograph shows an early embryonic stage of a human.

40x

ANIMAL EVOLUTION

Fossil evidence and genetic analyses indicate that the first animals evolved from colonies of flagella-equipped protists in the ancient seas approximately 600 million years ago. These first animals evolved to be multicellular and to consume other organisms. Around 540 million years ago, a rapid diversification of animals called the "Cambrian explosion" occurred; in a relatively short time span, a huge variety of animal forms evolved. All modern animals can trace their ancestry to the group of animals that evolved at that time.

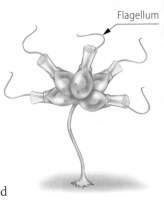

Flagellum

This hypothetical organism may resemble the colonies of protists that evolved into the first animals.

This artist's rendition of the variety of animals that lived in the Cambrian seas is based on fossils from that era.

Rock containing a number of trilobite (*Ellipsocephalus hoffi*) fossils from the middle Cambrian period, about 515 million years ago

ANIMAL PHYLOGENY

Of the 35 or so known animal phyla (the exact number remains a matter of debate), the vast majority of animals living today can be grouped into just nine phyla. The phylogenetic tree shown here presents a hypothesis about the evolutionary relationships between the nine major animal phyla based on structural features and genetic analyses. Each phylum is characterized by specific adaptations, but there are certain broad structural features (such as tissues, symmetry, and body cavities) that represent specific branch points in the evolution of animals.

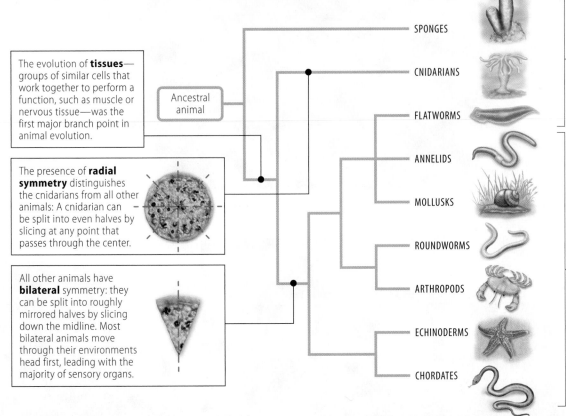

The evolution of **tissues**—groups of similar cells that work together to perform a function, such as muscle or nervous tissue—was the first major branch point in animal evolution.

The presence of **radial symmetry** distinguishes the cnidarians from all other animals: A cnidarian can be split into even halves by slicing at any point that passes through the center.

All other animals have **bilateral** symmetry: they can be split into roughly mirrored halves by slicing down the midline. Most bilateral animals move through their environments head first, leading with the majority of sensory organs.

Ancestral animal

SPONGES
CNIDARIANS
FLATWORMS
ANNELIDS
MOLLUSKS
ROUNDWORMS
ARTHROPODS
ECHINODERMS
CHORDATES

These phyla lack a **body cavity**, a fluid-filled space separating the digestive tract from the body's outer layer.

Most animals have a body cavity. In these animals, the internal organs can grow and move independently of the outer body wall.

Ninety-five percent of all animals are invertebrates, animals that lack a backbone.

CORE IDEA: Animals are multicellular eukaryotes that obtain nutrition by eating other organisms. Animals evolved from colonial protists around 600 million years ago. There are now nine major phyla of animals.

CORE QUESTION: What makes the sponges different from all other animals? (Hint: Look at the branch points near the top of the phylogenetic tree.)

ANSWER: Only the sponges lack tissues. The animals in all other major phyla have tissues.

Sponges and cnidarians have unusual body features

The vast majority of animals are **invertebrates**, animals without backbones. Some invertebrates have such odd body shapes that you might not even recognize them as animals. Two groups of aquatic invertebrates—the sponges and the cnidarians—have particularly unusual forms.

SPONGES

CNIDARIANS

FLATWORMS

ANNELIDS

MOLLUSKS

ROUNDWORMS

ARTHROPODS

ECHINODERMS

CHORDATES

SPONGES

Of the major animal groups living today, the **sponges** have the longest evolutionary history, and they have the simplest anatomy. Sponges come in a wide variety of shapes, sizes, and colors, but all share certain distinctive characteristics. First, they are asymmetrical (irregularly shaped, with no planes of symmetry). Second, they lack tissues, such as nerves and muscles. The lack of specialized tissue requires that every cell in a sponge be in contact with the environment. Third, unlike most animals, sponges are sessile, remaining anchored to one spot.

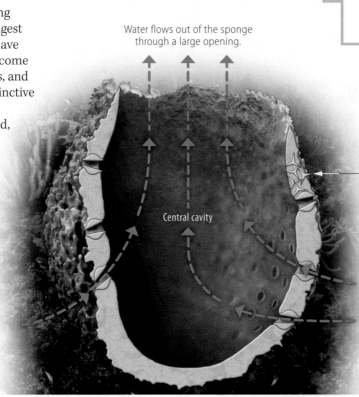

Water flows out of the sponge through a large opening.

Central cavity

Most sponges have skeletons made of small stiff structures.

The body of a sponge resembles a sac with pores. These holes draw nutrient-containing water into a central cavity.

An Australian cnidarian called the sea wasp is the deadliest organism on Earth: One animal contains enough poison to kill 60 people.

A GALLERY OF SPONGES
There are about 9,000 species of sponges. Most live in salt water, but about 100 species of sponges live in fresh water.

Orange elephant ear sponge (*Agelas clathrodes*)

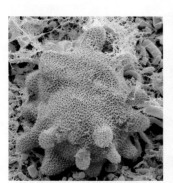

Freshwater sponge (*Spongilla*)

Orange sea sponge (*Tubastrea coccinea*)

Tube sponge (*Aplysina archeri*)

CNIDARIANS

There are around 10,000 species of **cnidarians** (pronounced with a silent "c"), almost all of which are marine (sea dwelling). Cnidarians include sea anemones, hydras, corals, and jellies (sometimes called "jellyfish," but biologists prefer the term "jellies" since they are not fish). Cnidarians have two distinct body forms: a stationary polyp and a floating medusa. Some cnidarians exist only as a polyp, some only as a medusa, and some live both ways at different times of their lives. Cnidarians—like all animals except sponges—have cells organized into tissues, such as nerve tissue that helps coordinate body movements.

The body of an adult cnidarian shows **radial symmetry**, with all body parts circularly arranged around a central axis.

This jelly is a **medusa**, the mobile form of a cnidarian. The medusa floats mouth down, either drifting or moving by contracting its body.

A sea anemone is a **polyp**, the stationary form of a cnidarian. Polyps grow while fixed to larger objects, waiting for prey as they extend their stinging tentacles.

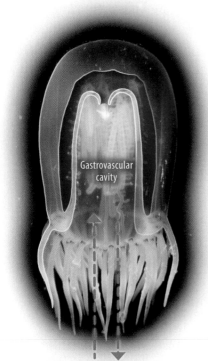

Cnidarians have a central digestive compartment called the **gastrovascular cavity**. Nutrient-containing food enters and waste exits the same opening, which therefore acts as both mouth and anus. Inside the cavity, fluid circulates nutrients and oxygen and helps remove wastes.

Gastrovascular cavity

Gastrovascular cavity

Water moves into and out of the gastrovascular cavity through a single opening.

10x

The most distinguishing feature of cnidarians is the stinging cells on their tentacles, which are used in defense and to paralyze prey (mostly small animals and protists), and to push food into the gastrovascular cavity. These spring-loaded cells contain threads that, when discharged, can cause painful stings to humans unfortunate enough to contact them.

A GALLERY OF CNIDARIANS

Sea anemone (*Alicia mirabilis*)

Mangrove upside-down jellyfish (*Cassiopea xamachana*)

Gray hydra (*Pelmatohydra oligactis*)

Coral animals, such as this boulder brain coral (*Colpophyllia natans*), secrete a hard external skeleton that, over generations, accumulates into the "rocks" we call coral.

CORE IDEA: Sponges are mostly marine organisms that lack body symmetry and tissues. Cnidarians are also primarily marine; they have tissues, radial body symmetry, and stinging cells.

CORE QUESTION: What is the characteristic body shape of a sponge? Of a cnidarian?

ANSWER: Sponges lack body symmetry. Cnidarians are radially symmetrical.

Mollusks and echinoderms primarily occupy marine habitats

Although the phyla of mollusks and echinoderms do not share a close evolutionary kinship, both primarily occupy marine (ocean) habitats. Indeed, all animal life first evolved in the oceans, with terrestrial adaptations coming along millions of years later. Each of these phyla has characteristics that distinguish it from other forms of animal life.

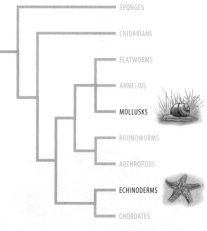

SPONGES

CNIDARIANS

FLATWORMS

ANNELIDS

MOLLUSKS

ROUNDWORMS

ARTHROPODS

ECHINODERMS

CHORDATES

MOLLUSKS

Mollusks are a phylum of soft-bodied animals, many of which are protected by a hard shell. There are 93,000 known species of mollusks, mostly marine. Despite their great variety, most mollusks have a similar body plan.

BILATERAL SYMMETRY
The adult body of all mollusks has a bilaterally symmetrical form, meaning that its body (like your own) can be divided down the middle into two nearly identical halves.

VISCERAL MASS
The **visceral mass** contains most of the internal organs.

SHELL
Many mollusks have an external shell. Some mollusks—such as slugs, squids, and octopi—have a small internal shell or no shell at all.

MANTLE
The **mantle** encloses the visceral mass and secretes the shell (if present).

FOOT
Most mollusks use a muscular **foot** to move.

CIRCULATORY SYSTEM
All mollusks have a circulatory system that pumps blood and distributes nutrients and oxygen through the body.

500x

RADULA
Many mollusks feed using a **radula**, a saw-like organ that can cut and scrape up food.

A GALLERY OF MOLLUSKS

GASTROPODS (SNAILS AND SLUGS)
Gastropods are the most numerous group of mollusks. They are found in salt- and freshwater habitats, and they include slugs, the only mollusks that live on land.

BIVALVES (CLAMS, MUSSELS, OYSTERS)
Bivalves are found in marine and freshwater environments. Most are sedentary and use their foot to dig into sand or mud. All bivalves have two shells connected by a hinge.

CEPHALOPODS (SQUIDS AND OCTOPI)
Cephalopods are fast and agile marine predators. Some (like the nautiluses) have large, heavy shells; some (like squids) have small internal shells; and some (like octopi) have no shells at all.

Blackmargin sea slug (*Glossodoris atromarginata*)

Bay scallop (*Argopecten irradians*)

European squid (*Loligo vulgaris*)

ECHINODERMS

All 7,000 species of **echinoderms** live in the ocean, and most move slowly. The phylum is named after the Greek words for "spiny skin" because all echinoderms have bumpy surfaces. Chordates (which include the vertebrates such as humans) share an evolutionary branch with echinoderms, making them the phylum most closely related to our own.

BODY SYMMETRY
Most echinoderms are radially symmetrical as adults—with parts radiating from the center of the body like spokes on a wheel—but they have bilateral symmetry as larvae.

SPINY SURFACE
Most echinoderms have a bumpy or spiky surface.

WATER VASCULAR SYSTEM
Echinoderms—and only echinoderms—have a circulatory system that uses seawater as blood. A network of water-filled canals allows the animal to exchange gases (O_2 in and CO_2 out) with the environment.

ENDOSKELETON
Echinoderms have a calcium-hardened **endoskeleton** (internal skeleton) that becomes visible when the organism is dead and dried out.

TUBE FEET
Sea stars and sea urchins move slowly across the seafloor using **tube feet**, extensions of the water-filled canals that end in tiny suction cups. Sea stars also use tube feet to grasp prey.

MOUTH
A sea star feeds by pulling apart the two shells of a bivalve and then pushing its stomach through the opening. The sea star's stomach digests its prey inside the mollusk shell.

REGENERATION
Sea stars and some other echinoderms can regenerate lost body parts, provided that a piece of the central body remains.

Not all sea stars have five legs; the crown-of-thorns sea star has up to 21 of them.

A GALLERY OF ECHINODERMS

Red sea urchin (*Astropyga radiata*)

Bennett's feather star (*Oxycomanthus bennetti*)

Brittle star (*Ophiothrix fragilis*)

Donkey dung sea cucumber (*Holothuria mexicana*)

CORE IDEA: Mollusks are a diverse group of bilaterally symmetrical soft-bodied animals, many with shells. Echinoderms are exclusively marine, have bumpy skin, a seawater circulatory system, and are often radially symmetrical as adults.

CORE QUESTION: Both cnidarians and echinoderms are primarily found in the ocean. Does that mean that these two groups are closely related?

ANSWER: No. Viewing the evolutionary tree at the top of the left page shows that these two phyla are distantly related.

Three phyla of worms have evolved unique structures

The animal kingdom contains three phyla of worms: flatworms, annelids, and roundworms. Although each has the same overall body shape (bilaterally symmetrical, long, and thin), the three different groups of worms vary considerably, spanning a size range from microscopic to several feet long. And, reflecting their evolutionary heritage, each group has distinctive features that set it apart from the others.

SPONGES
CNIDARIANS
FLATWORMS
ANNELIDS
MOLLUSKS
ROUNDWORMS
ARTHROPODS
ECHINODERMS
CHORDATES

FLATWORMS

Flatworms are members of the phylum Platyhelminthes. About 20,000 different species of flatworms can be found living in marine, freshwater, and damp terrestrial habitats, often attached to the undersurface of rocks. Their ribbonlike bodies can range in size from 1 mm to dozens of feet. Most flatworms have a **gastrovascular cavity** with a single opening that acts as both mouth and anus. Several species of flatworms—including tapeworms and blood flukes—are disease-causing parasites in humans.

BILATERAL SYMMETRY
Worms from all three phyla display bilateral symmetry, with nearly mirror-imaged body parts arranged on either side of the central axis.

EYE SPOTS
Some flatworms have crude sensory structures. The planarian shown here has eye spots that are able to distinguish light from dark. Planarians also have clusters of nerve cells that act as a simple brain.

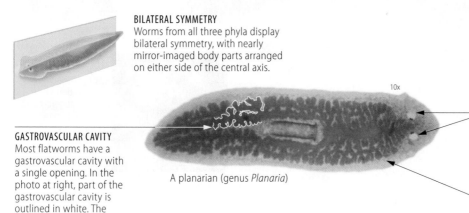

10x

GASTROVASCULAR CAVITY
Most flatworms have a gastrovascular cavity with a single opening. In the photo at right, part of the gastrovascular cavity is outlined in white. The cavity is highly branched, allowing the delivery of nutrients to every cell.

A planarian (genus *Planaria*)

NO BODY CAVITY
Flatworms, unlike other types of worms, have no internal body cavity (a fluid-filled space separating the digestive tract from the body's outer layer).

SEX ORGAN
Each unit of the tapeworm body contains eggs. The units break off and exit the host in its feces. If ingested by another host, the unit can hatch into a new tapeworm.

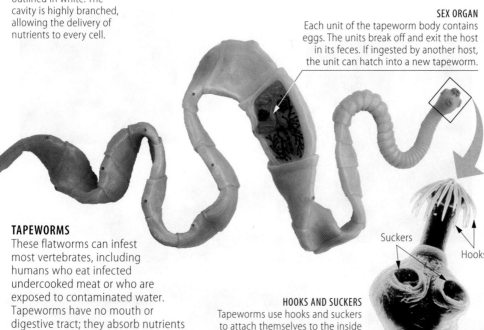

TAPEWORMS
These flatworms can infest most vertebrates, including humans who eat infected undercooked meat or who are exposed to contaminated water. Tapeworms have no mouth or digestive tract; they absorb nutrients directly through their body surface.

Suckers

Hooks

HOOKS AND SUCKERS
Tapeworms use hooks and suckers to attach themselves to the inside walls of the host's intestine.

15x

Adult tapeworm (*Taenia asiatica*) held in a gloved hand

ANNELIDS

Annelids display a key evolutionary adaptation: **segmentation**, the division of the body along its length into a series of repeated segments. Segmentation allows for greater flexibility and more complex movements. Additionally, annelids have two other characteristic features: a body cavity and a complete digestive tract. There are about 16,000 species of annelids found in marine, freshwater, and damp terrestrial habitats.

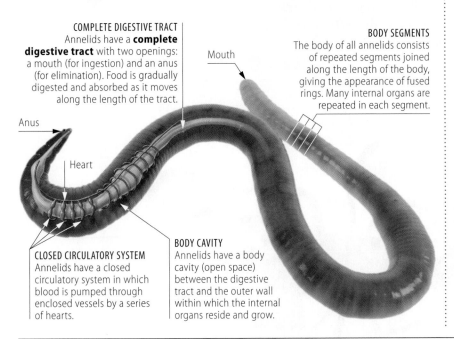

COMPLETE DIGESTIVE TRACT
Annelids have a **complete digestive tract** with two openings: a mouth (for ingestion) and an anus (for elimination). Food is gradually digested and absorbed as it moves along the length of the tract.

Mouth

Anus

Heart

CLOSED CIRCULATORY SYSTEM
Annelids have a closed circulatory system in which blood is pumped through enclosed vessels by a series of hearts.

BODY CAVITY
Annelids have a body cavity (open space) between the digestive tract and the outer wall within which the internal organs reside and grow.

BODY SEGMENTS
The body of all annelids consists of repeated segments joined along the length of the body, giving the appearance of fused rings. Many internal organs are repeated in each segment.

EARTHWORM
As an earthworm burrows underground, soil passes through its digestive tract. Nutrients are removed, and the wastes are excreted as worm castings. This action improves the soil for plants, making earthworms a valuable addition to any garden or farm.

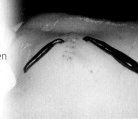

LEECH
Most leeches are free-living carnivores that eat insects and other small animals. Some species use teeth to create tiny slices in the skin of animals, then attach and suck blood. Leeches are used today to remove pooled blood from digits that have been reattached after accidental amputations.

POLYCHAETES
The polychaetes are mostly marine annelids. They have feathery appendages that can trap food in seawater.

ROUNDWORMS

Nematodes, or roundworms, are a phylum of small worms with cylindrical bodies that are tapered at both ends. Unlike annelids, roundworms are not segmented. There are over 25,000 known species of roundworms that live in a wide variety of habitats, and some biologists suspect that there are up to 10 times that many that remain undescribed. All nematodes have a complete digestive tract, with two openings (mouth and anus), through which food passes and is digested. Roundworms include many parasitic species that live within the bodies of other organisms. In soil, roundworms help recycle nutrients by breaking down decaying organic matter.

HEARTWORMS
The heartworm (*Dirofilaria immitis*) is a common parasite of dogs and cats. Spread by mosquitoes, it can be deadly if untreated.

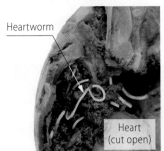

Heartworm

Heart (cut open)

TRICHINELLA
Humans are host to at least 50 species of roundworm parasites. *Trichinella spiralis*, usually acquired by eating infected undercooked pork, causes the disease trichinosis, which begins in the gastrointestinal tract but persists in muscle tissue.

800x

Trichinella roundworm

Muscle tissue

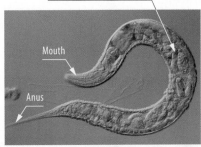

Complete digestive tract

Mouth

Anus

100x

C. ELEGANS
Many roundworms are free-living. This photo shows a *Caenorhabditis elegans*, a well-studied research organism—the lab rat of the worm world. *C. elegans* is a good research organism because its body is transparent and therefore well suited to study via microscope.

Parasitic tapeworms can grow to lengths of over 6 feet in the human intestine.

CORE IDEA: There are three major phyla of worms. Flatworms, some of which can parasitize humans, lack a digestive tract. Annelids—with bodies made of fused repeating segments—and roundworms both have complete digestive tracts.

CORE QUESTION: What is the fundamental difference between a gastrovascular cavity and a digestive tract?

ANSWER: A gastrovascular cavity has only one opening through which materials both enter and leave; materials move only one way through a digestive tract.

The arthropods are extremely diverse and numerous

With over a million identified species, and over a billion billion individuals, the **arthropods** are by far the most numerous and diverse phylum of animals. Despite such tremendous diversity, all arthropods share some basic characteristics: They all have segmented bodies, a tough exoskeleton, and several jointed appendages that perform specific functions such as walking, feeding, or swimming.

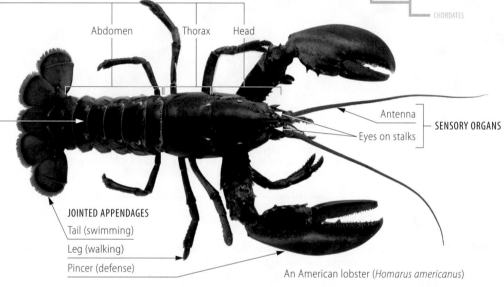

SPONGES
CNIDARIANS
FLATWORMS
ANNELIDS
MOLLUSKS
ROUNDWORMS
ARTHROPODS
ECHINODERMS
CHORDATES

FEATURES OF A TYPICAL ARTHROPOD

BODY SEGMENTS
An arthropod body is divided into segments along its length. Within the segments are specialized structures that perform specific functions.

EXOSKELETON
A hard exoskeleton made from protein and a polysaccharide called chitin provides protection and points of attachment for muscles. To grow in size, an arthropod must grow a new exoskeleton and then molt, shedding its old one.

Abdomen   Thorax   Head

Antenna
Eyes on stalks — **SENSORY ORGANS**

JOINTED APPENDAGES
Tail (swimming)
Leg (walking)
Pincer (defense)

An American lobster (*Homarus americanus*)

ARACHNIDS

Arachnids are a group of eight-legged arthropods that include spiders, scorpions, ticks, and mites. Most are terrestrial carnivores that use a pair of feeding appendages to capture prey. Some use fangs loaded with venom to paralyze prey and can thus be harmful or even deadly to humans.

Reproductive appendages

Feeding appendages with fangs

Four pairs of walking legs

Silk-spinning appendages

Yellow desert scorpion (*Leiurus quinquestriatus*), also know as the deathstalker

Chilean rose tarantula (*Grammostola rosea*)

Female Rocky Mountain wood tick (*Dermacentor andersoni*)
10x

Dust mite (*Dermatophagoides pteronyssinus*)
200x

For every human on Earth, there are about 100 million insects.

234

CRUSTACEANS

Nearly all **crustaceans** (except for pill bugs) live in the water. They are, in fact, the dominant group of arthropods in marine and freshwater habitats. Many of them (shrimp, lobster, crabs, etc.) are valuable food crops.

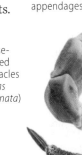

Brown crab (*Cancer pagurus*)

Two feeding appendages

Banded cleaner shrimp (*Stenopus hispidus*)

Goose-necked barnacles (*Lepas pectinata*)

Common rough woodlouse (*Porcellio scaber*), a type of pill bug

Crayfish (*Procambarus clarkii*)

Multiple pairs of locomotor appendages

INSECTS

Insects outnumber all other animals combined, by far. They are found on land and in freshwater. Most have easily recognizable segments (head, thorax, and abdomen), three pairs of legs, and one pair of antennae; many have two pairs of wings.

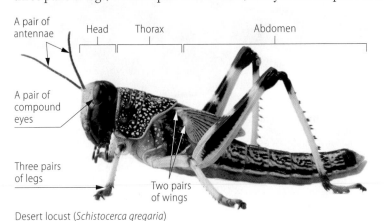

A pair of antennae

Head   Thorax   Abdomen

A pair of compound eyes

Three pairs of legs

Two pairs of wings

Desert locust (*Schistocerca gregaria*)

The tremendous diversity of insects is obvious in this collection of beetles at the Natural History Museum in London.

MILLIPEDES AND CENTIPEDES

Millipedes and **centipedes** are easily recognizable due to their obviously segmented bodies, which appear as many fused rings, each with jointed appendages attached. All are land dwelling. The two groups can be distinguished by the number of legs on each body segment.

One pair of legs per body segment

A pair of venomous claws on the head of an Amazonian giant centipede (*Scolopendra gigantea*) is used to subdue prey.

Two pairs of legs per body segment

A *Polydesmida* millipede

CORE IDEA: The arthropods include more individuals and species than any other animal phylum. Common groups of arthropods include crustaceans (such as lobsters), arachnids (such as spiders), insects (such as grasshoppers), and centipedes and millipedes.

CORE QUESTION: The body of many arthropods can be divided into what three sections?

ANSWER: Head, thorax, and abdomen.

Vertebrates belong to the chordate phylum

We humans belong to the phylum of **chordates**. Within this phylum are two broad divisions: the invertebrate chordates and the vertebrates (the sub-phylum to which we belong). The chordates in general and the vertebrates specifically can be recognized by several important structural features.

CHORDATES

Chordates are identified by four key features. All of these features are apparent in chordate embryos, but some of them are difficult to recognize in the adult.

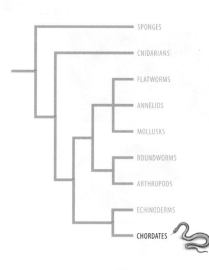

SPONGES

CNIDARIANS

FLATWORMS

ANNELIDS

MOLLUSKS

ROUNDWORMS

ARTHROPODS

ECHINODERMS

CHORDATES

A hollow **nerve cord** runs along the dorsal (top) surface of the back. In adult humans, this becomes part of the spinal cord.

All chordate embryos have a **tail** that extends backward from the anus. Human embryos have a tail, but in adults this persists only as the "tailbone" of the backbone.

A **notochord** is a flexible rod that provides support along the length of the back. In adult humans, this develops into a component of the spinal column.

Pharyngeal slits are grooves located just behind the mouth. In adult humans, part of these structures remain as the eustachian tubes, which link the ears to the throat.

INVERTEBRATE CHORDATES

Most chordates are vertebrates (animals with backbones). However, two groups of animals—lancelets and tunicates—are chordates but lack a skull and backbone.

INVERTEBRATE CHORDATES LACK SKULLS AND BACKBONES

Lancelets (such as this *Branchiostoma lanceolatum*) are small, sword-shaped animals that burrow into the sand and capture food by filtering water through their mouths.

Tunicates (such as these *Atriolum robustum* and *Clavelina*), also called sea squirts, adhere to stationary objects (rocks, coral, boats). They feed by filtering seawater.

VERTEBRATES

Except for lancelets and tunicates, all chordates are **vertebrates**, animals with backbones. In addition to the four features common to all chordates, all vertebrates have an **endoskeleton** (an internal skeleton) that includes a backbone and a skull. One group of vertebrates, the hagfishes, lack recognizable backbones as adults. But all other vertebrates have prominent backbones that provide strong support, as is apparent in this lizard skeleton.

Hagfishes have become endangered because their skin is used to make faux-leather "eel-skin" clothing.

Key characteristics of all vertebrates

Backbone (vertebrae)   Skull

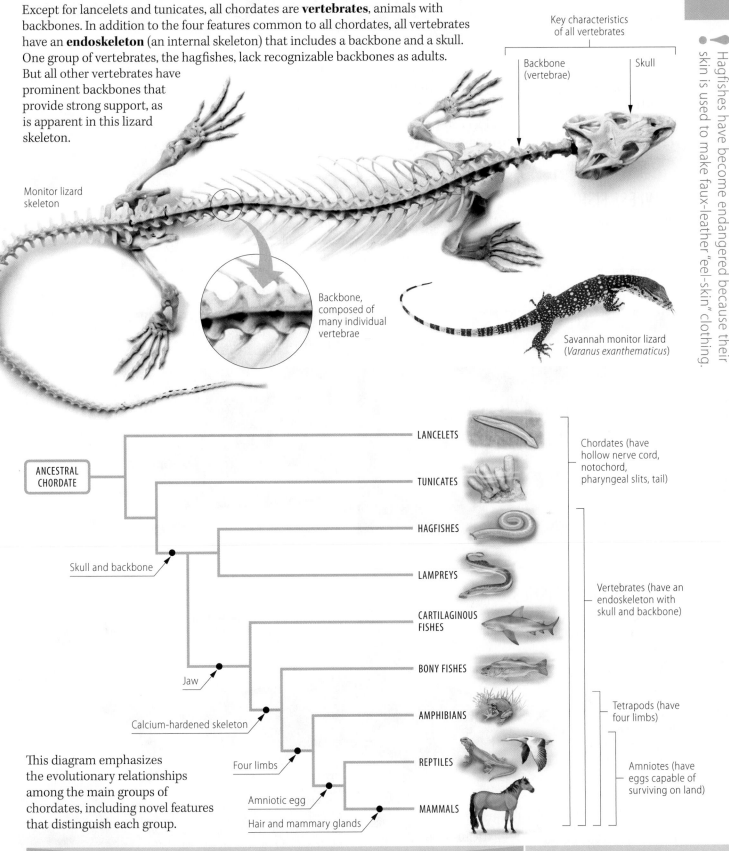

Monitor lizard skeleton

Backbone, composed of many individual vertebrae

Savannah monitor lizard (*Varanus exanthematicus*)

ANCESTRAL CHORDATE

Skull and backbone

Jaw

Calcium-hardened skeleton

Four limbs

Amniotic egg

Hair and mammary glands

LANCELETS

TUNICATES

HAGFISHES

LAMPREYS

CARTILAGINOUS FISHES

BONY FISHES

AMPHIBIANS

REPTILES

MAMMALS

Chordates (have hollow nerve cord, notochord, pharyngeal slits, tail)

Vertebrates (have an endoskeleton with skull and backbone)

Tetrapods (have four limbs)

Amniotes (have eggs capable of surviving on land)

This diagram emphasizes the evolutionary relationships among the main groups of chordates, including novel features that distinguish each group.

CORE IDEA: Chordates share four features (hollow nerve cord, notochord, pharyngeal slits, tail), all of which are visible in the embryo. Most chordates are vertebrates, meaning they also have an endoskeleton with a backbone and skull.

CORE QUESTION: Which group of animals includes the other: chordates or vertebrates?

ANSWER: The chordates include the vertebrates; all vertebrates are chordates, but there are some chordates that are not vertebrates.

The first vertebrates to evolve were fishes

The first vertebrates evolved around 540 million years ago (the early Cambrian period) in the oceans. They were so successful that the Devonian period (from about 416 to about 360 million years ago) is called the "Age of Fish." Today, the descendants of these early vertebrates are represented by several lineages, each with its own characteristic structures.

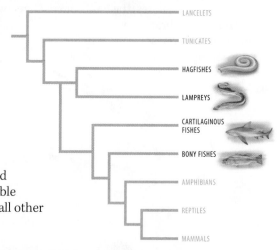

LANCELETS

TUNICATES

HAGFISHES

LAMPREYS

CARTILAGINOUS FISHES

BONY FISHES

AMPHIBIANS

REPTILES

MAMMALS

JAWLESS FISHES

Two groups of fishes living today have skulls but lack jaws: hagfishes and lampreys. Hagfishes (about 40 species) have a skull but lack a recognizable backbone in the adult. Lampreys have a skull and backbone, but unlike all other vertebrates, they lack a jaw.

FISHES WITH SKULLS BUT NO JAWS

NO RECOGNIZABLE BACKBONE

HAGFISH
Hagfishes (such as this *Eptatretus stoutii*) scavenge along the seafloor. They are nearly blind, but have excellent senses of touch and smell. When threatened, a hagfish can exude slime from special glands on the side of its body.

No recognizable backbone

Skull

No jaw

BACKBONE

LAMPREY
Lampreys, unlike most vertebrates, lack jaws. Most lamprey species live as parasites by attaching suckers to the sides of larger fish and feeding on blood.

Skull

No jaw

Backbone

Brook lamprey (*Lampetra planeri*)

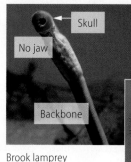

Close-up of a lamprey's sucking mouth

CARTILAGINOUS FISHES

The first vertebrates with jaws were **cartilaginous fishes**. They evolved around 470 million years ago and have changed little in the past 300 million years. Cartilaginous fishes have flexible skeletons made of cartilage (the same material in your nose) and use two pairs of fins and a tail to swim, making them fast and agile. Most of the 800 species of living cartilaginous fishes, including sharks and rays, are marine carnivores.

JAWS
Nearly all vertebrates have jaws supported by a hinged pair of bones.

LATERAL LINE SYSTEM
Sharks use a **lateral line system**, a row of sensory organs running along each side of the body, to detect changes in water pressure produced by nearby prey.

HAMMERHEAD SHARK
A hammerhead shark (*Sphyrna mokarran*) has strong jaws, keen senses, and powerful fins, making them fearsome predators.

SKELETON
The skeleton is made of flexible cartilage.

ELECTROSENSORS
Many sharks have electrosensors that can detect electric fields produced by muscle contractions of nearby prey.

GILLS
Cartilaginous fishes must constantly move in order to keep oxygenated water flowing through the gills.

STINGRAY
The bodies of rays (such as this southern stingray, *Dasyatis americana*) are adapted for life on the seafloor, with flat bodies and eyes on the top of the head. Their tails often bear sharp venomous spines that can be painful but are rarely fatal.

After lampreys invaded the Great Lakes via canals in the 1930s, they decimated many species of fishes.

BONY FISHES

There are more species of **bony fishes** than any other group of vertebrates. In fact, they are found in virtually every explored aquatic habitat, both freshwater and marine. Unlike cartilaginous fishes, bony fishes have skeletons that are reinforced with calcium, a swim bladder that helps maintain buoyancy, and an operculum that circulates water over the gills.

OPERCULUM
A protective flap called the **operculum** covers a chamber housing the gills just behind either side of the head. By moving the operculum, a bony fish can keep a steady supply of oxygenated water flowing over the gills even if the fish is not moving.

EYES
Bony fishes have keen senses of smell and eyesight.

SWIM BLADDER
The **swim bladder** is a gas-filled sac that helps keep a fish buoyant. A bony fish can therefore remain nearly motionless in the water, conserving energy.

GILLS
Gills are feathery organs that extract oxygen from the surrounding water.

SCALES
Most bony fishes have flattened scales covering their skin.

LATERAL LINE SYSTEM
Like the cartilaginous fishes, bony fishes have a lateral line system that can detect movement in the surrounding water.

Flying gurnard
(*Dactyloptena orientalis*)

Chevron barracuda
(*Sphyraena genie*)

Yellow-ribbon sweetlips
(*Plectorhinchus polytaenia*)

Thorny seahorse
(*Hippocampus histrix*)

LOBE-FINNED FISHES

The members of one evolutionary branch of the bony fishes, called the **lobe-finned fishes**, have muscular fins supported by rod-shaped bones that are homologous to amphibian limb bones. The lobefins include the lungfishes, several species of which can be found in the Southern Hemisphere. This evolutionary branch also includes a lineage that adapted to life on land and gave rise to amphibians.

COELACANTH
The coelacanth (*Latimeria chalumnae*) shown at left is a deep-sea dwelling lobe-finned fish once thought to be extinct until one was discovered by a biologist in a fish market in South Africa.

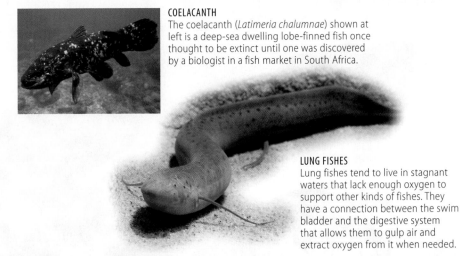

LUNG FISHES
Lung fishes tend to live in stagnant waters that lack enough oxygen to support other kinds of fishes. They have a connection between the swim bladder and the digestive system that allows them to gulp air and extract oxygen from it when needed.

CORE IDEA: There are several lineages of modern fishes, including hagfishes (which lack jaws and recognizable backbones), lampreys (which lack jaws), cartilaginous fishes (with cartilage skeletons), and bony fishes (with calcium-hardened skeletons).

CORE QUESTION: What two features allow a trout (but not a shark) to remain still in the water?

ANSWER: The buoyancy of the swim bladder allows bony fishes to stay still in the water, and an operculum allows bony fishes to extract oxygen as they do so.

Amphibians and reptiles were the first tetrapods to occupy land

All terrestrial (land-dwelling) vertebrates are **tetrapods** (literally "four feet"), animals with four limbs. The first tetrapods were amphibians descended from lobe-finned fishes, whose primitive lungs and muscular fins supported by strong bones enabled them to walk on land. All modern land tetrapods—including amphibians and reptiles—evolved from these first tetrapods and therefore have similar body plans.

AMPHIBIANS

Befitting their place as transitional organisms, **amphibians** exhibit a blend of aquatic and terrestrial adaptations. As the first land-dwelling vertebrates, the amphibians spread quite successfully across the ancient landscape, although they were limited by their need to breed in water. Today, most of the 6,000 species of amphibians are found primarily in moist habitats.

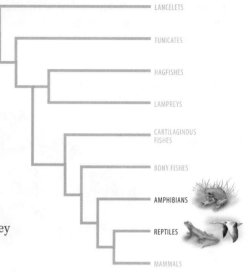

LANCELETS
TUNICATES
HAGFISHES
LAMPREYS
CARTILAGINOUS FISHES
BONY FISHES
AMPHIBIANS
REPTILES
MAMMALS

A GALLERY OF AMPHIBIANS

FROGS
Frogs have stout bodies, bulging eyes, and powerful hind legs that allow hopping.

TOADS
Toads are frogs that have rough skin and live entirely on land.

CAECILIANS
Because they lack limbs, this order of amphibians resembles worms or snakes.

SALAMANDERS
Some salamanders are entirely aquatic, but others walk on land with a distinctive side-to-side bending of the body.

AQUATIC ADAPTATIONS

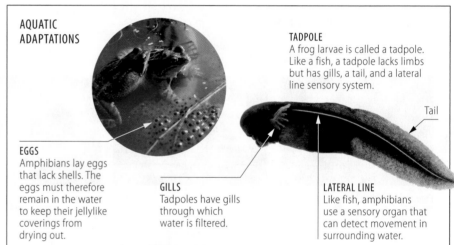

TADPOLE
A frog larvae is called a tadpole. Like a fish, a tadpole lacks limbs but has gills, a tail, and a lateral line sensory system.

Tail

EGGS
Amphibians lay eggs that lack shells. The eggs must therefore remain in the water to keep their jellylike coverings from drying out.

GILLS
Tadpoles have gills through which water is filtered.

LATERAL LINE
Like fish, amphibians use a sensory organ that can detect movement in surrounding water.

TERRESTRIAL ADAPTATIONS

LUNGS
Like all tetrapods, amphibians breathe air through lungs.

MOIST SKIN
Adult amphibians have moist skin that supplements the lungs by allowing gas exchange (O_2 in, CO_2 out).

MUSCULOSKELETAL SYSTEM
Amphibians, like all tetrapods, have muscular and skeletal systems that provide support and aid in locomotion.

Adult amphibians have limbs supported by strong bones.

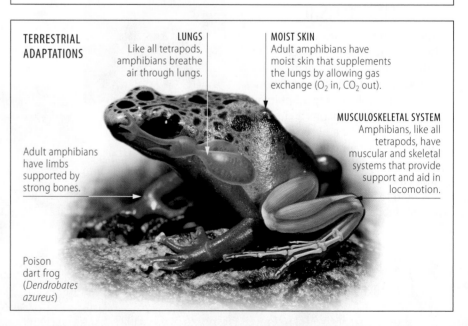

Poison dart frog (*Dendrobates azureus*)

REPTILES

The **reptiles** (which include birds) display some key evolutionary adaptations that enable them to complete their entire life cycle on land. Like the amphibians, reptiles are tetrapods. But reptiles (and mammals) are amniotes, animals that develop inside an amniotic egg that provides a life-support system for the embryo. The amniotic egg and other adaptations free reptiles from their ancestral ties to aquatic habitats, allowing them to occupy a much broader swath of the landscape.

AMNIOTES

Reptiles produce fluid-filled **amniotic eggs** that have a hard, waterproof shell surrounding a fluid-filled sac called the amnion. This provides an aquatic habitat (in a sense, a miniature pond) that encloses the developing embryo, allowing it to develop on land.

Grass snake (*Natrix natrix*) guarding its eggs

WATERPROOF SKIN

The skin of reptiles is covered with scales waterproofed by a protein called keratin. Scales help prevent dehydration. Unlike amphibians, reptiles cannot exchange gases across their skin and rely solely on their lungs.

Galapagos marine iguana (*Amblyrhynchus cristatus*)

The ostrich, the world's largest bird, can run over 40 miles per hour, covering up to 16 feet in a single stride.

ECTOTHERMS

Most non-bird reptiles are **ectotherms**, meaning they must obtain heat from the environment by, for example, basking in the sun. Because they don't expend energy on generating body heat, reptiles can survive on many fewer calories than other animals, but they are limited to warm climates.

Crocodilians include crocodiles (such as this Indian marsh crocodile, *Crocodylus palustris*) and alligators. They are the largest living reptiles. They spend most of their time in water, breathing through upturned nostrils. Saltwater crocodiles can be up to 20 feet long and weigh a ton.

ENDOTHERMS

Unlike other reptiles, birds are **endotherms**, meaning their metabolism provides heat to maintain a warm, constant body temperature.

FLIGHT

Nearly all birds fly, although there are a few species of flightless birds, including penguins, ostriches, and emus.

Emu (*Dromaius novaehollandiae*)

FEATHERS

Feathers are made of keratin, the same protein that makes up the scales of other reptiles. They aid in flight and provide insulation and waterproofing.

Powerful breast muscles power wings in flight.

Red-tailed Minla (*Minla ignotincta*)

Birds have strong, light, honeycombed bones.

CORE IDEA: Amphibians were the first tetrapods—four-legged vertebrates that live on land—but they are tied to the water to reproduce. Reptiles, including birds, reproduce via an amniotic egg that can survive on dry land.

CORE QUESTION: What is the key difference between a frog's eggs and a snake's eggs?

ANSWER: A frog's eggs must stay in water to survive; a snake's eggs can survive on land.

Mammals have hair and produce milk

The first mammals evolved around 200 million years ago. During the age of the dinosaurs, mammals were most likely small and nocturnal, surviving on a diet of insects. After the dinosaurs died in a mass extinction 65 million years ago, the mammals rapidly diversified. Today, large mammals dominate the modern landscape.

LANCELETS

TUNICATES

HAGFISHES

LAMPREYS

CARTILAGINOUS FISHES

BONY FISHES

AMPHIBIANS

REPTILES

MAMMALS

MAMMAL FEATURES

Mammals have two distinguishing features: mammary glands (after which the group is named) and hair. There are three major groups of mammals: monotremes, marsupials, and eutherians.

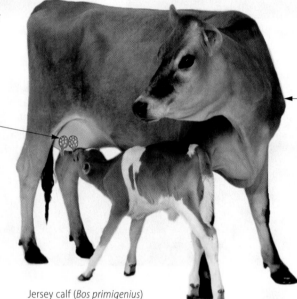

MAMMARY GLANDS
Mammary glands produce milk that nourishes newborns. Compared to most animals, mammals care for their children for a long time.

HAIR
Mammals are **endotherms**, animals whose warm, constant internal body temperature results from their own metabolism. Hair provides insulation, helping a mammal maintain its body temperature.

Jersey calf (*Bos primigenius*) suckling from his mother's udders

MONOTREMES

Monotremes are egg-laying mammals. Today, they are represented by the duckbilled platypus and four species of echidna.

DUCKBILLED PLATYPUS
The duckbilled platypus (*Ornithorhynchus anatinus*) is found exclusively in eastern Australia and the nearby island of Tasmania. Females typically build a leaf nest, lay two eggs, and then incubate them. After birth, the babies lick milk secreted by the mother onto her fur.

ECHIDNA
Found throughout Australia, the short-beaked echidna (*Tachyglossus aculeatus*) is also called the spiny anteater. When threatened, it curls into a spiked ball.

The blue whale is the largest animal that has ever lived.

MARSUPIALS

After a brief pregnancy, a mother **marsupial** gives birth to a tiny embryonic offspring that completes its development in its mother's pouch, attached to a nipple.

PLACENTA AND POUCH

A marsupial embryo starts developing in a simple placenta, an organ that physically joins the embryo to the mother within the uterus. After birth, the embryo migrates to its mother's pouch, where it nurses from a nipple.

Virginia opossum (*Didelphis virginiana*)

Opossum embryo attached to a nipple in its mother's pouch

MARSUPIAL DIVERSITY

Most marsupials are found in Australia, which split off from the other continents around 60 million years ago, before the evolution of later groups of mammals. While isolated on that island, marsupials diversified extensively, becoming the dominant animal group.

Nursing young are usually housed in a pouch, which extends from the mother's abdomen.

Eastern gray kangaroo (*Macropus giganteus*)

EUTHERIANS

An embryonic **eutherian**—also called a placental mammal, even though marsupials also have primitive placentas—spends a considerable amount of time (40 weeks for a human) attached to the placenta within its mother's uterus before being born as a fully developed animal. There are over 5,000 known species of eutherians, as opposed to just a few monotremes and a few hundred marsupials.

PLACENTA

The placenta of a eutherian is more complex than those of marsupials, and provides for a long physical association between mother and developing child.

AMNIOTIC SAC
Many mammals are born encased in an amniotic sac. The sac, attached to the placenta ("afterbirth"), is expelled shortly after the baby is born.

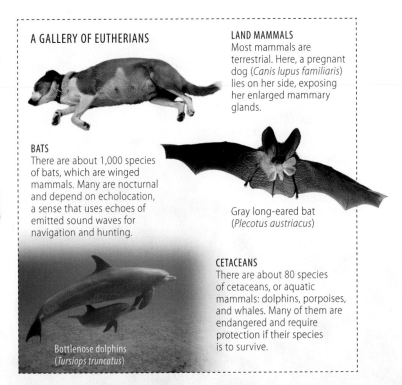

A GALLERY OF EUTHERIANS

LAND MAMMALS
Most mammals are terrestrial. Here, a pregnant dog (*Canis lupus familiaris*) lies on her side, exposing her enlarged mammary glands.

BATS
There are about 1,000 species of bats, which are winged mammals. Many are nocturnal and depend on echolocation, a sense that uses echoes of emitted sound waves for navigation and hunting.

Gray long-eared bat (*Plecotus austriacus*)

CETACEANS
There are about 80 species of cetaceans, or aquatic mammals: dolphins, porpoises, and whales. Many of them are endangered and require protection if their species is to survive.

Bottlenose dolphins (*Tursiops truncatus*)

CORE IDEA: All mammals have hair and produce milk in mammary glands. Modern mammals can be divided into three groups: monotremes (which lay eggs), marsupials (which mature in pouches), and eutherians (which mature in placentas).

CORE QUESTION: Which two groups of mammals have placentas? To which group do you belong?

ANSWER: The marsupials and the eutherians; eutherians.

Humans are primates

Where do humans sit on the tree of life on Earth? Like all animals, we are eukaryotes. Within our phylum of chordates, humans are mammals: We have hair and produce milk for developing young. The group of mammals to which humans and our closest kin belong are the **primates**. The first primates, which were small and lived in African trees, evolved around 65 million years ago. Today there are a variety of primates found throughout the world.

PRIMATES WITHIN THE TREE OF LIFE

Primates can be divided into two main groups. The first—which includes lemurs, lorises, pottos, and tarsiers—live exclusively in Madagascar, southern Asia, and Africa. The second group of primates is the **anthropoids**, which includes monkeys and apes, the latter being the group of primates to which we belong.

Of the 50 species of lemurs that lived on the island of Madagascar before the arrival of humans, 18 species have become extinct.

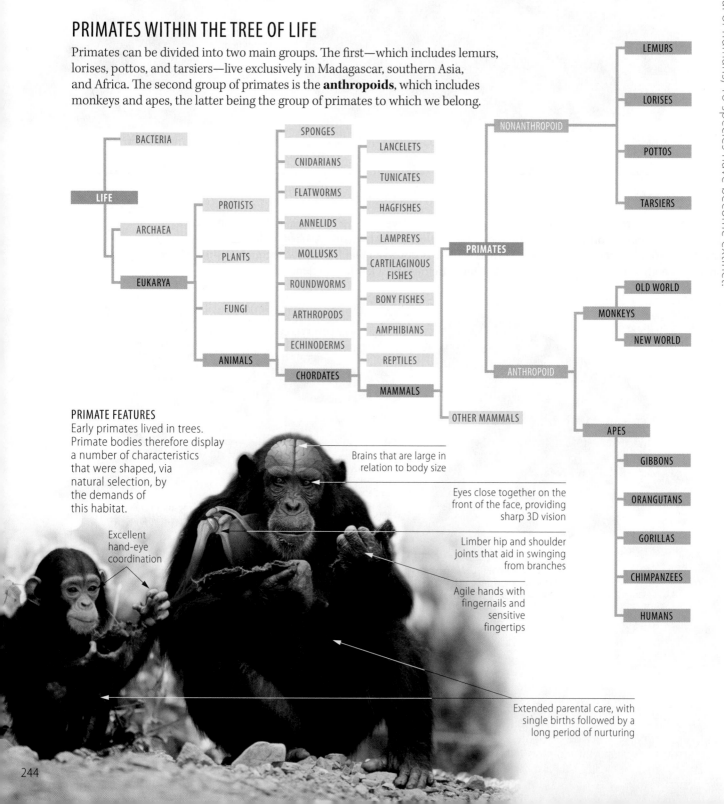

PRIMATE FEATURES
Early primates lived in trees. Primate bodies therefore display a number of characteristics that were shaped, via natural selection, by the demands of this habitat.

Excellent hand-eye coordination

Brains that are large in relation to body size

Eyes close together on the front of the face, providing sharp 3D vision

Limber hip and shoulder joints that aid in swinging from branches

Agile hands with fingernails and sensitive fingertips

Extended parental care, with single births followed by a long period of nurturing

PRIMATES
NONANTHROPOID

LEMURS
Lemurs, such as these Verreaux's sifakas (*Propithecus verreauxi*), are a diverse group of primates that live exclusively on the island of Madagascar. Different species of lemurs include individuals that range in size from 1 ounce to 17 pounds.

LORISES
Lorises, such as this Bornean slow loris (*Nycticebus menagensis*), are usually nocturnal and eat mostly insects.

POTTOS
Pottos, such as this *Perodicticus potto ibeanus*, hunt in the tops of trees in western central Africa.

TARSIERS
Tarsiers, such as this Philippine tarsier (*Tarsius syrichta*), are small, nocturnal tree-dwellers with flat faces and large, round eyes. Tarsiers are found only in Southern Asia.

ANTHROPOID

OLD WORLD MONKEYS
Most old world monkeys—including baboons and macaques—live in social bands on the ground rather than in trees.

Japanese macaque (*Macaca fuscata*)

Gelada baboon (*Theropithecus gelada*)

NEW WORLD MONKEYS
New world monkeys live in the Americas and can be recognized by a prehensile tail that acts as an extra grasping limb.

Brown capuchin (*Cebus apella*)

Wooly spider monkey (*Brachyteles hypoxanthus*)

APES

GIBBONS
Like all anthropoids, gibbons, such as this lar gibbon (*Hylobates lar*), have opposable thumbs that allow them to form fists with a strong grasp.

ORANGUTANS
Most apes are much larger than monkeys, have relatively long limbs, lack a tail, and live exclusively in the tropical areas of Africa and Asia. Of the apes, only gibbons and orangutans (such as this *Pongo abelii*) spend most of their time in trees.

GORILLAS
Gorillas, such as this mountain gorilla (*Gorilla beringei beringei*), are social and live in organized societies. They are the largest of the apes.

CHIMPANZEES
Chimpanzees, like this common chimpanzee (*Pan troglodyte*), have large brains and display complex behaviors.

HUMANS
Humans (*Homo sapiens*) are apes within the group of the anthropoid primates. Chimpanzees are our closest living relatives.

CORE IDEA: Having first evolved as tree-dwellers, all primates display features that are well adapted for that lifestyle, including large brains; keen senses; extended parenting; and limber joints, limbs, and digits.

CORE QUESTION: To which group and subgroup of primates do you belong?

ANSWER: Anthropoids and apes.

245

Humans evolved in the past few million years

On the vast phylogenetic tree that represents the evolutionary history of life on Earth, **humans** occupy a very recent branch. Fossil and DNA evidence suggests that the human lineage split from other primates around 5 to 7 million years ago. In the past 5 million years, many lineages of early humans evolved, but only one—ours—has survived. The emergence of modern humans involves several important anatomical features and a number of different species.

TIMELINE OF HUMAN EVOLUTION

Explorers have unearthed fossils from approximately 20 different species of extinct **hominins**, the human branch of the evolutionary tree of life. Although the evolutionary relationship of these various species remains a topic of active debate, it is clear that each of the features that distinguish modern humans first appeared long before our species, *Homo sapiens*.

mya = millions of years ago

7.0 to 6.0 million years ago:
Sahelanthropus tchadensis.
Based on a single skull discovered in Chad, this species is the oldest known hominin.

3.9 to 2.9 million years ago:
Australopithecus afarensis

AUSTRALOPITHECUS AFARENSIS

A 3.2-million-year-old female—nicknamed Lucy—unearthed in Ethiopia is the earliest known hominin to display a key human trait: bipedalism, or upright walking posture. Although she walked like a human, she wasn't a close relative: she was only about 3' tall, with a head the size of a softball, and a brain that was just a bit larger than a present-day chimp's brain, or about one-third the size of a modern human's brain.

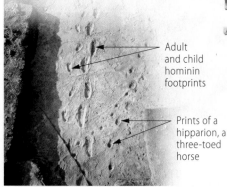

Adult and child hominin footprints

Prints of a hipparion, a three-toed horse

Trail of hominin footprints fossilized in volcanic ash, Tanzania

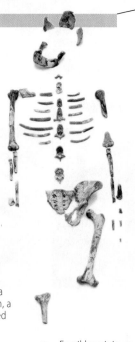

Fossil hominin skeleton known as Lucy

HOMO HABILIS

Some of the earliest fossils to display a second key feature of modern humans—enlarged brains—is found among *Homo habilis* ("handy man"), the oldest known member of our own genus. *Homo habilis* is just one of several species that have brain sizes intermediate between *Australopithecus* and modern humans. The fossil evidence suggests that by 2.4 million years ago—nearly 2 million years after hominins first walked upright—early humans evolved bigger brains (small relative to yours, but much bigger than previous species) that helped them invent tools on the African savannah.

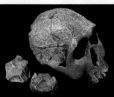

Shaped stones found with skull fossil of *Homo habilis* in Kenya

A model of *Homo habilis* using a stone as a tool

IMPORTANT POINTS ABOUT HUMAN EVOLUTION

Many people maintain mistaken notions about human evolution. Here, you can read the correct way to interpret three important points about human evolution.

HUMANS DID NOT EVOLVE FROM CHIMPS

There is a misconception that humans evolved from chimps. Although they share a common ancestor that lived 5–7 million years ago, humans and chimps have evolved separately for hundreds of thousands of generations. Modern humans are no more descended from modern chimps than you are descended from your distant cousin.

HUMAN EVOLUTION DID NOT PROCEED FROM ONE SPECIES TO THE NEXT, EACH ONE EVOLVING TO BE MORE MODERN

Another misconception is that human evolution proceeded from one species to the next. In fact, human evolution is not an orderly procession. Many hominin species represent evolutionary dead ends; they died out without leaving any modern descendants. Furthermore, many hominin species coexisted, living in the same times and places.

KEY HUMAN FEATURES DID NOT EVOLVE ALL AT ONCE

Finally, there is a misconception that key human features evolved all at once. However, the features that characterize modern humans—such as upright posture, the ability to speak, and large brains—each evolved separately and at different times over millions of years.

3 mya   2 mya   1 mya   0

2.4 to 1.7 million years ago: **Homo habilis**

1.8 to 1.3 million years ago: **Homo erectus**

350,000 years ago to 30,000 years ago: **Homo neanderthalensis**

200,000 years ago to the present day: **Homo sapiens**

HOMO ERECTUS

Homo erectus is the first hominin species known to have migrated out of Africa to other continents. Skeletons dating to 1.8 million years ago have been found in China, Indonesia, and the Asian country of Georgia. In comparison to earlier species, *H. erectus* was taller and had a larger brain, which may have aided its survival in colder climates. This species built huts and fires, made clothes, and designed stone tools.

Homo erectus skeleton

Part of stone handaxe used by *Homo erectus*

HOMO NEANDERTHALENSIS

Homo neanderthalensis, also called Neanderthals, were living in Europe by 350,000 years ago. By around 30,000 years ago—extremely recently in the history of life on Earth—they were extinct. They had brains larger than our own and hunted big game with tools made from stone and wood. Genetic analyses (using bits of DNA found inside fossil bones) suggest that Neanderthals and *H. sapiens* were not direct relatives but most likely interbred.

Reconstructed model of a Neanderthal man

HOMO SAPIENS

Older species within our genus (perhaps *Homo erectus*) gave rise to newer species and ultimately to our own: *Homo sapiens*. The oldest known fossils of our species were found in Ethiopia and date to (at most) 200,000 years ago; DNA analysis confirms this estimate and provides strong evidence that all living humans can trace their ancestry back to a single African woman who lived at that time. From Africa, our species spread into Asia approximately 50,000 years ago, and then to Europe and Australia.

Palaeolithic cave painting from Asturias, Spain from 30,000–10,000 years ago

CORE IDEA: Several known hominin species preceded the appearance of *Homo sapiens* around 200,000 years ago. These early species displayed key features such as bipedalism, enlarged brains, sophisticated tool making, and migration.

CORE QUESTION: Did modern humans ever coexist with any other species of hominin?

ANSWER: Yes, *Homo sapiens* and *Homo neanderthalensis* (Neanderthals) lived in the same times and places.

Animal bodies are organized into a structural hierarchy

The human body, like the body of all animals, can be studied at many levels of organization. The cell is the fundamental unit of life. In your body, individual cells are organized into tissues, which in turn constitute organs, which are grouped into organ systems, which work together to keep the whole organism functioning. Unique insights are gained by studying an organism at each level of the structural hierarchy.

FORM AND FUNCTION

The study of the human body reveals one of the fundamental principles of biology: the correlation of form and function. Studying the structure of a body part or system reveals information about what it does and how it works. Conversely, investigating the function of a system or part provides insight into its structure. For example, examining the structure of the human lung reveals a series of increasingly tiny branched passageways that end in millions of small sacs. This anatomy allows the lung to perform its function by providing a large surface area across which gases may be exchanged. The form and function of the lungs therefore each provide insight into the other.

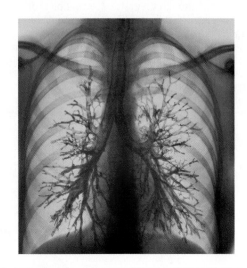

ANATOMY VERSUS PHYSIOLOGY

When studying animal bodies, you will notice that form and function are related. **Anatomy** is the study of the structure of an organism's body parts (its form). **Physiology** is the study of the functions of those parts. Each discipline involves a different way of studying the body—anatomical forms versus physiological functions. But the two disciplines complement each other in helping us to better understand how the body works.

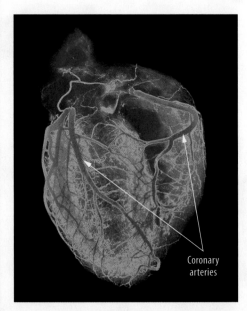

The anatomy of the heart and the arteries that surround it can be studied by injecting a dye that becomes visible under X-rays. This procedure produces an angiogram, which is particularly useful to detect blocked coronary arteries.

Coronary arteries

An electrocardiogram (EKG) can be used to study the physiology of the heart. Electrodes attached to the body can measure peaks and valleys of electrical activity that correspond to different stages of the heartbeat.

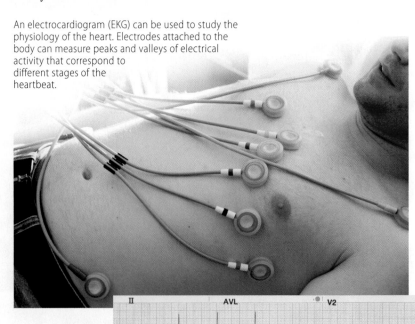

A STRUCTURAL HIERARCHY

All life can be viewed as a hierarchy of organization. In your body (and the bodies of nearly every animal), individual cells are organized into larger and larger working units. At each level of organization, new functions emerge that were not present in the structures at lower levels. For example, an organ performs functions that no individual tissue can carry out alone. In other words, your body is a whole, living unit that is greater than the sum of its parts.

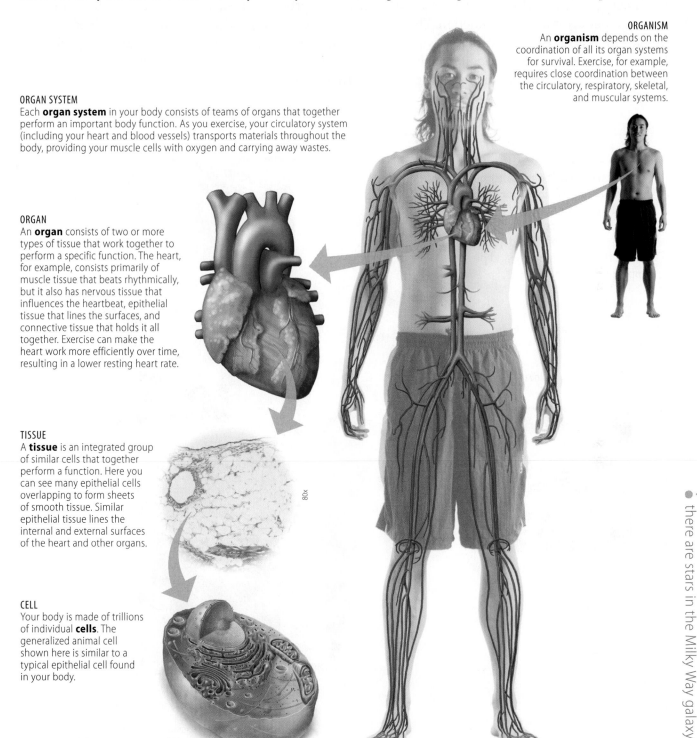

ORGANISM
An **organism** depends on the coordination of all its organ systems for survival. Exercise, for example, requires close coordination between the circulatory, respiratory, skeletal, and muscular systems.

ORGAN SYSTEM
Each **organ system** in your body consists of teams of organs that together perform an important body function. As you exercise, your circulatory system (including your heart and blood vessels) transports materials throughout the body, providing your muscle cells with oxygen and carrying away wastes.

ORGAN
An **organ** consists of two or more types of tissue that work together to perform a specific function. The heart, for example, consists primarily of muscle tissue that beats rhythmically, but it also has nervous tissue that influences the heartbeat, epithelial tissue that lines the surfaces, and connective tissue that holds it all together. Exercise can make the heart work more efficiently over time, resulting in a lower resting heart rate.

TISSUE
A **tissue** is an integrated group of similar cells that together perform a function. Here you can see many epithelial cells overlapping to form sheets of smooth tissue. Similar epithelial tissue lines the internal and external surfaces of the heart and other organs.

80x

CELL
Your body is made of trillions of individual **cells**. The generalized animal cell shown here is similar to a typical epithelial cell found in your body.

There are more cells in your body than there are stars in the Milky Way galaxy.

CORE IDEA: Form (anatomical structures) and function (physiological actions) relate to each other. Your body consists of a structural hierarchy running from cells to tissues to organs to organ systems to the whole organism.

CORE QUESTION: Your brain is an example of what level of the structural hierarchy?

ANSWER: An organ.

The human body contains several major types of tissues

The cell is the fundamental unit of life. However, in nearly all animals (including humans), individual cells rarely act alone. Instead, most cells work cooperatively as part of a **tissue**, an integrated group of similar cells that performs a specific function. The cells collected together in a tissue usually look similar to each other, with a structure that correlates with the tissue's function. Cells are joined into tissues by special junctions or sticky coatings that fuse cells together, or by being surrounded by a network of fibers. Two or more types of tissue are in turn joined together to form each organ.

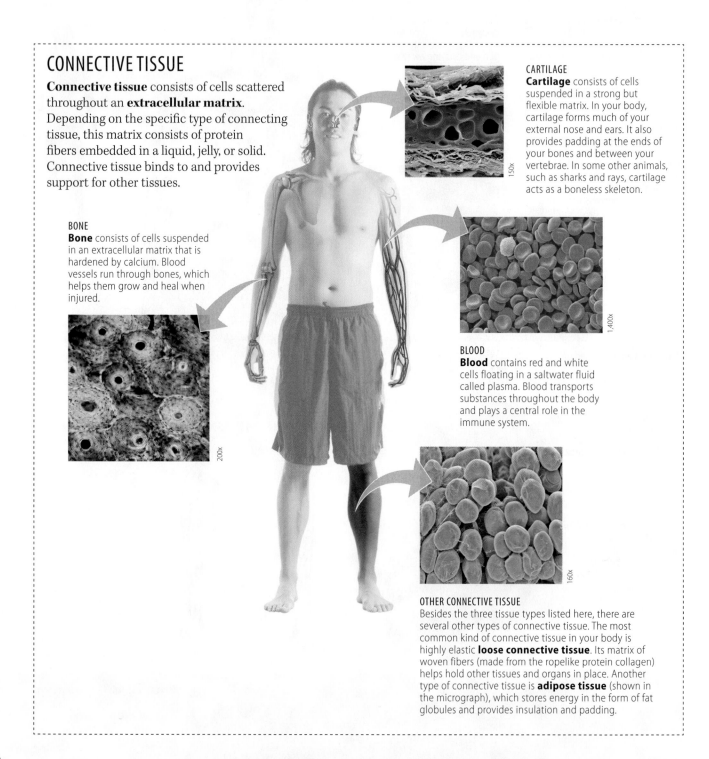

CONNECTIVE TISSUE

Connective tissue consists of cells scattered throughout an **extracellular matrix**. Depending on the specific type of connecting tissue, this matrix consists of protein fibers embedded in a liquid, jelly, or solid. Connective tissue binds to and provides support for other tissues.

BONE
Bone consists of cells suspended in an extracellular matrix that is hardened by calcium. Blood vessels run through bones, which helps them grow and heal when injured.

200x

CARTILAGE
Cartilage consists of cells suspended in a strong but flexible matrix. In your body, cartilage forms much of your external nose and ears. It also provides padding at the ends of your bones and between your vertebrae. In some other animals, such as sharks and rays, cartilage acts as a boneless skeleton.

150x

1,400x

BLOOD
Blood contains red and white cells floating in a saltwater fluid called plasma. Blood transports substances throughout the body and plays a central role in the immune system.

160x

OTHER CONNECTIVE TISSUE
Besides the three tissue types listed here, there are several other types of connective tissue. The most common kind of connective tissue in your body is highly elastic **loose connective tissue**. Its matrix of woven fibers (made from the ropelike protein collagen) helps hold other tissues and organs in place. Another type of connective tissue is **adipose tissue** (shown in the micrograph), which stores energy in the form of fat globules and provides insulation and padding.

EPITHELIUM

Most body and organ surfaces (such as the outer layer of your skin, and the inner and outer linings of your blood vessels and digestive tract) are covered with **epithelial tissue**, also called **epithelium**. Epithelium consists of sheets of tightly packed cells that are fused together. Your epidermis (the outermost layer of your skin) contains dense layers of tightly bound epithelial cells, forming a protective barrier that surrounds your body and helps protect its internal environment from external threats such as disease-causing microorganisms. Cells of the epidermis (and many other epithelial cells elsewhere in your body) continuously fall off and are renewed.

NERVOUS TISSUE

Nervous tissue communicates signals between your brain and the rest of the body. Within your brain, spinal cord, and nerves, individual cells called **neurons** can transmit rapid electrical signals along spindly extensions. Such signals allow you to sense the world, formulate a response, and act on it.

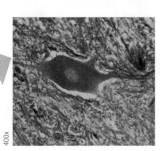

MUSCLE TISSUE

Muscle tissue is abundant in the majority of most animals, making up much of the "meat" that we consume. It consists of bundles of long cells called **muscle fibers**, each of which contains specialized proteins that allow it to contract (shorten).

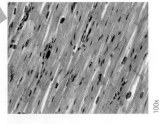

CARDIAC MUSCLE

The cells of **cardiac muscle**, found in heart tissue, branch and join one another to form a large interconnected tissue. This allows each beat of your heart to occur as one coordinated muscle contraction. Like smooth muscle, cardiac muscle is involuntary.

SMOOTH MUSCLE

Smooth muscle is found in many body systems, including the walls of the digestive tract and in blood vessels. Smooth muscle lacks any type of striation (striping) and is contracted through involuntary signals from the brain.

SKELETAL MUSCLE

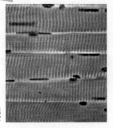

Skeletal muscle, attached to your bones by tendons, allows you to move. Under the microscope, skeletal muscle appears striated (striped). Skeletal muscle is, for the most part, under voluntary control.

Weight training does not increase the number of muscle cells but instead enlarges those already present.

CORE IDEA: Most animal bodies contain a number of tissues. Each tissue consists of similar cells that work together to perform a specific function. Organs are in turn made up of two or more types of tissue.

CORE QUESTION: Poke your forearm with your finger. What types of tissue can you feel?

ANSWER: You can probably feel epithelium (in your epidermis), adipose (fat), muscle, and bone.

An animal's internal environment remains relatively constant

One of the characteristics that distinguishes living organisms from nonliving matter is the ability to detect and react to environmental stimuli. The outside world is constantly changing—in temperature, humidity, salinity, and countless other ways. Yet the internal environment of your body is kept relatively constant. Your body contains many systems and mechanisms that work constantly to help maintain a relatively steady internal state despite wide fluctuations in the outside environment.

EXCHANGES WITH THE ENVIRONMENT

Although every animal is covered with a protective layer that separates the internal body from the external world, no animal is cut off from its environment. Every organism exchanges chemicals and energy with its surroundings. Such exchanges occur at every cell in the body through the help of several organ systems such as the circulatory, respiratory, and digestive systems.

HOMEOSTASIS

Animal bodies tend to maintain relatively constant internal conditions even when the external environment changes. Such a tendency is called **homeostasis**, a word that literally means "steady state." When the external environment changes drastically, the body uses various mechanisms that maintain internal systems within a narrow spectrum. Changes do occur, but they are small and tend to stay within a range that the body can tolerate. For example, your internal body temperature does not fluctuate more than a few degrees from average even if the temperature of your surroundings changes drastically.

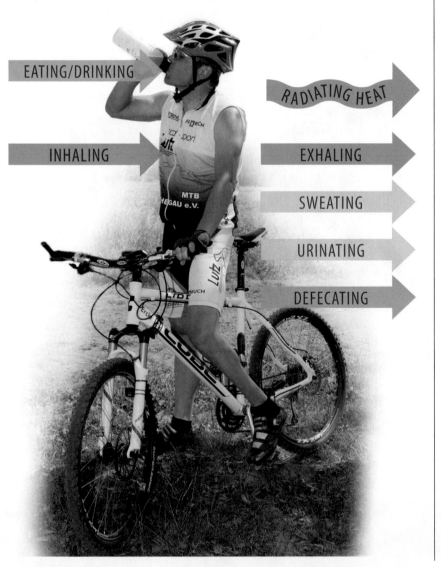

EATING/DRINKING

INHALING

RADIATING HEAT

EXHALING

SWEATING

URINATING

DEFECATING

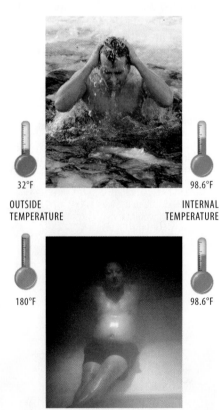

32°F
OUTSIDE TEMPERATURE

98.6°F
INTERNAL TEMPERATURE

180°F

98.6°F

NEGATIVE FEEDBACK

How does your body detect and respond to changes in the environment? The most common mechanism is called **negative feedback**, a form of regulation in which the result of a process inhibits that very process. A good example is a household thermostat. When the temperature falls below set point, the thermostat turns on the heater. This raises the temperature, which then causes the thermostat to turn off the heater. In other words, the result of the process (heat) inhibits the process (turns off the heater). The mechanism of negative feedback is very similar when regulating temperature (and many other processes) in your own body. The importance of homeostasis is apparent when you consider what happens if this mechanism breaks down: Both hypothermia (body temperature too low) and heat stroke (body temperature too high) can be deadly.

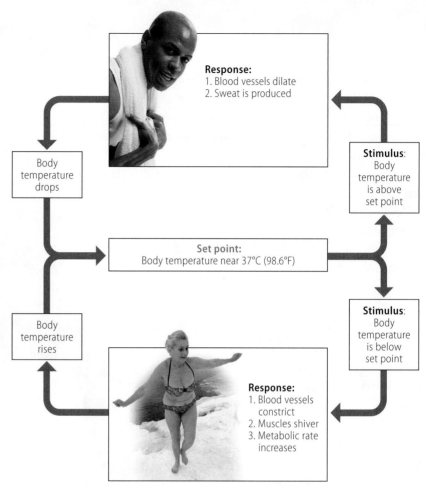

Response:
1. Blood vessels dilate
2. Sweat is produced

Body temperature drops

Stimulus: Body temperature is above set point

Set point: Body temperature near 37°C (98.6°F)

Body temperature rises

Stimulus: Body temperature is below set point

Response:
1. Blood vessels constrict
2. Muscles shiver
3. Metabolic rate increases

POSITIVE FEEDBACK

Positive feedback is a homeostatic mechanism in which the results of a process intensify that same process. This is like a chain reaction that powers a bomb: An explosion of some fuel ignites other fuel, leading to a climactic burst. Although less common in your body than negative feedback, this type of control plays some key roles, including the induction of labor at the end of pregnancy.

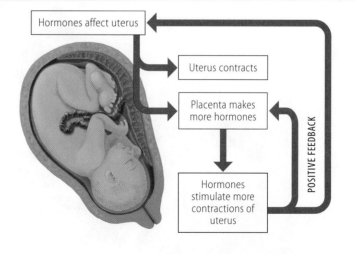

Hormones affect uterus

Uterus contracts

Placenta makes more hormones

Hormones stimulate more contractions of uterus

POSITIVE FEEDBACK

A very high body temperature disrupts the brain's temperature control system, leading to heat stroke.

CORE IDEA: Your body exchanges chemicals and energy with the environment, but it tends to maintain nearly constant internal conditions. Homeostasis is often maintained by negative feedback.

CORE QUESTION: Flushing a toilet causes water to rush into the tank, which lifts a float until it turns off the water. What kind of feedback is that?

ANSWER: Negative feedback, since the result of the process turns off that process.

Food is processed in a series of stages

Like all animals, humans consume other organisms to obtain nutrients and energy. Your body and your food consist of similar molecules; there is protein in a chicken leg, for example, that is nearly identical to the protein in your own muscles. But the larger nutrients in food are never used directly. Instead, all food is broken down into small molecular building blocks. These small molecules are absorbed and used to create the body's own molecules. The process of converting food to forms that the body can use is divided into four stages.

THE PROCESSING OF FOOD

To be used by your body, food must be broken down. Large pieces of food need to be deconstructed into individual molecules, which then must be dismantled into even smaller molecules. Small molecules that can be used by the body are absorbed, and unneeded substances are passed out as waste. A variety of body structures participate in this four-stage process.

If the folds of the small intestine were all flattened out, it would cover the area of a tennis court.

❶ INGESTION

Ingestion, or eating, is the first stage of food processing. All animals ingest other organisms, whole or by the piece.

❷ DIGESTION

Digestion is the breakdown of food into molecules that are small enough to be absorbed by the body. Because the body's cells cannot absorb large molecules, all nutrients must be broken into individual molecular building blocks before the body can use them.

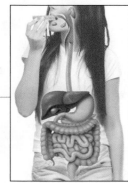

Mechanical digestion is the use of physical processes to break down food into smaller pieces. For example, your teeth slice, grind, mash, and chop food. In your stomach, muscles churn partially digested food, helping to break it down further.

Chemical digestion is the use of enzymes to perform hydrolysis, chemical reactions that use water to break bonds within large molecules (called polymers, such as proteins found in food), leaving behind smaller molecules (called monomers, such as amino acids).

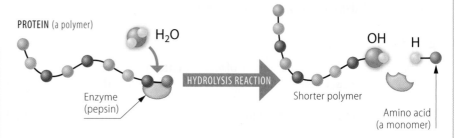

PROTEIN (a polymer)

H_2O

Enzyme (pepsin)

HYDROLYSIS REACTION

Shorter polymer

OH   H

Amino acid (a monomer)

❸ ABSORPTION

Only after extensive digestion are food molecules small enough for cells to absorb. **Absorption** is the uptake of these small nutrient molecules, primarily by the cells that line extensive folds of the small intestine. These nutrients enter blood vessels surrounding the small intestine. From there, they are transported to all the cells of the body. Cells use the molecular building blocks as a source of energy or in the construction of new molecules.

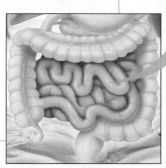

Intestinal villi

Villi are finger-like projections that line the surface of the small intestine and increase the surface area available for the absorption of nutrients from digested food.

1,080x

❹ ELIMINATION

Elimination is the disposal of undigested matter from the body. Food waste accumulates as feces in the rectum (the last section of the large intestine) and exits the digestive tract via the anus.

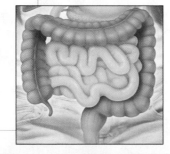

CORE IDEA: Food processing occurs via four stages: ingestion (also called eating), digestion (which can be mechanical or chemical), absorption (primarily by cells lining the small intestine), and elimination of undigested wastes.

CORE QUESTION: During which stage do the molecules in food actually enter body cells?

ANSWER: Absorption.

The human digestive system consists of an alimentary canal and accessory organs

You are what you eat: Maintaining the proper structure and function of your body depends on your ability to extract energy and nutrients from your food. This is the job of the **digestive system**, which consists of a long tube (called the alimentary canal) and a series of organs that secrete digestive chemicals into it. Your digestive system acts as a disassembly line, breaking down food into smaller and smaller bits until all that remains are molecules tiny enough to pass into your body cells. As you follow the path of food through the alimentary canal, try imagining the physical state of your food at each stage along the way.

ALIMENTARY CANAL

The human digestive system consists of an **alimentary canal** (a long tube sometimes called the "gut") that is divided along its length into specialized digestive organs. At each stop along the alimentary canal, specific steps in the processing of food occur.

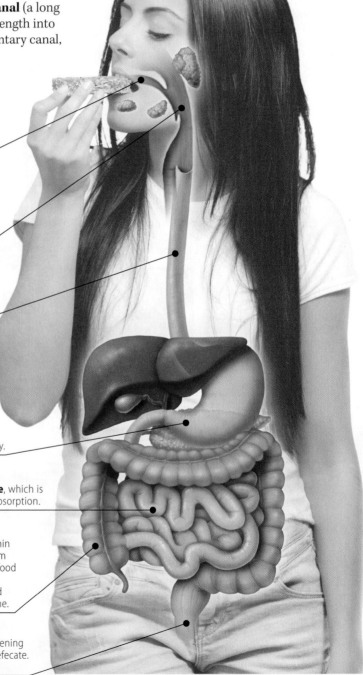

MOUTH
Also called the **oral cavity**, the mouth is the site of ingestion (eating), the start of mechanical digestion (via the teeth that cut, grind, and smash the food), and the start of chemical digestion (via saliva). The tongue provides the sensation of taste, helps to shape food into an easy-to-swallow ball, and moves food into the pharynx.

PHARYNX
Your **pharynx** is the intersection of your mouth, esophagus, and trachea (which leads to the lungs). Most of the time, air enters your trachea via your pharynx. But when you swallow, a flap called the **epiglottis** moves to cover the entrance to the trachea, directing food down the esophagus. Occasionally, food will "go down the wrong pipe," triggering a strong coughing reflex.

ESOPHAGUS
The **esophagus** is a muscular tube that connects the pharynx to the stomach. Food moves through the esophagus via muscle contractions called peristalsis.

STOMACH
Your stomach has elastic folds; it can stretch out to store large amounts of food and drink. Cells lining the stomach secrete **gastric juice**, an acidic fluid containing enzymes (such as **pepsin**) that help digest proteins and other molecules. The muscular walls of the stomach churn the food and the gastric juice together, producing chyme, a liquid with the consistency of thick soup. A sphincter (ring of muscle) periodically opens to send a small squirtful of chyme into the intestines. It takes about 2–6 hours after a meal for the stomach to empty.

SMALL INTESTINE
Chemical digestion is completed by enzymes within the **small intestine**, which is narrow but long. The small intestine is also the primary site of nutrient absorption.

LARGE INTESTINE
The **large intestine** measures about 5 feet long and 2 inches wide. Within the colon (the main portion of the large intestine), water is absorbed from the alimentary canal and returned to the bloodstream. What remains of food (primarily indigestible matter such as fiber), as well as huge numbers of prokaryotic organisms that normally inhabit the large intestine, is formed into feces and then stored in the **rectum**, the last 6 inches of the intestine.

ANUS
Two sphincter muscles, one voluntary and the other not, regulate the opening of the **anus**. Stimulation of nerves within the colon create the urge to defecate. When the voluntary sphincter muscle is relaxed, feces are expelled.

Your alimentary canal is, all together, about 30 feet long.

ACCESSORY ORGANS

Your digestive system contains a series of **accessory organs**, each of which secretes specific digestive chemicals into the alimentary canal via ducts.

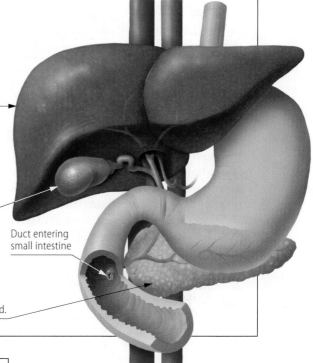

LIVER
The **liver** secretes a juice called **bile** via a duct into the small intestine. Bile helps the process of fat digestion.

GALLBLADDER
Bile produced by the liver can be stored in the **gallbladder** and later secreted via a duct into the small intestine.

PANCREAS
Liquid secreted from the **pancreas** via a duct into the small intestine helps neutralize stomach acid and continues the chemical digestion of food.

Duct entering small intestine

SALIVARY GLANDS
Several salivary glands secrete saliva into the mouth. Saliva contains the digestive enzyme salivary amylase, which breaks down starch into simple sugars.

NUTRIENT ABSORPTION IN THE SMALL INTESTINE

The structure of the small intestine correlates with its function of nutrient absorption. Epithelial tissue lining the inside of the small intestine is extensively folded into tiny finger-like extensions called **villi**. In turn, each cell along the villi has microscopic projections called **microvilli**. This structure—folds with villi with microvilli—gives the small intestine a huge surface area across which absorption can occur.

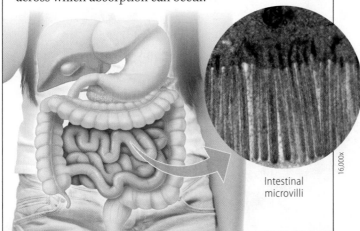

Intestinal microvilli

16,000x

PERISTALSIS

Waves of contractions in the muscles that line the alimentary canal, called **peristalsis**, propel food through your gut. You probably know that you can swallow food even if you are upside down, proving that food is squeezed into your stomach, rather than falling into it.

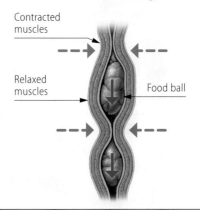

Contracted muscles

Relaxed muscles

Food ball

CORE IDEA: The human digestive system consists of an alimentary canal (divided into organs along its length) and various accessory organs, including salivary glands and organs that release digestive juices into the small intestine.

CORE QUESTION: If the most important function of the digestive system is the absorption of nutrients, which one organ is primarily responsible?

ANSWER: The small intestine.

Proper nutrition provides energy and building materials

Like any animal, you require a proper diet. Eating a well-balanced diet helps your body in two ways: It provides the building materials needed for the body's structures, and it provides the energy needed to maintain the body's functions. A well-balanced diet is rich in whole grains, fresh fruits and vegetables, and vitamins and minerals, with moderate amounts of lean meat and small quantities of fats, sugars, and salt. Together, such foods provide your body with the molecular building blocks of good health.

ESSENTIAL NUTRIENTS

To maintain proper health, an animal's diet must contain sufficient quantities of the **essential nutrients**, materials that the body cannot make itself. The absence of any essential nutrient causes disease. For example, your cells can manufacture 12 different kinds of amino acids. There are another eight, however, that cannot be made and so must be ingested to avoid malnutrition. As you can see in the table and in the meal pictured at right, there are four classes of essential nutrients.

ESSENTIAL NUTRIENTS: Required in the diet; absence causes disease			
Minerals	Vitamins	Essential fatty acids	Essential amino acids
Inorganic chemical elements	Organic compounds	Required to build several important molecules	Required to build proteins

FOOD AS FUEL

In order to perform its functions, every cell in your body requires a constant supply of energy in the form of the molecule **ATP** (adenosine triphosphate). Within each cell's mitochondria, the process of **cellular respiration** uses oxygen (O_2) to "burn" the molecule glucose ($C_6H_{12}O_6$), producing carbon dioxide (CO_2), water (H_2O), and several dozen molecules of ATP. The glucose that serves as the input to cellular respiration is obtained by digestion of the food you eat.

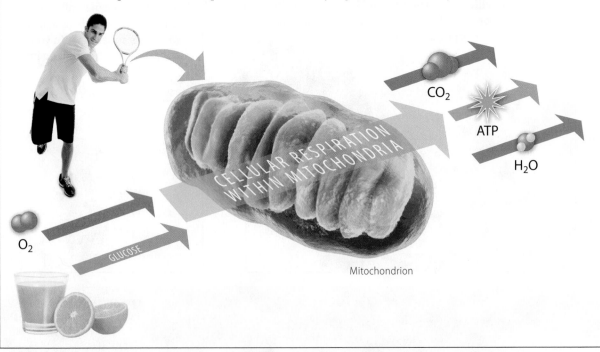

CELLULAR RESPIRATION WITHIN MITOCHONDRIA

O_2

GLUCOSE

CO_2

ATP

H_2O

Mitochondrion

One tablespoon of vitamin B_{12} can provide the daily requirement of nearly 1 million people.

FOUR CATEGORIES OF ESSENTIAL NUTRIENTS

MINERALS
Minerals are elements required to maintain health. Some, such as calcium, are required in relatively large amounts. Just over one cup of broccoli provides the daily requirement of calcium. Other minerals, such as iron, are required in tiny quantities. Rice is a good source of iron, which is required to build molecules of hemoglobin, the oxygen-carrying molecule found in red blood cells. Too little iron can result in anemia.

VITAMINS
A **vitamin** is an organic (carbon-containing) nutrient required in your diet, but only in very small amounts. All vitamins are necessary to your health: lack of any vitamin in the diet leads to disease. Some vitamins (C and the B-complex vitamins) are water soluble, while others (A, D, E, and K) are fat soluble. Sugar snap peas and lemon are rich in vitamin C, which is required for the production of connective tissue. A lack of vitamin C in the diet causes the disease scurvy.

ESSENTIAL FATTY ACIDS
Your cells use many kinds of fatty acids to make fats and other lipids. Some fatty acids can be produced from scratch by the body. Those that cannot, the **essential fatty acids**, must be obtained from food. For example, many fish are rich in linoleic acid, an omega-6 fatty acid used to make the phospholipid molecules of cell membranes. In industrialized nations, deficiencies in essential fatty acids are rare.

ESSENTIAL AMINO ACIDS
All proteins are built from 20 different kinds of amino acids. Eight of these are **essential amino acids** that must be obtained from the diet because human cells cannot make them. (Infants require a ninth.) Different foods contain different essential amino acids. Many animal proteins (meat, fish, dairy) contain all the essential amino acids, but most plant proteins have only a subset. However, a combination of plant proteins (such as a grain plus a legume) can provide all the essential amino acids.

CORE IDEA: A proper diet provides raw materials and energy to cells. Essential nutrients cannot be produced by the body itself. They include inorganic minerals, organic vitamins, essential fatty acids, and essential amino acids.

CORE QUESTION: Vitamins and minerals are often discussed together. What is the difference between them?

ANSWER: Vitamins are organic (carbon-containing) compounds while minerals are inorganic (non-carbon-containing) elements.

An unbalanced diet or malfunctioning digestive system can lead to health problems

Two broad categories of ailments are associated with the digestive system: those caused by a malfunctioning or diseased digestive organ, and those caused by improper diet. You can see examples of both types of ailments on these pages.

WHAT CAN GO WRONG

Many illnesses are caused by infection or malfunction of the organs of the digestive system.

ACID REFLUX
Acid reflux (commonly but erroneously called "heartburn") is caused by backflow of partially digested food into the esophagus. Frequent, severe reflux that harms the lining of the esophagus is called gastroesophageal reflux disease, or GERD. Acid reflux can often be treated by lifestyle changes and antacids.

Antacids

GALLSTONES
Gallstones, solid crystals of bile, can obstruct the gallbladder or its ducts. Often, the only cure is surgical removal of the gallbladder.

CONSTIPATION
If peristalsis slows the movement of feces through the colon, too much water is reabsorbed and the feces become compacted. Constipation most commonly results from a lack of exercise or a fiber-poor diet.

APPENDICITIS
The appendix is a small finger-shaped extension of the large intestine. Infection of the appendix by bacteria is called appendicitis and is indicated by sharp pains on the right side of the abdomen. Most people are cured by surgical removal of the appendix.

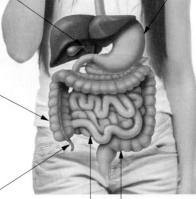

ULCER
A coating of mucus prevents gastric juice from harming the lining of the stomach. Infection by the acid-tolerant bacterium *Helicobacter pylori* damages the mucus, leading to an erosion of the lining called a gastric ulcer. Most gastric ulcers are treated with medication containing bismuth and, for serious cases, antibiotics.

12,700x

Helicobacter pylori bacterium

Medication containing bismuth

GASTRIC BYPASS
The most common weight loss surgery in the United States is gastric bypass: the small intestine about 18 inches downstream is attached to the stomach, which has been reduced by stapling. Patients quickly fill up, and absorption is reduced.

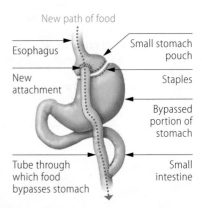

New path of food

Esophagus

Small stomach pouch

New attachment

Staples

Bypassed portion of stomach

Tube through which food bypasses stomach

Small intestine

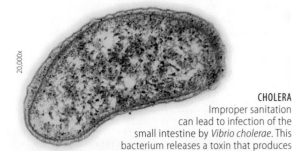

20,000x

Vibrio cholerae bacterium

CHOLERA
Improper sanitation can lead to infection of the small intestine by *Vibrio cholerae*. This bacterium releases a toxin that produces diarrhea (inefficient water reabsorption) so severe that it can quickly lead to dehydration and, if untreated, death.

INFLAMMATORY BOWEL DISEASE
Several intestinal disorders—including Crohn's disease and ulcerative colitis—are caused by painful swelling (inflammation) of the intestinal wall.

NUTRITIONAL IMBALANCES

Dietary problems can have severe health consequences. A number of different nutritional disorders can arise, each with different causes.

OBESITY

In America, the nutritional imbalance of greatest concern is **obesity**, defined as an inappropriately high **body mass index (BMI)**, a ratio of weight to height. About one-third of all Americans are overweight, and another one-third are obese. Obesity contributes to a number of health problems, including type 2 diabetes, cancer of the colon and breast, and cardiovascular disease.

PERCENTAGE OF OBESE ADULTS IN THE UNITED STATES

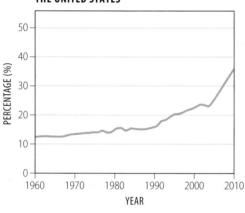

CALCULATING BMI

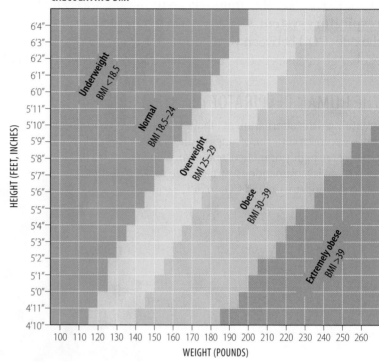

MALNUTRITION

Hunger affects nearly 1 billion people around the world. **Malnutrition** is caused by a diet that lacks sufficient calories or essential nutrients. It can be caused by inadequate intake (hunger) or medical problems. The most common form of malnutrition is protein deficiency, the insufficient intake of essential amino acids. Although many animal products provide sufficient protein, people who rely on a single plant staple—just rice or just potatoes, for example—will suffer protein deficiencies. In developing nations, protein-deficient children often have swollen bellies (due to fluid accumulation in the abdomen), a condition called kwashiorkor. This child, a refugee in northern Uganda, is suffering from malnutrition due to lack of food.

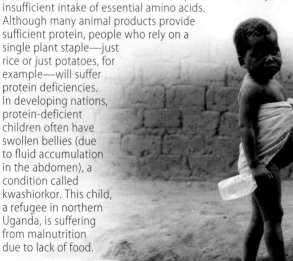

EATING DISORDERS

Malnutrition is not always caused by a lack of access to food. Millions of Americans (mostly female) have an eating disorder. **Anorexia nervosa** is caused by self-starvation due to a fear of gaining weight, even when the person is underweight. **Bulimia** is a pattern of binge eating followed by purging through induced vomiting, abuse of laxatives, or excessive exercise. The causes of eating disorders are unknown. Treatment includes counseling and antidepressant medications.

Paula Abdul, spokesperson for the National Eating Disorders Association

CORE IDEA: Health problems can result from a number of abnormalities in the human digestive system. Additionally, improper diet—resulting in obesity, malnutrition, or eating disorders—can cause significant health problems.

CORE QUESTION: Is it possible for someone to be both obese and malnourished?

ANSWER: Yes, if someone eats too many calories but not enough nutrients.

Each day, 14,000 children under the age of 5 starve to death worldwide.

The respiratory system exchanges gases between the environment and the body

You can survive a week without food, and several days without water. But if you don't breathe for 10 minutes, you'll die. Your minute-to-minute survival thus depends on a properly functioning **respiratory system**, several organs that facilitate the exchange of oxygen (O_2) and carbon dioxide (CO_2) between your body and the environment. Many structures contribute, but the actual exchange of gases occurs between blood capillaries and alveoli (tiny air sacs) in your lungs.

THE HUMAN RESPIRATORY SYSTEM

Breathing—the alternation of inhalation (in) and exhalation (out)—exposes your inner lungs to air. Because your airways are highly branched, ending in a huge number of sacs, there is a tremendous surface area for O_2 to diffuse into the bloodstream and CO_2 to diffuse out.

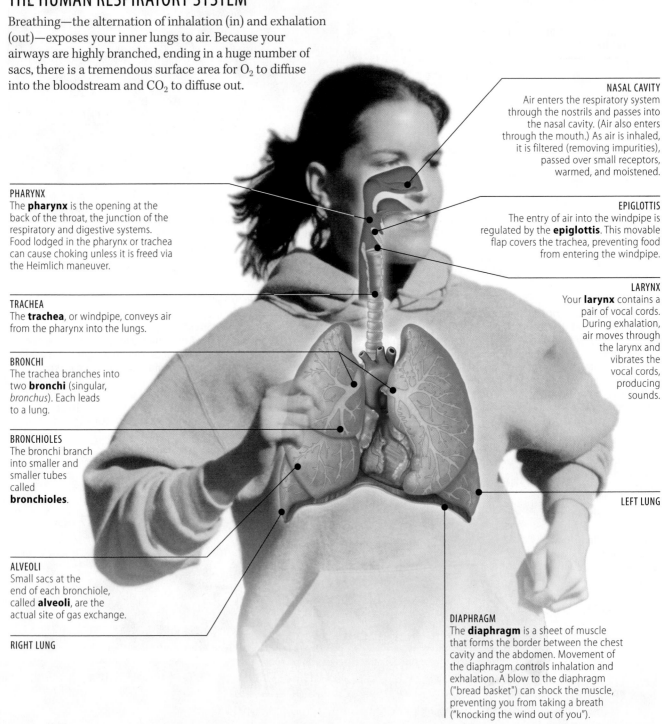

NASAL CAVITY
Air enters the respiratory system through the nostrils and passes into the nasal cavity. (Air also enters through the mouth.) As air is inhaled, it is filtered (removing impurities), passed over small receptors, warmed, and moistened.

EPIGLOTTIS
The entry of air into the windpipe is regulated by the **epiglottis**. This movable flap covers the trachea, preventing food from entering the windpipe.

LARYNX
Your **larynx** contains a pair of vocal cords. During exhalation, air moves through the larynx and vibrates the vocal cords, producing sounds.

PHARYNX
The **pharynx** is the opening at the back of the throat, the junction of the respiratory and digestive systems. Food lodged in the pharynx or trachea can cause choking unless it is freed via the Heimlich maneuver.

TRACHEA
The **trachea**, or windpipe, conveys air from the pharynx into the lungs.

BRONCHI
The trachea branches into two **bronchi** (singular, *bronchus*). Each leads to a lung.

BRONCHIOLES
The bronchi branch into smaller and smaller tubes called **bronchioles**.

ALVEOLI
Small sacs at the end of each bronchiole, called **alveoli**, are the actual site of gas exchange.

RIGHT LUNG

LEFT LUNG

DIAPHRAGM
The **diaphragm** is a sheet of muscle that forms the border between the chest cavity and the abdomen. Movement of the diaphragm controls inhalation and exhalation. A blow to the diaphragm ("bread basket") can shock the muscle, preventing you from taking a breath ("knocking the wind out of you").

GAS EXCHANGE

The primary function of the respiratory system is to exchange gases, bringing O_2 from the air into the body via inhalation and then removing CO_2 from the body via exhalation. The circulatory system transports O_2 (via the blood) from the lungs to all body cells, and transports CO_2 from body cells back to the lungs. The two systems thus work in close cooperation.

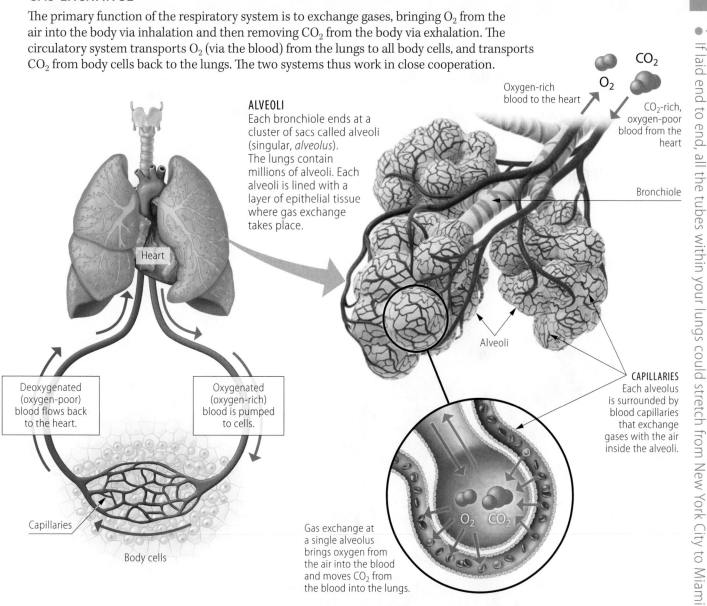

ALVEOLI
Each bronchiole ends at a cluster of sacs called alveoli (singular, *alveolus*). The lungs contain millions of alveoli. Each alveoli is lined with a layer of epithelial tissue where gas exchange takes place.

Oxygen-rich blood to the heart

CO_2

O_2

CO_2-rich, oxygen-poor blood from the heart

Bronchiole

Heart

Deoxygenated (oxygen-poor) blood flows back to the heart.

Oxygenated (oxygen-rich) blood is pumped to cells.

Alveoli

CAPILLARIES
Each alveolus is surrounded by blood capillaries that exchange gases with the air inside the alveoli.

Capillaries

Body cells

O_2 CO_2

Gas exchange at a single alveolus brings oxygen from the air into the blood and moves CO_2 from the blood into the lungs.

If laid end to end, all the tubes within your lungs could stretch from New York City to Miami.

WHAT CAN GO WRONG

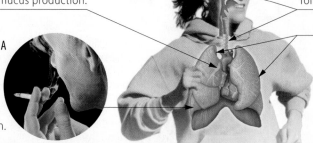

BRONCHITIS
Bronchitis is most commonly caused by infection of the bronchioles by viruses. Symptoms include a persistent cough and mucus production.

UPPER RESPIRATORY INFECTIONS
URIs include a wide variety of illnesses of the nose, pharynx, or larynx. Viral infection is the most common cause, followed by bacterial infection.

EMPHYSEMA
Most often caused by long-term exposure to tobacco smoke or air pollution, emphysema causes progressively worsening shortness of breath.

ASTHMA
Asthma is a long-term inflammation of the airway. Environmental irritants may cause difficulty in breathing.

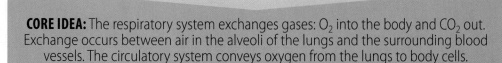

CORE IDEA: The respiratory system exchanges gases: O_2 into the body and CO_2 out. Exchange occurs between air in the alveoli of the lungs and the surrounding blood vessels. The circulatory system conveys oxygen from the lungs to body cells.

CORE QUESTION: Where are gases actually exchanged between the body and the environment?

ANSWER: In the alveoli of the lungs.

The circulatory system transports materials throughout the body

Every cell in your body needs to exchange materials—energy, nutrients, gases, wastes—with the outside environment. But the vast majority of your cells are located well inside your body, far from the outside world. Your cells therefore rely on your **circulatory system**, a collection of organs and tissues that acts as an internal transport network. Circulating blood passes near every body cell, allowing materials to diffuse in and out.

THE HUMAN CIRCULATORY SYSTEM

Your **cardiovascular system** consists of your heart ("cardio") and your blood vessels ("vascular"). The heart pumps **blood** through a series of branching tubes (arteries, arterioles, capillaries, venules, and veins). The blood carries nutrients and wastes between different bodily locations.

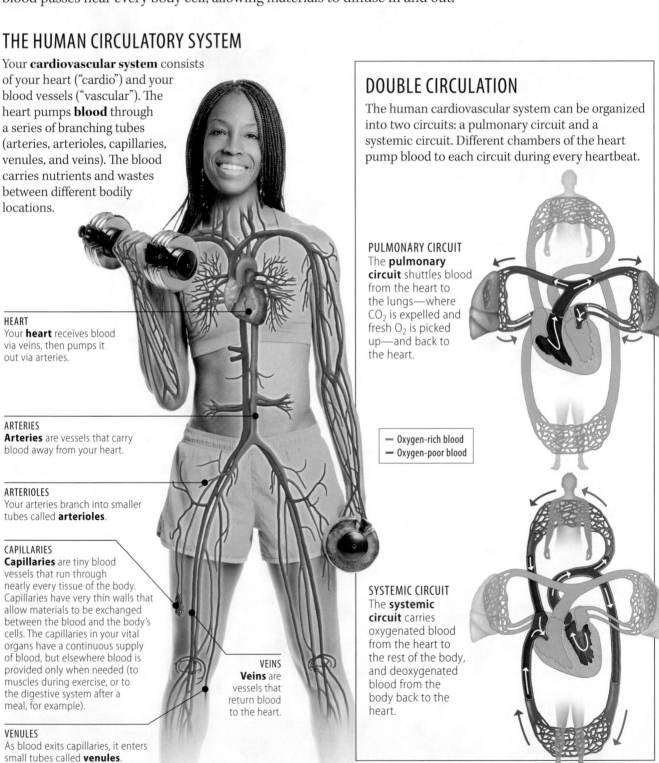

HEART
Your **heart** receives blood via veins, then pumps it out via arteries.

ARTERIES
Arteries are vessels that carry blood away from your heart.

ARTERIOLES
Your arteries branch into smaller tubes called **arterioles**.

CAPILLARIES
Capillaries are tiny blood vessels that run through nearly every tissue of the body. Capillaries have very thin walls that allow materials to be exchanged between the blood and the body's cells. The capillaries in your vital organs have a continuous supply of blood, but elsewhere blood is provided only when needed (to muscles during exercise, or to the digestive system after a meal, for example).

VEINS
Veins are vessels that return blood to the heart.

VENULES
As blood exits capillaries, it enters small tubes called **venules**.

DOUBLE CIRCULATION

The human cardiovascular system can be organized into two circuits: a pulmonary circuit and a systemic circuit. Different chambers of the heart pump blood to each circuit during every heartbeat.

PULMONARY CIRCUIT
The **pulmonary circuit** shuttles blood from the heart to the lungs—where CO_2 is expelled and fresh O_2 is picked up—and back to the heart.

— Oxygen-rich blood
— Oxygen-poor blood

SYSTEMIC CIRCUIT
The **systemic circuit** carries oxygenated blood from the heart to the rest of the body, and deoxygenated blood from the body back to the heart.

BLOOD VESSELS

Three types of blood vessels make up the "plumbing" of your circulatory system. Arteries (and smaller arterioles) carry blood away from the heart. Veins (and smaller venules) carry blood to the heart. Capillaries join arterioles to venules, providing the site for exchange of materials between the blood and the body's cells. This local chemical exchange—between the blood and cells near capillaries—is the key step in the functioning of the circulatory system.

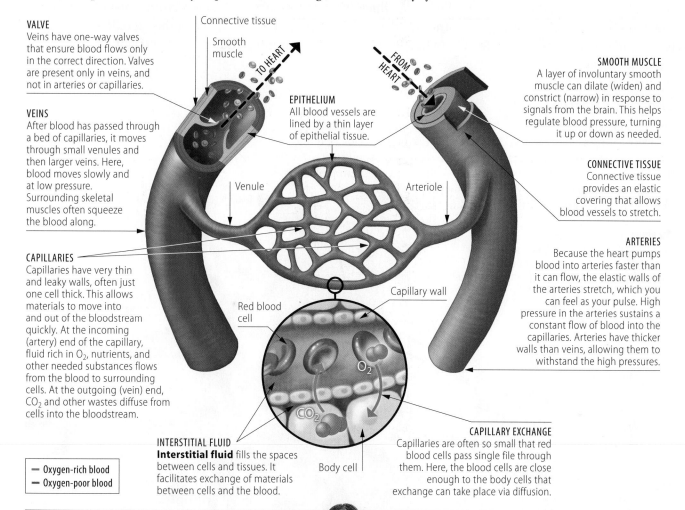

VALVE
Veins have one-way valves that ensure blood flows only in the correct direction. Valves are present only in veins, and not in arteries or capillaries.

VEINS
After blood has passed through a bed of capillaries, it moves through small venules and then larger veins. Here, blood moves slowly and at low pressure. Surrounding skeletal muscles often squeeze the blood along.

CAPILLARIES
Capillaries have very thin and leaky walls, often just one cell thick. This allows materials to move into and out of the bloodstream quickly. At the incoming (artery) end of the capillary, fluid rich in O_2, nutrients, and other needed substances flows from the blood to surrounding cells. At the outgoing (vein) end, CO_2 and other wastes diffuse from cells into the bloodstream.

Connective tissue

Smooth muscle

TO HEART

FROM HEART

EPITHELIUM
All blood vessels are lined by a thin layer of epithelial tissue.

SMOOTH MUSCLE
A layer of involuntary smooth muscle can dilate (widen) and constrict (narrow) in response to signals from the brain. This helps regulate blood pressure, turning it up or down as needed.

CONNECTIVE TISSUE
Connective tissue provides an elastic covering that allows blood vessels to stretch.

ARTERIES
Because the heart pumps blood into arteries faster than it can flow, the elastic walls of the arteries stretch, which you can feel as your pulse. High pressure in the arteries sustains a constant flow of blood into the capillaries. Arteries have thicker walls than veins, allowing them to withstand the high pressures.

Venule

Arteriole

Capillary wall

Red blood cell

O_2

CO_2

Body cell

CAPILLARY EXCHANGE
Capillaries are often so small that red blood cells pass single file through them. Here, the blood cells are close enough to the body cells that exchange can take place via diffusion.

INTERSTITIAL FLUID
Interstitial fluid fills the spaces between cells and tissues. It facilitates exchange of materials between cells and the blood.

— Oxygen-rich blood
— Oxygen-poor blood

WHAT CAN GO WRONG

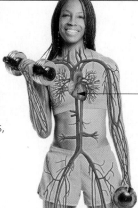

HYPERTENSION
Hypertension, or high blood pressure, affects about one-quarter of American adults. Although it significantly increases the risk of heart attack, heart disease, and stroke, many people do not know they have hypertension. Regular check-ups, eating a healthy diet, avoiding smoking, regular exercise, and medications can decrease your risk of hypertension.

HEART DISEASE
Heart disease is the most common cause of death among Americans. It often results from fatty deposits blocking the arteries that supply the heart muscle with oxygen.

ANEMIA
Anemia occurs when there is an abnormally low amount of hemoglobin (the oxygen-carrying molecule within red blood cells) or a low number of red blood cells. An anemic person feels constantly tired and rundown because the body cells do not get enough oxygen.

CORE IDEA: The human circulatory system contains a heart that pumps blood through a series of blood vessels. A double circulation system conveys blood from the heart to the lungs and back, then to the rest of the body and back.

CORE QUESTION: Do arteries always carry oxygenated blood?

ANSWER: No. That is true in the systemic circuit, but the arteries that leave the heart and connect to the lungs (pulmonary circuit) carry oxygen-poor blood.

The heart is the hub of the human circulatory system

The **heart**, a fist-sized muscular organ located under your sternum (breastbone), is the central hub of the human circulatory system. The heart conveys blood to the lungs. There, the blood picks up oxygen (O_2) and returns to the heart, which pumps the oxygenated blood to the rest of the body. At the body's cells, O_2 and other needed substances are delivered, and carbon dioxide (CO_2) and other wastes are collected. In this way, the heart ensures a steady flow of nutrients and waste removal.

CARDIAC ANATOMY

The human heart contains four chambers that control the movement of blood: one atrium (plural, *atria*) and one ventricle on each side (left and right). The atria collect blood returning to the heart and squeeze it a short distance into the ventricles. The thick, muscular ventricles pump blood out of the heart to other body organs. Here you can follow the path of blood as it makes one complete trip around the body.

① FROM THE BODY
Two large veins called the inferior vena cava and superior vena cava (collectively called vena cavae) bring oxygen-depleted blood from the body to the heart.

② RIGHT ATRIUM
Oxygen-poor blood enters the heart at the right atrium.

③ RIGHT VENTRICLE
The right atrium pumps the blood directly to the right ventricle, which then pumps the oxygen-depleted blood to the lungs via pulmonary arteries.

④ LUNGS
As blood flows through lung capillaries, CO_2 diffuses out of the blood and O_2 diffuses in. The O_2-rich blood then flows through pulmonary veins to the left atrium.

⑤ LEFT ATRIUM
O_2-rich blood from the lungs enters the heart at the left atrium, which pumps it directly to the left ventricle.

⑥ LEFT VENTRICLE
The left ventricle uses powerful contractions to pump O_2-rich blood through the aorta and out to the body.

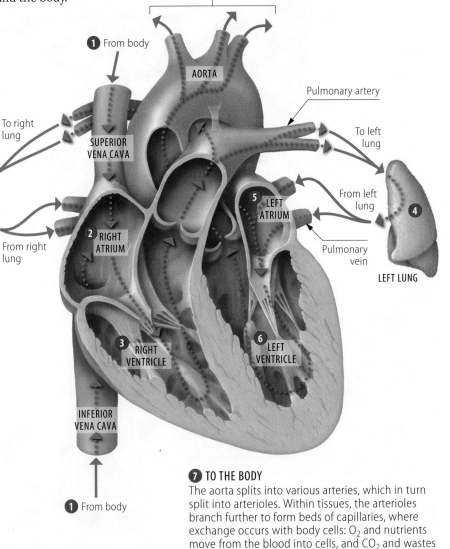

⑦ TO THE BODY
The aorta splits into various arteries, which in turn split into arterioles. Within tissues, the arterioles branch further to form beds of capillaries, where exchange occurs with body cells: O_2 and nutrients move from the blood into cells, and CO_2 and wastes move from cells into the blood. The O_2-poor blood then flows into venules, which flow into veins, which flow into the vena cavae, and enter the heart once again at the right atrium.

CARDIAC CYCLE

The heartbeat is maintained by the **cardiac cycle**, a rhythmic contraction and relaxation of the heart muscles. The normal heart rate for an adult is about 72 heartbeats per minute. A natural electrical pacemaker within the heart coordinates the steps in each cardiac cycle. The pacemaker interfaces with the nervous system, allowing the cardiac cycle to be sped up or slowed down in response to signals from the brain.

SINOATRIAL NODE
The **SA (sinoatrial) node** is the pacemaker of the heart. Within the wall of the right atrium, the SA node sends out electrical impulses that spread through the walls of both atria, causing them to contract simultaneously.

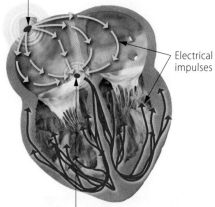

Electrical impulses

ATRIOVENTRICULAR NODE
When the electrical impulses reach the **AV (atrioventricular) node**, they pause for about 0.1 second, allowing the atria to empty. The impulses then cause the ventricles to contract strongly and in unison, pushing blood out of the heart.

1 DIASTOLE
During **diastole**, the heart muscles relax, allowing blood to flow into all four chambers of the heart.

DIASTOLE

SYSTOLE

3 VENTRICULAR SYSTOLE
At the end of systole, the AV node stimulates the ventricles to contract, pumping blood out of the heart and into the aorta and pulmonary arteries.

2 ATRIAL SYSTOLE
At the start of systole, the SA node stimulates the atria to contract, moving blood into the ventricles.

Through extensive conditioning, endurance athletes can achieve resting heart rates that are almost one-third slower than normal.

CORONARY ARTERIES

Like all the cells in your body, the cells of your heart require a steady supply of oxygen-rich blood. When blood exits the heart via the aorta, several coronary arteries immediately branch off, running along the outside of the heart, and supply the heart muscle itself with oxygenated blood. You can see the coronary arteries in this angiogram, an X-ray image created by injecting dye into the circulatory system. If one or more of the coronary arteries become blocked, heart muscle cells quickly die from lack of O$_2$. This is called a myocardial infarction, or **heart attack**. Although the symptoms may appear suddenly, the blockage of coronary arteries is usually the result of a gradual process called atherosclerosis. Plaque (fatty deposits) clogs up the inner walls of arteries, restricting blood flow. Treatment options for atherosclerosis include cholesterol-lowering drugs, angioplasty (a procedure in which a tiny balloon is used to widen clogged arteries), stents (wire mesh tubes that prop open clogged arteries), or bypass surgery (in which a section of blood vessel from the leg is used to shunt around the clog).

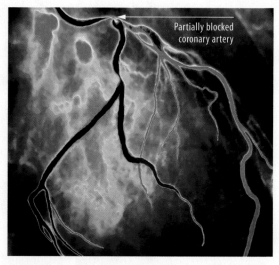
Partially blocked coronary artery

CORE QUESTION: What is the cause of a heart attack?

CORE IDEA: The heart lies at the center of the circulatory system. Blood enters the heart at the atria, which pump it a short distance to the ventricles. The ventricles then pump it out of the heart to the rest of the body.

ANSWER: A poor supply of O$_2$-rich blood to the muscle cells of the heart.

267

Blood contains cells in liquid

Blood is the fluid of the human circulatory system. Your body contains about 11 pints (5 liters) of blood. Your blood is about half water. The other half consists of a variety of cells (mostly red blood cells), proteins, salts, and a wide variety of substances that are being transported, such as nutrients, wastes, and hormones. Altogether, your blood is a complex mixture that helps maintain the working body.

COMPOSITION OF THE BLOOD

Your blood consists of many small molecules and several types of cells dissolved in a liquid called plasma. Blood provides long-distance transportation—for oxygen, carbon dioxide, nutrients, wastes, hormones, and countless other substances—throughout the body.

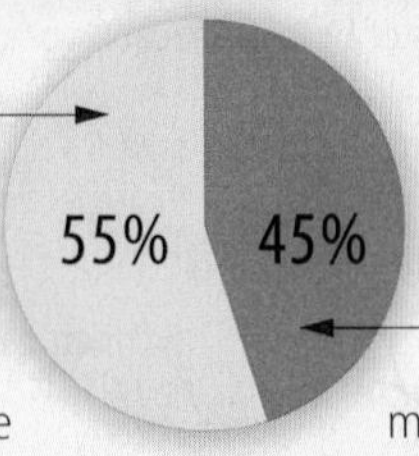

Plasma is a straw-colored liquid that makes up over half (55%) of the volume of blood. Plasma's share of blood volume consists of:

• Water (52%)

• Proteins, electrolytes (such as sodium, potassium, calcium), and substances being transported (such as O_2, CO_2, nutrients, wastes, hormones) (3%)

55%   45%

Cellular elements make up just under half (45%) of the volume of blood. They include:

• Red blood cells (44%)

• White blood cells and platelets (1%)

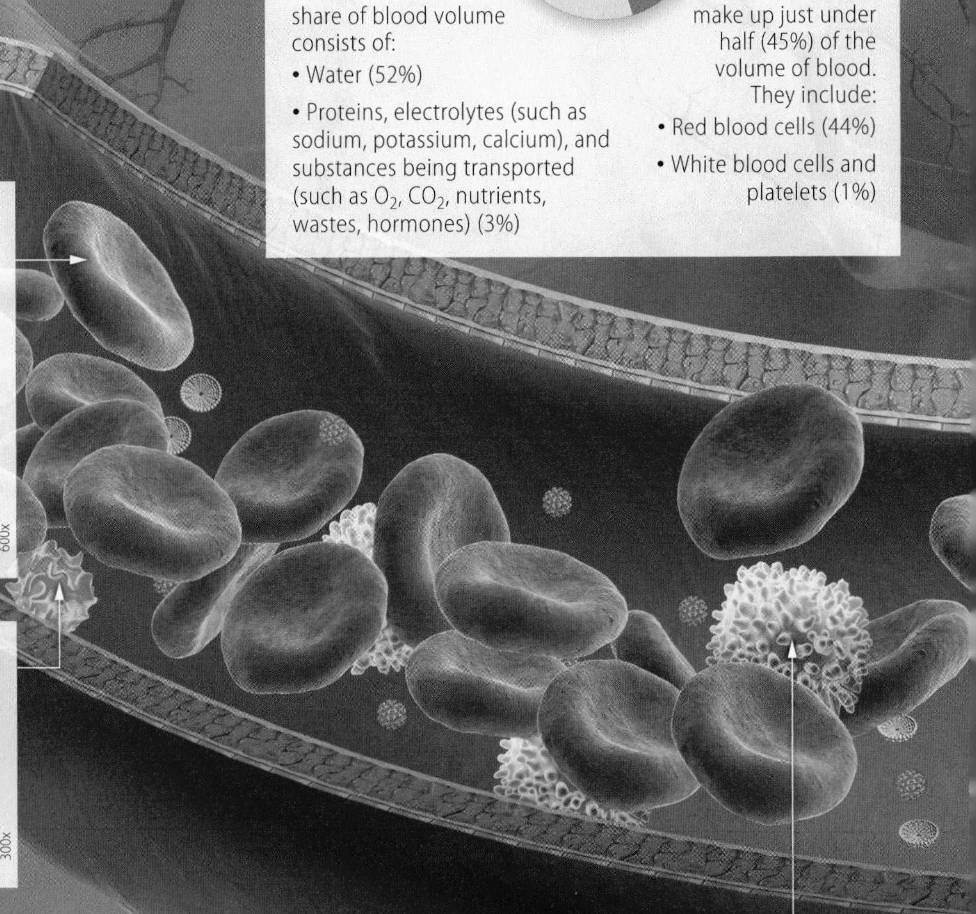

RED BLOOD CELLS (ERYTHROCYTES)

Red blood cells are by far the most numerous type of blood cell. Each one is small, lacks many organelles normally found in animal cells (such as a nucleus), and is shaped like a disk with indentations. Red blood cells are responsible for binding oxygen in the lungs and releasing it to body cells.

600x

PLATELETS (THROMBOCYTES)

Platelets are cellular fragments that aid in blood clotting.

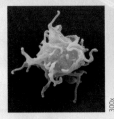

300x

WHITE BLOOD CELLS (LEUKOCYTES)

White blood cells, which make up less than 1% of all blood cells, fight infections as part of the immune system. There are many different kinds of white blood cells.

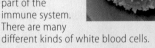

500x

11.11

You have over a trillion red blood cells in your bloodstream.

HEMOGLOBIN

Each red blood cell contains about a quarter billion molecules of **hemoglobin**, a protein that binds and transports oxygen. Each molecule of hemoglobin contains four polypeptide chains (two each of two different types), and each chain has an iron-containing chemical group called heme. Within the capillaries surrounding your lungs, oxygen diffuses from the alveoli into red blood cells and binds to the heme group of hemoglobin. In the capillaries of your body tissues, the process is reversed: Hemoglobin unloads its O_2 cargo, which diffuses out of the blood and into body cells.

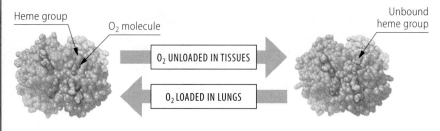

Heme group

O_2 molecule

Unbound heme group

O_2 UNLOADED IN TISSUES

O_2 LOADED IN LUNGS

OXYHEMOGLOBIN
In the capillaries surrounding the lungs, a molecule of O_2 binds to each heme group (four in total), forming oxyhemoglobin.

DEOXYHEMOGLOBIN
In the capillaries surrounding body cells, O_2 exits oxyhemoglobin, forming deoxyhemoglobin. Oxygen-poor blood circulates to the lungs, where it picks up a new supply of O_2.

CLOTTING

We all get cuts and scrapes frequently, but we don't bleed to death because our blood contains self-sealing substances that become activated when blood vessels are injured. **Platelets**, bits of cells pinched off from larger cells in the bone marrow, begin the healing process.

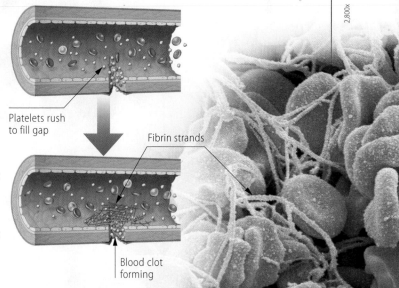

PLATELETS
Almost immediately after the tissue lining a blood vessel is damaged, platelets form a sticky plug that can seal a minor break.

Platelets rush to fill gap

Fibrin strands

FIBRIN
Platelets release molecules that convert a blood protein called fibrinogen into a thread-like form called fibrin. Molecules of fibrin cross-link to form a clot which, if on your skin, is called a scab.

Blood clot forming

2,800x

CORE IDEA: Your blood is mostly water, but it also contains red blood cells (which transport oxygen using hemoglobin), white blood cells (which fight infections), and platelets (which produce clots).

CORE QUESTION: Which cellular component of blood is the most like the majority of other body cells?

ANSWER: White blood cells, since cellular fragments and red blood cells lack many organelles.

The immune system contains a huge number of defensive elements

Your environment teems with **pathogens**, disease-causing viruses and microorganisms. Yet, you are not constantly ill, thanks to your **immune system**, the body's system of protection against infectious diseases. A vast array of immune components helps defend you from a wide variety of potentially harmful agents.

EXTERNAL DEFENSES

Your first line of defense against infections is a set of barriers that prevent pathogens from penetrating deep inside the body. These defenses are innate; that is, they are always deployed and ready to act.

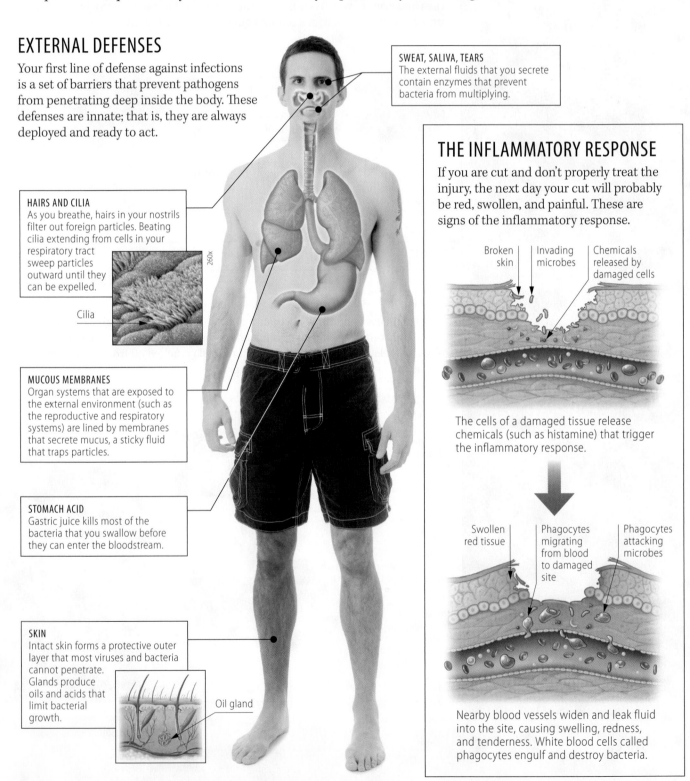

SWEAT, SALIVA, TEARS
The external fluids that you secrete contain enzymes that prevent bacteria from multiplying.

HAIRS AND CILIA
As you breathe, hairs in your nostrils filter out foreign particles. Beating cilia extending from cells in your respiratory tract sweep particles outward until they can be expelled.

260x

Cilia

MUCOUS MEMBRANES
Organ systems that are exposed to the external environment (such as the reproductive and respiratory systems) are lined by membranes that secrete mucus, a sticky fluid that traps particles.

STOMACH ACID
Gastric juice kills most of the bacteria that you swallow before they can enter the bloodstream.

SKIN
Intact skin forms a protective outer layer that most viruses and bacteria cannot penetrate. Glands produce oils and acids that limit bacterial growth.

Oil gland

THE INFLAMMATORY RESPONSE

If you are cut and don't properly treat the injury, the next day your cut will probably be red, swollen, and painful. These are signs of the inflammatory response.

Broken skin | Invading microbes | Chemicals released by damaged cells

The cells of a damaged tissue release chemicals (such as histamine) that trigger the inflammatory response.

Swollen red tissue | Phagocytes migrating from blood to damaged site | Phagocytes attacking microbes

Nearby blood vessels widen and leak fluid into the site, causing swelling, redness, and tenderness. White blood cells called phagocytes engulf and destroy bacteria.

COMPLEMENT SYSTEM

If an invader breaches the body's external barriers and enters the bloodstream, a series of proteins called the **complement system** can provide protection.

Bacterium

Proteins

Proteins of the complement system assemble on the surface of an invading bacterial cell, forming a hole.

Once a hole is formed, water and ions rush inside the foreign cell.

The target cell swells and eventually bursts.

When your doctor feels around your neck or under your armpits, she or he is checking for swollen lymph nodes, a sign of ongoing infection.

WHITE BLOOD CELLS

Your immune system contains a vast number of **white blood cells**, which come in two main types.

WHITE BLOOD CELLS			
INNATE		LYMPHOCYTES	
These cells are innate: They are pre-made and ready to attack.		Each type of **lymphocyte** is not normally available in great numbers, but can be rapidly produced by the immune system after contact with a specific invader.	
Phagocytic cells	Natural killer cells	B cells	T cells
Phagocytic cells engulf cells or molecules from foreign invaders or from dead cells. This process is called phagocytosis.	When a **natural killer cell** recognizes a body cell that has been infected by viruses or has become cancerous, it releases chemicals that target and may destroy the cell.	After being formed by stem cells in the bone marrow, **B cells** remain there to mature. Mature B cells produce and release defensive proteins called antibodies.	After being produced in the bone marrow, **T cells** mature in the thymus gland. T cells attack abnormal body cells.

320x

440x

THE LYMPHATIC SYSTEM

When your body fights an infection, the **lymphatic system** kicks into high gear. Consisting of a branching network of fluid-filled vessels and numerous small organs called **lymph nodes**, the lymphatic system produces huge numbers of white blood cells when needed. Fluid flows from blood capillaries into intermingled lymphatic capillaries. Invading microbes are picked up by this fluid, called **lymph**, and are swept into lymph nodes, where they are attacked by **lymphocytes** (white blood cells). Lymph drains into larger and larger lymphatic vessels, eventually reentering the circulatory system through two large lymphatic vessels that fuse with veins near the shoulders.

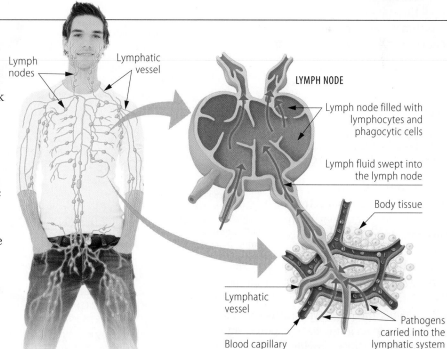

Lymph nodes

Lymphatic vessel

LYMPH NODE

Lymph node filled with lymphocytes and phagocytic cells

Lymph fluid swept into the lymph node

Body tissue

Lymphatic vessel

Blood capillary

Pathogens carried into the lymphatic system

CORE IDEA: Your immune system consists of a variety of defenses that protect against pathogens. These defenses include external barriers, the inflammatory response, complement system proteins, and white blood cells.

CORE QUESTION: What is the difference between the two types of lymphocytes, B cells and T cells?

ANSWER: They differ in where they mature (bone marrow versus thymus) and how they act (producing antibodies versus attacking abnormal body cells).

271

The immune system mounts highly specific attacks against invaders

Your immune system contains many elements that help prevent infection or mount a generalized defense. If these fail and infection does occur, the **adaptive defenses** come into play. Unlike innate defenses such as physical barriers or inflammation—which are always at the ready—the adaptive defenses must first be primed by exposure to an **antigen**, a molecule that elicits an immune response. Examples of antigens include molecules on the surfaces of microorganisms. The adaptive defenses depend on two types of **lymphocytes**, white blood cells that reside in the lymphatic system: B cells and T cells. The ability to recognize specific invaders is at the heart of the adaptive immune response.

ANTIBODIES: RECOGNIZING INVADERS

Every lymphocyte (B cell or T cell) has, on its surface, many copies of a specific protein that is capable of binding to one kind of antigen. The immune system contains a tremendous variety of lymphocytes, enough to bind to just about every possible antigen to which the body might be exposed. Once a particular antigen appears, it binds to its matching B cell. This binding activates the B cell. The activated B cell then gives rise to short-lived cells that secrete **antibodies**, proteins that circulate in the blood that are specific for that same antigen.

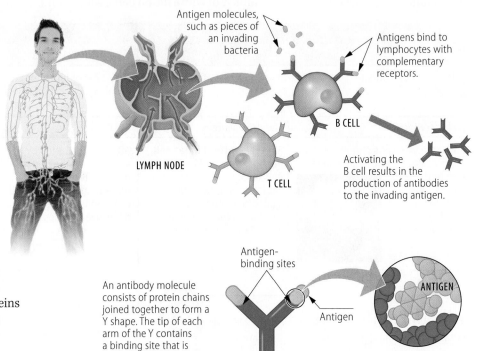

CLONAL SELECTION: BATTLING INVADERS

Upon entering the lymphatic system, an antigen will activate only those few lymphocytes that have matching receptors. Once activated, those lymphocytes will multiply, producing a large population of lymphocytes that are all specific for that same invading antigen. The produced lymphocytes are of two types: **effector cells**, which provide an immediate response, and **memory cells**, which provide long-lasting protection. This process, called **clonal selection**, allows the immune system to maintain a vast army of cells but to produce reinforcements only when they are needed.

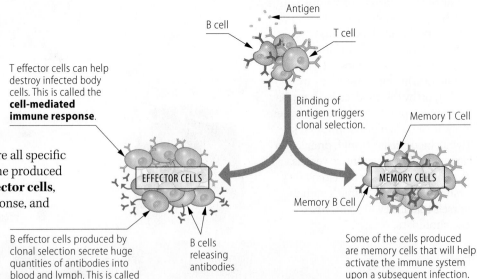

T CELLS: DESTROYING THE INVADERS

Once antigens are recognized, they must be destroyed. The first step occurs when antibodies produced via clonal selection bind the antigen. Sometimes, mere binding of the antibodies neutralizes the antigen. Other times, antibodies marshal several types of cells and molecules that work to destroy the invader.

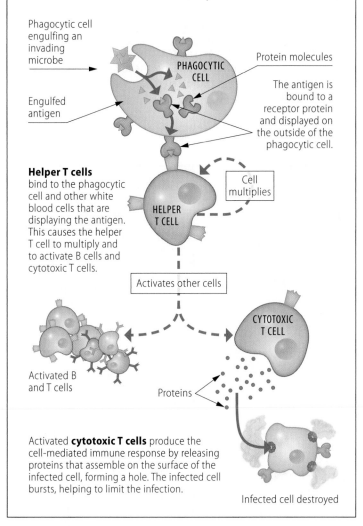

Phagocytic cell engulfing an invading microbe

Engulfed antigen

PHAGOCYTIC CELL

Protein molecules

The antigen is bound to a receptor protein and displayed on the outside of the phagocytic cell.

Helper T cells bind to the phagocytic cell and other white blood cells that are displaying the antigen. This causes the helper T cell to multiply and to activate B cells and cytotoxic T cells.

HELPER T CELL

Cell multiplies

Activates other cells

Activated B and T cells

CYTOTOXIC T CELL

Proteins

Activated **cytotoxic T cells** produce the cell-mediated immune response by releasing proteins that assemble on the surface of the infected cell, forming a hole. The infected cell bursts, helping to limit the infection.

Infected cell destroyed

MEMORY CELLS: REMEMBERING THE INVADER

The first time lymphocytes encounter an antigen, clonal selection is triggered. This is called the **primary immune response**. It produces effector cells that work to limit infection as well as memory cells that can live for decades. The primary immune response usually takes several days, during which time the invading cells are multiplying and perhaps causing illness. However, if the same invader is encountered again, the memory cells will consequently instigate a vigorous and rapid **secondary immune response** that neutralizes the invader before it causes illness. Thus, once exposed to an infectious disease, you may have lifetime immunity against it.

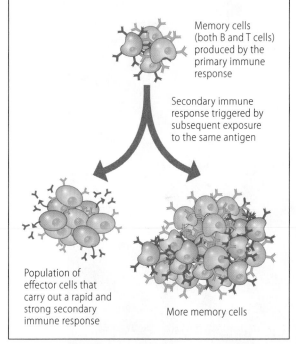

Memory cells (both B and T cells) produced by the primary immune response

Secondary immune response triggered by subsequent exposure to the same antigen

Population of effector cells that carry out a rapid and strong secondary immune response

More memory cells

VACCINATION

Vaccination (also called *immunization*) involves purposefully exposing the immune system to an antigen (often a deactivated, harmless version of a pathogen), which stimulates the production of memory cells. If that antigen is ever encountered again by exposure to the natural, harmful pathogen, the body can quickly fight it off with a secondary immune response. Vaccinations—which offer the only medical help for many viral diseases—have successfully eradicated many previously feared diseases such as polio and smallpox in industrialized nations. However, such diseases remain a threat in much of the developing world where immunizations are less available.

CORE IDEA: A lymphocyte binding an antigen triggers clonal selection, the humoral immune response (via antibodies produced by B cells), the cell-mediated immune response (via T cells), and immunological memory (via memory cells).

CORE QUESTION: Why does getting chicken pox as a child normally prevent you from getting it as an adult?

ANSWER: Because the first exposure will result in the production of memory cells that will quickly destroy the virus should you be exposed to it again.

Immune system malfunctions cause a variety of disorders

Your immune system is a finely tuned network that protects you against a vast array of pathogens in the environment. But if the intricate interplay of immune system components goes awry, a variety of problems may result, ranging from mild irritations to deadly diseases.

ALLERGIES

Allergies are sensitivities to allergens, components of the environment that trigger an allergic reaction but are otherwise harmless. Common allergens include protein molecules on pollen grains, on the feces of tiny mites that feed on dust, and in shed animal skin. An allergic reaction—essentially a hyperactive immune response—involves two stages, only the second of which produces allergy symptoms.

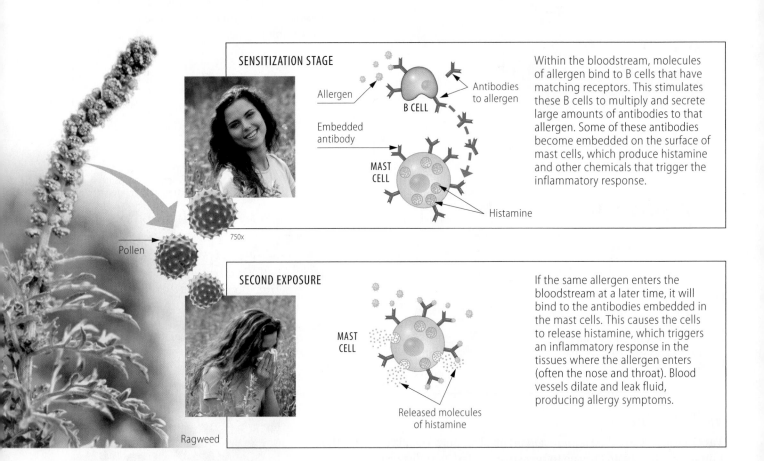

Pollen

750x

Ragweed

SENSITIZATION STAGE

Allergen

Antibodies to allergen

B CELL

Embedded antibody

MAST CELL

Histamine

Within the bloodstream, molecules of allergen bind to B cells that have matching receptors. This stimulates these B cells to multiply and secrete large amounts of antibodies to that allergen. Some of these antibodies become embedded on the surface of mast cells, which produce histamine and other chemicals that trigger the inflammatory response.

SECOND EXPOSURE

MAST CELL

Released molecules of histamine

If the same allergen enters the bloodstream at a later time, it will bind to the antibodies embedded in the mast cells. This causes the cells to release histamine, which triggers an inflammatory response in the tissues where the allergen enters (often the nose and throat). Blood vessels dilate and leak fluid, producing allergy symptoms.

AUTOIMMUNE DISEASES

Each cell in your body has "self" proteins on the surface that identify it as your own and therefore off limits to attack by your immune system. But sometimes this self-recognition system breaks down. **Autoimmune diseases** occur when the immune system improperly turns against the body's own molecules. For example, rheumatoid arthritis is an autoimmune disease that leads to damage and painful inflammation of the cartilage and bones of joints. In type 1 (insulin-dependent) diabetes, the insulin-producing cells of the pancreas wrongly become the targets of immune cells. In multiple sclerosis (MS), T cells wrongly attack proteins in nerve cells.

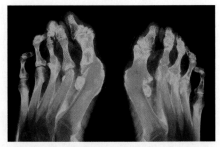

Colored X-ray image of the deformed feet of a patient suffering from rheumatoid arthritis.

IMMUNODEFICIENCIES

Immunodeficiency diseases result when one or more components of the immune system are missing or defective. Immunodeficient people are susceptible to a wide variety of infections that would normally be fought off. These frequent and recurrent opportunistic infections may eventually be fatal.

Severe combined immunodeficiency (SCID) is a rare genetic disease that destroys or deactivates both B cells and T cells. Because they are extremely susceptible to even minor infections, people with SCID (such as this boy from Texas, who lived until age 12 in a sterile environment) must remain behind protective barriers or receive bone marrow transplants. Gene therapy techniques have been successful in treating some people with SCID.

AIDS (acquired immunodeficiency syndrome) is a worldwide pandemic caused by **HIV (human immunodeficiency virus)**. When HIV enters the bloodstream—usually through unprotected sex or sharing needles—it destroys helper T cells (shown in the micrograph). This severely impairs immunity, causing the patient to die not from HIV itself but from other infectious agents or from cancer. The name AIDS refers to the fact that the disease is *acquired* through an infection that results in severe *immunodeficiency* and presents as a *syndrome*, a combination of symptoms.

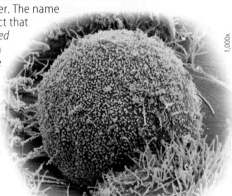

1,000×

Many people who are allergic to cats and dogs are actually allergic to proteins in the animal's saliva that get deposited on the fur when the animal licks itself.

IMMUNE REJECTION

The immune system's ability to recognize and attack foreign molecules does not always work in our favor. When a person receives an organ transplant, the immune system may recognize the donated tissue as foreign and attack it. To minimize rejection, proteins from the donor and recipient should match as closely as possible by using the recipient's own cells, those of a close relative, or those of a matching stranger. In graft-versus-host disease, foreign immune cells from transplanted bone marrow may attack the host's body cells. To avoid this, the patient will usually be given drugs that suppress the immune response.

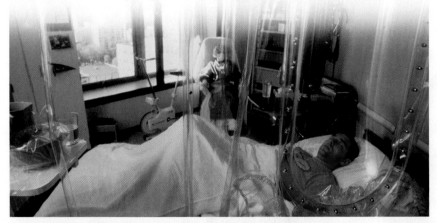

IMMUNITY AND STRESS

Evidence suggests that emotional stress can trigger the release of hormones that may suppress the immune system. Conversely, during times of happiness, the brain releases chemicals that may enhance immunity. The link between mood and health continues to be an active area of research.

CORE IDEA: When the immune response is too strong, allergies, autoimmune diseases, or organ rejection may occur. When the immune response is not strong enough, the result may be immunodeficiency diseases such as SCID and AIDS.

CORE QUESTION: Would a person be expected to have an allergic reaction to bee sting venom the first time they are bit?

ANSWER: No. Normally, the first sting primes the allergic reaction to occur upon the second sting.

The endocrine system regulates the body via hormones

The human **endocrine system** consists of several **endocrine glands** and other tissues that make and secrete hormones. **Hormones** are chemical signals that are produced by endocrine tissue and then transported by the circulatory system to other parts of the body, where they communicate regulatory messages. Because hormones reach all parts of the body, the endocrine system is important in controlling whole-body activities, such as metabolism, growth, and reproduction. A tiny amount of a hormone is often sufficient to influence many cells.

THE HUMAN ENDOCRINE SYSTEM

Of the major endocrine glands shown here, some—such as the thyroid and pituitary glands—are endocrine specialists, performing no other functions. Others—including the pancreas, which secretes hormones and digestive enzymes—have both endocrine and nonendocrine functions. And there are other organs (not shown here, such as the stomach) that secrete hormones but primarily serve other functions. Many of the glands of the endocrine system are controlled by the hypothalamus within the brain.

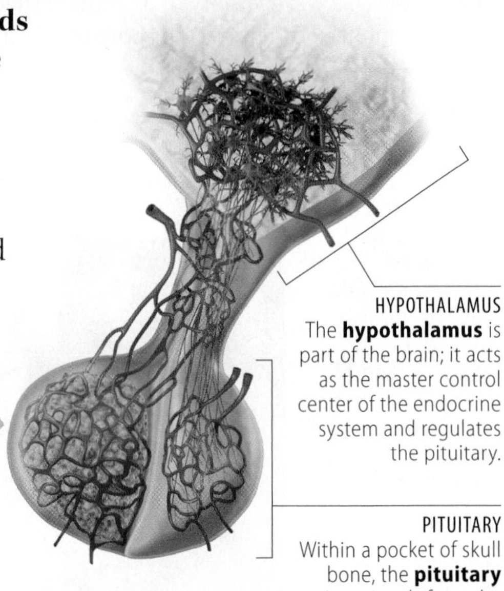

HYPOTHALAMUS
The **hypothalamus** is part of the brain; it acts as the master control center of the endocrine system and regulates the pituitary.

PITUITARY
Within a pocket of skull bone, the **pituitary** receives signals from the hypothalamus. The pituitary produces a wide variety of hormones that regulate many body functions.

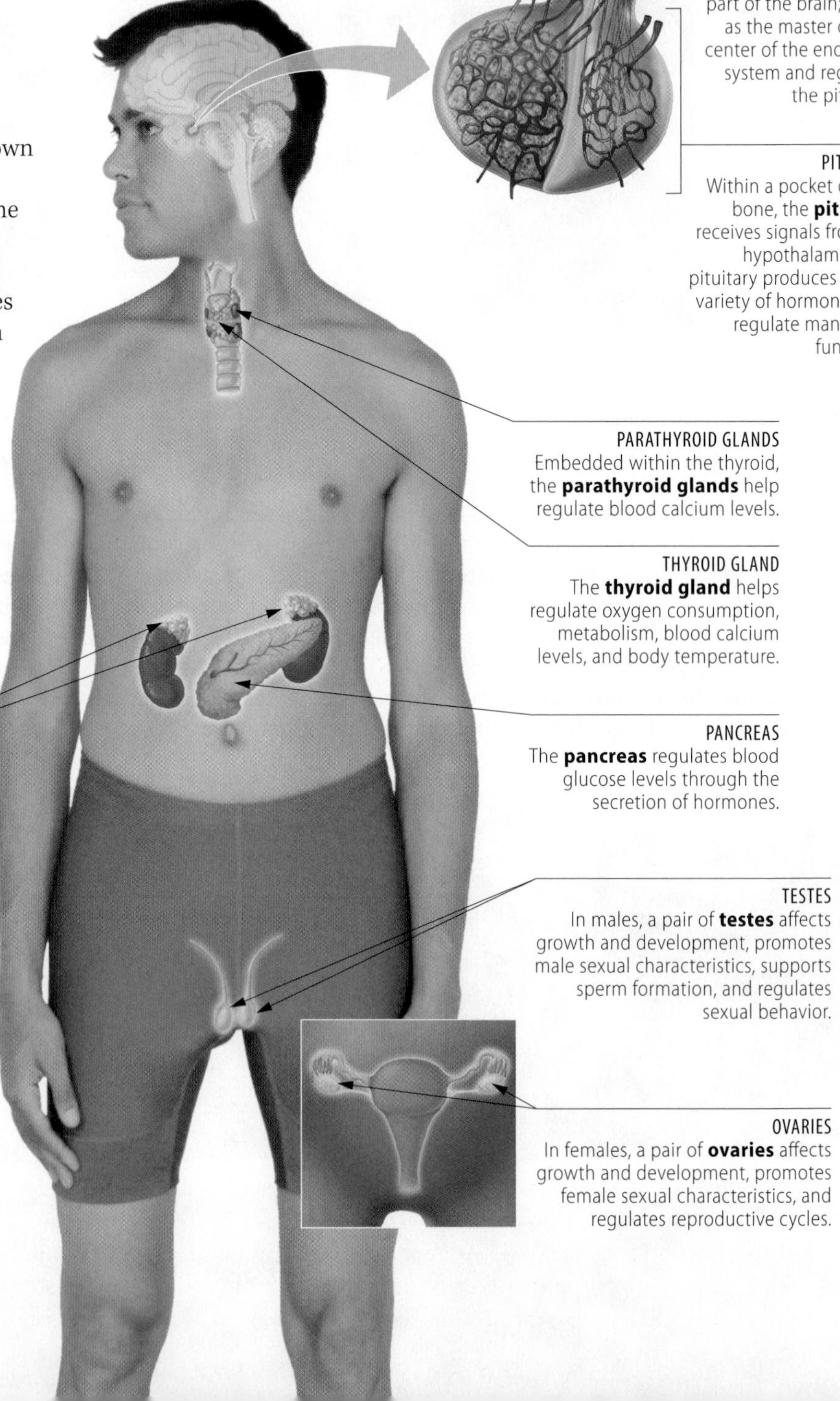

PARATHYROID GLANDS
Embedded within the thyroid, the **parathyroid glands** help regulate blood calcium levels.

THYROID GLAND
The **thyroid gland** helps regulate oxygen consumption, metabolism, blood calcium levels, and body temperature.

PANCREAS
The **pancreas** regulates blood glucose levels through the secretion of hormones.

ADRENAL GLANDS
There are two **adrenal glands**, one sitting atop each kidney; each one consists of two fused glands (the adrenal medulla and the adrenal cortex) that together regulate metabolism and responses to stress.

TESTES
In males, a pair of **testes** affects growth and development, promotes male sexual characteristics, supports sperm formation, and regulates sexual behavior.

OVARIES
In females, a pair of **ovaries** affects growth and development, promotes female sexual characteristics, and regulates reproductive cycles.

HOW HORMONES WORK

Within an endocrine cell, hormones are stored in membrane-enclosed vesicles. When the vesicles fuse with the cell's plasma membrane, the hormone is released from the cell into the circulatory system. Although a hormone may travel through the entire body, only cells with receptors for that specific hormone—called target cells—are affected. Cells without the matching receptor ignore the hormone. There are two main types of hormones that bring about changes within target cells in different ways: water-soluble hormones and fat-soluble hormones.

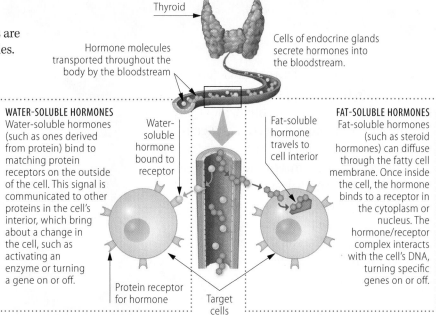

Thyroid

Hormone molecules transported throughout the body by the bloodstream

Cells of endocrine glands secrete hormones into the bloodstream.

WATER-SOLUBLE HORMONES
Water-soluble hormones (such as ones derived from protein) bind to matching protein receptors on the outside of the cell. This signal is communicated to other proteins in the cell's interior, which bring about a change in the cell, such as activating an enzyme or turning a gene on or off.

Water-soluble hormone bound to receptor

FAT-SOLUBLE HORMONES
Fat-soluble hormones (such as steroid hormones) can diffuse through the fatty cell membrane. Once inside the cell, the hormone binds to a receptor in the cytoplasm or nucleus. The hormone/receptor complex interacts with the cell's DNA, turning specific genes on or off.

Fat-soluble hormone travels to cell interior

Protein receptor for hormone

Target cells

About 6 million Americans have undiagnosed diabetes.

THE PANCREAS AND GLUCOSE REGULATION

The **pancreas**, located in the abdomen, performs two very different but vital body functions: it produces digestive enzymes that are secreted into the small intestine, and it produces two water-soluble hormones that regulate the body's metabolism of glucose. These two hormones, insulin and glucagon, are antagonistic hormones, meaning each one counteracts the effects of the other in a feedback circuit. Together, they help maintain a homeostatic balance, keeping the concentration of glucose in the bloodstream relatively near a balanced value.

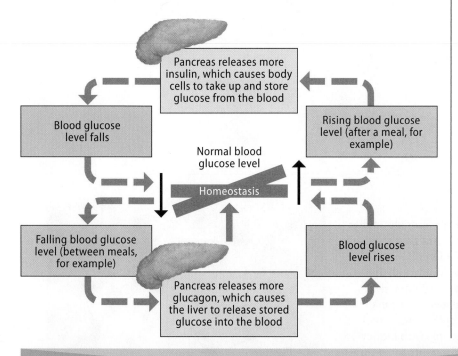

Pancreas releases more insulin, which causes body cells to take up and store glucose from the blood

Blood glucose level falls

Rising blood glucose level (after a meal, for example)

Normal blood glucose level

Homeostasis

Falling blood glucose level (between meals, for example)

Blood glucose level rises

Pancreas releases more glucagon, which causes the liver to release stored glucose into the blood

DIABETES

In a person with **diabetes mellitus**, the body either fails to produce enough insulin (as in type 1, or insulin-dependent diabetes) or target cells do not respond normally to the insulin that is produced (as in type 2, or non-insulin-dependent diabetes). Without proper regulation by insulin, cells of an individual with diabetes cannot obtain enough glucose from the blood, even though there is plenty. Furthermore, glucose concentrations in the blood can fluctuate widely, becoming too high or too low, requiring frequent monitoring. Type 1 patients require strict monitoring of blood glucose levels and regular injections of insulin. Type 2 diabetes is usually associated with obesity and can often be controlled through diet and exercise.

A home blood glucose monitor

CORE IDEA: The endocrine system consists of organs, glands, and other tissues that secrete hormones, chemical signals that are transported by the bloodstream and affect target cells throughout the body.

CORE QUESTION: What is the primary difference between water- and fat-soluble hormones in terms of where they act?

ANSWER: Water-soluble hormones bind to the outside of cells, whereas fat-soluble hormones enter cells and act there.

The urinary system regulates water and rids the body of wastes

The survival of humans and all animals depends on **osmoregulation**, the control of the gain or loss of water and dissolved ions. If too much water enters or leaves your body, you cannot survive. You gain water by eating and drinking, and you lose water through urinating, defecating, breathing, and perspiring. Water balance in your body is maintained by your **urinary system**. It excretes waste-carrying urine while keeping the proper amounts of water and the substances dissolved within it. The kidneys are the central hub of the urinary system.

THE HUMAN URINARY SYSTEM

The human urinary system produces and excretes urine. It also reclaims water and vital substances (such as glucose, important ions, and minerals) and returns them to circulation. The main processing centers are the two kidneys, located on either side of the abdomen, each a bit smaller than a fist.

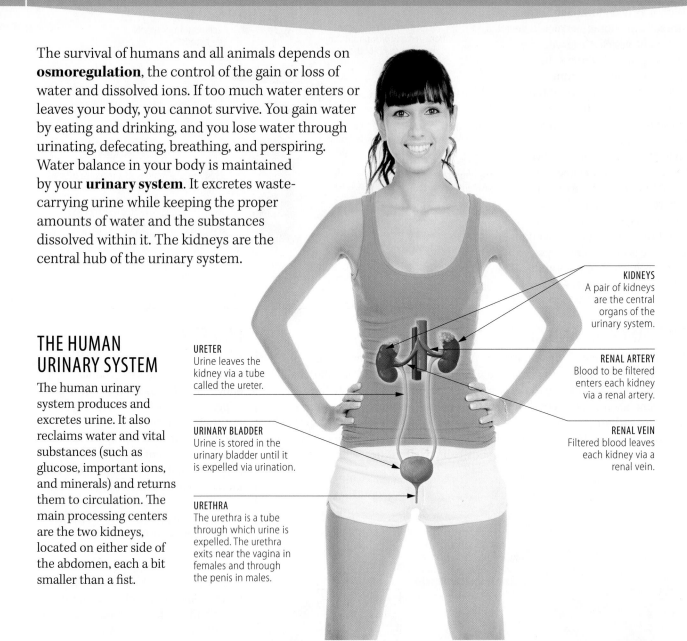

URETER
Urine leaves the kidney via a tube called the ureter.

URINARY BLADDER
Urine is stored in the urinary bladder until it is expelled via urination.

URETHRA
The urethra is a tube through which urine is expelled. The urethra exits near the vagina in females and through the penis in males.

KIDNEYS
A pair of kidneys are the central organs of the urinary system.

RENAL ARTERY
Blood to be filtered enters each kidney via a renal artery.

RENAL VEIN
Filtered blood leaves each kidney via a renal vein.

DIALYSIS

Kidney failure, the inability of the kidneys to filter blood, can be caused by injury, illness, or prolonged use of pain relievers, alcohol, or other drugs. A person with one functioning kidney can lead a normal life, but if both kidneys fail, the buildup of toxic wastes will lead to certain and rapid death. One option is to place the person on **dialysis**, filtration of the blood by a machine that mimics the action of a kidney. Dialysis treatment is life sustaining for people with kidney failure, but it is costly, requires about 4 to 6 hours three times a week, and must be continued for life. Another option is a kidney transplant, but the average wait for a kidney donation in the United States is three to five years.

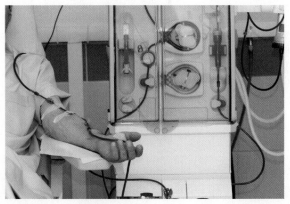

People suffering from kidney failure can have their blood filtered by a dialysis machine.

THE KIDNEYS

Blood flows continuously through the kidneys. As it does, some substances within the blood are filtered out and enter kidney tubules (small tubes). As the filtered liquid passes through the kidney tubules, needed substances (such as water and glucose) are reclaimed back into the bloodstream, while wastes (such as urea) are concentrated. The actual filtering of blood and formation of urine takes place within nephrons. The end result is urine, which passes out of the kidney via the ureter and is stored in the urinary bladder until it is excreted.

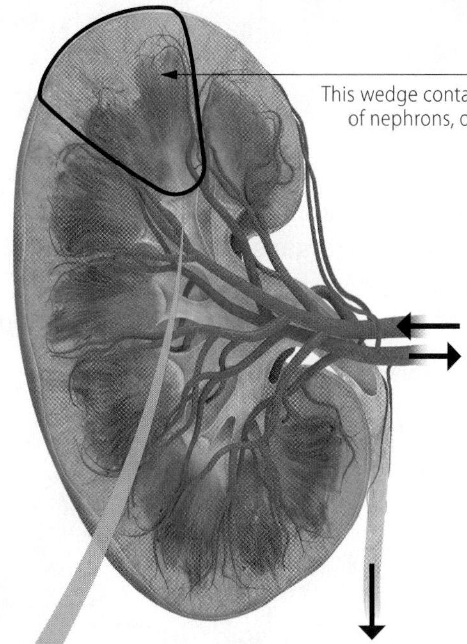

NEPHRONS

There are millions of nephrons in a kidney. A single **nephron**, consisting of one tubule and surrounding capillaries, is where the blood is actually filtered.

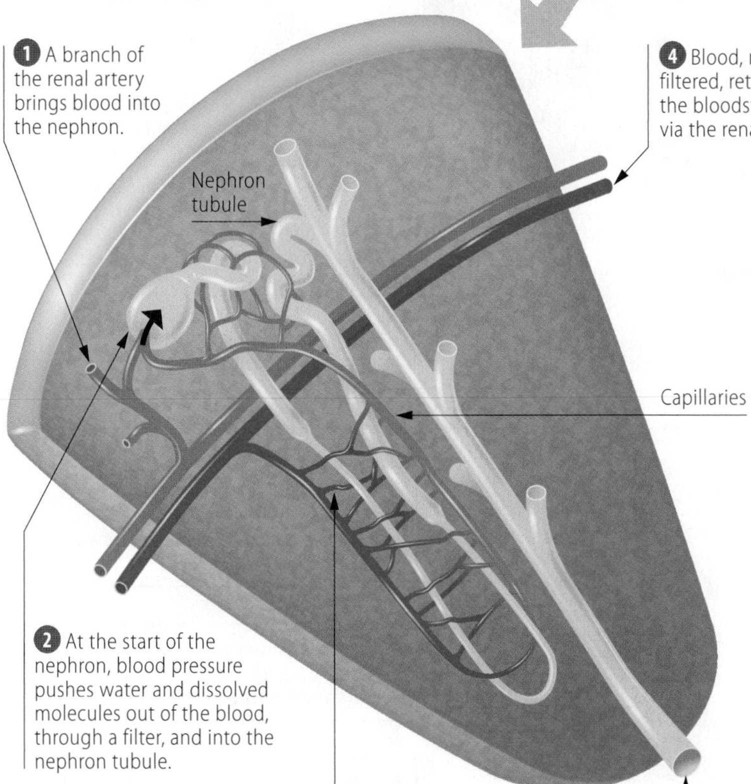

This wedge contains thousands of nephrons, one of which is shown below.

Blood to be filtered enters the kidneys via a renal artery, which branches into millions of tiny blood vessels within the kidney.

Filtered blood leaves the kidney via a renal vein.

Urine exits the kidney via the ureter and collects in the urinary bladder.

1 A branch of the renal artery brings blood into the nephron.

Nephron tubule

4 Blood, now filtered, returns to the bloodstream via the renal vein.

Capillaries

2 At the start of the nephron, blood pressure pushes water and dissolved molecules out of the blood, through a filter, and into the nephron tubule.

3 Water and valuable solutes are reclaimed and returned to the blood via tiny capillaries that surround the tubules, while unneeded substances are moved from the blood via these same capillaries into the kidney.

5 Concentrated urine travels through a connecting duct, exits the nephron via the ureter, and is stored in the urinary bladder until it is expelled.

COMPOSITION OF URINE
Urine is mostly water. The remainder consists of dissolved solutes and urea, a nitrogen-containing waste product.

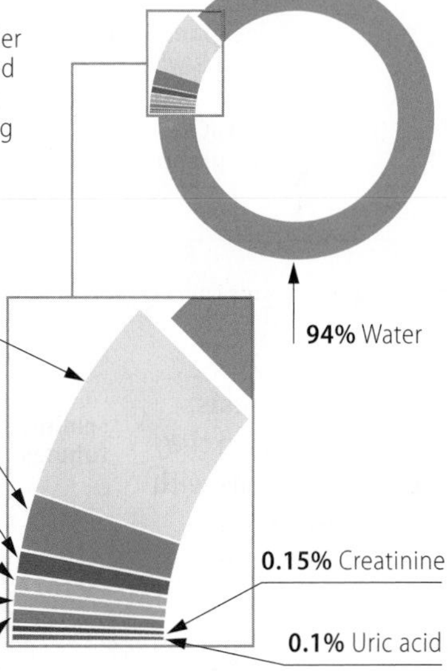

3.5% Urea
1% Chloride
0.5% Sodium
0.25% Potassium
0.25% Phosphate
0.25% Sulphate
94% Water
0.15% Creatinine
0.1% Uric acid

CORE IDEA: The urinary system disposes of wastes and helps regulate the concentration of water and dissolved substances within the body. Inside kidney nephrons, needed substances are retained and wastes are collected, forming urine.

CORE QUESTION: What is the functional unit of the urinary system, where urine is actually formed?

Males and females produce, store, and deliver gametes

Although we tend to focus on the anatomical differences between human males and females, the **reproductive systems** of both sexes have several important similarities. For example, both sexes have a pair of **gonads**, the organs that produce gametes: sperm in males and eggs in females. Both sexes have a system of ducts that store and deliver the gametes for reproduction. And both sexes have structures that facilitate copulation (sexual intercourse).

HUMAN MALE REPRODUCTIVE ANATOMY

→ PATH OF SPERM

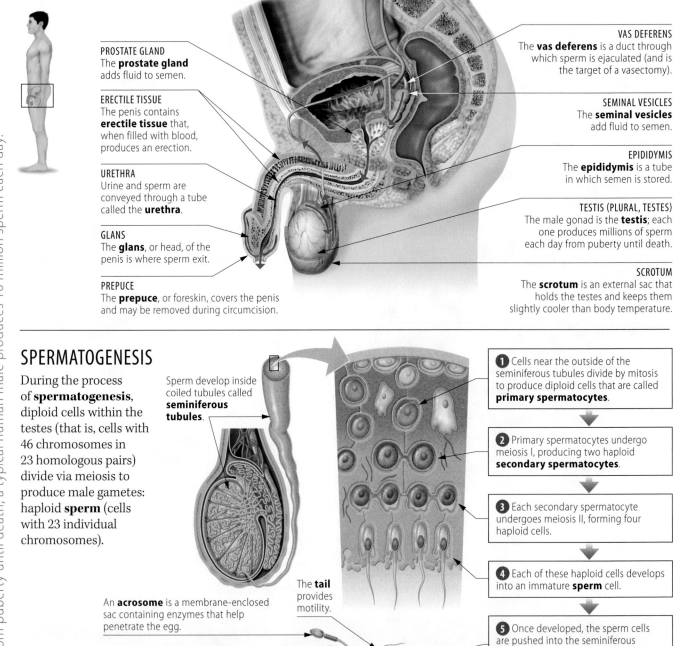

PROSTATE GLAND
The **prostate gland** adds fluid to semen.

ERECTILE TISSUE
The penis contains **erectile tissue** that, when filled with blood, produces an erection.

URETHRA
Urine and sperm are conveyed through a tube called the **urethra**.

GLANS
The **glans**, or head, of the penis is where sperm exit.

PREPUCE
The **prepuce**, or foreskin, covers the penis and may be removed during circumcision.

VAS DEFERENS
The **vas deferens** is a duct through which sperm is ejaculated (and is the target of a vasectomy).

SEMINAL VESICLES
The **seminal vesicles** add fluid to semen.

EPIDIDYMIS
The **epididymis** is a tube in which semen is stored.

TESTIS (PLURAL, TESTES)
The male gonad is the **testis**; each one produces millions of sperm each day from puberty until death.

SCROTUM
The **scrotum** is an external sac that holds the testes and keeps them slightly cooler than body temperature.

SPERMATOGENESIS

During the process of **spermatogenesis**, diploid cells within the testes (that is, cells with 46 chromosomes in 23 homologous pairs) divide via meiosis to produce male gametes: haploid **sperm** (cells with 23 individual chromosomes).

Sperm develop inside coiled tubules called **seminiferous tubules**.

The **tail** provides motility.

An **acrosome** is a membrane-enclosed sac containing enzymes that help penetrate the egg.

The **head** contains a haploid nucleus with 23 chromosomes.

❶ Cells near the outside of the seminiferous tubules divide by mitosis to produce diploid cells that are called **primary spermatocytes**.

❷ Primary spermatocytes undergo meiosis I, producing two haploid **secondary spermatocytes**.

❸ Each secondary spermatocyte undergoes meiosis II, forming four haploid cells.

❹ Each of these haploid cells develops into an immature **sperm** cell.

❺ Once developed, the sperm cells are pushed into the seminiferous tubule and then the epididymis, where they mature and are stored.

From puberty until death, a typical human male produces 10 million sperm each day.

HUMAN FEMALE REPRODUCTIVE ANATOMY

OVARY
The **ovary** is the female gonad, where eggs are produced and released.

CERVIX
The **cervix** is a narrow neck at the bottom of the uterus.

VAGINA
The **vagina**, or birth canal, is where sperm enters and a baby exits.

URETHRA
The **urethra** is a tube through which urine is excreted.

OVIDUCT
The **oviduct** (also called the fallopian tube) is the site where egg meets sperm.

UTERUS, OR WOMB
The **uterus** is the site of pregnancy where an embryo develops into a baby.

CLITORIS
The **clitoris** contains erectile tissue that swells during arousal.

LABIA MINORA
The **labia minora** are a pair of skin folds that border the opening of the vagina.

LABIA MAJORA
The **labia majora** are a pair of thick ridges that protect underlying structures.

VULVA
The **vulva** is the collective name for all of the external female reproductive structures.

FEMALE REPRODUCTIVE CYCLE

The human female reproductive cycle is a series of recurring 28-day (approximately) events that prepares the body for reproduction. Day 0 marks the start of a woman's menstruation (period). **Ovulation**, the release of an egg cell from the ovaries, occurs around day 14. A series of hormones regulates the formation of a blood-rich uterine lining and—if pregnancy has not occurred—its release during menstruation.

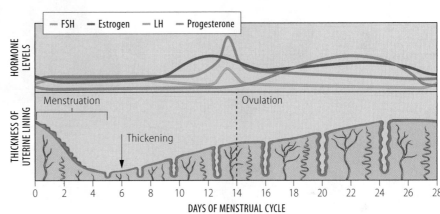

— FSH   — Estrogen   — LH   — Progesterone

HORMONE LEVELS

THICKNESS OF UTERINE LINING

Menstruation   Ovulation

Thickening

DAYS OF MENSTRUAL CYCLE

OOGENESIS

Here, you can follow the events in **oogenesis**, the development of a mature egg (also called an **ovum**; plural, *ova*) within the ovary. An egg is a haploid cell formed by the division of a diploid cell. A woman is born with many primary oocytes. Each month, one (or, rarely, more) primary oocyte within the ovary develops further and is released.

OVARY

❶ A **primary oocyte** is a diploid cell that is paused in meiosis I. A hormone can trigger the dormant cell to complete its development.

❷ A **secondary oocyte** is formed when a primary oocyte completes meiosis I and enters meiosis II.

❸ **Ovulation** is the release of a secondary oocyte from the ovary into the oviduct. It will complete meiosis II to form a mature egg only if it contacts sperm in the oviduct.

❹ The **corpus luteum** develops from the ruptured follicle. If the oocyte is not fertilized, the corpus luteum degenerates.

CORE QUESTION: What organs are the actual male and female gonads?

CORE IDEA: The reproductive systems of both males and females produce, store, and release gametes (sperm and eggs, respectively). Haploid gametes are produced from diploid cells within the gonads (testes and ovaries, respectively).

ANSWER: Testes in males and ovaries in females.

A human develops from a single cell

Although you may not wish to think about the details, your biological story began when one of your father's sperm fused with one of your mother's eggs. Starting at that point, a typical human pregnancy lasts 266 days (or 38 weeks), although it is often measured as 40 weeks (or 9 months) from the start of the last menstrual cycle. To follow the journey from fertilized egg to free-living baby, start at the bottom of this page.

3 CLEAVAGE

Starting about 36 hours after fertilization, the zygote undergoes **cleavage**, a series of rapid divisions that continues for several days. As it continues to move down the oviduct, the embryo (this name is used after the first cell division) forms a ball with many cells, although the ball remains the same overall size of the original egg.

DAYS 2–6

4 BLASTOCYST

By day 7, the early embryo has reached the uterus and begun to implant in the endometrium lining. The embryo is now a fluid-filled ball of about 100 cells called a **blastocyst**. It contains an inner cell mass that will eventually form the fetus and an outer layer called the trophoblast that aids implantation.

DAYS 7–9

Trophoblast

Inner cell mass containing embryonic stem cells

Endometrium

5 GASTRULA

Starting on day 9, the cells begin to migrate within the blastocyst, organizing themselves into three main layers. The layers of the resulting **gastrula** will each develop into specific structures (for example, the middle layer eventually gives rise to the muscles, heart, and other organs).

DAYS 9–14

Gastrula

2 ZYGOTE

If a sperm meets an egg in the oviduct, the sperm's membrane fuses with the egg's membrane, and the sperm nucleus enters the egg. The two nuclei fuse, with 23 chromosomes from the sperm joining 23 chromosomes from the egg to form a diploid nucleus in the zygote. This first cell switches into high metabolic gear, preparing for the tremendous growth that will soon begin.

DAY 1

The **oviduct**, also called the fallopian tube, contains cilia that sweep the egg toward the uterus.

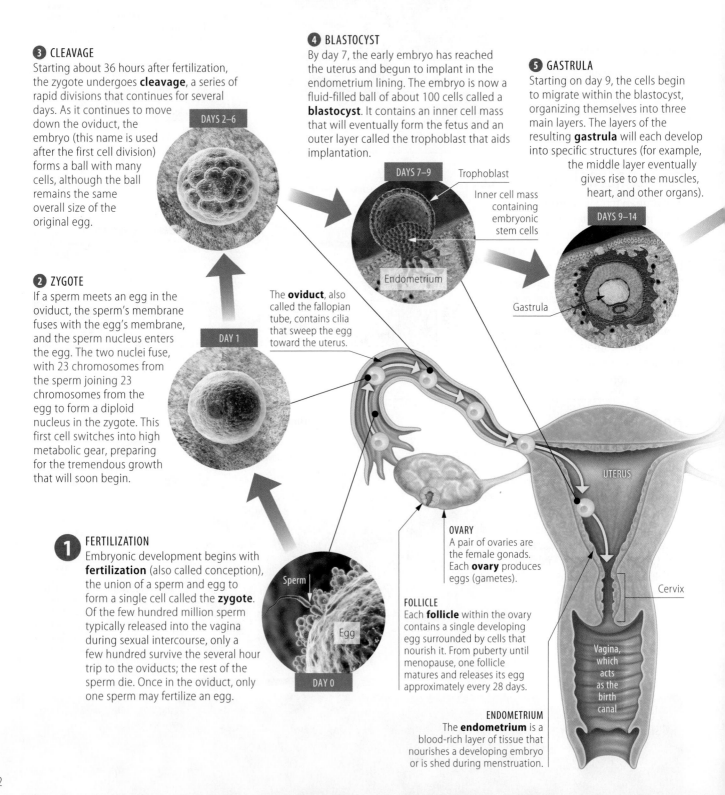

UTERUS

Cervix

Vagina, which acts as the birth canal

1 FERTILIZATION

Embryonic development begins with **fertilization** (also called conception), the union of a sperm and egg to form a single cell called the **zygote**. Of the few hundred million sperm typically released into the vagina during sexual intercourse, only a few hundred survive the several hour trip to the oviducts; the rest of the sperm die. Once in the oviduct, only one sperm may fertilize an egg.

Sperm

Egg

DAY 0

OVARY

A pair of ovaries are the female gonads. Each **ovary** produces eggs (gametes).

FOLLICLE

Each **follicle** within the ovary contains a single developing egg surrounded by cells that nourish it. From puberty until menopause, one follicle matures and releases its egg approximately every 28 days.

ENDOMETRIUM

The **endometrium** is a blood-rich layer of tissue that nourishes a developing embryo or is shed during menstruation.

6 LIFE-SUPPORT EQUIPMENT

By the third week after fertilization, the embryo is attached to the wall of the uterus, and several important early structures have begun to develop. Although substances can pass back and forth via diffusion, the embryo now has its own blood supply that does not mix with the mother's.

7 FIRST TRIMESTER: ORGAN FORMATION

Pregnancy is often divided into three trimesters of about 3 months each. At a fraction of an inch long, a month-old embryo is not yet distinctly human in appearance. Just after the end of the second month, the embryo has developed into a 2-inch-long **fetus** (this term is used from the ninth week until birth). By the end of the third month, the fetus looks like a miniature (but disproportioned) human being.

PLACENTA
The **placenta** provides nourishment and energy to the embryo and helps dispose of its waste.

DAY 21

SPINAL CORD
The spinal cord and brain have begun to form.

The eyes are starting to form.

4 WEEKS

These gill-like structures will form into parts of the head.

The limb buds will develop into arms and legs.

Tail

AMNION
The **amnion** is a fluid-filled sac that encloses and protects the embryo.

UMBILICAL CORD
The **umbilical cord** connects the embryo to the placenta.

8 WEEKS

The joints are developing into limbs.

The muscles of the back and ribs have formed.

The digits are visible.

20 WEEKS

8 SECOND TRIMESTER: GROWTH

During this stage, the fetus grows, and its features become more distinct. At 20 weeks, the fetus is about 8 inches long, weighs about a pound, has many recognizable features and a measurable heartbeat, and can usually be felt moving.

35 WEEKS

9 THIRD TRIMESTER: PREPARING FOR BIRTH

During this stage, the fetus grows rapidly and develops the organs (such as the lungs and heart) required to survive outside the womb. Its muscles grow and strengthen, and it usually turns head-down as it fills the uterus.

40 WEEKS

Placenta

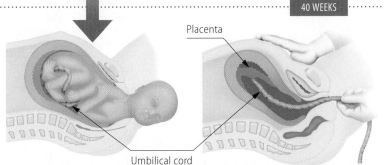

Umbilical cord

10 BIRTH

Childbirth is brought about by **labor**, a series of contractions by the muscles that line the uterus. Labor is often preceded by breaking of the amnion, which releases amniotic fluid ("water breaking"). The cervix dilates (widens) and a positive feedback loop of hormones produces increasingly strong contractions. After the baby is pushed through the uterus and vagina, the umbilical cord is clamped and cut, and the placenta ("afterbirth") is delivered.

An elephant's pregnancy lasts 22 months, while a mouse's lasts 3 weeks.

CORE IDEA: The time from human fertilization to birth takes about 38 weeks. During this time, a zygote develops into an embryo and then into a fetus.

CORE QUESTION: Where does fertilization take place?

ANSWER: In the oviduct (fallopian tube).

Issues of reproductive health affect us all

Knowledge of the anatomy and physiology of the human reproductive system is important to everyone. Issues such as having a baby (or not), sexually transmitted diseases, and infertility will arise in your life and the lives of those you care about.

CONTRACEPTION

Contraception is the deliberate prevention of pregnancy. The various types of contraception interfere with different steps in the process of becoming pregnant.

FEMALE INFERTILITY

About 15% of couples are **infertile**—that is, they are unable to conceive a child despite one year of trying. There are many potential causes of infertility, and several possible solutions that can help infertile couples produce a baby.

INTRAUTERINE DEVICE (IUD)
An IUD is a T-shaped device placed within the uterus by a health-care provider. IUDs can prevent pregnancy for up to 12 years and can be safely removed at any time.

TUBAL LIGATION
Also called "having your tubes tied," tubal ligation is a form of sterilization in which a section of each oviduct is removed and the remaining ends are tied (ligated).

RHYTHM METHOD
The rhythm method, or natural family planning, depends on refraining from intercourse during the days around ovulation. Because ovulation is hard to detect with certainty, this method is unreliable.

ORAL CONTRACEPTIVES
Oral contraceptives (birth control pills) contain synthetic hormones that prevent development of an egg in the ovaries. These hormones are also available as a shot or a skin patch.

ABSTINENCE
Abstinence (avoiding intercourse) is the only totally effective method of contraception.

DIAPHRAGM
A diaphragm is a rubber cap that covers the cervix, preventing sperm from entering the uterus. It requires a doctor's visit and is most effective when combined with spermicide, sperm-killing chemicals.

ECTOPIC PREGNANCY
An embryo may become lodged within an oviduct. Such a situation, called an ectopic pregnancy, is inevitably fatal to the embryo and can be dangerous to the mother, requiring immediate medical attention.

BLOCKED OVIDUCT
Several sexually transmitted diseases damage the oviducts. The resulting scarring can block them, preventing sperm from reaching eggs.

OVULATION PROBLEMS
Female infertility can result from a lack of eggs or a failure to ovulate, often due to hormonal irregularities. Fertility drugs may offer a cure, but they frequently cause multiple eggs to be released at once, therefore resulting in multiple pregnancies (twins, triplets, or beyond). Alternatively, a woman may conceive using eggs from a donor.

MISCARRIAGES
Some women can conceive but are unable to support a growing embryo in the uterus, resulting in a miscarriage or stillbirth. Some women turn to surrogate mothers, a woman who has entered into a legal contract to carry someone else's baby.

From 1980 to 2003, the increased use of fertility drugs caused the rate of "supertwin" births (triplets, quadruplets, or more) to rise by 600%.

CONTRACEPTION

VASECTOMY
A form of male sterilization, a vasectomy involves cutting out a section of each vas deferens, thereby preventing sperm from reaching the urethra.

CONDOM
Condoms are barriers, usually made of latex, that physically prevent sperm from entering the vagina. Condoms also provide protection against STDs.

WITHDRAWAL
Withdrawal of the penis before ejaculation is often ineffective because sperm may exit the penis before orgasm.

MALE INFERTILITY

IMPOTENCE
Impotence (also called erectile dysfunction) is the inability to maintain an erection. Impotence can be temporary due to alcohol, drugs, or psychological problems, or it can be permanent due to nervous or circulatory system problems. Drug therapies (such as Viagra) can offer help.

LOW SPERM COUNT
If a man's testes fail to produce enough sperm, or if the sperm produced are not vigorous, a change of underwear from briefs (which keep the testes next to the body and therefore warm) to boxers (which reduce the temperature of the testes, increasing sperm production) may offer a cure. If a man is unable to produce sperm on his own, sperm may be collected from a donor, concentrated, and injected into a woman's uterus to promote pregnancy.

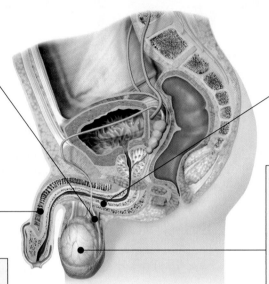

SEXUALLY TRANSMITTED DISEASES

Sexually transmitted diseases (STDs), also called sexually transmitted infections (STIs), are contagious diseases spread by sexual contact. This table shows the most common STDs in the United States, organized by the type of organism that causes the infection. While many STDs can be cured via antibiotics, STDs caused by viruses last a lifetime; there is no cure for viral STDs. Sexually active individuals are encouraged to be screened for STDs annually.

	Bacteria	Viruses	Protists	Fungi
Example of infectious agent	*Neisseria gonorrhoeae*	Herpes virus	*Trichomonas vaginalis*	*Candida albicans*
Diseases	• Gonorrhea often causes painful, burning urination in men, but may produce no symptoms in women. • Chlamydia is the most common STD. It often presents no symptoms but can lead to pelvic inflammatory disease and sterility.	• The virus that causes genital herpes can remain dormant for years, but then reappear and cause a new outbreak. • Some cases of genital warts are caused by the human papillomavirus, for which there is a vaccine. • AIDS, caused by HIV, is fatal unless treated.	• Trichomoniasis is caused by infection of the urethra or vagina by a single-celled protozoan parasite. Symptoms in women include burning, but men may be carriers without symptoms.	• Candidiasis, also called a yeast infection, is caused by a fungal infection of the mucous membranes of the vagina. About 75% of women will have candidiasis during their lives. Luckily, it is easily treatable.

CORE IDEA: Issues of reproductive health—contraception, STDs, and infertility—can be understood in relation to the anatomy of the human reproductive system.

CORE QUESTION: Of the methods of contraception listed, which one(s) actually prevent gametes from being formed?

ANSWER: Only female oral contraceptives.

The brain is the hub of the human nervous system

Your **nervous system** forms a communication and coordination network that runs throughout your body. The brain acts as the central hub while the spinal cord acts as the main conduit to the rest of the body.

THE HUMAN NERVOUS SYSTEM

Like the nervous systems of all other vertebrates, the human nervous system can be divided into two distinct divisions: the central nervous system (CNS) and the peripheral nervous system (PNS).

WHAT CAN GO WRONG

Diseases and injuries that affect the nervous system can have a profound impact.

MENINGES
The CNS is protected by a layer of connective tissue called the **meninges**.

BRAIN
The **brain** receives and integrates sensory information, keeps the body functioning, controls the muscles, and is the center of emotion and intellect.

CENTRAL NERVOUS SYSTEM
The **central nervous system (CNS)** receives incoming signals from the senses (such as touch), integrates them to formulate a response, and transmits signals that produce reactions, such as moving a muscle.

SPINAL CORD
The **spinal cord** is a jellylike bundle of millions of nerve fibers protected inside the hard spine (backbone). The spinal cord is the central communication conduit between the brain and the body.

CEREBROSPINAL FLUID
Both the brain and the spinal cord contain spaces filled with **cerebrospinal fluid**. It provides a cushion and helps supply nutrients, hormones, and white blood cells.

AUTONOMIC NERVOUS SYSTEM
The autonomic nervous system controls many internal body organs systems, such as the circulatory, excretory, and endocrine systems. This control is generally involuntary (unconscious).

PERIPHERAL NERVOUS SYSTEM
The **peripheral nervous system (PNS)** contains the nerves that convey information into and out of the CNS.

SOMATIC NERVOUS SYSTEM
The somatic nervous system is mostly under conscious control, although it also includes involuntary reflexes. For example, motor neurons carry signals to skeletal muscles, allowing you to move as you wish.

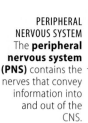

DEPRESSION
Nearly 20 million Americans, about two-thirds of them women, are affected by depressive illness: major depression (extreme and persistent sadness that leaves the sufferer unable to live a normal life) or bipolar disorder (also called manic-depressive disorder, involving extreme mood swings).

ALZHEIMER'S DISEASE
Alzheimer's disease is a form of dementia (mental deterioration) that causes confusion, memory loss, and personality changes. Nearly 35% of Americans who reach age 80 show symptoms. The disease is progressive and hard to diagnose definitively.

PARALYSIS
A traumatic blow to the spine can crush the delicate nerve bundles within, preventing signals from passing through. All parts of the body controlled by nerves beneath the site of injury may become paralyzed. Paraplegia is paralysis of the lower half of the body, while quadriplegia is paralysis from the neck down. Such injuries are often permanent because the nerves of the spinal cord cannot be repaired.

SPINAL INFECTIONS
Infections of the cerebrospinal fluid can be detected by a spinal tap, a procedure that involves carefully inserting a narrow needle through gaps between back bones and removing a sample of cerebrospinal fluid.

THE BRAIN

The human brain is an incredibly sophisticated system for receiving, processing, and communicating information. It consists of over 100 billion intricately networked neurons and even more supporting cells.

CEREBRAL CORTEX
The cerebral cortex is the outermost 0.2 inch of the cerebrum. Because it is highly folded, the cortex contains about 80% of the total brain mass. It produces our most distinctively human traits: reasoning, mathematical ability, language skills, imagination, artistic talent, sensory perceptions, and personality.

HYPOTHALAMUS
The hypothalamus controls the secretion of hormones via the pituitary gland and regulates many body responses such as internal temperature, the biological clock, hunger, thirst, feelings of pleasure, and emotions.

Meninges

CEREBRUM
The largest and most complex part of the brain, the cerebrum consists of right and left cerebral hemispheres.

MOTOR CORTEX
This area of the brain sends commands to skeletal muscles.

CORPUS CALLOSUM
The corpus callosum is a thick band of nerve fibers that connects the two cerebral hemispheres, allowing them to act in coordination.

CEREBELLUM
The cerebellum uses sensory information to plan and coordinate body movements.

PONS
The pons controls breathing.

MEDULLA OBLONGATA
The medulla oblongata controls breathing, circulation, swallowing, and digestion.

THALAMUS
The thalamus sorts and relays information to the cerebral cortex.

MIDBRAIN
The midbrain receives and integrates auditory information, controls visual reflexes, and passes sensory information to other parts of the brain.

BRAINSTEM
These three structures make up the brainstem, which acts as a sensory filter, selecting information to be passed on to other brain regions. The brainstem also regulates sleep and helps coordinate body movements.

LEFT AND RIGHT HEMISPHERES
The cerebrum is divided into left and right hemispheres, each of which receives information from and controls the movement of the opposite side of the body.

WHAT CAN GO WRONG

MENINGITIS
If the cerebrospinal fluid becomes infected, the meninges may become inflamed. Viral meningitis is generally not harmful, whereas bacterial meningitis can be serious but is usually cured by antibiotics.

CONCUSSION
The most common form of traumatic brain injury, a concussion causes a temporary breakdown of one or more brain functions. Concussions are often caused by the brain impacting the skull with more force than can be cushioned by the cerebrospinal fluid, after a blow to the head, for example. Although symptoms often disappear within a few weeks, repeated concussions can cause permanent brain injury.

ADDICTION
Within the brain there are "pleasure centers" that are strongly affected by certain addictive drugs, such as amphetamines and cocaine. The effect of these drugs on the brain helps explain their highly addictive qualities.

The brain of a sperm whale weighs 17 pounds; the brain of a goldfish weighs 1/5,000th of a pound.

CORE IDEA: The nervous system consists of the central nervous system (brain and spinal cord)—which receives, processes, and sends out information—and the peripheral nervous system (nerves)—which conveys information to and from the CNS.

CORE QUESTION: Do the nerves in your fingers connect directly to your brain?

ANSWER: No. Nerves send signals to neurons of the spinal cord, which is directly connected to the brain.

The nervous system receives input, processes it, and sends output

The **nervous system** forms a communication and coordination network throughout an animal's body. Your nervous system consists of your brain, spinal cord, and many nerves. All three of the structures contain **neurons**, individual nerve cells that carry electrical signals from one part of the body to another. Each neuron may communicate with thousands of others, forming networks that enable us to move, perceive our surroundings, learn, and remember.

CELL BODY
The **cell body** is the central hub of a neuron, housing the nucleus and other organelles.

NEURON ANATOMY
A **nerve** is a communication line made from cable-like bundles of neuron fibers tightly wrapped in connective tissue.

NERVE

Single neuron

Bundle of neuron fibers

Ring of connective tissue

Blood vessels

DENDRITES
Usually numerous, short, and highly branched, **dendrites** receive signals from other neurons and convey them toward the cell body.

NERVE SIGNAL

MYELIN SHEATH
The myelin sheath is a chain of bead-like supporting cells that insulate the axon and help speed the electrical signal.

AXON
This entire long extension from the cell body is a single **axon**. A signal travels from the cell body to the tip of the axon.

ACTION POTENTIALS
At rest, an axon has more positive charge outside the cell than inside. This electrical charge difference across the axon membrane of a neuron at rest is called the resting potential. A nerve signal involves a temporary reversal of the electric charge, caused by ions flowing into and out of the axon.

NEURON MEMBRANE
Proteins in the plasma membrane regulate the passage of ions into and out of the neuron.

At rest, proteins (not shown) establish an electrical charge difference across the membrane.

The nerve signal consists of a flow of positive ions into the interior of the atom, followed by a flow back out.

INTERIOR OF AXON

NERVE SIGNAL

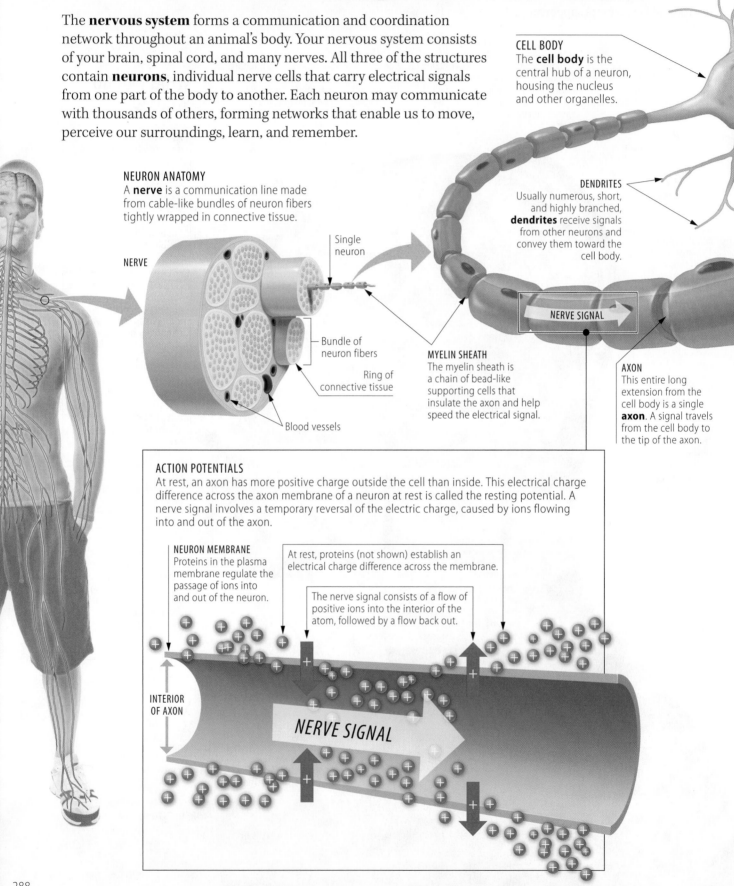

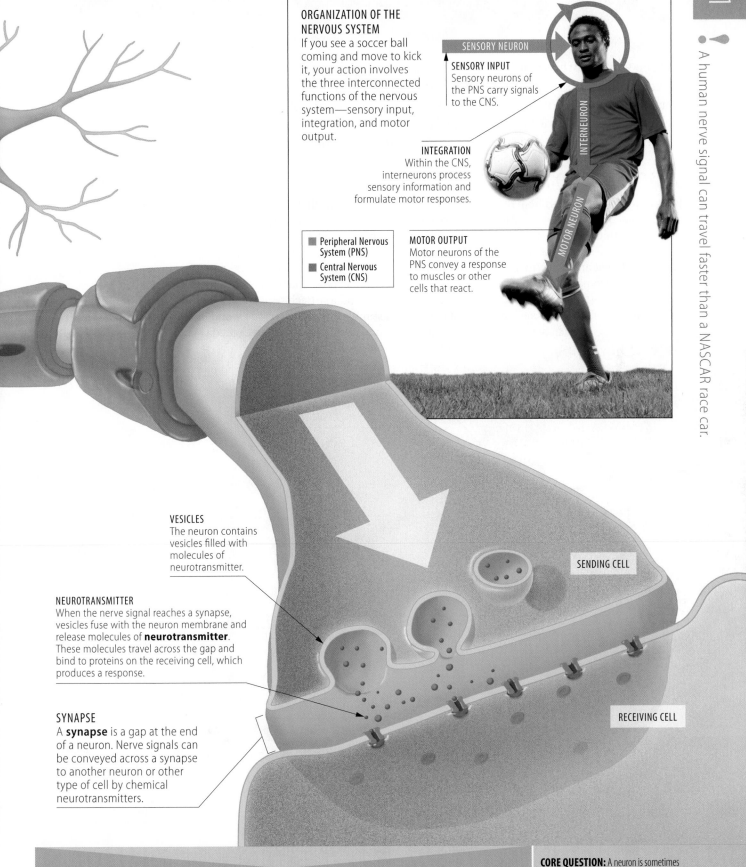

ORGANIZATION OF THE NERVOUS SYSTEM

If you see a soccer ball coming and move to kick it, your action involves the three interconnected functions of the nervous system—sensory input, integration, and motor output.

SENSORY NEURON

SENSORY INPUT
Sensory neurons of the PNS carry signals to the CNS.

INTERNEURON

INTEGRATION
Within the CNS, interneurons process sensory information and formulate motor responses.

MOTOR NEURON

■ Peripheral Nervous System (PNS)
■ Central Nervous System (CNS)

MOTOR OUTPUT
Motor neurons of the PNS convey a response to muscles or other cells that react.

VESICLES
The neuron contains vesicles filled with molecules of neurotransmitter.

SENDING CELL

NEUROTRANSMITTER
When the nerve signal reaches a synapse, vesicles fuse with the neuron membrane and release molecules of **neurotransmitter**. These molecules travel across the gap and bind to proteins on the receiving cell, which produces a response.

SYNAPSE
A **synapse** is a gap at the end of a neuron. Nerve signals can be conveyed across a synapse to another neuron or other type of cell by chemical neurotransmitters.

RECEIVING CELL

CORE IDEA: Nerve signals travel along the length of an axon. The movement of ions into and out of the axon generates the signal. At the synapse, the signal may be communicated to another cell via chemical neurotransmitters.

CORE QUESTION: A neuron is sometimes described as having "many ears, but one mouth." What does that mean in terms of neuron structure?

ANSWER: A neuron has many dendrites ("ears," since they "hear" incoming signals), but only one axon ("mouth"), since it "talks" to other cells.

Your senses use receptors to convey information about the outside world

To keep your body functioning, your nervous system must receive information about the outside world. This information is obtained by your sensory organs and conveyed via sensory neurons to your central nervous system. Once there, the brain can formulate and convey responses. Underlying the wide variety of senses you experience—sight, hearing, touch, taste, smell—there is unity in the way that sensory receptors function.

SENSORY RECEPTOR CELLS

Each of your senses arises from the action of sensory receptors, cells that are tuned to the condition of the outside world. After a sensory receptor detects its stimulus, it conveys this information to the central nervous system as an action potential, a traveling electrical signal. There are several categories of sensory receptors: pain receptors, thermoreceptors (that detect heat and cold), mechanoreceptors (touch, pressure, motion, sound, body position), electromagnetic receptors (energy, including photoreceptors that detect light), and chemoreceptors (chemicals, such as those found in foods and odors). Each type of receptor works in a manner similar to the smell receptors shown here: a stimulus from the outside world creates a change in potential within a receptor cell; the signal is then carried via the nervous system to the brain, where it is processed.

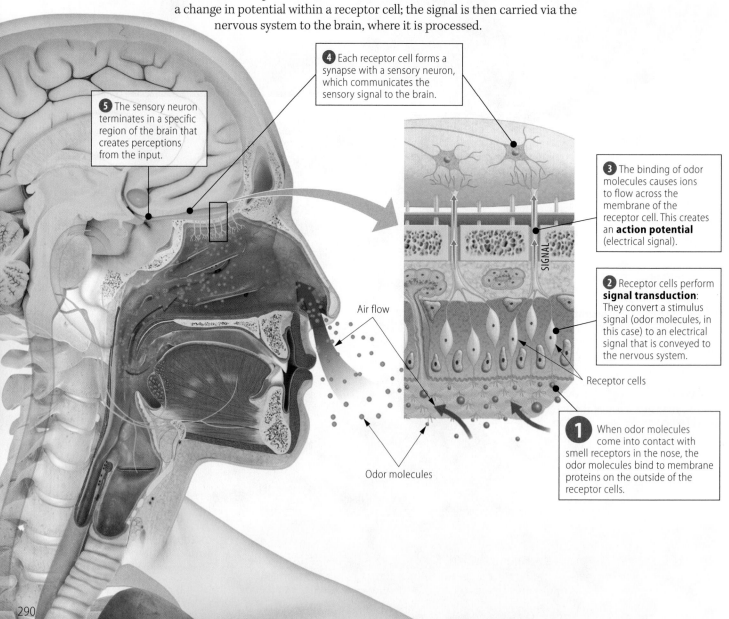

4 Each receptor cell forms a synapse with a sensory neuron, which communicates the sensory signal to the brain.

5 The sensory neuron terminates in a specific region of the brain that creates perceptions from the input.

3 The binding of odor molecules causes ions to flow across the membrane of the receptor cell. This creates an **action potential** (electrical signal).

2 Receptor cells perform **signal transduction**: They convert a stimulus signal (odor molecules, in this case) to an electrical signal that is conveyed to the nervous system.

SIGNAL

Air flow

Odor molecules

Receptor cells

1 When odor molecules come into contact with smell receptors in the nose, the odor molecules bind to membrane proteins on the outside of the receptor cells.

THE HUMAN EYE

The human eye is a remarkable example of a sensory organ. It is able to detect light and colors, form images of objects near and far, and respond to even tiny amounts of stimulus. Many parts of the eye work together to focus light energy on the sensory receptor cells that convert this stimulus into nerve signals sent to the brain.

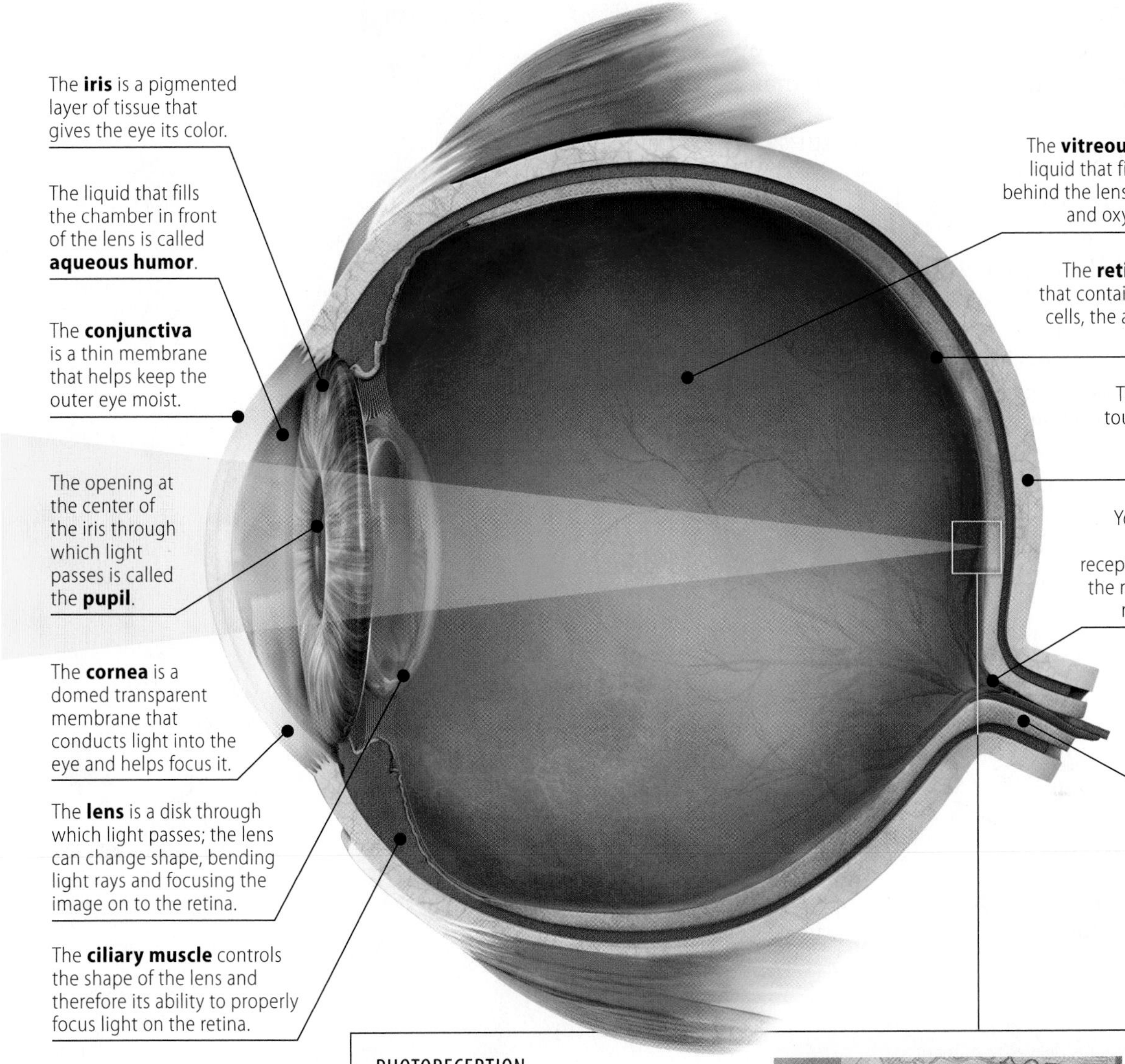

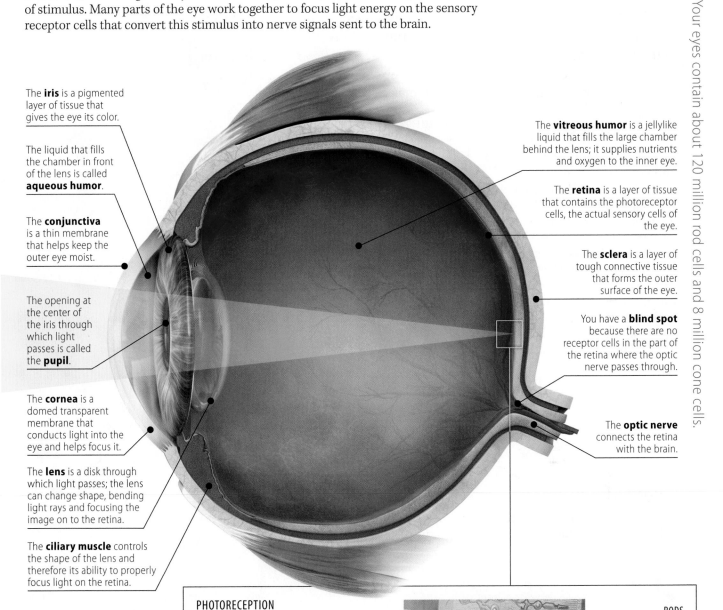

The **iris** is a pigmented layer of tissue that gives the eye its color.
The liquid that fills the chamber in front of the lens is called **aqueous humor**.
The **conjunctiva** is a thin membrane that helps keep the outer eye moist.
The opening at the center of the iris through which light passes is called the **pupil**.
The **cornea** is a domed transparent membrane that conducts light into the eye and helps focus it.
The **lens** is a disk through which light passes; the lens can change shape, bending light rays and focusing the image on to the retina.
The **ciliary muscle** controls the shape of the lens and therefore its ability to properly focus light on the retina.
The **vitreous humor** is a jellylike liquid that fills the large chamber behind the lens; it supplies nutrients and oxygen to the inner eye.
The **retina** is a layer of tissue that contains the photoreceptor cells, the actual sensory cells of the eye.
The **sclera** is a layer of tough connective tissue that forms the outer surface of the eye.
You have a **blind spot** because there are no receptor cells in the part of the retina where the optic nerve passes through.
The **optic nerve** connects the retina with the brain.

PHOTORECEPTION

When light strikes photoreceptor cells embedded in the back of the retina, it causes a chemical change in pigment molecules located within the rods and cones. This change triggers a change in potential. The signal travels along the optic nerve and into the cerebral cortex of the brain, which integrates the signal into a visual perception.

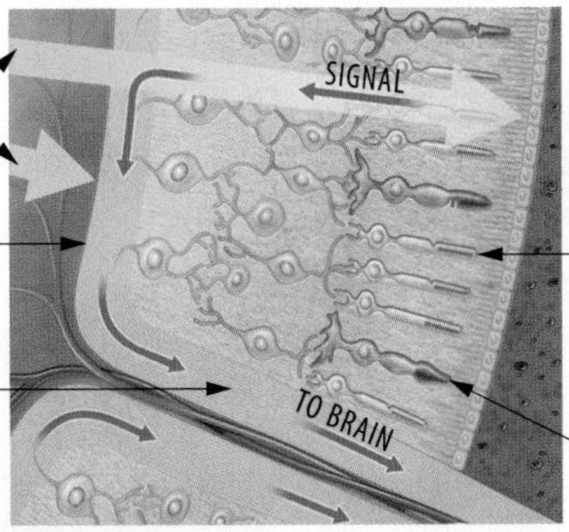

Light
Retina
Optic nerve
SIGNAL
TO BRAIN
RODS
Rods are photoreceptor cells that are very sensitive to light, but only in shades of gray.
CONES
Cones require bright light, but can distinguish color.

Your eyes contain about 120 million rod cells and 8 million cone cells.

CORE IDEA: Senses are created when sensory receptor cells detect a stimulus and convert it to an electrical nerve signal that is communicated to the brain. Smell receptors detect chemicals, whereas photoreceptors detect light.

CORE QUESTION: Why do you have a "blind spot" in your eye?

ANSWER: At the "blind spot," there are no light-detecting sensory cells because the optic nerve passes through that location.

The human skeleton contains 206 bones

Your **skeletal system** supports your body, protects your vital organs, and anchors your internal structure. A study of the skeletal system should include a map of the bones, examination of their internal structure, a survey of different joints, and consideration of problems that may arise.

THE SKELETAL SYSTEM

Like all vertebrates, you have an **endoskeleton**, a bony skeleton located inside your body that is intermingled with your soft tissues. (Other animals, such as insects and lobsters, have a hard exoskeleton on the outside.) Your skeleton consists of 206 bones organized into an axial skeleton (which supports the axis, or trunk, of your body) and an appendicular skeleton (all the rest of your bones), plus cartilage that provides flexibility and cushioning.

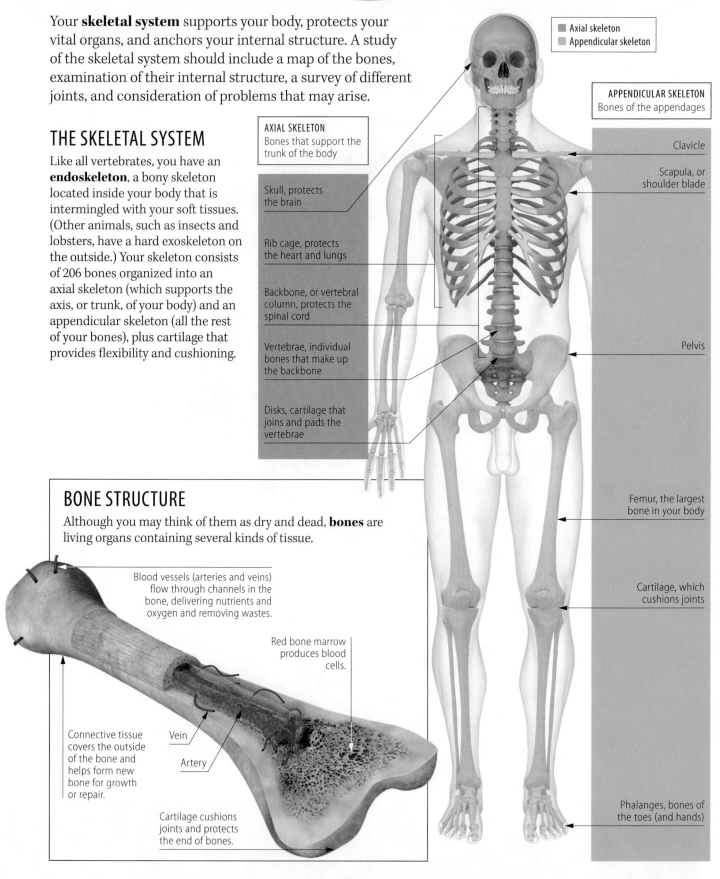

■ Axial skeleton
■ Appendicular skeleton

AXIAL SKELETON
Bones that support the trunk of the body

Skull, protects the brain

Rib cage, protects the heart and lungs

Backbone, or vertebral column, protects the spinal cord

Vertebrae, individual bones that make up the backbone

Disks, cartilage that joins and pads the vertebrae

APPENDICULAR SKELETON
Bones of the appendages

Clavicle

Scapula, or shoulder blade

Pelvis

Femur, the largest bone in your body

Cartilage, which cushions joints

Phalanges, bones of the toes (and hands)

BONE STRUCTURE

Although you may think of them as dry and dead, **bones** are living organs containing several kinds of tissue.

Blood vessels (arteries and veins) flow through channels in the bone, delivering nutrients and oxygen and removing wastes.

Red bone marrow produces blood cells.

Connective tissue covers the outside of the bone and helps form new bone for growth or repair.

Vein

Artery

Cartilage cushions joints and protects the end of bones.

JOINTS

Joints allow for complex skeletal movements. **Ligaments** (a strong fibrous connective tissue) hold bones together at joints.

BALL AND SOCKET
Ball-and-socket joints enable us to rotate our arms and legs in several directions.

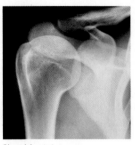

Shoulder joint

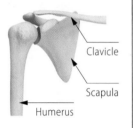

Clavicle

Scapula

Humerus

PIVOT
A pivot joint enables you to rotate your head around in a circle.

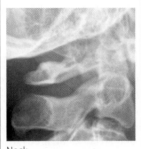

Neck

Atlas (top vertebra)

Axis (second vertebra)

HINGE
Hinge joints permit movement in a single plane, as in the way you can bend your elbow or knee in just one direction.

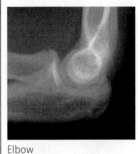

Elbow

Humerus

Radius

Ulna

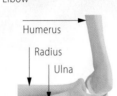

SADDLE
Consisting of two U-shaped surfaces, and found only at the base of the thumb, this joint permits movement in two planes.

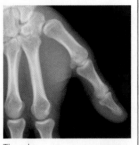

Thumb

Trapezium (wrist bone)

First metacarpal of thumb

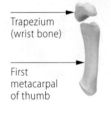

WHAT CAN GO WRONG

OSTEOPOROSIS
Although most common in postmenopausal women, anyone can develop osteoporosis, characterized by a low bone mineral density. Because the bones are thinner, they are more easily broken.

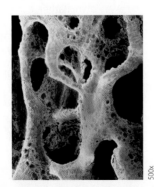

500x

FRACTURES
Forces that exceed a bone's ability to flex will result in a fracture, or broken bone. Treatment involves putting the bone back into its normal shape and then immobilizing it with a splint, cast, or plate. This allows fresh bone to be rebuilt.

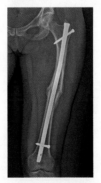

ARTHRITIS
Arthritis, inflammation of the joints, is part of the normal aging process as the cartilage between bones wears down. Rheumatoid arthritis results from the body's immune system improperly destroying joint tissue. Other types of arthritis can be caused by injuries or infections.

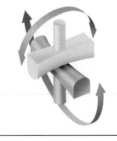

The average American will break two bones during his or her lifetime.

CORE QUESTION: Why is the saying "dry as a bone" inaccurate?

CORE IDEA: Your skeletal system consists of 206 bones and cartilage. Your bones contain living tissue that can regenerate after an injury. Several types of joints permit movement.

ANSWER: Living bones contain a blood supply and marrow.

Skeletal muscles produce movement

All animals move thanks to an interplay of the nervous, skeletal, and muscular systems. The nervous system issues commands. In response, **skeletal muscles** contract (shorten) and exert a force against the stationary skeleton. Together, this provides movement. Regardless of the location, all muscles act through similar mechanisms in which proteins within small segments of a muscle fiber slide across one another.

STRUCTURE OF A MUSCLE

Each skeletal muscle consists of bundles of parallel muscle fibers. Each fiber is a single long, cylindrical cell that has many nuclei. Most of a muscle cell consists of myofibrils, which are bundles of proteins. The proteins of a myofibril give skeletal muscle a striped appearance, with alternating light and dark bands. Hence, skeletal muscle is called striated (striped) muscle.

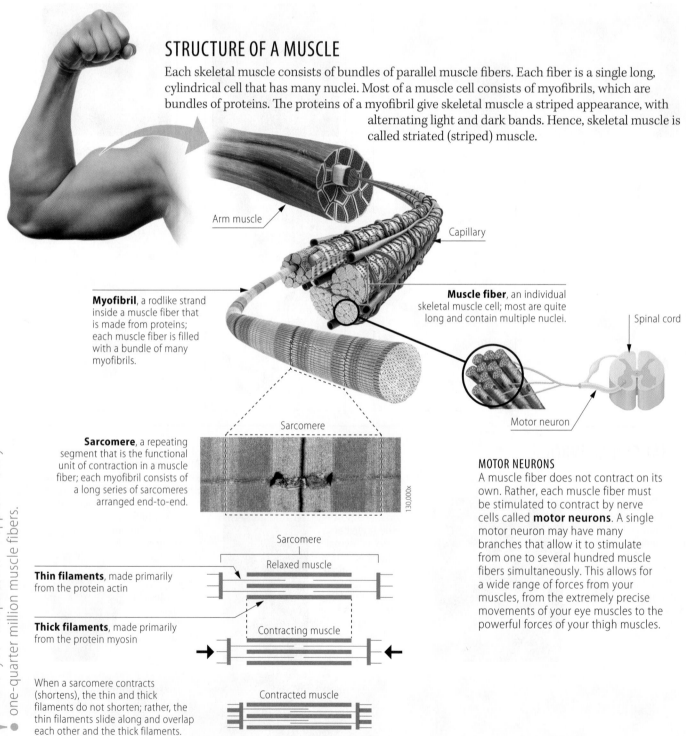

Arm muscle

Capillary

Myofibril, a rodlike strand inside a muscle fiber that is made from proteins; each muscle fiber is filled with a bundle of many myofibrils.

Muscle fiber, an individual skeletal muscle cell; most are quite long and contain multiple nuclei.

Spinal cord

Sarcomere

Motor neuron

Sarcomere, a repeating segment that is the functional unit of contraction in a muscle fiber; each myofibril consists of a long series of sarcomeres arranged end-to-end.

130,000x

MOTOR NEURONS

A muscle fiber does not contract on its own. Rather, each muscle fiber must be stimulated to contract by nerve cells called **motor neurons.** A single motor neuron may have many branches that allow it to stimulate from one to several hundred muscle fibers simultaneously. This allows for a wide range of forces from your muscles, from the extremely precise movements of your eye muscles to the powerful forces of your thigh muscles.

Sarcomere

Relaxed muscle

Thin filaments, made primarily from the protein actin

Thick filaments, made primarily from the protein myosin

Contracting muscle

When a sarcomere contracts (shortens), the thin and thick filaments do not shorten; rather, the thin filaments slide along and overlap each other and the thick filaments.

Contracted muscle

Each of your biceps contains approximately one-quarter million muscle fibers.

MUSCLE CONTRACTION

All muscles move by the contracting (shortening) of sarcomeres. Contraction occurs when the thick filaments (myosin, shown in red) and thin filaments (actin, blue) slide over each other. As long as ATP molecules are available to provide energy, contraction may continue.

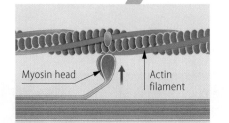

Myosin head   Actin filament

1 The myosin head attaches to the actin filament.

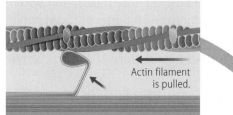

Actin filament is pulled.

2 The myosin head bends, shortening the sarcomere by pulling the actin filament toward the center.

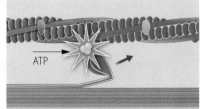

ATP

3 A molecule of ATP binds to the myosin head, causing it to detach from the actin filament.

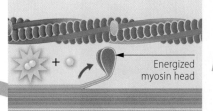

Energized myosin head

4 The energy released by the breakdown of ATP causes the myosin filament to move to a high-energy position, where it can bind to the actin filament once again.

MUSCLES ACT IN PAIRS

Many skeletal muscles work in antagonistic pairs. That is, two muscles perform opposite tasks that together allow movement in opposing directions at the joint.

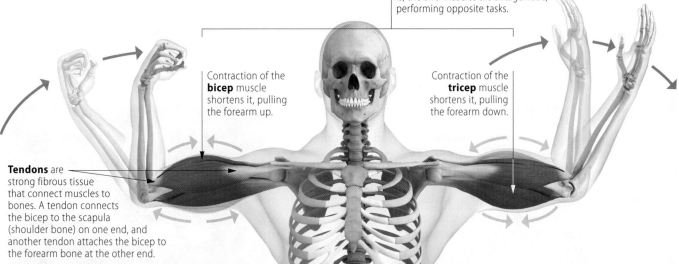

Notice that when the bicep is contracted, the tricep is relaxed, and vice versa. That is, the two muscles are antagonistic, performing opposite tasks.

Contraction of the **bicep** muscle shortens it, pulling the forearm up.

Contraction of the **tricep** muscle shortens it, pulling the forearm down.

Tendons are strong fibrous tissue that connect muscles to bones. A tendon connects the bicep to the scapula (shoulder bone) on one end, and another tendon attaches the bicep to the forearm bone at the other end.

CORE IDEA: Your muscular system contains skeletal muscles that are anchored to bones by tendons. Contraction of the sarcomeres within muscle fibers allows muscles to shorten, producing body movements.

CORE QUESTION: Can you work out your biceps and your triceps simultaneously?

ANSWER: No. Because they are antagonistic, one must be relaxing while the other is contracting.

Ecology is the study of organisms in their environments

Ecology is the scientific study of how organisms interact with each other and with their environment. Biologists sometimes refer to the study of life on this scale as "skin-out" biology, meaning that ecology is generally concerned with the levels of biological organization from an organism outward into its environment.

DISCOVERY AND HYPOTHESIS-DRIVEN ECOLOGY

Humans have always observed other organisms and the environments in which they live. To learn about ecosystems, scientists make use of two broad means of study: discovery science and hypothesis-driven science.

DISCOVERY SCIENCE
Making verifiable observations of organisms within their environments is at the heart of the discovery approach to ecology. For example, these students are investigating the water quality of a stream by cataloging the invertebrate organisms found within it.

HYPOTHESIS-DRIVEN SCIENCE
Observations of the natural world can prompt questions, which lead to the formation of hypotheses and the design of experiments. This manner of hypothesis-driven science may be conducted in the field or in the lab under controlled conditions. For example, these students are using a grid to survey the plants growing in a field; the data they collect will be used to formulate testable hypotheses, which will lead to further experiments.

ENVIRONMENTALISM

Ecology is distinct from **environmentalism**, a broad philosophy and social movement that seeks to maintain environmental quality. As the human population and its consumption of resources grows, so will our impact on the biosphere and its various habitats. The science of ecology can inform an understanding of our environment—a necessary first step toward solving environmental problems. But solutions to environmental problems require broad participation among a wide swath of society—including politicians, community leaders, and ordinary citizens—and a personal commitment on the part of many. In this photo, environmental activists from Indonesia protest government plans to replace natural forests with palm oil plantations.

ECOLOGY CAN BE STUDIED ON MANY LEVELS

When we study the interactions between organisms and their environments, it is convenient to divide ecology into five increasingly comprehensive levels.

The deepest part of the biosphere is the Mariana Trench, which is nearly 7 miles deep.

ORGANISMAL ECOLOGY

An **organism** is an individual living being. Organismal ecology focuses on the ways that organisms adapt to their environments through physiology and behavior. For example, how does this wolf change its hunting habits as the surrounding environment changes with the seasons?

POPULATION ECOLOGY

A **population** is a group of individuals of the same species living in the same place at the same time. Population ecology is concerned with the environmental factors that affect population size, growth, and density. For example, how many wolves can be supported by the resources available in the area?

COMMUNITY ECOLOGY

A **community** consists of all the populations (of multiple species) living in a particular place. Questions in community ecology focus on interactions between species and how this affects the makeup and organization of the community. For example, how do the food-gathering behaviors of different species affect the community within Alaska's Denali National Park?

ECOSYSTEM ECOLOGY

An **ecosystem** is all the life living in a particular area together with all the nonliving components of that environment, including water, minerals, and physical factors such as light and air. Ecosystem ecology is concerned with questions of energy flow and chemical cycling. For example, how does the element nitrogen move between the plants and animals within Denali National Park?

BIOSPHERE

The **biosphere** is the global ecosystem—all life and all of life's environments. The biosphere is the sum of all ecosystems on Earth. This includes the lower atmosphere (to a height of over one mile), the land (to a depth of about one mile), and all water.

CORE IDEA: Ecology is the use of discovery or hypothesis-driven science to study the interactions of organisms and their environments. Ecology can be studied at levels ranging from individual organisms to the whole biosphere.

CORE QUESTION: If you are studying the predation of deer by wolves in a particular habitat, what level of ecology are you studying?

ANSWER: Community ecology (since you are investigating multiple populations).

Ecosystems include a variety of abiotic factors

Ecology is the scientific study of organisms within their ecosystems. An ecosystem includes the life found within an area and also the nonliving factors that affect that life. The organisms of an ecosystem constitute its **biotic factors**, or living components. The **abiotic factors** of an ecosystem are its nonliving components, which include several different physical and chemical factors.

ENERGY

All organisms require energy to survive. Therefore, all ecosystems require a source of energy to sustain life. Most ecosystems on Earth are powered by solar energy via sunlight. The energy in sunlight is used to drive photosynthesis in plants and other producers such as algae. The sugars created via photosynthesis are used directly by the plant that made them, and also provide food for other organisms in the ecosystem. In rare cases, ecosystems can be powered by non-solar sources.

HYDROTHERMAL VENT
In rare cases, ecosystems can be completely dark and receive their energy from non-solar sources. For example, volcanic gases occasionally seep through cracks in the deep ocean floor. Near these cracks, called hydrothermal vents, certain species of prokaryotes are able to use the chemicals as an energy source. These prokaryotes become the basis of a food chain within these unusual ecosystems.

TERRESTRIAL SUNLIGHT
The amount of available sunlight is often the most important factor in determining how much life can be sustained in a given area. For example, shading by tall trees creates intense competition for light among low-growing plants.

AQUATIC SUNLIGHT
Light cannot penetrate deeply into water, so most photosynthesis occurs near the surface. Photosynthetic kelp thrives in sunlight and is eaten by many other organisms in the shallow oceans.

NUTRIENTS

The availability of inorganic nutrients—elements required for growth, such as nitrogen and phosphorus—can affect the ability of an ecosystem to sustain life. In terrestrial habitats, the soil structure—its moisture content, pH, and nutrient content—plays a major role in determining which plants (and how many) can grow in a given habitat. In marine habitats, low levels of nutrients in the water can limit growth of algae and bacteria.

Corn living in nutrient-rich soil (left) can grow much larger than corn growing in nitrogen-deficient soil (right).

The pitcher plant (*Sarracenia purpurea*) and other carnivorous plants grow in nitrogen-poor habitats and so must obtain nitrogen by ingesting insects.

WIND

Wind is an important abiotic factor in some terrestrial ecosystems. Wind can affect an animal's ability to regulate its internal temperature (by increasing evaporative cooling) and can affect the growth of plants.

Strong, consistent wind can influence the growth habits of a tree, such as this hawthorn. Wind also significantly affects pollination and water loss via evaporation.

A lichen growing near the South Pole is so well adapted to extreme cold that it can photosynthesize even when the temperature drops to −4°F.

TEMPERATURE

Temperature is an important abiotic factor that limits the distribution of organisms that can survive in a given habitat. Most organisms can function only when the temperatures are between about 32°F (0°C) and 110°F (43°C). Many terrestrial organisms display adaptations that allow them to regulate body temperature, such as fur or the ability to sweat.

Because lizards (like this ocellated lizard, *Timon lepidus*) depend on environmental heat to keep themselves warm, fewer lizard species are found in more northerly regions.

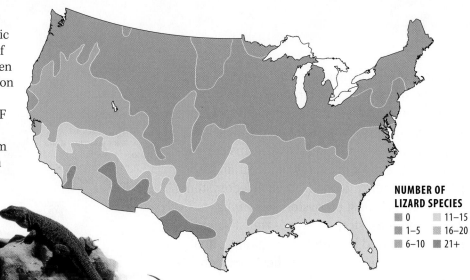

NUMBER OF LIZARD SPECIES
- 0
- 1–5
- 6–10
- 11–15
- 16–20
- 21+

WATER

All living cells are mostly water. Therefore, it is no surprise that water is essential to all life and that all species display evolutionary adaptations that help regulate their internal water content. Terrestrial organisms have many adaptations to prevent drying out. Aquatic organisms, although surrounded by water, must maintain a proper balance of water in their cells in order to survive.

The outside of most exposed plant structures includes a waxy cuticle that helps seal in water.

To compensate, the fish's kidneys excrete large amounts of urine.

Water enters the fish through the body surface.

Freshwater fish balance the water that flows into the body from their surroundings by excreting large amounts of dilute urine.

FIRE

In some ecosystems, such as prairie grasslands, fires periodically act as an important abiotic factor. Some plants have evolved mechanisms that take advantage of these disturbances, such as seeds that sprout only after being scorched.

A 1988 fire in Yellowstone National Park consumed pine forests.

Regrowth after a fire in Yellowstone

CORE IDEA: Abiotic factors (nonliving components) can have a profound effect on the life within an ecosystem. Examples of abiotic factors include the supply of energy and nutrients, wind, temperature, water, and fire.

CORE QUESTION: Of the abiotic factors discussed here, which apply to both marine organisms and terrestrial organisms?

ANSWER: The availability of energy and nutrients, and the need to regulate water and temperature.

Energy flows through ecosystems

An ecosystem consists of both living organisms (called the biotic factors) and nonliving components that affect them (called the abiotic factors). The most important abiotic factor in any ecosystem is energy. All organisms depend on a steady supply of energy to maintain life's processes. Energy flows through an ecosystem. Energy enters most ecosystems as sunlight; within the ecosystem, energy is transformed, used, and reused by various inhabitants; finally, energy exits the ecosystem as heat. Such energy flow has several important consequences to ecosystem ecology.

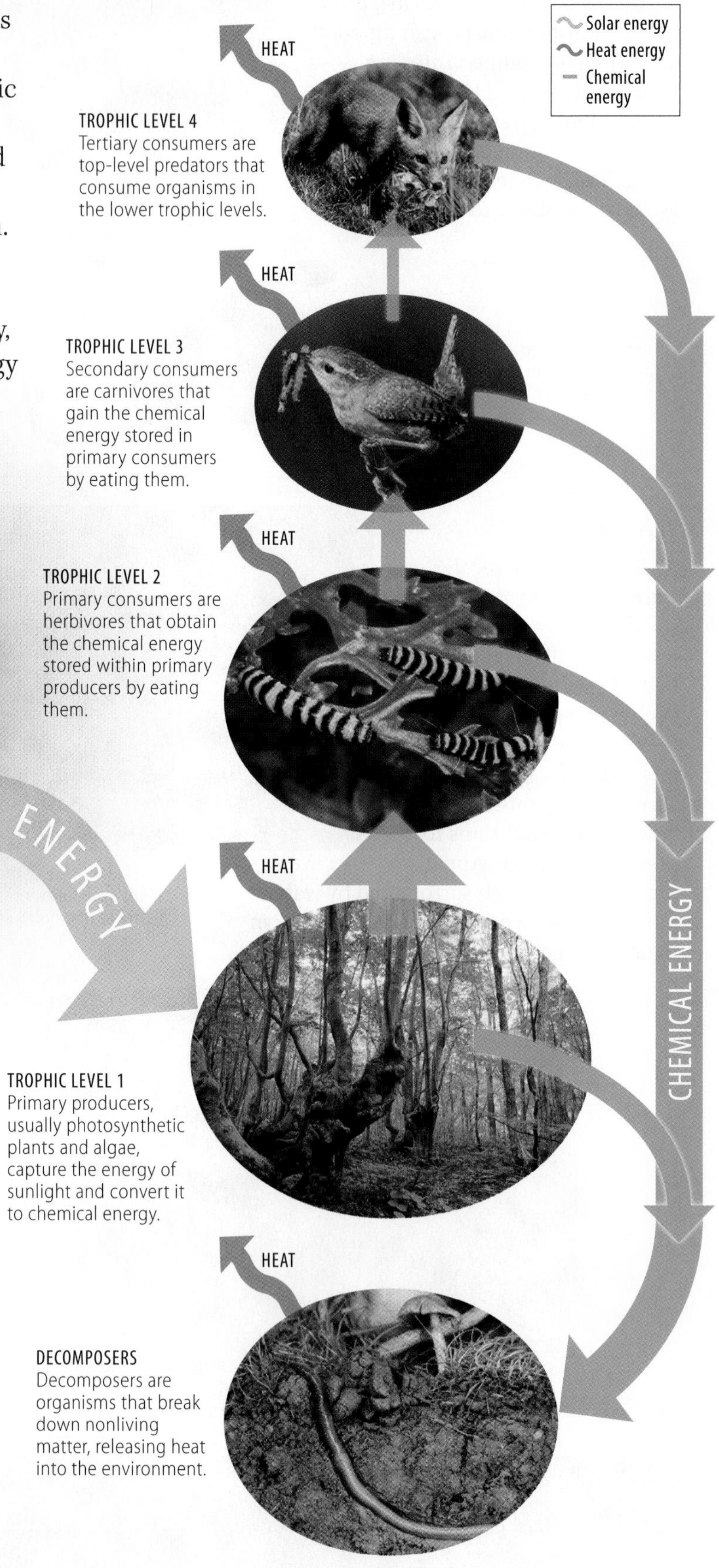

ENERGY FLOW

After energy enters an ecosystem as sunlight, **primary producers** (such as plants and algae) convert the solar energy to chemical energy via photosynthesis. **Primary consumers** are able to use some of this chemical energy when they eat plants, but much of the energy is lost as heat. Higher-level predators prey on other consumers, again gaining some of the chemical energy as food, but losing much as heat. **Decomposers** (including worms, insects, fungi, and bacteria) break down the dead remains of other organisms, again releasing heat. Overall, the constant input of energy from sunlight into an ecosystem balances the constant output of energy as heat. The **trophic structure** of an ecosystem—such as the one shown to the right—describes the feeding relationships that determine how chemical energy transfers from one level to another. The width of the arrows signifies the relative amount of chemical energy transferred between each trophic level.

~ Solar energy
~ Heat energy
— Chemical energy

HEAT

TROPHIC LEVEL 4
Tertiary consumers are top-level predators that consume organisms in the lower trophic levels.

HEAT

TROPHIC LEVEL 3
Secondary consumers are carnivores that gain the chemical energy stored in primary consumers by eating them.

HEAT

TROPHIC LEVEL 2
Primary consumers are herbivores that obtain the chemical energy stored within primary producers by eating them.

HEAT

SOLAR ENERGY

CHEMICAL ENERGY

TROPHIC LEVEL 1
Primary producers, usually photosynthetic plants and algae, capture the energy of sunlight and convert it to chemical energy.

HEAT

DECOMPOSERS
Decomposers are organisms that break down nonliving matter, releasing heat into the environment.

PRIMARY PRODUCTION

Biomass is the total amount of living material in an ecosystem; it can serve as a measure of the capacity of an ecosystem to support life. **Primary production** is a measure of the rate at which solar energy is converted to biomass within a given ecosystem during a given time period. Different ecosystems have different rates of primary production. The deep ocean (where sunlight penetrates only a small fraction of the total volume) has the lowest rate of biomass production. But because it is so large, the deep ocean contributes more than any other ecosystem to Earth's total biomass.

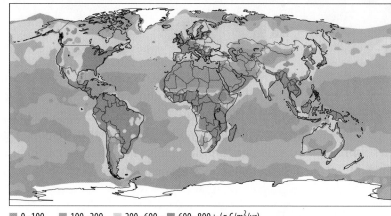

■ 0–100 ■ 100–200 ■ 200–600 ■ 600–800+ (g C/m²/yr)

This map is shaded to show the primary productivity of different ecosystems. On land, primary productivity increases with temperature and precipitation, which is why the tropics are so productive. In the sea, the primary productivity increases where light and nutrients are abundant, mostly near the continents.

ECOLOGICAL PYRAMIDS

Each time chemical energy is passed from one trophic level to the next, most of it is lost as heat. For example, only about 1% of the energy in sunlight that strikes a region is converted to chemical energy by plants, and herbivores in an ecosystem will gain only about 10% of that energy by eating plant material produced via photosynthesis. The cumulative loss of energy means that each level of trophic structure can support fewer organisms than the last. A given ecosystem is therefore able to support many more vegetarians (primary consumers) than meat eaters (secondary consumers). This also explains why top-level predators are so rare; each one requires many prey organisms from the trophic levels below.

TROPHIC LEVELS

SECONDARY CONSUMERS
2,000 Calories of beef feed one person for one day.

PRIMARY CONSUMERS
20,000 Calories eaten by 10 cattle produce 2,000 Calories worth of beef.

PRIMARY PRODUCERS
20,000 Calories of organic material made by primary producers

2,000,000 Calories of sunlight enter the ecosystem.

TROPHIC LEVELS

PRIMARY CONSUMERS
20,000 Calories feed 10 people for one day (2,000 Calories per person per day).

PRIMARY PRODUCERS
20,000 Calories of organic material made by primary producers

2,000,000 Calories of sunlight enter the ecosystem.

CORE IDEA: Energy enters an ecosystem as sunlight, is transformed to chemical energy via photosynthesis by primary producers, and is then passed through various trophic levels of consumers. Each transfer loses most of the energy as heat.

CORE QUESTION: Why will prey always greatly outnumber their predators?

ANSWER: Because so much energy is lost between levels of trophic structure, each predator requires a large number of prey to gain the energy it needs to live.

Elements cycle through the biosphere

Although the sun provides the Earth with a constant input of energy, there is no significant extraterrestrial source of chemical elements. Instead, life depends on the continuous recycling of chemicals. All organisms "borrow" chemical elements from the ecosystem: Nutrients are acquired and used, but eventually all of the elements in living matter are returned to the environment. A particular chemical element—such as carbon (C) or nitrogen (N)—can be traced through an ecosystem by following its exchange between the living and nonliving components.

CHEMICAL CYCLING

Imagine that you are tracking a single atom of an element as it moves through an ecosystem over a long period of time. You would find that it passes between biotic (living) and abiotic (nonliving) components. Such an overall flow through an ecosystem is called a **biogeochemical cycle**. These cycles occur both locally (within a particular ecosystem) and globally (within the entire biosphere).

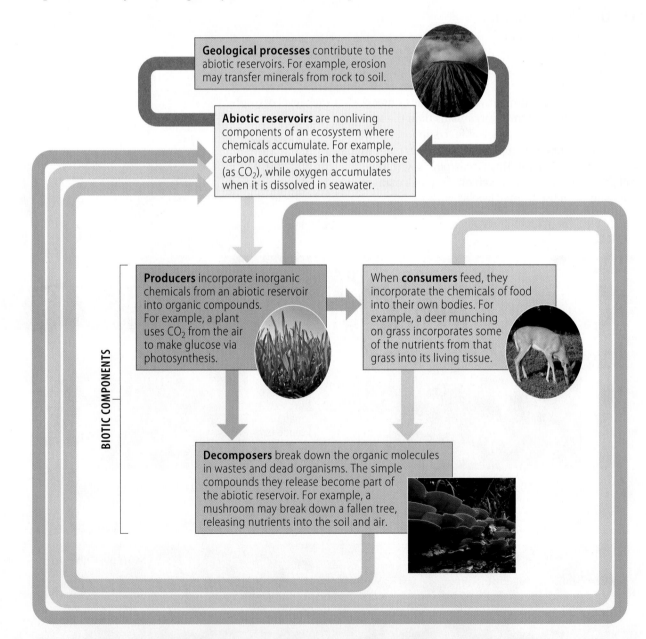

Geological processes contribute to the abiotic reservoirs. For example, erosion may transfer minerals from rock to soil.

Abiotic reservoirs are nonliving components of an ecosystem where chemicals accumulate. For example, carbon accumulates in the atmosphere (as CO_2), while oxygen accumulates when it is dissolved in seawater.

BIOTIC COMPONENTS

Producers incorporate inorganic chemicals from an abiotic reservoir into organic compounds. For example, a plant uses CO_2 from the air to make glucose via photosynthesis.

When **consumers** feed, they incorporate the chemicals of food into their own bodies. For example, a deer munching on grass incorporates some of the nutrients from that grass into its living tissue.

Decomposers break down the organic molecules in wastes and dead organisms. The simple compounds they release become part of the abiotic reservoir. For example, a mushroom may break down a fallen tree, releasing nutrients into the soil and air.

THE CARBON CYCLE

Carbon is the major ingredient of all organic molecules and therefore all life on Earth. The major abiotic reservoir of carbon is the atmosphere, which is rich in CO_2. Other abiotic reservoirs include fossil fuels, limestone rock, and carbon compounds dissolved in seawater. Photosynthesis transfers carbon from the abiotic reservoir (the atmosphere) to living producers (for example, plants). Conversely, cellular respiration moves carbon from living cells back to the atmosphere. On a global scale, the uptake of CO_2 from the atmosphere by photosynthesis roughly balances the return of CO_2 via cellular respiration. But this balance is affected by human activities (shown in red)—primarily the burning of fossil fuels, which releases trapped carbon into the atmosphere.

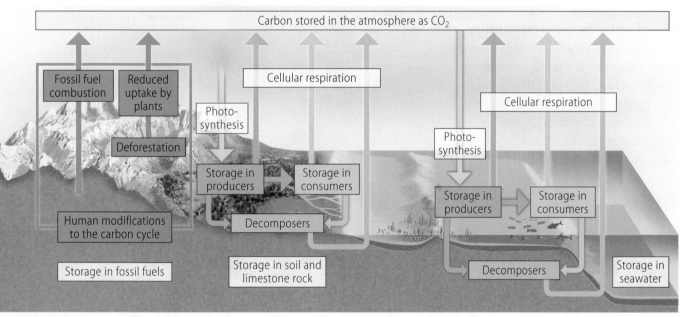

THE NITROGEN CYCLE

As an ingredient of proteins and nucleic acids, nitrogen is essential to all living cells. Abiotic reservoirs of nitrogen include the soil and the atmosphere, which is made up primarily of N_2, nitrogen gas. To be used by living organisms, N_2 must first undergo nitrogen fixation, a process whereby soil bacteria convert N_2 to compounds that plants can take up, such as nitrates (NO_3^-) and ammonium (NH_4^+). As with the carbon cycle, you can see that human effects (shown in red) also play a large role.

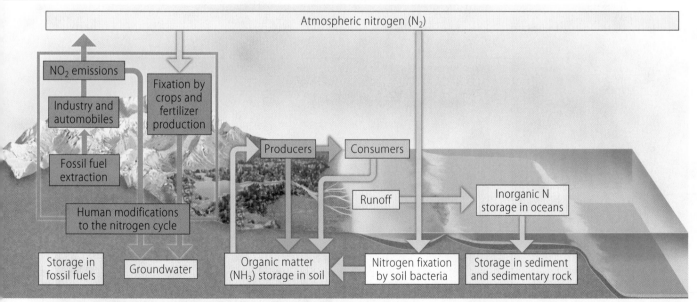

12.4

Globally, humans apply 165 million tons of nitrogen fertilizer to the soil each year.

CORE IDEA: Chemical elements—such as carbon and nitrogen—cycle through the biosphere, moving from abiotic (nonliving) reservoirs such as the air, soil, and water, to biotic (living) components of ecosystems.

CORE QUESTION: What process moves carbon from its primary abiotic reservoir to the living components of an ecosystem?

ANSWER: Photosynthesis.

All water on Earth is interconnected in a global cycle

Water is vital to all life on Earth. It flows within and between all ecosystems. In other words, all of the ecosystems on Earth are interconnected by a **global water cycle**. **Precipitation** (including rain and snow) transfers water from the atmosphere to the land. Conversely, **evaporation** from bodies of water and **transpiration** (the movement of water up and out of plants) move water from terrestrial sources to the atmosphere.

Solar energy drives the global patterns of air and water movement.

Winds carry water vapor in clouds across the land.

WATER VAPOR OVER THE OCEAN

PRECIPITATION OVER THE OCEAN

Over the oceans, the rate of evaporation exceeds the rate of precipitation, resulting in a net movement of water from the oceans into clouds as water vapor.

EVAPORATION FROM THE OCEAN

Surface water and groundwater eventually flow back into the oceans.

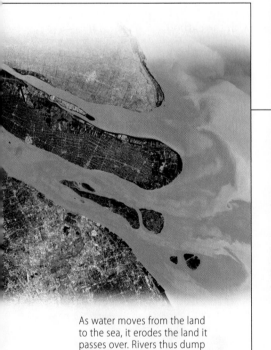

As water moves from the land to the sea, it erodes the land it passes over. Rivers thus dump a continuous supply of silt and minerals into the oceans.

Irrigation moves water from subterranean groundwater sources to the surface. This increases the amount of evaporation over land and can deplete groundwater supplies.

Because transpiration from plants is a major step in the water cycle, deforestation can have global effects, reducing the amount of water vapor in the air. Much of the world's rainforests are being burned for agriculture, and this is changing the global water cycle.

WATER VAPOR OVER THE LAND

TRANSPIRATION
Water evaporates from plants during the process of transpiration, which pulls water from the soil into the roots, up the stems, and out the leaves.

Over land, the rate of precipitation exceeds the rate of evaporation, resulting in a net movement of water from cloud vapor to the land.

EVAPORATION OVER THE LAND

PRECIPITATION OVER THE LAND

SURFACE WATER AND GROUNDWATER
The precipitation that falls to the ground may stay on the surface, or it may trickle through the soil to become groundwater.

CORE IDEA: All water on Earth is interconnected in a global water cycle. Over the oceans, the overall net movement of water is from the sea into the atmosphere. Over land, the net movement is from clouds down to the land.

CORE QUESTION: After water evaporates from the surface of the ocean into clouds, where does most of it go?

ANSWER: After the clouds are blown over land, the water vapor in the clouds falls back to the land as precipitation.

Aquatic biomes cover much of Earth's surface

Ecologists categorize the Earth's surface (and near surface) into a series of biomes. Each **biome** is a type of ecological community that occupies a particular zone. **Aquatic biomes** are defined by their abiotic factors, primarily salinity. Marine biomes (such as oceans or coral reefs) generally have water that is about 3% salt by weight, while freshwater biomes (such as lakes and rivers) generally have salt concentrations of less than 1%. You may recognize many of the aquatic biomes shown here.

FRESHWATER BIOMES

Although **freshwater biomes** cover less than 1% of the Earth's surface, and they contain only 0.01% of the Earth's water, they are home to nearly 6% of Earth's species. We depend on these biomes for drinking water, irrigation, and many of the conveniences of modern civilization. Freshwater biomes can be standing (such as lakes and ponds) or moving (such as rivers and streams).

LAKES AND PONDS
Standing bodies of freshwater range from small ponds to giant lakes. Communities of organisms (including algae, plants, and animals) are distributed according to depth of the water and distance from the shore. The growth of **phytoplankton**, photosynthetic aquatic algae and cyanobacteria, depends largely on the availability of nitrogen and phosphorus. This is why fertilizer runoff—and the resulting explosive growth of phytoplankton—can have a profound impact on the health of a lake or pond.

RIVERS AND STREAMS
Flowing bodies of water, such as rivers and streams, are usually cold and low in nutrients near their source. Downstream, the water is often warmer and has accumulated nutrients, supporting a more diverse community of life.

WETLANDS
Wetlands, such as swamps, bogs, and marshes, are biomes that occur where aquatic and terrestrial biomes meet. Some have permanent water, others seasonal, but all contain a rich supply of nutrients and a diverse community of species. Because they can absorb and filter large amounts of water, wetlands reduce flooding and improve water quality. Recognition of their value has led to efforts to protect and restore wetlands.

INTERMEDIATE BIOMES

Where land and freshwater biomes meet the ocean, unique intermediate biomes are formed.

INTERTIDAL ZONES
An **intertidal zone**, where the ocean meets land, can be a sandy or a rocky habitat that provides shelter for many sedentary organisms such as algae, barnacles, and mussels. The intertidal zone is periodically pounded by surf and exposed to the sun, creating unique conditions.

ESTUARIES
An **estuary** is a transitional biome between a freshwater river and a saltwater ocean. Laden with nutrient-rich sediment, estuaries support highly productive and diverse communities.

MARINE BIOMES

Marine biomes include coral reefs and, within the ocean, several distinct zones and realms.

CORAL REEFS

Coral reefs occur in the photic zone of warm tropical waters above the continental shelf. Successive generations of coral animals with hard skeletons build up the reef. Coral reefs provide shelter and food that support a huge variety of animals.

The **continental shelf** is a shallow region where a continental plate is submerged in the ocean.

0 m
50 m
100 m
150 m
200 m

Low tide line

High tide line

Intertidal zone

The **pelagic realm** is the open water region of the ocean.

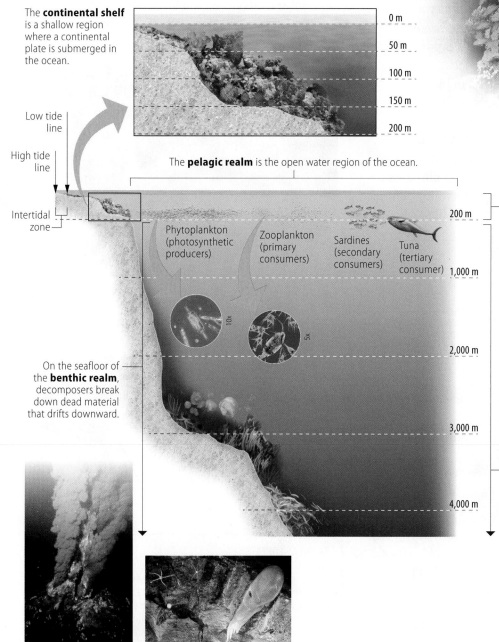

Phytoplankton (photosynthetic producers)

Zooplankton (primary consumers)

Sardines (secondary consumers)

Tuna (tertiary consumer)

200 m

1,000 m

2,000 m

3,000 m

4,000 m

10x

5x

On the seafloor of the **benthic realm**, decomposers break down dead material that drifts downward.

The **photic zone** includes waters where light can penetrate to drive photosynthesis, generally to depths of about 650 feet (200 meters). Here, phytoplankton (photosynthetic producers) form the basis of a diverse food chain. Zooplankton (primary consumers, which include protists and small crustaceans) eat phytoplankton, and are in turn eaten by fish, who are eaten by larger fish, including the blue shark shown below.

Light to power photosynthesis does not extend into the **aphotic zone**. Many animals here—such as the deep sea squid—consume other animals. Other aphotic dwellers rise to the surface to feed, or consume dead organic matter found on the seafloor. In general, the density of life in the aphotic zone is low.

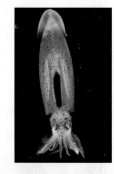

A typical coral reef grows at a rate of 1 inch per year.

DEEP SEA

Where the Earth's tectonic plates meet, more than a mile below the ocean's surface, lie hydrothermal vents. There, scalding water and hot gases spew from the Earth's interior. Some prokaryotes can derive energy directly from these chemicals, providing the basis for a unique food chain. Even under these extreme conditions, life thrives. At this hydrothermal vent, you can see sea stars and even an octopus that ultimately derive energy from the chemicals released by the vent.

CORE IDEA: Aquatic biomes include freshwater biomes—rivers, streams, lakes, ponds, and wetlands—and marine biomes. Intertidal zones lie between the ocean and land, and estuaries lie between fresh and salt water.

CORE QUESTION: What is the only biome zone that can support coral reefs?

ANSWER: The photic zone of the ocean.

There are a variety of terrestrial biomes

A biome is a community of organisms found living in a particular type of climatic zone. **Terrestrial biomes** are identified primarily by the types of vegetation found within them, since primary producers (plants) provide food and organic material for most members of an ecosystem. Around the globe, the distribution of biomes depends largely on two measures of climate: temperature and rainfall. A few biomes are also characterized by periodic wildfires. Although there is considerable variation within and between biomes, it is generally true that similar climates in different parts of the globe support similar biomes, and therefore similar communities of life.

POLAR ICE
Polar ice is found in the northernmost and southernmost parts of the Earth. These regions are so cold and dry that few species can survive. But small plants, most particularly mosses and lichens, can grow in some regions.

TUNDRA AND ALPINE TUNDRA
Tundra occupies the latitudes between coniferous forests and polar ice. It is characterized by dryness, bitter cold temperatures, and high winds that together produce permafrost, permanently frozen soil. The landscape is dominated by low-growing grasses, shrubs, mosses, and lichens. High winds and cold temperatures create plant communities called alpine tundra on very high mountaintops at all latitudes, including the tropics. Although these communities are similar to arctic tundra, there is no permafrost beneath alpine tundra.

TEMPERATE GRASSLANDS
Temperate grasslands are found in regions with cold winters, low rainfall, and periodic drought. Trees grow poorly, so the landscape is dominated by grasses. Much of the prairie of the American heartland is temperate grassland, although little of the original prairie remains.

TROPICAL FORESTS
Tropical forests occur in wet, warm climates near the equator. They can be rain forests or, if the region has rainy seasons punctuated by dry spells, tropical dry forests. In rain forests, a dense forest canopy towers over smaller trees and shrubs, often thick with vines. The dense vegetation provides a rich and varied habitat that supports a wide variety of animals.

CHAPARRAL

Chaparral is found in coastal areas where ocean currents produce mild, rainy winters and hot, dry summers. The landscape is dominated by evergreen shrubs, with annual flowers common during the winter and spring. Plant species of the chaparral have adaptations that help them survive periodic wildfires.

Temperature range | Precipitation

Fires

CONIFEROUS FORESTS

Coniferous forests are dominated by cone-bearing evergreens such as pine, hemlock, and redwood. A huge coniferous forest that stretches across northern North America, Europe, and Asia—the largest terrestrial biome on Earth—is characterized by long, cold winters and short, wet summers.

Temperature range | Precipitation

TEMPERATE BROADLEAF FORESTS

Temperate broadleaf forests occur in regions of hot summers, cold winters, and relatively frequent rain. This supports the growth of dense forests of deciduous (leaf-bearing) trees. Virtually all of the original ("old growth") temperate broadleaf forests of North America were destroyed by logging and development well over a century ago. Today's forest is dominated by young "new growth" species.

Temperature range | Precipitation

SAVANNA

Savannas are warm, fairly dry climates that primarily contain grasses with scattered, isolated trees. During the wet season, rapidly growing vegetation provides rich grazing for large herbivores, many species of which are found only in the African savanna. Because of periodic dry seasons, the plant life on savannas is adapted to survive fire. Some have belowground growing shoots, while others have hardy seeds that sprout quickly after a fire.

Temperature range | Precipitation

Fires

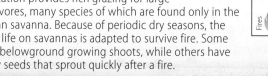

DESERT

Deserts have very low rainfall. Some, like the Sahara, are hot, but others, like the Gobi, can be cold. Most plants that survive in the desert have evolved adaptations that help them store water. Many cacti, for example, swell as they store water internally.

Temperature range | Precipitation

A large, fully grown saguaro cactus can store more water than a hot tub.

CORE IDEA: Earth's land can be divided into terrestrial biomes. Each is characterized by a particular climate that differs in temperature, rainfall, and prevalence of wildfires, and each has a dominant form of vegetation.

CORE QUESTION: Which highly productive biome is found almost exclusively near the equator?

ANSWER: Tropical forests.

Interactions between species play important roles in communities

A biological **community** is the sum of all the populations of organisms (plants, animals, fungi, and even microbes) that live in the same area and so are able to interact. Organisms interact with members of their community as they compete for food, water, sunlight, or living space. Such interactions between populations affect the composition of the community.

SPECIES INTERACTIONS

A community ecologist typically studies **interspecies interactions**, relationships between species. Such interactions can be classified according to how they affect each population involved. Different types of interspecies interactions can be helpful, harmful, or neutral to each population involved.

Type of interaction	Effect on population 1	Effect on population 2	Example
Competition	⊖ (negative)	⊖	Different species of forest plants compete for sunlight.
Mutualism	⊕ (positive)	⊕	Flowers and their pollinators benefit each other.
Predation	⊕	⊖	Cheetahs hunt gazelles.
Herbivory	⊕	⊖	Deer eat forest plants.
Parasitism/pathogens	⊕	⊖	Heartworms multiply inside the cardiac tissue of dogs.
Commensalism	⊕	⓪ (neutral)	Egrets eat insects stirred up during cattle grazing, but the egrets don't affect the cattle at all.

COMPETITION

Each species in a biological community has a set of resources that it needs for survival, such as food, sunlight, water, and living space. When the required resources of two species overlap, they compete for a limited supply of these resources. Such interspecies competition means that reproductive success by one species harms the chances of survival of the other, and vice versa. In fact, if the resources required by two species are too similar, they cannot coexist. This principle is called the **competitive exclusion principle**.

In the forests of the American northeast, black bears and chipmunks compete to feed on a limited supply of acorns. If one species benefits by eating more, the other species suffers.

MUTUALISM

Mutualism is a form of interspecies interaction in which both populations benefit. Mutualism often occurs among species that are **symbiotic**, living in close physical association with one another. (Don't confuse symbiosis with mutualism; two organisms may live in contact but both may not benefit, as in a parasitic relationship.)

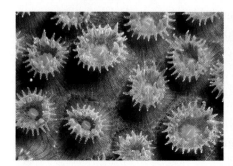

CORAL
Coral reefs contain many individual coral polyps. Each polyp is an animal that contains millions of unicellular algae living within it. The algae provide sugar to the coral via photosynthesis while receiving living space and access to sunlight.

POLLINATORS
Flowers provide sugar (in the form of nectar) to pollinators, such as bees. At the same time, the pollinators help the flowers reproduce by carrying sperm-containing pollen from one flower to another.

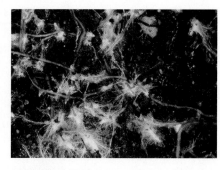

MYCORRHIZAE
Mycorrhizae are mutualistic associations of plant roots and fungal filaments. The plant supplies organic nutrients to the fungus, and the fungus supplies mineral nutrients to the plant.

PREDATION, HERBIVORY, PARASITISM, AND PATHOGENS

These four types of interspecies interactions involve distinct behaviors, but they all benefit one species at the expense of another.

PREDATORS
Predation is an interaction in which a predator species kills and eats a prey species. Prey species evolve adaptations through natural selection that help them avoid predators. For example, this impala uses speed and agility to avoid cheetahs. Other prey adaptations include camouflage, shells, quills, and chemical defenses.

HERBIVORES
Herbivory is the eating of plant parts by an animal. Plants evolve adaptations that provide defense against herbivores, such as the spines of this barrel cactus, or chemical toxins.

PARASITES
A **parasite** lives on or in (but does not kill) a host, from which it obtains nutrients. Both plants and animals may have parasites. Animal parasites may live internally (flukes, tapeworms, roundworms) or externally (ticks, lice, mites, leeches). This mouse flea feeds on its host's blood.

5x

PATHOGENS
Pathogens are disease-causing microorganisms—most often bacteria, viruses, fungi, or protists. Humans have many pathogens that cause disease (such as the viruses that cause colds or the bacteria that cause pneumonia), and so do most other large organisms. These chestnut trees were killed by blight due to protist infection.

CORE IDEA: Within a community, species interactions can be mutually harmful (competition), mutually beneficial (mutualism), or harmful for one population but helpful for another (predation, herbivory, parasitism, and pathogens).

CORE QUESTION: What relationship does a human have to a pet dog? To a chicken raised for meat? To an apple tree?

ANSWER: Mutualism; predation; herbivory.

Food webs describe multiple trophic structures

When considering a biological community (a group of populations living in the same place and the same time), we often think of feeding relationships: Who eats whom? A **trophic structure** describes the feeding relationships within a community. Understanding trophic structure allows an ecologist to describe the flow of energy and nutrients through an ecosystem. Trophic structures can be quite complex, involving many organisms and many different trophic levels.

FOOD CHAINS

A **food chain** is a simplified description of one part of the trophic structure, following the passage of food energy from one individual to another. Here, you can see two specific examples of food chains, one from a terrestrial ecosystem and one from an aquatic ecosystem. Read each one starting from the bottom.

TERRESTRIAL	TROPHIC LEVELS	AQUATIC
Quaternary consumers on land include hawks, coyotes, and other top-level predators.  Ferruginous hawk	**QUATERNARY CONSUMERS** Some but not all ecosystems have **quaternary consumers**, top-level predators that eat tertiary consumers.	 Orca In the ocean, top-level consumers include sharks and killer whales.
On land, tertiary consumers—such as snakes and mammals—eat other carnivores.  Bull snake	**TERTIARY CONSUMERS** **Tertiary consumers** eat secondary consumers.	 King salmon In aquatic environments, tertiary consumers include large fish that prey on bait fish.
On land, secondary consumers include a great variety of small animals as well as large animals that hunt grazers.  Chipmunk	**SECONDARY CONSUMERS** All trophic levels above primary consumers contain predatory carnivores that eat consumers from the level below. **Secondary consumers** eat primary consumers.	 Flagfin shiner In the ocean, secondary consumers are mainly small fishes (sometimes called bait fish) that eat zooplankton.
On land, primary consumers include herbivores such as insects, snails, grazing animals, and birds.  Cricket	**PRIMARY CONSUMERS** All trophic levels above producers, called consumers, depend on the output of producers. **Primary consumers** eat producers directly.	 Squid larva In the ocean, primary consumers include a variety of zooplankton, microscopic animals and protists that eat phytoplankton.
Plants are the main producers on land.  Grass	**PRODUCERS** **Producers** support all other levels of the food chain. Most producers use photosynthesis to convert the energy in sunlight to organic molecules that serve as food.	 Ceratium Marine producers include phytoplankton (photosynthetic single-celled protists and cyanobacteria) and seaweed (multicellular algae).

FOOD WEBS

Although a food chain is a useful tool for understanding the trophic levels of an ecosystem, a food chain may greatly simplify real-world biological communities. Several different species occupy most trophic levels and compete with each other for the same prey. Furthermore, some predators consume prey from multiple trophic levels. Actual feeding relationships within ecosystems must thus be expressed as a multibranched **food web** that interconnects multiple food chains.

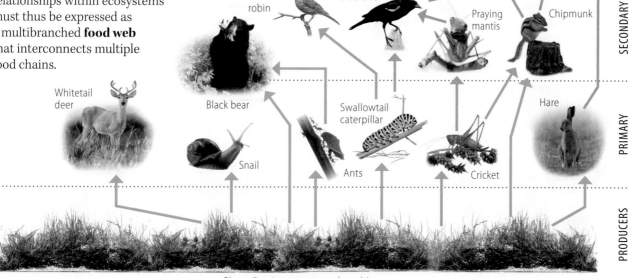

FOOD WEB OF A HYPOTHETICAL FOREST COMMUNITY

Great horned owl

Black rat snake

Peregrine falcon

Barn owl

Gray fox

American robin

Red-winged blackbird

Praying mantis

Chipmunk

Whitetail deer

Black bear

Snail

Swallowtail caterpillar

Ants

Cricket

Hare

Plants (berries, grasses, and seeds)

QUATERNARY CONSUMERS

TERTIARY CONSUMERS

SECONDARY CONSUMERS

PRIMARY CONSUMERS

PRODUCERS

BIOLOGICAL MAGNIFICATION

Organic material passes through the trophic levels of an ecosystem in a stepwise fashion, from producers to primary consumers and so on. Because so much energy is lost as heat during each transfer, the organisms at each level must consume many organisms at the lower levels to gain the required energy to live. Many toxins, once they enter the food chain (typically via industrial pollution), cannot be digested and so are passed from one trophic level to another. This causes **biological magnification**, the tendency of toxins to become concentrated as they pass through a food chain. Thus, top-level predatory fish are often significantly more contaminated than bait fish.

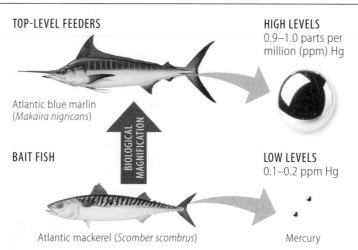

TOP-LEVEL FEEDERS

Atlantic blue marlin (*Makaira nigricans*)

BAIT FISH

BIOLOGICAL MAGNIFICATION

Atlantic mackerel (*Scomber scombrus*)

HIGH LEVELS
0.9–1.0 parts per million (ppm) Hg

LOW LEVELS
0.1–0.2 ppm Hg

Mercury

Mercury (Hg) is a toxin that is passed from one trophic level to another without being broken down. Consequently, higher trophic-level carnivores have higher concentrations of mercury, and so are less safe for human consumption. For example, every marlin eats many mackerel, causing mercury to accumulate in the body of the predator. You should limit your consumption of top-level consumers to avoid high levels of accumulated toxins.

Albacore tuna has about three times more mercury per can than light tuna.

CORE IDEA: Food chains and food webs describe the transfer of organic material from one trophic level to the next, starting with producers and ending with top predators. Toxins can accumulate in higher levels of a food web.

CORE QUESTION: Which would you expect to have higher levels of lead in its bloodstream: a mouse or a hawk?

ANSWER: A hawk, because it occupies a higher trophic level.

12.10

Several factors affect species diversity

When comparing one community to another, or tracing changes in a community over time, ecologists often measure **species diversity**, the variety of species that live within a community. One community may be calculated to be more diverse than another. For example, modern agricultural fields have little diversity: crops usually consist of just a single species grown over a wide area. In contrast, coral reefs and rain forests have extremely high species diversity.

SPECIES RICHNESS AND RELATIVE ABUNDANCE

The species diversity of a community has two components. The **species richness** is the number of different species in a community. A community with few different species has low species richness, while a community with many different species has high species richness. The **relative abundance** is the fraction of the total life in a community accounted for by each species. If an ecosystem is dominated by one species, that species has high relative abundance. Here you can see these two measures for a pair of hypothetical stands of trees in a northeastern U.S. deciduous forest.

FOREST A

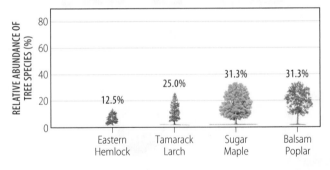

FOREST B

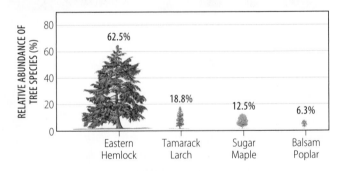

These two hypothetical communities have the identical species richness: four species each. However, in Forest B, one species dominates, yielding a community in which that one species has a much higher relative abundance than the others. Hence, Forest B is less diverse than Forest A.

KEYSTONE SPECIES

Within a community, one species may have a disproportionate effect on the overall species diversity. Such a species is called a **keystone species** (named for a wedge-shaped stone at the top of an arch that, if removed, causes the entire arch to collapse). For example, ecologists studying the Alaskan coast discovered that a decline in the sea otter population allowed sea urchins (the otters' main food source) to quickly multiply. The urchins devoured the kelp forests that supported a wide diversity of marine life. Removing the keystone species (sea otters) thereby affected the richness and diversity of the entire community.

DISTURBANCES AND ECOLOGICAL SUCCESSION

Communities are rarely static. Periodic disturbances (such as fires, floods, storms, volcanic eruptions, and droughts) kill organisms and affect the availability of resources. After a disturbance, a habitat will be colonized by a series of different species. This process, called **ecological succession**, may take hundreds or thousands of years.

1980

2011

1985

2009

DISTURBANCE
A severe disturbance, such as the 1980 eruption of the Mt. St. Helens volcano in Washington, can drastically alter an ecosystem.

PRIMARY SUCCESSION
Primary succession occurs when an area has been rendered virtually lifeless with no soil. The first multicellular life to return is often photosynthetic lichens and mosses, followed by other small plants.

As soil begins to accumulate from the breakdown of rock and the decomposition of early colonizers, grasses, shrubs, and finally trees will return to the area via windblown seeds.

Eventually, the area will be colonized by the community's dominant form of vegetation. The presence of abundant producers allows consumers to return.

When Mt. St. Helens erupted, it created the largest rock avalanche ever recorded, instantly destroying over 230 square miles.

SECONDARY SUCCESSION

Secondary succession occurs after a disturbance that kills much of the life in an area but leaves the soil intact, allowing plant life to return more rapidly than during primary succession.

A giant bushfire raging in Victoria, Australia

Regrowth six weeks after bushfires

Fire-damaged trees showing new growth a year after bushfires

CORE IDEA: A community's diversity can be measured by the number of species it contains (species richness) and how prevalent each species is (relative abundance). After a disturbance, an area will be reoccupied by a succession of communities.

CORE QUESTION: When did Mt. St. Helens have a greater species diversity: in 1985 or in 2009?

ANSWER: 2009; at the start of primary succession in 1985, only a few species were present, yielding a low species diversity.

Invasive species can disrupt ecosystems

A biological community can undergo rapid change after introduction of an **invasive species**, a non-native species that occupies a foreign habitat and spreads quickly, causing environmental damage. For example, when a species introduced by humans into a new habitat lacks predators, the invasive population may quickly multiply and displace native species. Although not every introduced population is successful, the ones that are can be quite damaging. In fact, invasive species are a leading cause of extinctions of local populations.

A ROGUE'S GALLERY OF INVASIVE SPECIES

Around the world, ecologists have documented thousands of invasive species, including hundreds within the United States. They come in all varieties: plants, mammals, birds, fishes, mollusks, and even insects.

KUDZU
A member of the pea family, kudzu (*Pueraria lobata*) is a climbing vine native to Japan and China that was introduced into America at the 1876 Exposition in Philadelphia. It spreads rapidly, overgrowing native species, depriving them of light. It is now common along roadsides in the southern United States.

BURMESE PYTHON
Burmese pythons (*Python molurus*) are large snakes that were set loose in Florida. The warm, wet climate of the Everglades—similar to their native Asian habitat—and bountiful prey (including birds, mammals, amphibians, lizards, and even household pets) have allowed them to thrive. Biologists estimate there are now tens of thousands of pythons living in Florida, and their range is expanding.

LIONFISH
Released from hobbyists' aquariums into the Caribbean, the lionfish (*Pterois volitans*), a native of the Indian and South Pacific Oceans, spread rapidly. Lionfish consume other fish voraciously, including juveniles of economically important species such as snapper. Although beautiful, they have venomous spines and are causing considerable havoc to tropical reef ecosystems.

RABBITS
In 1859, 24 rabbits were released on an Australian ranch for game hunting. By 1900, the rabbit population had increased to several hundred million, spanning most of the continent. They ate crops, destroyed grazing lands, and became a significant economic nuisance. The government has been battling to control rabbit populations ever since.

ZEBRA MUSSELS
The zebra mussel (*Dreissena polymorpha*) is a native of Europe. It was introduced to the U.S. Great Lakes in 1988 through the dumping of ballast water from ocean-crossing cargo ships. Once introduced, the mussels spread quickly and prolifically. They now cause an estimated $250 million worth of damage annually to power plants and water treatment facilities.

Kudzu spreads over 150,000 new acres each year.

BIOLOGICAL CONTROL

Why might an invasive species be rapidly successful in a new habitat? One reason could be an absence of predators, herbivores (if the invasive species is a plant), or pathogens (infectious microorganisms). Efforts to control invasive species therefore often focus on **biological control**, the intentional release of a natural enemy that attacks the invader. Biological control can also be used as a tool against noninvasive pests that attack crops, such as insects and weeds. Although often less harmful to an ecosystem than chemical control, the effect of biological control on a given habitat is difficult to predict.

STINGLESS WASPS

The European core borer (*Ostrinia nubilalis*; see the photo below) is a worm that devastates corn and other valuable crops throughout North America. Introduction of a small Chinese stingless wasp, *Trichogramma ostriniae*, helps keep populations of the borer in check.

MONGOOSE

Asian rats have become invasive in many islands of the Western world. In Hawaii and the Caribbean, they cause considerable damage to crops. In an attempt at biological control, growers introduced the Indian mongoose (*Herpestes edwardsii*; see below), a fierce hunter of rats. Unfortunately, the mongooses are such good hunters that they also decimated native reptiles, amphibians, and birds, as well as flocks of domestic fowl and other crops. This example serves as a cautionary tale that biological control often has unintended consequences.

INTEGRATED PEST MANAGEMENT

Modern agriculture—which usually involves growing a large population of genetically identical individuals—presents unique challenges of biological control and management. Farmers (and home gardeners) constantly battle weeds, pests, and pathogens that, if they affect one individual, may wipe out all of them. Although chemicals may be effective, they often have problems of their own. **Integrated pest management** is a method that utilizes several strategies, including planting pest-resistant crop varieties, careful monitoring, judicious use of chemicals, release of sterile pests, and other biological and behavioral changes. The goal is not eradication of pests, but the maintenance of a constant, low population. Such techniques are being tried around the world, and have met with some success. For example, integrated pest management has been used effectively for cotton crops in India.

CORE IDEA: Invasive species spread rapidly and cause damage after being introduced to a new environment. Invasive species are often combated with biological methods, such as the introduction of natural predators, with mixed success.

CORE QUESTION: Name the practice that the following example illustrates: Some farmers have introduced goats that eat kudzu in their fields.

ANSWER: Biological control.

Biodiversity is measured on many levels

Biodiversity is a general term for the variety of living things on Earth. Biodiversity can be studied on many levels. The term may refer to the variety of genes (inherited DNA sequences) within a given population. On a somewhat larger scale, it may refer to the variety of different species that can be found in a given area. On an even larger scale, it can be a measure of the diversity of ecosystems found within the biosphere. On any scale, the preservation of biodiversity requires conscious effort on all our parts.

GENETIC BIODIVERSITY

Genetic biodiversity refers to the collection of genes within a population. Genes are the raw material upon which evolution by natural selection acts. Genes that confer a survival advantage in a particular environment will be passed along in greater proportion than genes that do not; the result is evolutionary adaptation. As members of a population are lost, so are their genes and the potential they represent. Severely reducing genetic variation makes the population less able to adapt to a changing environment, thereby threatening the survival of the species. A classic example is the Irish Potato Famine (1845–1852). Because virtually all the potatoes in the country were genetically identical, a change in the environment (in the form of a new pathogen) caused widespread crop loss. The subsequent starvation and emigration of Irish citizens to other countries affected the fate of people in several countries on multiple continents.

Potato infected with blight

A memorial to those who died in the Irish Potato Famine

SPECIES BIODIVERSITY

Species biodiversity refers to the number of different species in an ecosystem or in the biosphere as a whole. In the long term, new species appear as a result of evolutionary change and old species undergo **extinction**, the irreversible loss of all populations of a species. At present, planet-wide species diversity is declining rapidly. If this rate of species loss continues, half of all currently living plant and animal species on Earth will be extinct by the end of this century. Examples of recently extinct American species include the Catahoula salamander (an amphibian), the Carolina parakeet (a bird), the blue walleye of the Great Lakes (a fish), and the Sampson's pearly mussel (a mollusk).

The last Carolina parakeet (*Conuropsis carolinensis*) died in a zoo in 1918.

ECOSYSTEM BIODIVERSITY

A tropical reef damaged by a drag net.

Ecosystem biodiversity refers to the variety of ecosystems found on Earth. This is biodiversity in its broadest context. The degradation of ecosystems threatens **ecosystem services**, benefits that ecosystems provide to people. For example, wetlands absorb large quantities of rainfall; the loss of wetlands results in more frequent flooding of surrounding areas. Among the most threatened ecosystems are coral reefs, which provide shelter for other species, protect shorelines by acting as breakwaters, and provide recreation. About 20% of the coral reefs in the world have been destroyed by human activities, and 75% of the remaining reefs are threatened.

Nearly one-third of all known amphibian species are either near extinction or endangered.

CAUSES OF BIODIVERSITY LOSS

Ecologists have identified a variety of factors that threaten biodiversity. In modern times, the root cause is usually the growing size of the human population and the activities required to sustain it.

INVASIVE SPECIES

When a non-native plant or animal is introduced by humans into a new habitat, its population size will sometimes expand very rapidly. The invader may have no natural predators and can thus multiply unchecked, preying upon, outcompeting, and displacing native species. The havoc this plays on a biological community can lead to significant loss of biodiversity. For example, when brown tree snakes (*Boiga irregularis*) were introduced to the island of Guam, they drove three bird species to extinction, and two others have no living members in the wild.

GLOBAL CLIMATE CHANGE

A changing global climate is having a significant impact on biodiversity. Changes in patterns of temperature and rainfall are altering ecosystems faster than life within them can adapt. For example, earlier arrival of warm temperatures in spring has been shown to affect communities by disrupting the timing of seasonal events: Flowers may emerge before their pollinators have matured, threatening both species. In the oceans, high temperatures cause coral to expel their symbiotic algae. This phenomenon, called coral bleaching, threatens many tropical reef ecosystems.

HABITAT DESTRUCTION

The single greatest threat to biodiversity—and the cause of the vast majority of extinctions—is the destruction of habitat due to development, agriculture, forestry, mining, and dam construction. Deforestation is occurring quite rapidly and—particularly in the tropical rain forest—threatens many species. Coastal development also threatens many delicate terrestrial and aquatic ecosystems. This photo shows Brazilian rain forest being cleared for agricultural use.

POLLUTION

Pollution of the air and water contributes to biodiversity loss at the local, regional, and global levels. Local ecological disasters, such as oil spills, can decimate a habitat. On a larger scale, pollutants released into the air in one part of the world may be carried by winds thousands of miles before falling back and polluting the ground and water of a distant area. This photo shows the effects of the 2010 *Deepwater Horizon* oil rig explosion in the Gulf of Mexico.

OVERHARVESTING

If people harvest wildlife from an ecosystem faster than the wildlife can naturally replenish, the result may be a loss of biodiversity. On land, overhunting has drastically reduced wild populations of American bison, and overcutting has endangered many hardwood tree species. In the ocean, overfishing has decimated the wild populations of many fish and mammal species.

CORE IDEA: Biodiversity—the variety of living things—can be considered on many levels: genetic (within a population), species, and ecosystem-wide. Biodiversity at every level is threatened by human activities.

CORE QUESTION: Mountaintop removal—a method of mining that strips off large areas of soil, rock, and vegetation—disrupts biodiversity at which level?

ANSWER: Ecosystem biodiversity.

Populations vary in age structure, survivorship, density, and dispersion

A **population** consists of members of the same species living in an ecosystem at the same time. The definition of who does and does not belong to a population is somewhat loose, but it generally represents those individuals who may breed with one another. Members of a population compete for the same resources and are influenced by the same environmental factors. **Population ecology** is the study of changes in populations over time. Topics of study for a population ecologist include factors that affect population size, density, and growth.

AGE STRUCTURE

An important measure of a population is a demographic tool called **age structure**, the number of individuals in different age groups. Comparing age structures of different human populations can provide important insights into social conditions. Graphs comparing the age structures of the United States and Ethiopia, for example, indicate a higher need for elderly health care in the United States, but a higher need for schools and employment opportunities in Ethiopia.

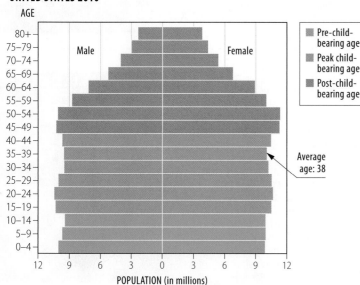

UNITED STATES 2010

AGE

Male — Female

Pre-child-bearing age
Peak child-bearing age
Post-child-bearing age

Average age: 38

POPULATION (in millions)

In the United States today, birth rates are low but survivorship is high. This is reflected in the average age: 38.

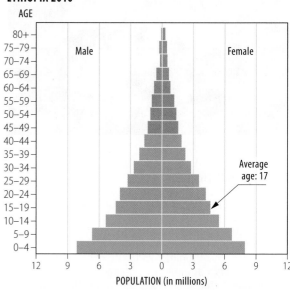

ETHIOPIA 2010

AGE

Male — Female

Average age: 17

POPULATION (in millions)

In the African country of Ethiopia, the broad base of the age structure reflects high birth rates but low survivorship. Thus, the population has a high proportion of children: Over half of Ethiopians are 17 years old or younger, versus one-fifth of Americans.

The birth rate in the United States is about twice that of China but about one-third that of Somalia.

SURVIVORSHIP CURVES

Survivorship is the chance that an individual member of a given population will live to a particular age. Gathering data on survivorship can help ecologists understand the dynamics of a population over time. Such data may be presented as a **survivorship curve**, a graph showing the percentage of individuals alive at each age, with the age often expressed as a percentage of the life span.

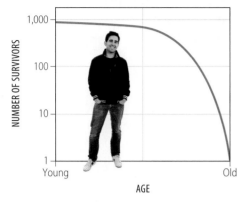

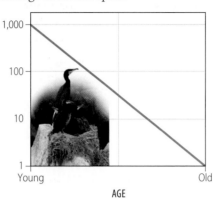

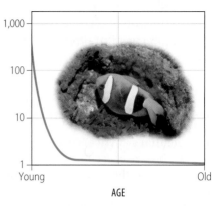

Large animals (such as humans) tend to have few babies, but most survive to old age, after which they die off fairly rapidly.

Some species (such as birds, lizards, and rodents) have approximately constant survivorship over their lifetimes. Individuals are equally likely to die at any age.

Many species (such as fish, ferns, and insects) produce huge numbers of offspring, few of which survive to adulthood. But those that do survive childhood often live their full life span.

POPULATION DENSITY AND DISPERSION

One important measure to ecologists is the **population density**, the number of members of a species per unit area or volume of the habitat. Population density varies over time as individuals enter (via birth or immigration) and leave (via death or emigration). The size of the population may be counted directly (the number of ferns in a forest, for example) or estimated indirectly (for example, by counting some sample areas and then extrapolating to the total habitat). Population ecologists also record the **dispersion pattern**, the way individuals are spaced within a habitat.

CLUMPED
The most common dispersion pattern in nature is clumped, with individuals found in small groups with spaces between groups. Clumping may result from an unequal distribution of resources or result from behaviors that reduce the risk of predation.

UNIFORM
A uniform dispersion pattern often results from competition between individuals for the same resource. For example, some plants secrete chemicals that inhibit the growth of nearby plants, and some animals display territorial behavior. Spacing allows each individual access to needed resources.

RANDOM
Rarely, individuals in a population are dispersed with no pattern. Such a random distribution may arise, for example, when windblown seeds are randomly dispersed.

CORE IDEA: By studying populations (interacting members of the same species) in terms of their age structure, survivorship curves, density, and dispersion, ecologists can gain insights into how a population changes over time.

CORE QUESTION: In which country—Ethiopia or the United States—is there a greater need to build schools?

ANSWER: Ethiopia; notice that the green region of its age structure graph accounts for a greater percentage of the population than the green region of the U.S. graph.

Growth models can predict changes in population size

The size of a population—a group of individuals of the same species that may interact in the same habitat—changes over time. Although some population sizes remain relatively constant over time, others change dramatically. Population ecologists have developed idealized models that can be used to predict and explain changes in populations in the real world.

EXPONENTIAL GROWTH MODEL

Exponential growth occurs when the population size of each new generation is a multiple of the previous generation—for example, a population that doubles every generation. Under such conditions, the rate of population growth continuously rises with population size. Exponential growth requires unlimited resources and a nearly vacant environment. This may happen after a disturbance (such as fire, drought, or flood) wipes out most life in an area, or when a new area is occupied. However, no natural environment can possibly sustain exponential growth for very long.

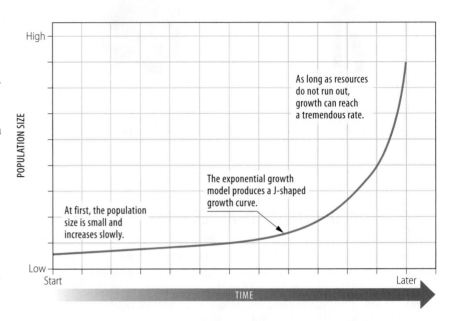

As long as resources do not run out, growth can reach a tremendous rate.

The exponential growth model produces a J-shaped growth curve.

At first, the population size is small and increases slowly.

LOGISTIC GROWTH MODEL

Most natural environments cannot support unlimited growth for long. In reality, most environments have **limiting factors**, environmental constraints that put a cap on the size of a population. The **carrying capacity** is the maximum population size that can survive in an environment, given the limiting factors that exist. **Logistic growth** is where the size of a population grows rapidly until it nears its carrying capacity for that environment. At that point, the rate of population growth slows. Once the carrying capacity is met, the population size does not change because the birth rate and death rate equal each other. Many real-world populations exhibit logistic population growth when resources are limited.

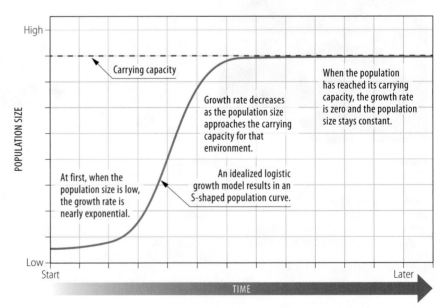

Carrying capacity

When the population has reached its carrying capacity, the growth rate is zero and the population size stays constant.

Growth rate decreases as the population size approaches the carrying capacity for that environment.

At first, when the population size is low, the growth rate is nearly exponential.

An idealized logistic growth model results in an S-shaped population curve.

LIMITING FACTORS

As a population approaches the carrying capacity, its growth slows due to limiting factors in the environment. Two broad categories of limiting factors are found in the natural world.

DENSITY-DEPENDENT LIMITING FACTORS

Factors that increase or decrease their effect depending on the density of the population are called **density-dependent limiting factors**. The most common is competition between individuals of the same species for limited resources. As the population grows, competition intensifies, lowering birth rates. Limits on food, water, and living space all foster competition within a population in a manner proportional to the existing population density.

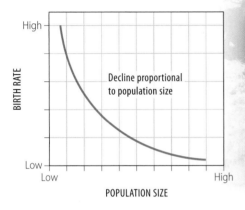

DENSITY-INDEPENDENT LIMITING FACTORS

Factors for which the effect does not relate to population size are called **density-independent limiting factors**. Abiotic factors such as weather or disturbances may affect population size by causing a rapid decline. The decline is expected to be just as drastic whether the population size is large or small. For example, many insect populations, regardless of their size, experience drastic die-offs when the first winter frost arrives.

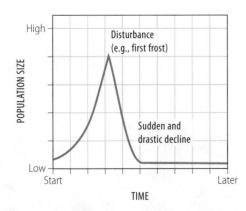

If not limited by the environment, a single bacterium could give rise to enough bacteria to cover the Earth in just 36 hours.

CORE QUESTION: What kind of limiting factor is a volcanic eruption? After an eruption, what growth model would you predict among the survivors?

CORE IDEA: Population growth can be understood by considering hypothetical growth models. Limiting factors (which may or may not depend on the number of individuals) affect the maximum population size in an ecosystem.

ANSWER: It is a density-independent limiting factor. Afterward, the surviving population will grow exponentially.

Human population growth has been exponential

Population ecologists study changes in populations over time. Because humans have a disproportionately high impact on nearly all ecosystems, the trends in human population growth are particularly relevant to the study of ecology.

THE HUMAN POPULATION GROWTH CURVE

Throughout most of human history, parents bore many children, but the infant mortality rate was high. During this period, population growth was slow. Advances in nutrition, sanitation, and medical care caused the death rate to decline while the birth rate remained the same. The result was a population explosion during the past 100 years. The birth rate, the number of new people born each year, recently peaked. Since that time, the birth rate has declined, and the rate of growth has slowed. However, the total human population—currently about 7.1 billion—continues to rise.

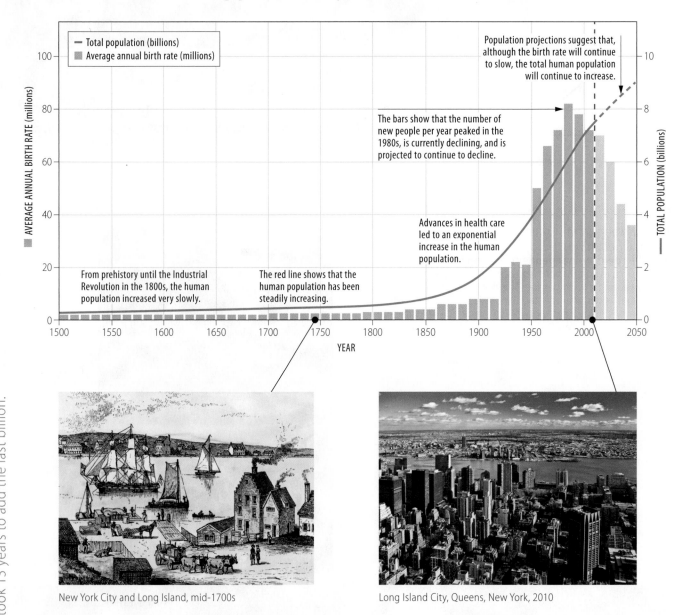

New York City and Long Island, mid-1700s

Long Island City, Queens, New York, 2010

It took until 1810 to add the first billion humans; it took 13 years to add the last billion.

324

HUMAN AGE STRUCTURES

An important measure of a population is its age structure, the number of individuals in different age groups. The age structure of a population can be used to predict that population's future growth. For example, consider the age structures for the United States and Mexico in 1985, 2010, and as projected for 2035, and the significant impact these changes will have on their respective societies. Notice that the present birth rate is affected by the birth rate 15 to 30 years ago, since that is when today's child-bearers were born.

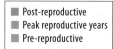

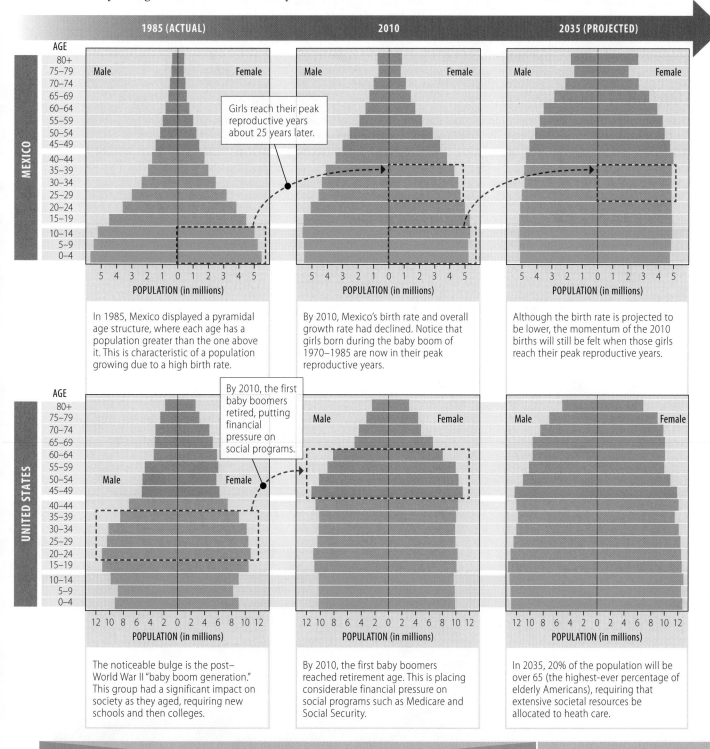

CORE IDEA: The human population has exhibited exponential growth for the last 250 years. Examining the age structure of a population can give insight into future societal needs.

CORE QUESTION: What accounts for the bulge in 15- to 24-year-olds in the United States in 2010?

ANSWER: These are the children of the baby boomers, offspring of the bulge from 25 years prior.

Humans cause many ecological problems

Throughout our history, humans have become increasingly adept at gathering and producing food, extracting minerals and other resources from the environment, and in general using the Earth to sustain our population growth. As a result, our civilization has flourished. But this progress comes at a price: Human activities are changing the biosphere in ways both negative and unpredictable. No part of the biosphere remains unchanged by the collective influence of over 7 billion humans and our unquenchable thirst for resources.

ECOLOGICAL FOOTPRINT

Based on current data, the population of most nations on Earth will continue to increase for the foreseeable future, although the rate of increase will be slower than the past. How many people does the planet have resources to sustain? An **ecological footprint** is an estimate of the amount of land and water required to sustain one person. Measured this way, the ecological footprint of the average American is well beyond what the planet can support; we are running an enormous ecological deficit because we consume a disproportionate amount of food and fuel and own an overabundance of goods. Indeed, the largest footprints all belong to the wealthiest nations.

AVERAGE ECOLOGICAL FOOTPRINTS

Citizens of different countries have vastly different ecological footprints, expressed here as global hectares (a unit of area) per person. Notice that the average citizen of the world uses more resources than the planet can provide in the long term.

BANGLADESH AVERAGE: 0.7

US AVERAGE: 7.2

WORLD AVERAGE: 2.7

WORLD AVERAGE REQUIRED FOR SUSTAINABILITY: 1.8

The United States alone consumes 23% of the planet's ecological capacity.

This map shows the per-person ecological footprint of every country in the world. Notice that the footprints of the vast majority of countries exceeds long-term sustainable levels.

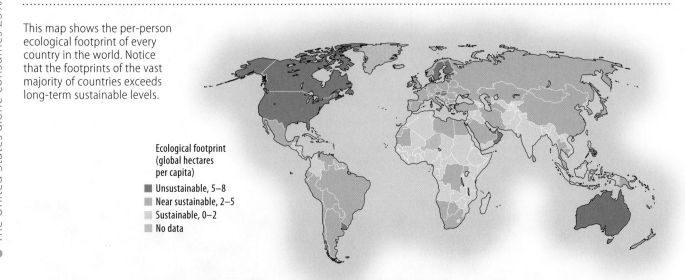

Ecological footprint (global hectares per capita)

- Unsustainable, 5–8
- Near sustainable, 2–5
- Sustainable, 0–2
- No data

HUMAN IMPACTS ON THE ENVIRONMENT

Humans have a tremendous impact on ecological resources. Several different types of problems result from our collective effects.

FOREST DESTRUCTION

We depend on the land as a source of food, fuel, and shelter. The Earth's land also provides ecosystem services such as nutrient cycling, flood control, recreation, and reducing atmospheric levels of greenhouse gases. But about three-quarters of the Earth's land surface has been altered by humans through agriculture and urban development. Forests have been particularly hard hit. The primary cause of forest destruction is clearing for agriculture. Other forests are being destroyed by logging, mining, road construction, or pollution. Present rates of forest destruction cannot be sustained for long, or soon there will be no forests at all.

SANTA CRUZ, BOLIVIA, 1986

SANTA CRUZ, BOLIVIA, 1998

These satellite images of one region of South American rain forest show how quickly forest landscapes can be altered. In just 12 years, roads for agriculture and logging spread throughout the region, destroying much of the natural habitat.

FRESHWATER DEPLETION

Humans cannot survive more than a few days without fresh water, so the health of freshwater ecosystems is clearly important to our species. The biggest threats to freshwater ecosystems include fertilizer runoff, pollution, depletion due to development in dry climates, and overuse of groundwater for irrigation. Some regions of the world are already experiencing dire shortages, and the problem will probably get worse in the near future.

Polluted creek in Orange County, California, collects urban runoff that contains fertilizers, detergents, and other pollutants.

Depleted water levels at Arrowrock Dam near Boise, Idaho

RAPIDLY DECLINING SPECIES

The Hawksbill sea turtle (*Eretmochelys imbricata*) shown here is being tagged as part of a conservation program. This species was once common throughout the Indian, Atlantic, and Pacific Oceans, but the Hawksbill is now an **endangered species**, one with a rapidly declining population due (almost always) to human activity such as habitat destruction. A **threatened species**, such as the Canada lynx (*Lynx canadensis*), is one that is likely to become endangered soon.

The Hawksbill turtle (*Eretmochelys imbricata*), an endangered species

The Canada lynx (*Lynx canadensis*), a threatened species

CORE IDEA: Humans depend on the continued health of Earth's ecosystems for our own survival, but we threaten the health of ecosystems in several ways.

CORE QUESTION: What is the difference between an endangered and a threatened species?

ANSWER: An endangered species has a rapidly declining population; a threatened species is one that is likely to become endangered soon.

Although human activities are the cause of many environmental problems, we also have the potential to repair and reverse much of the damage. Ecology is not just useful for pointing out problems; it is also the foundation of our solutions. **Conservation biology** is a branch of ecology that seeks to investigate and reverse the loss of biodiversity by sustaining ecosystems and maintaining the genetic diversity that is the raw material for evolutionary adaptation. The field of **restoration ecology** uses ecological principles to help repair degraded areas. Both of these important and expanding fields are making significant progress.

BIODIVERSITY HOT SPOTS

Biodiversity hot spots are relatively small areas with unusually high concentrations of endemic species (ones that are found nowhere else), endangered species, and threatened species. Although such hot spots account for less than 1.5% of Earth's land surface, they are home to over 30% of all species of plants and vertebrates. By identifying such biodiversity hot spots and establishing protected zones around them, conservation biologists can help save much biodiversity at a relatively low cost.

■ Biodiversity hot spots

The southernmost tip of Florida is a biodiversity hot spot that includes many unique creatures, such as manatees (*Trichechus manatus*).

The ring-tailed lemur (*Lemur catta*) is just one of about 100 species of lemur, all of which are endemic to the island of Madagascar.

SPECIES RECOVERY

Although human activities are the most common causes of species extinction today, humans can also act to preserve endangered or threatened species. When people are motivated to act, recovery can be dramatic. Bald eagles, for example, declined dramatically in the lower 48 U.S. states due to habitat loss and environmental toxins, primarily the insecticide DDT. Due to preservation efforts, the few hundred bald eagle pairs that were alive in 1963 have given rise to over 10,000 breeding pairs today.

Bald eagles (*Haliaeetus leucocephalus*) perched on driftwood in the Kachemak Bay, Alaska

● Madame Berthe's mouse lemur, a native of Madagascar, weighs just over 1 ounce, making it the world's smallest primate.

CONSERVING ECOSYSTEMS

Beyond saving individual species, conservation biologists aim to preserve the biodiversity of entire ecosystems. One of the most harmful types of habitat destruction is fragmentation, the splitting and isolation of habitats that causes small populations to become isolated from each other. One strategy to remedy fragmentation involves building movement corridors, narrow strips of suitable habitat that connect otherwise isolated patches. Corridors can assist with migration through heavily developed areas, particularly for animals that migrate seasonally.

A bridge in Banff, Canada, allows animals to access areas otherwise separated by a road.

BIOREMEDIATION

Restoration ecologists often turn to **bioremediation**, the use of living organisms to detoxify polluted ecosystems. For example, microorganisms naturally capable of degrading oil played a significant role in cleaning up the 2010 *Deepwater Horizon* oil spill in the Gulf of Mexico.

After the 2011 Fukushima nuclear disaster in Japan, workers planted sunflowers that naturally remove heavy metal toxins (such as lead) from topsoil.

SUSTAINABLE DEVELOPMENT

While modern ecological practices seek to preserve wildlife and the ecosystems in which they live, few biologists would suggest that all wildlife be locked away, protected from any use. A more realistic goal is to achieve **sustainable development**, the responsible management and conservation of the Earth's resources. The goal of sustainable development is to maintain the productivity of Earth's ecosystems indefinitely. Although the biosphere is in peril, ecologists hope that sensible, science-driven policies can help us take corrective action now, while much hope remains.

These young trees are being harvested as part of a biomass research project with the goal of developing sustainable fuel sources.

Through education, students are becoming increasingly aware of the need to conserve our biosphere.

CORE IDEA: Conservation biologists and restoration ecologists seek to maintain biodiversity by identifying and protecting hot spots, aiding species recovery, conserving and restoring ecosystems, and promoting sustainable development.

CORE QUESTION: Biodiversity hot spots are relatively small, so what makes them so important?

ANSWER: They house a disproportionately high percentage of endemic, endangered, and threatened species.

The accumulation of greenhouse gases is causing global climate change

Greenhouse gases are airborne chemicals that capture and hold heat within Earth's atmosphere. Greenhouse gases include carbon dioxide (CO_2), methane (CH_4), nitrous oxide (N_2O), and water (H_2O) vapor. The concentration of greenhouse gases in the atmosphere is rising, and this is changing global patterns of weather and climate. There is no debate among the vast majority of scientists: Global climate change is occurring.

GLOBAL CLIMATE CHANGE: THE DATA

The average temperature over the surface of the Earth has been rising since the late 1800s—including a rise of about 1°F in the past 30 years—a trend called **global warming**. During that same period, the atmospheric concentration of greenhouse gases has also been rising. The concentration of CO_2, the dominant greenhouse gas, is higher now than it has been at any time in the past 800,000 years. The consensus among scientists is that these rising concentrations—and, thus, global warming—are the result of human activities.

Due to warming temperatures, Montana's Glacier National Park has 27 glaciers today, versus 150 in 1910.

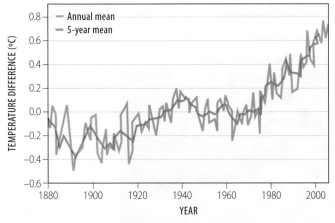

GLOBAL TEMPERATURES

The temperature of the Earth's surface was measured at thousands of weather stations throughout the planet and compared to the average temperature at that location. Combining all the data yields the global temperature relative to average. The blue line shows that the temperature varies considerably from year to year, but the red line (a five-year running average) makes the steady increase apparent.

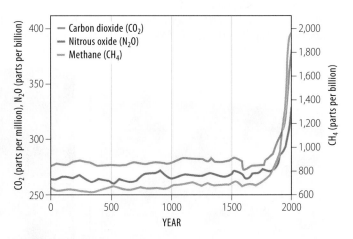

GREENHOUSE GASES

The levels of the three greenhouse gases shown here have all risen sharply since the late 1800s, the start of the Industrial Revolution. And the increase in these emissions is accelerating: From 2000 to 2005, global CO_2 emissions increased four times faster than in the preceding 10-year span.

CO_2 IN THE ATMOSPHERE

Humans add to the amount of CO_2 in the atmosphere in several ways, no one of which accounts for the majority. The human activities that contribute the most CO_2 to the atmosphere are power generation and industry.

This graph shows the contribution of various human activities to atmospheric CO_2 levels.

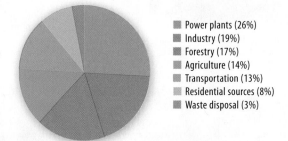

- Power plants (26%)
- Industry (19%)
- Forestry (17%)
- Agriculture (14%)
- Transportation (13%)
- Residential sources (8%)
- Waste disposal (3%)

THE GREENHOUSE EFFECT

Why are temperatures on Earth increasing? As solar radiation (sunlight) strikes Earth, much of it is reflected back into the atmosphere as heat. Greenhouse gases in the atmosphere reflect some of this heat back downward. A similar effect operates within a greenhouse: sunlight enters, warms surfaces, and then this heat is trapped by the inside air, warming the greenhouse. In essence, greenhouse gases act as an atmospheric blanket, trapping heat. The more greenhouse gases there are, the more heat is trapped. The greenhouse effect is vital to life on Earth; without it, most of Earth's surface would be too cold to support life. But the accumulation of human-made greenhouse gases is gradually warming the Earth, causing global climate change.

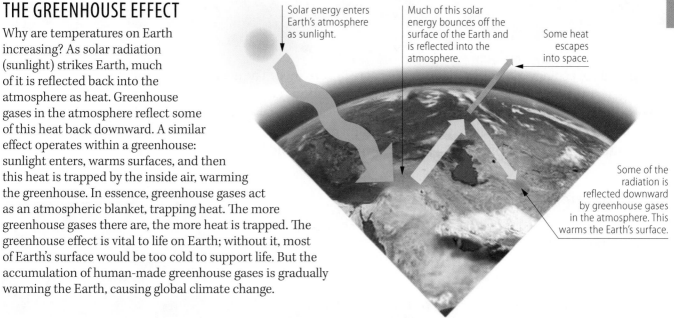

Solar energy enters Earth's atmosphere as sunlight.

Much of this solar energy bounces off the surface of the Earth and is reflected into the atmosphere.

Some heat escapes into space.

Some of the radiation is reflected downward by greenhouse gases in the atmosphere. This warms the Earth's surface.

EFFECTS OF GLOBAL CLIMATE CHANGE

Global climate change results in new patterns of temperature and rainfall. This is having a significant effect on the distribution of life. Melting permafrost is causing the boundary of tundra biomes to shift northward. Additionally, the borders of some deserts are expanding. Such changes affect food production and the availability of fresh drinking water. Make no mistake about it: We humans are susceptible to the effects of a changing climate.

HABITAT CHANGE

With rising temperatures, the ranges of many species are shifting toward the poles or to higher elevations. Ecologists have identified dozens of species of birds and butterflies that have migrated north or to higher altitudes in the last few decades.

POLAR CLIMATES

As the globe warms, the ice-covered hunting grounds of the far northern Arctic polar bears (*Ursus maritimus*) are melting away. Similarly, the disappearance of southern sea ice is blamed for decreases in the populations of Antarctic penguins.

CORAL BLEACHING

Warming oceans cause coral to lose their symbiotic algae. When the algae flee, the coral turns white, a phenomenon called coral bleaching. Coral cannot survive long this way, threatening entire reef ecosystems.

FIRES

Earlier melting of snow has extended the dry season in much of western North America, creating conditions that promote widespread, devastating wildfires.

CORE IDEA: Increasing concentrations of greenhouse gases in the atmosphere are causing rising temperatures. This global climate change has many far-reaching effects on ecosystems and the life within them, including human life.

CORE QUESTION: According to the data presented, what is the largest source of greenhouse gases in our atmosphere?

ANSWER: Power plants.

Organisms adapt to their environments

The fields of ecology and evolutionary biology are intertwined. The geographic distribution of organisms shows that ecosystems shape how populations adapt via natural selection. Populations evolve in response to both living (biotic) factors such as predators as well as nonliving (abiotic) factors such as temperature and rainfall. The range of possible adaptations falls into three broad categories: physiological, anatomical, and behavioral responses.

The arctic tern, a seabird, migrates over 44,000 miles each year.

PHYSIOLOGICAL RESPONSES

Many organisms respond to changing environmental conditions by changing their physiology, the functioning of body components.

GOOSE BUMPS
In most mammals, contraction of muscles attached to hairs beneath the skin creates a temporary layer of insulation. Since humans lack fur, the result is goose bumps.

FEATHERS
Feathers provide insulation. On a cold day, a bird contracts skin muscles that raise its feathers so they appear unusually puffy. This response traps pockets of air, which help warm the animal.

ACCLIMATION
Acclimation is a gradual, reversible physiological change in response to a changing environment. For example, if you move to a high-altitude area, the concentration of red blood cells in your bloodstream will gradually increase over a period of weeks.

ANATOMICAL RESPONSES

In response to changing environments, organisms may change their anatomy, adjusting body shape or structure. Anatomical responses are particularly important for plants, which can only grow in (not move through) their environment.

WIND
Wind can shape plants. For example, strong sustained wind can cause "flagging," the growth of some trees along the direction of the wind.

PHOTOTROPISM
Plants change their patterns of growth to orient themselves toward a source of light.

FUR
Many mammals, such as these marmots (*Marmota marmota*), grow heavy coats of fur before the arrival of winter, and then shed them when summer comes.

BEHAVIORAL RESPONSES

Many animals respond to changes in the environment by altering their behavior.

CLOTHING
We humans have a wide range of behavioral responses to varying environmental conditions, such as changing our clothing.

BASKING
Many desert ectotherms (animals that obtain body heat from the environment) move between sun and shade to maintain a proper body temperature. For example, these Galapagos land iguanas (*Conolophus subcristatus*) are taking advantage of the midday sun by basking on rocks, warming their bodies.

MIGRATION
Some animals, including many birds and ocean-going turtles and whales, migrate great distances in response to changing environmental cues. The migration of these blue wildebeests (*Connochaetes taurinus*) in the African Serengeti is a particularly striking example.

CORE IDEA: Organisms adapt to environmental conditions in a variety of ways: physiological responses, anatomical responses, and behavioral responses.

CORE QUESTION: Natives of the Himalayas have a greater capacity to absorb oxygen into their bloodstream than nearby lowland dwellers. This is an example of what type of adaptive response?

ANSWER: Physiological.

Appendix A Metric Conversion Table

MEASUREMENT	UNIT AND ABBREVIATION	METRIC EQUIVALENT	APPROXIMATE METRIC-TO-ENGLISH CONVERSION FACTOR	APPROXIMATE ENGLISH-TO-METRIC CONVERSION FACTOR
Length	1 kilometer (km)	= 1,000 (10^3) meters	1 km = 0.6 mile	1 mile = 1.6 km
	1 meter (m)	= 100 (10^2) centimeters	1 m = 1.1 yards	1 yard = 0.9 m
		= 1,000 millimeters	1 m = 3.3 feet	1 foot = 0.3 m
			1 m = 39.4 inches	
	1 centimeter (cm)	= 0.01 (10^{-2}) meter	1 cm = 0.4 inch	1 foot = 30.5 cm
				1 inch = 2.5 cm
	1 millimeter (mm)	= 0.001 (10^{-3}) meter	1 mm = 0.04 inch	
	1 micrometer (μm)	= 10^{-6} meter (10^{-3} mm)		
	1 nanometer (nm)	= 10^{-9} meter (10^{-3} μm)		
	1 angstrom (Å)	= 10^{-10} meter (10^{-4} μm)		
Area	1 hectare (ha)	= 10,000 square meters	1 ha = 2.5 acres	1 acre = 0.4 ha
	1 square meter (m^2)	= 10,000 square centimeters	1 m^2 = 1.2 square yards	1 square yard = 0.8 m^2
			1 m^2 = 10.8 square feet	1 square foot = 0.09 m^2
	1 square centimeter (cm^2)	= 100 square millimeters	1 cm^2 = 0.16 square inch	1 square inch = 6.5 cm^2
Mass	1 metric tonne (t)	= 1,000 kilograms	1 t = 1.1 tons	1 ton = 0.91 t
	1 kilogram (kg)	= 1,000 grams	1 kg = 2.2 pounds	1 pound = 0.45 kg
	1 gram (g)	= 1,000 milligrams	1 g = 0.04 ounce	1 ounce = 28.35 g
			1 g = 15.4 grains	
	1 milligram (mg)	= 10^{-3} gram	1 mg = 0.02 grain	
	1 microgram (μg)	= 10^{-6} gram		
Volume *(solids)*	1 cubic meter (m^3)	= 1,000,000 cubic centimeters	1 m^3 = 1.3 cubic yards	1 cubic yard = 0.8 m^3
	1 cubic centimeter (cm^3 or cc)	= 10^{-6} cubic meter	1 m^3 = 35.3 cubic feet	1 cubic foot = 0.03 m^3
	1 cubic millimeter (mm^3)	= 10^{-9} cubic meter (10^{-3} cubic centimeter)	1 cm^3 = 0.06 cubic inch	1 cubic inch = 16.4 cm^3
Volume *(liquids and gases)*	1 kiloliter (kL or kl)	= 1,000 liters	1 kL = 264.2 gallons	1 gallon = 3.79 L
	1 liter (L)	= 1,000 milliliters	1 L = 0.26 gallon	1 quart = 0.95 L
			1 L = 1.06 quarts	
	1 milliliter (mL or ml)	= 10^{-3} liter	1 mL = 0.03 fluid ounce	1 quart = 946 mL
		= 1 cubic centimeter	1 mL = 1/4 teaspoon	1 pint = 473 mL
			1 mL = 15−16 drops	1 fluid ounce = 29.6 mL
	1 microliter (μL or μl)	= 10^{-6} liter (10^{-3} milliliter)		1 teaspoon = 5 mL
Time	1 second (s)	= 1/60 minute		
	1 millisecond (ms)	= 10^{-3} second		
Temperature	Degrees Celsius (°C)		$°F = \dfrac{9}{5}°C + 32$	$°C = \dfrac{5}{9}(°F - 32)$

Appendix B The Periodic Table

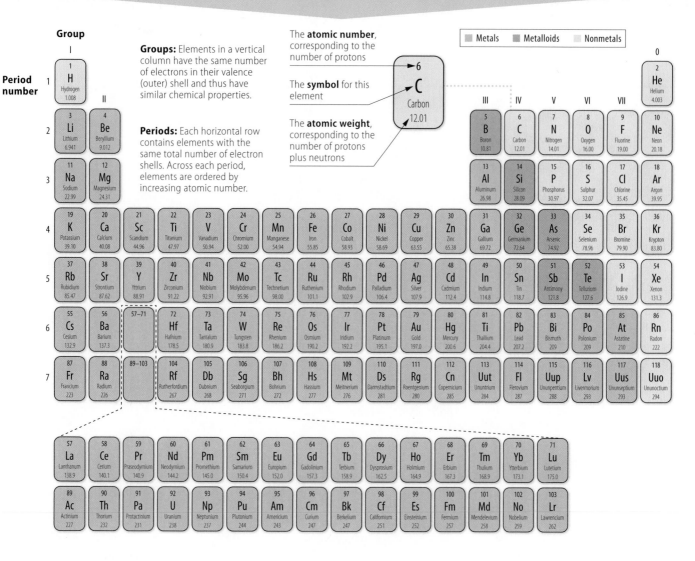

Group I — Elements in a vertical column have the same number of electrons in their valence (outer) shell and thus have similar chemical properties.

Periods: Each horizontal row contains elements with the same total number of electron shells. Across each period, elements are ordered by increasing atomic number.

The **atomic number**, corresponding to the number of protons

The **symbol** for this element

The **atomic weight**, corresponding to the number of protons plus neutrons

Metals ▪ Metalloids ▪ Nonmetals

NAME (SYMBOL)	ATOMIC NUMBER	NAME (SYMBOL)	ATOMIC NUMBER	NAME (SYMBOL)	ATOMIC NUMBER	NAME (SYMBOL)	ATOMIC NUMBER	NAME (SYMBOL)	ATOMIC NUMBER
Actinium (Ac)	89	Copper (Cu)	29	Iron (Fe)	26	Phosphorus (P)	15	Sulfur (S)	16
Aluminum (Al)	13	Curium (Cm)	96	Krypton (Kr)	36	Platinum (Pt)	78	Tantalum (Ta)	73
Americium (Am)	95	Darmstadtium (Ds)	110	Lanthanum (La)	57	Plutonium (Pu)	94	Technetium (Tc)	43
Antimony (Sb)	51	Dubnium (Db)	105	Lawrencium (Lr)	103	Polonium (Po)	84	Tellurium (Te)	52
Argon (Ar)	18	Dysprosium (Dy)	66	Lead (Pb)	82	Potassium (K)	19	Terbium (Tb)	65
Arsenic (As)	33	Einsteinium (Es)	99	Lithium (Li)	3	Praseodymium (Pr)	59	Thallium (Tl)	81
Astatine (At)	85	Erbium (Er)	68	Livermorium (Lv)	116	Promethium (Pm)	61	Thorium (Th)	90
Barium (Ba)	56	Europium (Eu)	63	Lutetium (Lu)	71	Protactinium (Pa)	91	Thulium (Tm)	69
Berkelium (Bk)	97	Fermium (Fm)	100	Magnesium (Mg)	12	Radium (Ra)	88	Tin (Sn)	50
Beryllium (Be)	4	Flerovium (Fl)	114	Manganese (Mn)	25	Radon (Rn)	86	Titanium (Ti)	22
Bismuth (Bi)	83	Fluorine (F)	9	Meitnerium (Mt)	109	Rhenium (Re)	75	Tungsten (W)	74
Bohrium (Bh)	107	Francium (Fr)	87	Mendelevium (Md)	101	Rhodium (Rh)	45	Uranium (U)	92
Boron (B)	5	Gadolinium (Gd)	64	Mercury (Hg)	80	Roentgenium (Rg)	111	Ununtrium (Uut)	113
Bromine (Br)	35	Gallium (Ga)	31	Molybdenum (Mo)	42	Rubidium (Rb)	37	Ununoctium (Uuo)	118
Cadmium (Cd)	48	Germanium (Ge)	32	Neodymium (Nd)	60	Ruthenium (Ru)	44	Ununpentium (Uup)	115
Calcium (Ca)	20	Gold (Au)	79	Neon (Ne)	10	Rutherfordium (Rf)	104	Ununseptium (Uus)	117
Californium (Cf)	98	Hafnium (Hf)	72	Neptunium (Np)	93	Samarium (Sm)	62	Vanadium (V)	23
Carbon (C)	6	Hassium (Hs)	108	Nickel (Ni)	28	Scandium (Sc)	21	Xenon (Xe)	54
Cerium (Ce)	58	Helium (He)	2	Niobium (Nb)	41	Seaborgium (Sg)	106	Ytterbium (Yb)	70
Cesium (Cs)	55	Holmium (Ho)	67	Nobelium (No)	102	Selenium (Se)	34	Yttrium (Y)	39
Chlorine (Cl)	17	Hydrogen (H)	1	Osmium (Os)	76	Silicon (Si)	14	Zinc (Zn)	30
Chromium (Cr)	24	Indium (In)	49	Oxygen (O)	8	Silver (Ag)	47	Zirconium (Zr)	40
Cobalt (Co)	27	Iodine (I)	53	Palladium (Pd)	46	Sodium (Na)	11		
Copernicium (Cn)	112	Iridium (Ir)	77			Strontium (Sr)	38		

Glossary

A

abiotic factor
A nonliving component of an ecosystem, such as air, water, light, minerals, or temperature.

abiotic reservoir
The part of an ecosystem where a chemical, such as carbon or nitrogen, accumulates or is stockpiled outside of living organisms.

absorption
The uptake of small nutrient molecules by an organism's own body. In animals, absorption is the third main stage of food processing, following digestion; in fungi, it is acquisition of nutrients from the surrounding medium.

accessory organ
Organ (such as the salivary glands, pancreas, gallbladder, and liver) that interfaces with the alimentary canal of the digestive system. Accessory organs add digestive chemicals to the alimentary canal.

acid
A substance that increases the hydrogen ion (H^+) concentration in a solution.

acrosome
A membrane-enclosed sac at the tip of a sperm. The acrosome contains enzymes that help the sperm penetrate an egg.

action potential
A self-propagating change in the voltage across the plasma membrane of a neuron; a nerve signal.

activation energy
The amount of energy that reactants must absorb before a chemical reaction will start.

active site
The part of an enzyme molecule where a substrate molecule attaches (by means of weak chemical bonds); typically, a pocket or groove on the enzyme's surface.

active transport
The movement of a substance across a biological membrane against its concentration gradient, aided by specific transport proteins and requiring the input of energy (often as ATP).

adaptation
An inherited character that enhances an organism's ability to survive and reproduce in a particular environment.

adaptive defense
An immune system defense that is found only in vertebrates and that must be activated by exposure to a specific invader.

adipose tissue
A type of connective tissue in which the cells contain fat.

ADP (Adenosine diphosphate)
A molecule composed of adenosine and two phosphate groups. The molecule ATP is made by combining a molecule of ADP with a third phosphate in an energy-consuming reaction.

adrenal gland
One of a pair of endocrine glands, located atop each kidney in mammals, composed of an outer cortex and a central medulla.

age structure
The relative number of individuals of each age in a population.

AIDS
Acquired immunodeficiency syndrome; the late stages of HIV infection, characterized by a reduced number of T cells; usually results in death caused by opportunistic infections.

alga (plural, **algae**)
An informal term that describes a great variety of protists, most of which are unicellular or colonial photosynthetic autotrophs with chloroplasts. Heterotrophic and multicellular protists closely related to unicellular autotrophs are also regarded as algae.

alimentary canal
A digestive tube running between a mouth and an anus; also called a digestive tract.

allele
An alternative version of a gene.

allergy
An exaggerated sensitivity to an antigen. Symptoms are triggered by histamines released from mast cells.

allopatric speciation
The formation of a new species in populations that are geographically isolated from one another. *See also* sympatric speciation.

alveolus (plural, **alveoli**)
One of millions of tiny sacs within the vertebrate lungs where gas exchange occurs.

amino acid
An organic molecule containing a carboxyl group, an amino group, a hydrogen atom, and a variable side group (also called a radical group or R group); serves as the monomer of proteins.

amnion
In vertebrate animals, the extraembryonic membrane that encloses the fluid-filled amniotic sac containing the embryo.

amniotic egg
A shelled egg in which an embryo develops within a fluid-filled amniotic sac and is nourished by yolk. Produced by reptiles (including birds) and egg-laying mammals, it enables them to complete their life cycles on dry land.

amoeba
A general term for a protozoan (animal-like protist) characterized by great structural flexibility and the presence of pseudopodia.

amphibian
Member of a class of vertebrate animals that includes frogs and salamanders.

anabolic steroid
A synthetic variant of the male hormone testosterone that mimics some of its effects.

anatomy
The study of the structure of an organism and its parts.

anemia
A condition in which an abnormally low amount of hemoglobin or a low number of red blood cells results in the body cells not receiving enough oxygen.

angiosperm
A flowering plant, which forms seeds inside a protective chamber called an ovary.

annelid
A segmented worm. Annelids include earthworms, polychaetes, and leeches.

anorexia nervosa
An eating disorder that results in self-starvation due to an intense fear of gaining weight, even when the person is underweight.

anthropoid
Member of a primate group made up of the apes (gibbons, orangutans, gorillas, chimpanzees, and bonobos), monkeys, and humans.

antibiotic
A naturally occurring substance that inhibits the growth of or kills bacteria.

antibody
A protein that is secreted by a B cell and attaches to a specific kind of antigen, helping counter its effects.

anticodon
On a tRNA molecule, a specific sequence of three nucleotides that is complementary to a codon triplet on mRNA.

antigen
Any molecule that elicits a response from a lymphocyte.

anus
The opening through which undigested materials are expelled.

aphotic zone
The region of an aquatic ecosystem beneath the photic zone, where light levels are too low for photosynthesis to take place.

aquatic biome
A major life zone found in water. Each aquatic biome is characterized by specific environmental conditions.

aqueous humor
A liquid in the space between the lens and cornea in the vertebrate eye that helps maintain the shape of the eye, supplies nutrients and oxygen to its tissues, and disposes of its wastes.

aqueous solution
A solution in which water is the solvent.

arachnid
Member of a major arthropod group that includes spiders, scorpions, ticks, and mites.

Archaea
One of two prokaryotic domains of life, the other being Bacteria.

arteriole
A small vessel that conveys blood between an artery and a capillary bed.

artery
A vessel that carries blood away from the heart to other parts of the body.

arthropod
Member of the most diverse phylum in the animal kingdom; includes the horseshoe crab, arachnids (for example, spiders, ticks, scorpions, and mites), crustaceans (for example, crayfish, lobsters, crabs, and barnacles), millipedes, centipedes, and insects. Arthropods are characterized by a chitinous exoskeleton, molting, jointed appendages, and a body formed of distinct groups of segments.

artificial selection
The selective breeding of domesticated plants and animals to promote the occurrence of desirable traits.

asexual reproduction
The creation of genetically identical offspring by a single parent, without the participation of gametes (sperm and egg).

atom
The smallest unit of matter that retains the properties of an element.

atomic number
The number of protons in each atom of a particular element.

atomic weight
Also called the atomic mass; the total mass of an atom, which is the mass in grams of 1 mole of the atom.

ATP (adenosine triphosphate)
A molecule composed of adenosine and three phosphate groups; the main energy source for cells. A molecule of ATP can be broken down to a molecule of ADP (adenosine diphosphate) and a free phosphate; this reaction releases energy that can be used for cellular work.

autoimmune disease
An immunological disorder in which the immune system improperly attacks the body's own molecules.

autosome
A chromosome not directly involved in determining the sex of an organism; in mammals, for example, any chromosome other than X or Y.

AV (atrioventricular) node
A region of muscle tissue between the heart's right atrium and right ventricle. It generates electrical impulses that primarily cause the ventricles to contract.

axon
A neuron fiber that extends from the cell body and conducts signals to another neuron or to an effector cell.

B

B cell
A type of lymphocyte that matures in the bone marrow and later produces antibodies; responsible for the humoral immune response. *See also* T cell.

bacillus (plural, **bacilli**)
A rod-shaped prokaryotic cell.

Bacteria
One of two prokaryotic domains of life, the other being Archaea.

bacteriophage
A virus that infects bacteria; also called a phage.

Barr body
A dense body formed from a deactivated X chromosome found in the nuclei of female mammalian cells.

base
A substance that decreases the hydrogen ion (H^+) concentration in a solution.

base pair
Two DNA nucleotides, one from each polynucleotide strand in a DNA double helix, joined by hydrogen bonds. In DNA, the two types of base pairs are A/T and C/G.

benign tumor
An abnormal mass of cells that remains at its original site in the body.

benthic realm
A seafloor or the bottom of a freshwater lake, pond, river, or stream. The benthic realm is occupied by communities of organisms known as benthos.

bicep
A muscle of the arm that, when contracted, causes the arm to bend at the elbow.

bilateral symmetry
An arrangement of body parts such that an organism can be divided equally by a single cut passing longitudinally through it. A bilaterally symmetric organism has mirror-image right and left sides.

bile
A solution of salts secreted by the liver that emulsifies fats and aids in their digestion.

binary fission
A means of asexual reproduction in which a parent organism, often a single cell, divides into two individuals of about equal size.

biodiversity
The variety of living things; includes genetic diversity, species diversity, and ecosystem diversity.

biodiversity hot spot
A small geographic area that contains a large number of threatened or endangered species and an exceptional concentration of endemic species (those found nowhere else).

biofilm
A surface-coating cooperative colony of prokaryotes.

biogenesis
The principle that all life arises by the reproduction of preexisting life.

biogeochemical cycle
Any of the various chemical circuits occurring in an ecosystem, involving both biotic and abiotic components of the ecosystem.

biogeography
The study of the geographic distribution of species.

bioinformatics
The use of computers, software, and mathematical models to analyze biological information.

biological control
The intentional release of a natural enemy to attack a pest population.

biological magnification
The accumulation of persistent chemicals in the living tissues of consumers in food chains.

biomass
The amount, or mass, of living organic material in an ecosystem.

biome
A major terrestrial or aquatic life zone, characterized by vegetation type in terrestrial biomes and the physical environment in aquatic biomes.

bioremediation
The use of living organisms to detoxify and restore polluted and degraded ecosystems.

biosphere
The global ecosystem; the entire portion of Earth inhabited by life; all of life and where it lives.

biotechnology
The manipulation of living organisms to perform useful tasks. Today, biotechnology often involves DNA technology.

biotic factor
A living component of a biological community; any organism that is part of an individual's environment.

bivalve
Member of a group of molluscs that includes clams, mussels, scallops, and oysters.

blastocyst A
mammalian embryo (equivalent to an amphibian blastula) made up of a hollow ball of cells that results from cleavage and that implants in the mother's endometrium.

blind spot
The location on the retina where the optic nerve passes through. Because there are no photoreceptor cells at this location, any image that strikes this particular spot will not be perceived by the brain.

blood
A type of connective tissue with a fluid matrix called plasma in which blood cells are suspended.

body cavity
A fluid-filled space separating the digestive tract from the outer body wall.

body mass index (BMI)
A ratio of weight to height used as a measure of obesity.

bone
A type of connective tissue consisting of living cells held in a rigid matrix of collagen fibers embedded in calcium salts.

bony fish
A fish that has a stiff skeleton reinforced by calcium salts.

bottleneck effect
Genetic drift resulting from a drastic reduction in population size.

brain
The master control center of the central nervous system, which is involved in regulating and controlling body activity and interpreting information from the senses.

bronchiole
A thin breathing tube that branches from a bronchus within a lung.

bronchus (plural, **bronchi**)
One of a pair of breathing tubes that branch from the trachea into the lungs.

bryophyte
A type of plant that lacks xylem and phloem; a nonvascular plant. Bryophytes include mosses and their close relatives.

buffer
A chemical substance that resists changes in pH by accepting hydrogen ions from or donating hydrogen ions to solutions.

bulimia
An eating disorder characterized by episodic binge eating followed by purging through induced vomiting, abuse of laxatives, or excessive exercise.

C

Calvin cycle
The second of two stages of photosynthesis; a cyclic series of chemical reactions that occur in the stroma of a chloroplast, using the carbon in CO_2 and the ATP and NADPH produced by the light reactions to make the energy-rich sugar molecule G3P, which is later used to produce glucose.

cancer
A malignant growth or tumor caused by abnormal and uncontrolled cell division.

capillary
A microscopic blood vessel that conveys blood between an artery and a vein or between an arteriole and a venule; enables the exchange of nutrients and dissolved gases between the blood and interstitial fluid.

capsid
The protein shell that encloses a viral genome.

capsule
A dense layer of polysaccharide or protein that surrounds the cell wall of some prokaryotes and is sticky, protecting the cell and enabling it to adhere to substrates or other cells.

carbohydrate
A biological molecule consisting of a simple sugar (a monosaccharide), two monosaccharides joined into a double sugar (a disaccharide), or a chain of monosaccharides (a polysaccharide).

carcinogen
A cancer-causing agent, either high-energy radiation (such as X-rays or UV light) or a chemical.

cardiac cycle
The alternating contractions and relaxations of the heart.

cardiac muscle
Striated muscle that forms the contractile tissue of the heart.

cardiovascular system
A closed circulatory system, found in vertebrates, with a heart and a branching network of arteries, capillaries, and veins.

carrier
An individual who is heterozygous for a recessively inherited disorder and who therefore does not show symptoms of that disorder.

carrying capacity
The maximum population size that a particular environment can sustain.

cartilage
A type of connective tissue consisting of living cells embedded in a rubbery matrix with collagen fibers.

cartilaginous fish
A fish that has a flexible skeleton made of cartilage.

cell
A basic unit of living matter separated from its environment by a plasma membrane; the fundamental structural unit of life.

cell body
The part of a cell, such as a neuron, that houses the nucleus and other organelles.

cell cycle
An ordered sequence of events (including interphase and the mitotic phase) that extends from the time a eukaryotic cell is first formed from a dividing parent cell until its own division into two cells.

cell cycle control system
A cyclically operating set of proteins that triggers and coordinates events in the eukaryotic cell cycle.

cell division
The reproduction of a cell.

cell plate
A membranous disk that forms across the midline of a dividing plant cell. During cytokinesis, the cell plate grows outward, accumulating more cell wall material and eventually fusing into a new cell wall.

cell theory
The theory that all living things are composed of cells and that all cells come from other cells.

cell wall
A protective layer external to the plasma membrane in plant cells, bacteria, fungi, and some protists; protects the cell and helps maintain its shape.

cell-mediated immune response
The type of adaptive immune response that involves T cells, lymphocytes that fight body cells infected with pathogens. *See also* humoral immune response.

cellular respiration
The aerobic harvesting of energy from food molecules; the energy-releasing chemical breakdown of food molecules, such as glucose, and the storage of potential energy in a form that cells can use to perform work; involves glycolysis, the citric acid cycle, the electron transport chain, and chemiosmosis.

cellulose
A large polysaccharide composed of many glucose monomers linked into cable-like fibrils that provide structural support in plant cell walls. Because cellulose cannot be digested by animals, it acts as fiber, or roughage, in the diet.

centipede
A carnivorous terrestrial arthropod that has one pair of long legs for each of its numerous body segments, with the front pair modified as poison claws.

central nervous system (CNS)
The integration and command center of the nervous system including the brain and, in vertebrates, the spinal cord.

centromere
The region of a chromosome where two sister chromatids are joined and where spindle microtubules attach during mitosis and meiosis. The centromere divides at the onset of anaphase during mitosis and anaphase II of meiosis.

cephalopod
Member of a group of molluscs that includes squids and octopuses.

cerebrospinal fluid
Fluid that surrounds, cushions, and nourishes the brain and spinal cord and protects them from infection.

cervix
The narrow neck at the bottom of the uterus, which opens into the vagina.

chaparral
A terrestrial biome limited to coastal regions where cold ocean currents circulate offshore, creating mild, rainy winters and long, hot, dry summers. Chaparral vegetation is adapted to fire.

character
A heritable feature that varies among individuals within a population, such as flower color in pea plants.

chemical bond
An attraction between two atoms resulting from a sharing of outer-shell electrons or the presence of opposite charges on the atoms. The bonded atoms gain complete outer electron shells.

chemical digestion
The breakdown by enzymes of large food molecules into the smaller molecules that make them up. Chemical digestion begins in the mouth and continues throughout the alimentary canal.

chemical energy
Energy stored in the chemical bonds of molecules; a form of potential energy.

chemical reaction
A process leading to chemical changes in matter, involving the making and/or breaking of chemical bonds.

chemistry
The scientific study of matter.

chemotherapy
Treatment for cancer in which drugs are administered to disrupt cell division of the cancer cells.

chlorophyll
A light-absorbing pigment in chloroplasts that plays a central role in converting solar energy to chemical energy.

chloroplast
An organelle found in plants and photosynthetic protists. Enclosed by two concentric membranes, a chloroplast absorbs sunlight and uses it to power the synthesis of organic food molecules (sugars).

cholesterol
A steroid that is an important component of animal cell membranes and that acts as a precursor molecule for the synthesis of other steroids, such as hormones.

chordate
An animal that at some point during its development has a dorsal, hollow nerve cord, a notochord, pharyngeal slits, and a post-anal tail. Chordates include lancelets, tunicates, and vertebrates.

chromatin
The combination of DNA and proteins that constitutes chromosomes; often used to refer to the diffuse, very extended form taken by the chromosomes when a eukaryotic cell is not dividing.

chromosome
A gene-carrying structure found in the nucleus of a eukaryotic cell and most visible during mitosis and meiosis; also, the main gene-carrying structure of a prokaryotic cell. Each chromosome consists of one very long threadlike DNA molecule and associated proteins. *See also* chromatin.

ciliary muscle
A muscle within the eye that controls the shape of the lens, allowing light to be focused on the retina.

cilium (plural, **cilia**)
A short appendage that propels some protists through the water and moves fluids across the surface of many tissue cells in animals.

circulatory system
The organ system that transports materials such as nutrients, O_2, and hormones to body cells and transports CO_2 and other wastes from body cells.

citric acid cycle
The metabolic cycle that is fueled by acetyl CoA formed after glycolysis in cellular respiration. Chemical reactions in the cycle complete the metabolic breakdown of glucose molecules to carbon dioxide. The cycle occurs in the matrix of mitochondria and supplies most of the NADH molecules that carry energy to the electron transport chains. Also referred to as the Krebs cycle.

clade
An ancestral species and all its descendants—a distinctive branch in the tree of life.

cladistics
The study of evolutionary history; specifically, an approach to systematics in which organisms are grouped by common ancestry.

cleavage
(1) Cytokinesis in animal cells and in some protists, characterized by pinching in of the plasma membrane. (2) In animal development, the succession of rapid cell divisions without cell growth, converting the animal zygote to a ball of cells.

cleavage furrow
The first sign of cytokinesis during cell division in an animal cell; a shallow groove in the cell surface near the old metaphase plate.

clitoris
An organ in the female that engorges with blood and becomes erect during sexual arousal.

clonal selection
The production of a population of genetically identical lymphocytes (white blood cells) that recognize and attack the specific antigen that stimulated their proliferation. Clonal selection is the mechanism that underlies the immune system's specificity and memory of antigens.

clone
As a verb, to produce genetically identical copies of a cell, organism, or DNA molecule. As a noun, the collection of cells, organisms, or molecules resulting from cloning; also (colloquially), a single organism that is genetically identical to another because it arose from the cloning of a somatic cell.

cnidarian
An animal characterized by cnidocytes, radial symmetry, a gastrovascular cavity, and a polyp or medusa body form. Cnidarians includes hydras, jellies, sea anemones, and corals.

coccus (plural, **cocci**)
A spherical prokaryotic cell.

coconut oil
A high-fat liquid extract of the coconut fruit.

codon
A three-nucleotide sequence in mRNA that specifies a particular amino acid or polypeptide termination signal; the basic unit of the genetic code.

cohesion
The attraction between molecules of the same kind.

colony
Collection of members of the same species, usually prokaryotes or protists, living in close contact.

community
All the organisms inhabiting and potentially interacting in a particular area; an assemblage of populations of different species.

competitive exclusion principle
The concept that populations of two species cannot coexist in a community if their niches are nearly identical. Using resources more efficiently and having a reproductive advantage, one of the populations will eventually outcompete and eliminate the other.

complement system
A family of innate defensive blood proteins that cooperate with other components of the vertebrate defense system to protect against microbes; can enhance phagocytosis, directly lyse pathogens, and amplify the inflammatory response.

complementary DNA (cDNA)
A DNA molecule made in vitro using mRNA as a template and the enzyme reverse transcriptase. A cDNA molecule therefore corresponds to a gene but lacks the introns present in the DNA of the genome.

complete digestive tract
A digestive tube with two openings, a mouth and an anus.

compound
A substance containing two or more elements in a fixed ratio; for example, table salt (NaCl) consists of one atom of the element sodium (Na) for every atom of chlorine (Cl).

cone
(1) In vertebrates, a photoreceptor cell in the retina, stimulated by bright light and enabling color vision. (2) In conifers, a reproductive structure bearing pollen or ovules.

coniferous forest
A terrestrial biome characterized by conifers, cone-bearing evergreen trees.

conjugation
The union (mating) of two bacterial cells or protist cells and the transfer of DNA between the two cells.

conjunctiva
A thin mucous membrane that lines the inner surface of vertebrate eyelids.

connective tissue
Tissue consisting of cells held in an abundant extracellular matrix.

conservation biology
A goal-oriented science that seeks to understand and counter the loss of biodiversity.

consumer
An organism that obtains its food by eating plants or by eating animals that have eaten plants.

continental shelf
The submerged part of a continent.

contraception
The deliberate prevention of pregnancy.

controlled experiment
A component of the process of science whereby a scientist carries out two parallel tests, an experimental test and a control test. The experimental test differs from the control by one factor, the variable.

coral reef
Tropical marine biome characterized by hard skeletal structures secreted primarily by the resident cnidarians.

cornea
The transparent front portion of the sclera that admits light into the vertebrate eye.

corpus luteum
A small body of endocrine tissue that develops from an ovarian follicle after ovulation. The corpus luteum secretes progesterone and estrogen during pregnancy.

cotyledon
The first leaf that appears on an embryo of a flowering plant; a seed leaf. Monocot embryos have one cotyledon; dicot embryos have two.

covalent bond
An attraction between atoms that share one or more pairs of outer-shell electrons.

crossing over
The exchange of segments between chromatids of homologous chromosomes during prophase I of meiosis.

crustacean
Member of a major arthropod group that includes lobsters, crayfish, crabs, shrimps, and barnacles.

cuticle
(1) In animals, a tough, nonliving outer layer of the skin. (2) In plants, a waxy coating on the surface of stems and leaves that helps retain water.

cytokinin
Any of a family of plant hormones that promote cell division, retard aging in flowers and fruits, and may counter the effects of auxins in regulating plant growth and development.

cytoplasm
Everything inside a eukaryotic cell between the plasma membrane and the nucleus; consists of a semifluid medium and organelles; can also refer to the interior of a prokaryotic cell.

cytoskeleton
A meshwork of fine fibers in the cytoplasm of a eukaryotic cell; includes microfilaments, intermediate filaments, and microtubules.

cytosol
The fluid part of the cytoplasm, in which organelles are suspended.

cytotoxic T cell
A type of lymphocyte that directly attacks body cells infected by pathogens.

D

Darwinian fitness
The contribution an individual makes to the gene pool of the next generation, relative to the contributions of other individuals in the population.

decomposer
An organism that secretes enzymes that digest molecules in organic material and convert them to inorganic form.

dehydration synthesis reaction
A chemical reaction in which a monomer is joined to another monomer or polymer, forming a larger molecule and releasing a molecule of water.

dendrite
A short, branched neuron fiber that receives signals and conveys them from its tip inward, toward the rest of the neuron.

density-dependent limiting factor
A limiting factor whose effects intensify with increasing population density.

density-independent limiting factor
A limiting factor whose occurrence and effects are not related to population density.

deoxyribonucleic acid (DNA)
The genetic material that organisms inherit from their parents; a double-stranded helical macromolecule consisting of nucleotide monomers with deoxyribose sugar, a phosphate group, and the nitrogenous bases adenine (A), cytosine (C), guanine (G), and thymine (T). *See also* gene.

desert
A terrestrial biome characterized by low and unpredictable rainfall (less than 30 cm per year).

development
The growth and specialization of cells and tissues that occur as the cells of a zygote multiply and differentiate to form a multicellular organism.

diabetes mellitus
A human hormonal disease in which body cells cannot absorb enough glucose from the blood and become energy starved. Body fats and proteins are then consumed for their energy. Insulin-dependent diabetes results when the pancreas does not produce insulin. Non-insulin-dependent diabetes results when body cells fail to respond to insulin. A third type of diabetes, called gestational diabetes, can affect any pregnant woman, even one who has never shown symptoms of diabetes before.

dialysis
Separation and disposal of metabolic wastes from the blood by mechanical means; an artificial method of performing the functions of the kidneys.

diaphragm
(1) The sheet of muscle separating the chest cavity from the abdominal cavity in mammals. Its contraction expands the chest cavity, and its relaxation reduces it. (2) A dome-shaped rubber cap that covers a woman's cervix, serving as a barrier method of contraception.

diastole
The stage of the cardiac cycle in which the heart muscle is relaxed, allowing the chambers to fill with blood.

dicot
A flowering plant whose embryos have two seed leaves, or cotyledons.

diffusion
The spontaneous movement of particles of any kind down a concentration gradient; that is, movement of particles from where they are more concentrated to where they are less concentrated.

digestion
The mechanical and chemical breakdown of food into molecules small enough for the body to absorb; the second stage of food processing, following ingestion.

digestive system
The organ system involved in ingestion and digestion of food, absorption of nutrients, and elimination of wastes.

dihybrid cross
A mating of individuals differing at two genetic loci.

diploid
Containing two sets of chromosomes (pairs of homologous chromosomes) in each cell, one set inherited from each parent; referring to a 2*n* cell.

disaccharide
A sugar molecule consisting of two monosaccharides linked by a dehydration reaction.

discovery science
The process of scientific inquiry that focuses on using observations to describe nature.

dispersion pattern
The manner in which individuals in a population are spaced within their area. Three types of dispersion patterns are clumped (individuals are aggregated in patches), uniform (individuals are evenly distributed), and random (unpredictable distribution).

DNA (deoxyribonucleic acid)
The genetic material that organisms inherit from their parents; a double-stranded helical macromolecule consisting of nucleotide monomers with deoxyribose sugar, a phosphate group, and the nitrogenous bases adenine (A), cytosine (C), guanine (G), and thymine (T). *See also* gene.

DNA ligase
An enzyme, essential for DNA replication, that catalyzes the covalent bonding of adjacent DNA nucleotides; used in genetic engineering to paste a specific piece of DNA containing a gene of interest into a bacterial plasmid or other vector.

DNA polymerase
An enzyme that assembles DNA nucleotides into polynucleotides using a preexisting strand of DNA as a template.

DNA profiling
A procedure that analyzes an individual's unique collection of genetic markers using PCR and gel electrophoresis. DNA profiling can be used to determine whether two samples of genetic material were derived from the same individual.

DNA replication
The process by which a DNA molecule is copied.

DNA technology
Methods used to study or manipulate genetic material.

domain
A taxonomic category above the kingdom level. The three domains of life are Archaea, Bacteria, and Eukarya.

dominant allele
In a heterozygote, the allele that determines the phenotype with respect to a particular gene.

double bond
A double covalent bond; the sharing of two pairs of valence electrons by two atoms.

double helix
The form assumed by DNA in living cells, referring to its two adjacent polynucleotide strands wound into a spiral shape.

Down syndrome
A human genetic disorder resulting from a condition called trisomy 21, the presence of an extra chromosome 21; characterized by heart and respiratory defects and varying degrees of mental retardation.

E

echinoderm
Member of a group of slow-moving or stationary marine animals characterized by a rough or spiny skin, a water vascular system, typically an endoskeleton, and radial symmetry in adults. Echinoderms include sea stars, sea urchins, and sand dollars.

ecological footprint
An estimate of the amount of land and water area required to provide the resources an individual or nation consumes and to absorb the waste it generates.

ecological succession
The process of biological community change resulting from disturbance; transition in the species composition of a biological community. *See also* primary succession; secondary succession.

ecology
The scientific study of how organisms interact with their environment.

ecosystem
All the organisms in a given area, along with the nonliving (abiotic) factors with which they interact; a biological community and its physical environment.

ecosystem biodiversity
A measure of the variety of ecosystems found on Earth.

ecosystem service
Function performed by an ecosystem that directly or indirectly benefits people.

ectotherm
An animal that warms itself mainly by absorbing heat from its surroundings.

effector cell
A short-lived lymphocyte that has an immediate effect against a specific pathogen; a cell capable of carrying out some action in response to a command from the nervous system.

egg
A female gamete.

electron
A subatomic particle with a single unit of negative electrical charge. One or more electrons move around the nucleus of an atom.

electron transport chain
A series of electron carrier molecules that shuttle electrons during the redox reactions that release energy used to make ATP; located in the inner membrane of mitochondria, the thylakoid membrane of chloroplasts, and the plasma membrane of prokaryotes.

element
A substance that cannot be broken down into other substances by chemical means. Scientists recognize 92 chemical elements occurring in nature.

elimination
The passing of undigested material out of the digestive compartment; the fourth stage of food processing, following absorption.

emerging virus

A virus that has appeared suddenly or has recently come to the attention of medical scientists.

endangered species

As defined in the U.S. Endangered Species Act, a species that is in danger of extinction throughout all or a significant portion of its range.

endocrine gland

A gland that synthesizes hormone molecules and secretes them directly into the bloodstream.

endocrine system

The body's main system for internal chemical regulation, consisting of all hormone-secreting cells; cooperates with the nervous system in regulating body functions and maintaining homeostasis.

endocytosis

The movement of materials from the external environment into the cytoplasm of a cell via vesicles or vacuoles.

endomembrane system

A network of organelles that partitions the cytoplasm of eukaryotic cells into functional compartments. Some of the organelles are structurally connected to each other, whereas others are structurally separate but functionally connected by the traffic of vesicles among them.

endometrium

The inner lining of the uterus in mammals, richly supplied with blood vessels that provide the maternal part of the placenta and nourish the developing embryo.

endoplasmic reticulum (ER)

An extensive membranous network in a eukaryotic cell, continuous with the outer nuclear membrane and composed of ribosome-studded (rough) and ribosome-free (smooth) regions. *See also* rough ER; smooth ER.

endoskeleton

A hard interior skeleton located within the soft tissues of an animal; found in all vertebrates and a few invertebrates (such as echinoderms).

endospore

A thick-coated, protective cell produced within a prokaryotic cell exposed to harsh conditions.

endosymbiosis

Symbiotic relationship in which one species resides within another species. The mitochondria and chloroplasts of eukaryotic cells probably evolved from symbiotic associations between small prokaryotic cells living inside larger ones.

endotherm

An animal that derives most of its body heat from its own metabolism.

energy

The capacity to cause change, or to move matter in a direction it would not move if left alone.

entropy

A measure of disorder, or randomness. One form of disorder is heat, which is random molecular motion.

environmentalism

A broad philosophical and social movement that seeks to improve and maintain the quality of the natural environment through conservation and restoration.

enzyme

A protein that serves as a biological catalyst, changing the rate of a chemical reaction without itself being changed in the process.

epididymis

A long coiled tube into which sperm pass from the testis and are stored until mature and ejaculated.

epiglottis

A flap of elastic cartilage that protects the entrance to the trachea. Normally, the epiglottis is positioned to allow air to enter the trachea; it changes position when food is swallowed, allowing food to enter the esophagus and preventing food from entering the trachea.

epithelial tissue

A sheet of tightly packed cells lining organs and cavities; also called epithelium.

epithelium (plural, **epithelia**)

See epithelial tissue.

erectile tissue

Tissue within the penis of human males and the clitoris of humans females that fills with blood and swells during sexual arousal.

esophagus

The channel through which food passes in a digestive tube, connecting the pharynx to the stomach.

essential amino acid

An amino acid that an animal cannot synthesize itself and must obtain from food. Eight amino acids are essential for the human adult.

essential fatty acid

An unsaturated fatty acid that an animal needs but cannot make.

essential nutrient

A substance that an organism must absorb in preassembled form because it cannot synthesize the nutrient from any other material. Humans require vitamins, minerals, essential amino acids, and essential fatty acids.

estuary

The area where a freshwater stream or river merges with seawater.

eukaryote

An organism characterized by eukaryotic cells. *See also* eukaryotic cell.

eukaryotic cell

A type of cell that has a membrane-enclosed nucleus and other membrane-enclosed organelles. All organisms except bacteria and archaea are composed of eukaryotic cells.

eutherian

Mammal whose young complete their embryonic development in the uterus, nourished via the mother's blood vessels in the placenta; also called a placental mammal.

evaporation

The process by which a liquid changes to a gas. For example, water evaporates from the ocean, entering the atmosphere as vapor.

evolution

Descent with modification; genetic change in a population or species over generations; the heritable changes that have produced Earth's diversity of organisms.

exocytosis

The movement of materials out of the cytoplasm of a cell via membranous vesicles or vacuoles.

exon

In eukaryotes, a coding portion of a gene. *See also* intron.

exponential growth

A continuous increase in the size of a population by a multiple of the previous generation.

extinction

The irrevocable loss of a species.

extracellular matrix

The meshwork that surrounds animal cells, consisting of a web of protein and polysaccharide fibers embedded in a liquid, jelly, or solid.

F

F_1 generation

The offspring of two parental (P generation) individuals. F_1 stands for first filial.

facilitated diffusion

The passage of a substance across a biological membrane down its concentration gradient, aided by specific transport proteins.

fatty acid

A carboxylic acid with a long carbon chain. Fatty acids vary in length and in the number and location of double bonds; three fatty acids linked to a glycerol molecule form a fat molecule, also known as a triglyceride.

feathers

Structures that facilitate flight in birds. Feathers, made of the protein keratin, also aid in waterproofing and provide insulation.

fermentation

The anaerobic harvest of food by some cells.

fertilization

The union of a haploid sperm cell with a haploid egg cell, producing a zygote.

fetus

A developing human from the 9th week of pregnancy until birth. The fetus has all the major structures of an adult.

flagellum (plural, **flagella**)

A long appendage that propels protists through the water and moves fluids across the surface of many tissue cells in animals. A cell may have one or more flagella.

flatworm

A bilateral animal with a thin, flat body form, a gastrovascular cavity with a single opening, and no body cavity. Flatworms include planarians, flukes, and tapeworms.

flower

In an angiosperm, a short stem with four sets of modified leaves, bearing structures that function in sexual reproduction.

fluid mosaic

A description of membrane structure, depicting a cellular membrane as a mosaic of diverse protein molecules embedded in a fluid bilayer of phospholipid molecules.

follicle

A cluster of cells surrounding, protecting, and nourishing a developing egg cell in the ovary. The follicle also secretes estrogen.

food chain

The sequence of food transfers between the trophic levels of a community, beginning with the producers.

food web

A network of interconnecting food chains.

foot

In an invertebrate animal, a structure used for locomotion or attachment, such as the muscular organ extending from the ventral side of a mollusc.

fossil record

The ordered sequence of fossils as they appear in the rock layers, marking the passing of geologic time.

founder effect

The genetic drift resulting from the establishment of a small, new population whose gene pool differs from that of the parent population.

freshwater biome

An aquatic biome that occurs in fresh water, such as a lake, pond, river, or stream.

fructose

A monosaccharide with the chemical formula $C_6H_{12}O_6$ found in many fruits; also called "fruit sugar."

fruit

A ripened, thickened ovary of a flower, which protects dormant seeds and aids in their dispersal.

functional group

A group of atoms that form the chemically reactive part of an organic molecule. A particular functional group usually behaves similarly in different chemical reactions.

fungus (plural, **fungi**)

A chemo-heterotrophic eukaryote that digests its food externally and absorbs the resulting small nutrient molecules. Most fungi consist of a netlike mass of filaments called hyphae. Molds, mushrooms, and yeasts are examples of fungi.

G

gallbladder

An organ that stores bile and releases it as needed into the small intestine.

gamete

A sex cell; a haploid egg or sperm. The union of two gametes of opposite sex (fertilization) produces a zygote.

gametophyte

The multicellular haploid form in the life cycle of organisms undergoing alternation of generations; results from a union of spores and mitotically produces haploid gametes that unite and grow into the sporophyte generation.

gastric juice

The collection of fluids secreted by the epithelium lining the stomach.

gastropod

Member of the largest group of molluscs, including snails and slugs.

gastrovascular cavity

A digestive compartment with a single opening that serves as both the entrance for food and the exit for undigested wastes; may also function in circulation, body support, and gas exchange. Jellies and hydras are examples of animals with a gastrovascular cavity.

gastrula

The embryonic stage resulting from gastrulation in animal development. Most animals have a gastrula made up of three layers of cells: ectoderm, endoderm, and mesoderm.

gel electrophoresis

A technique for sorting macromolecules. A mixture of molecules is placed on a gel between a positively charged electrode and a negatively charged one; negative charges on the molecules are attracted to the positive electrode, and the molecules migrate toward that electrode. The molecules separate in the gel according to their rates of migration.

gene

A unit of inheritance in DNA (or RNA, in some viruses) consisting of a specific nucleotide sequence that programs the amino acid sequence of a polypeptide. Most of the genes of a eukaryote are located in its chromosomal DNA; a few are carried by the DNA of mitochondria and chloroplasts.

gene cloning
The production of multiple copies of a gene.

gene expression
The process whereby genetic information flows from genes to proteins; the flow of genetic information from the genotype to the phenotype: DNA → RNA → protein.

gene flow
The gain or loss of alleles from a population by the movement of individuals or gametes into or out of the population.

gene pool
All the genes in a population at any one time.

gene regulation
The turning on and off of specific genes within a living organism.

gene therapy
A treatment for a disease in which the patient's defective gene is supplemented or altered.

genetic biodiversity
A measure of biodiversity based on the collection of different genes within a population. Reduced genetic biodiversity can leave a population susceptible to catastrophic loss when the environment changes suddenly and drastically.

genetic cross
A mating of two sexually reproducing individuals.

genetic drift
A change in the gene pool of a population due to chance.

genetic engineering
The direct manipulation of genes for practical purposes.

genetically modified (GM) organism
An organism that has acquired one or more genes by artificial means. If the gene is from another organism, typically of another species, the recombinant organism is also known as a transgenic organism.

genetics
The scientific study of heredity (inheritance).

genome
The complete set of genetic material of an organism or virus.

genomic library
The entire collection of DNA segments from an organism's genome. Each segment is usually carried by a plasmid or phage.

genomics
The study of whole sets of genes and their interactions.

genotype
The genetic makeup of an organism.

geological process
Process (such as erosion and volcanic eruptions) that transfers chemical elements from the Earth itself to an abiotic reservoir such as the atmosphere.

gill
An extension of the body surface of an aquatic animal, specialized for gas exchange and/or suspension feeding.

glans
The rounded, highly sensitive head of the clitoris in females and penis in males.

global warming
Increase in temperature all around the planet, due mostly to increasing atmospheric CO_2 levels from the burning of fossil fuels. Global warming is a major aspect of global climate change.

global water cycle
The flow of water—by, among other processes, evaporation and precipitation—among all the ecosystems of Earth.

glucose
A six-carbon monosaccharide that serves as a building block for many polysaccharides and whose oxidation in cellular respiration is a major source of ATP for cells.

glycogen
A complex, extensively branched polysaccharide made up of many glucose monomers; serves as an energy-storage molecule in liver and muscle cells.

glycolysis
The multistep chemical breakdown of a molecule of glucose into two molecules of pyruvic acid; the first stage of cellular respiration in all organisms; occurs in the cytoplasmic fluid.

Golgi apparatus
An organelle in eukaryotic cells consisting of stacks of membranous sacs that modify, store, and ship products of the endoplasmic reticulum.

gonad
An animal sex organ that produces gametes; an ovary or a testis.

granum (plural, **grana**)
A stack of hollow disks formed of thylakoid membrane in a chloroplast. Grana are the sites where light energy is trapped by chlorophyll and converted to chemical energy during the light reactions of photosynthesis.

greenhouse gas
Any of the gases in the atmosphere that absorb heat radiation, including CO_2, methane, water vapor, and synthetic chlorofluorocarbons.

growth factor
A protein secreted by certain body cells that stimulates other cells to divide.

gymnosperm
A naked-seed plant. Its seed is said to be naked because it is not enclosed in an ovary.

H

halophile
Prokaryotic microorganism that lives in a highly salty environment, such as the Great Salt Lake or the Dead Sea.

haploid
Containing a single set of chromosomes; referring to an *n* cell.

head
With respect to human sperm, the end of the cell that contains the chromosomes.

heart
(1) The chambered muscular organ in vertebrates that pumps blood received from the veins into the arteries, thereby maintaining the flow of blood through the entire circulatory system. (2) A similarly functioning structure in invertebrates.

heart attack
Damage or death of heart muscle tissue and the resulting failure of the heart to deliver enough blood to the body.

heat
The amount of kinetic energy contained in the movement of the atoms and molecules in a body of matter. Heat is energy in its most random form.

helper T cell
A type of lymphocyte that helps activate other T cells and helps stimulate B cells to produce antibodies.

hemoglobin
An iron-containing protein in red blood cells that reversibly binds O_2 and transports it to body tissues.

herbivory
The consumption of plant parts or algae by an animal.

heredity
The transmission of traits from one generation to the next.

heterotroph
An organism that cannot make its own organic food molecules from inorganic ingredients and must obtain them by consuming other organisms or their organic products; a consumer or a decomposer in a food chain.

heterozygous
Having two different alleles for a given gene.

hexokinase
An example of an enzyme, a protein that promotes a chemical reaction.

HIV (human immunodeficiency virus)
The retrovirus that attacks the human immune system and causes AIDS.

homeostasis
The steady state of body functioning; the tendency to maintain relatively constant conditions in the internal environment even when the external environment changes.

homeotic gene
A master control gene that determines the identity of a body structure of a developing organism, presumably by controlling the developmental fate of groups of cells. (In plants, such genes are called organ identity genes.)

hominin
Any anthropoid on the human branch of the evolutionary tree, more closely related to humans than to chimpanzees.

homologous chromosomes
The two chromosomes that make up a matched pair in a diploid cell. Homologous chromosomes are of the same length, centromere position, and staining pattern and possess genes for the same characters at corresponding loci. One homologous chromosome is inherited from the organism's father, the other from the mother.

homozygous
Having two identical alleles for a given gene.

hormone
In multicellular organisms, a regulatory chemical that travels in body fluids from its production site to other sites, where target cells respond to the regulatory signal.

human
Member of the species *Homo sapiens*.

Human Genome Project
An international collaborative effort that sequenced the DNA of the entire human genome.

humoral immune response
The type of adaptive immune response that involves secretion of antibodies into the blood and lymph by effector B cells. These antibodies fight bacteria and viruses in body fluids. *See also* cell-mediated immune response.

hydrogen bond
A type of weak chemical bond formed when a partially positive hydrogen atom from one polar molecule is attracted to the partially negative atom in another molecule (or in another part of the same molecule).

hydrogenation
The process of converting unsaturated fats to saturated fats by adding hydrogen.

hydrolysis reaction
Chemical reaction in which macromolecules are broken down by the chemical addition of water molecules to the bonds linking their monomers. A hydrolysis reaction is the opposite of a dehydration reaction.

hydrophobic
"Water-fearing"; pertaining to nonpolar molecules (or parts of molecules), which do not dissolve in water.

hypertension
Abnormally high blood pressure consisting of a persistent systolic blood pressure higher than 140 and/or diastolic blood pressure higher than 90. This condition can lead to a variety of serious cardiovascular disorders.

hypha (plural, **hyphae**)
One of many filaments making up the body of a fungus.

hypothalamus
The main control center of the endocrine system, located in the vertebrate forebrain. The hypothalamus functions in maintaining homeostasis, especially in coordinating the endocrine and nervous systems. It synthesizes hormones secreted by the posterior pituitary and regulates the secretion of hormones by the anterior pituitary.

hypothesis (plural, **hypotheses**)
A tentative explanation that a scientist proposes for a specific phenomenon that has been observed.

I

immune system
The body's system of defenses against infectious disease.

immunodeficiency disease
An immunological disorder in which the body lacks one or more components of the immune system, making a person susceptible to infectious agents that would not ordinarily cause a problem.

incomplete dominance
A type of inheritance in which the phenotype of a heterozygote (*Aa*) is intermediate between the phenotypes of the two types of homozygotes (*AA* and *aa*).

induction
During embryonic development, the influence of one group of cells on an adjacent group of cells.

infertile
The result of infertility, the inability to conceive after one year of trying.

ingestion
The act of eating; the first stage of food processing.

inhibitor
A substance that reduces the activity of an enzyme by binding to it.

insect
An arthropod that usually has three body segments (head, thorax, and abdomen), three pairs of legs, and one or two pairs of wings.

integrated pest management
The maintenance of a low population of a pest through a variety of methods meant to promote a healthy environment.

interphase

The phase in the eukaryotic cell cycle when the cell is not actually dividing. During interphase, cellular metabolic activity is high, chromosomes and organelles are duplicated, and cell size may increase. Interphase accounts for 90% of the cell cycle. *See also* mitosis.

interspecies interaction

Interaction between members of different species. The effect on each species may be beneficial, neutral, or harmful.

interstitial fluid

An aqueous solution that surrounds body cells and through which materials pass back and forth between the blood and the body tissues.

intertidal zone

A shallow zone where the waters of an estuary or ocean meet land.

intron

In eukaryotes, a nonexpressed (noncoding) portion of a gene that is excised from the RNA transcript. *See also* exon.

invasive species

A non-native species that has spread far beyond the original point of introduction and causes environmental or economic damage by colonizing and dominating suitable habitats.

invertebrate

An animal that does not have a backbone.

ion

An atom or molecule that has gained or lost one or more electrons, thus acquiring an electrical charge.

ionic bond

An attraction between two ions with opposite electrical charges. The electrical attraction of the opposite charges holds the ions together.

iris

The colored part of the vertebrate eye, formed by the anterior portion of the pigmented layer called the choroid.

isomer

One of two or more molecules with the same molecular formula but different structures and thus different properties.

isotope

A variant form of an atom. Isotopes of an element have the same number of protons and electrons but different numbers of neutrons.

K

karyotype

A display of micrographs of the metaphase chromosomes of a cell, arranged by size and centromere position.

keratin

Any member of a family of fibrous proteins; keratin is a key component of human skin, nails, and hair.

keystone species

A species whose impact on its community is much larger than its biomass or abundance indicates.

kinetic energy

Energy of motion. Moving matter performs work by transferring its motion to other matter, such as leg muscles pushing bicycle pedals.

kingdom

In classification, the broad taxonomic category above phylum.

L

labia majora

A pair of outer thickened folds of skin that protect the female genital region.

labia minora

A pair of inner folds of skin bordering and protecting the female genital region.

labor

A series of strong, rhythmic uterine contractions that expels a baby out of the uterus and vagina during childbirth.

large intestine

The tubular portion of the vertebrate alimentary canal between the small intestine and the anus.

larynx

The voice box, containing the vocal cords.

lateral line system

A row of sensory organs along each side of a fish's body. Sensitive to changes in water pressure, it enables a fish to detect minor vibrations in the water.

law of independent assortment

A general rule of inheritance, first proposed by Gregor Mendel, that states that when gametes form during meiosis, each pair of alleles for a particular character segregate (separate) independently of each other pair.

law of segregation

A general rule of inheritance, first proposed by Gregor Mendel, that states that the two alleles in a pair segregate (separate) into different gametes during meiosis.

lens

The disklike structure in an eye that focuses light rays onto the retina.

life

The set of common characteristics that distinguish living organisms from nonliving matter, including such properties and processes as order, regulation, growth and development, energy utilization, response to the environment, reproduction, and the capacity to evolve over time.

life cycle

The entire sequence of stages in the life of an organism, from the adults of one generation to the adults of the next.

ligament

A type of fibrous connective tissue that joins bones together at joints.

light reactions

The first of two stages in photosynthesis, the steps in which solar energy is absorbed and converted to chemical energy in the form of ATP and NADPH. The light reactions power the sugar-producing Calvin cycle but produce no sugar themselves.

lignin

A chemical that hardens the cell walls of plants. Lignin makes up most of what we call wood.

limiting factor

An environmental factor that restricts the number of individuals that can occupy a particular habitat, thus holding population growth in check.

linked genes

Genes located close enough together on a chromosome that they are usually inherited together.

lipid

An organic compound consisting mainly of carbon and hydrogen atoms linked by nonpolar covalent bonds and therefore mostly hydrophobic and insoluble in water. Lipids include fats, waxes, phospholipids, and steroids.

liver
The largest organ in the vertebrate body. The liver performs diverse functions, such as producing bile, preparing nitrogenous wastes for disposal, and detoxifying poisonous chemicals in the blood.

lobe-finned fish
A bony fish with strong, muscular fins supported by bones.

logistic growth
A description of the increase in a size of a population in which the increase slows as the population approaches its carrying capacity.

loose connective tissue
The most widespread connective tissue in the vertebrate body. It binds epithelia to underlying tissues and functions as packing material, holding organs in place.

lymph
A fluid similar to interstitial fluid that circulates in the lymphatic system.

lymph node
A small organ that is located along a lymph vessel and that filters lymph.

lymphatic system
The organ system through which lymph circulates; includes lymph vessels, lymph nodes, and several other organs. The lymphatic system helps remove toxins and pathogens from the blood and interstitial fluid and returns fluid and solutes from the interstitial fluid to the circulatory system.

lymphocyte
A type of white blood cell that carries out adaptive defenses—recognizing and responding to specific invading pathogens. There are two types of lymphocytes: B cells and T cells. *See also* B cell; T cell.

lysosome
A digestive organelle in eukaryotic cells; contains enzymes that digest the cell's food and wastes.

M

macroevolution
Evolutionary change above the species level, including the origin of evolutionary novelty and new taxonomic groups and the impact of mass extinctions on the diversity of life and its subsequent recovery.

macromolecule
A giant molecule formed by joining smaller molecules. Examples of macromolecules include proteins, polysaccharides, and nucleic acids.

malignant tumor
An abnormal tissue mass that spreads into neighboring tissue and to other parts of the body; a cancerous tumor.

malnutrition
The absence of one or more essential nutrients from the diet.

mammal
Member of a class of endothermic amniotes that possesses mammary glands and hair.

mammary gland
An characteristic gland of mammals that secretes milk to nourish the young.

mantle
In molluscs, the outgrowth of the body surface that drapes over the animal. The mantle produces the shell and forms the mantle cavity.

marsupial
A pouched mammal, such as a kangaroo, opossum, or koala. Marsupials give birth to embryonic offspring that complete development while housed in a pouch and attached to nipples on the mother's abdomen.

matter
Anything that occupies space and has mass.

mechanical digestion
Physical processes, such as chewing, that help break food down into smaller molecules.

medusa (plural, **medusae**)
One of two types of cnidarian body forms; a floating, umbrella-like body form; also called a jelly.

meiosis
In a sexually reproducing organism, the process of cell division that produces haploid gametes from diploid cells within the reproductive organs.

memory cell
A long-lived lymphocyte that responds to subsequent exposures to a specific pathogen. A memory cell is formed during the primary immune response and is activated by exposure to the same antigen that triggered its formation. When activated, a memory cell forms large clones of effector cells and memory cells that mount the secondary immune response.

meninges
Layers of connective tissue that enwrap and protect the brain and spinal cord.

meristem
Plant tissue consisting of undifferentiated cells that divide and generate new cells and tissues.

messenger RNA (mRNA)
The type of ribonucleic acid that encodes genetic information from DNA and conveys it to ribosomes, where the information is translated into amino acid sequences.

metabolism
The total of all the chemical reactions in an organism.

metastasis
The spread of cancer cells beyond their original site.

methanogen
An archaean that produces methane as a metabolic waste product.

microevolution
A change in a population's gene pool over a succession of generations; evolutionary changes in species over relatively brief periods of geologic time.

microRNA (miRNA)
A small, single-stranded RNA molecule that associates with one or more proteins in a complex that can degrade or prevent translation of an mRNA with a complementary sequence.

microvillus (plural, **microvilli**)
One of many microscopic projections on the epithelial cells in the lumen of the small intestine. Microvilli increase the surface area of the small intestine.

millipede
A terrestrial arthropod that has two pairs of short legs for each of its numerous body segments and that eats decaying plant matter.

mineral
In nutrition, a simple inorganic nutrient that an organism requires in small amounts for proper body functioning.

missense mutation

A change in the nucleotide sequence of a gene that alters the amino acid sequence of the resulting polypeptide. In a missense mutation, a codon is changed from encoding one amino acid to encoding a different amino acid.

mitochondrion (plural, **mitochondria**)

An organelle in eukaryotic cells where cellular respiration occurs. Enclosed by two concentric membranes, it is where most of the cell's ATP is made.

mitosis

The division of a single nucleus into two genetically identical daughter nuclei. Mitosis and cytokinesis make up the mitotic phase of the cell cycle.

mitotic phase

The phase of the cell cycle when mitosis divides the nucleus and distributes its chromosomes to the daughter nuclei and cytokinesis divides the cytoplasm, producing two daughter cells.

mitotic spindle

A spindle-shaped structure formed of microtubules and associated proteins that is involved in the movement of chromosomes during mitosis and meiosis. (A spindle is shaped roughly like a football.)

molecule

A group of two or more atoms held together by covalent bonds.

mollusc

A soft-bodied animal characterized by a muscular foot, mantle, mantle cavity, and radula. Molluscs include gastropods (snails and slugs), bivalves (clams, oysters, and scallops), and cephalopods (squids and octopuses).

monocot

A flowering plant whose embryos have a single seed leaf, or cotyledon.

monohybrid cross

A mating of individuals differing at one genetic locus.

monomer

A chemical subunit that serves as a building block of a polymer.

monosaccharide

The smallest kind of sugar molecule; a single-unit sugar; also known as a simple sugar. Monosaccharides are the building blocks of more complex sugars and polysaccharides.

monotreme

An egg-laying mammal, such as the duck-billed platypus.

motor neuron

A nerve cell that conveys command signals from the central nervous system to effector cells, such as muscle cells or gland cells.

multiple alleles

The existence of more than two common versions of a gene. For example, the gene that determines human blood type (ABO) comes in three common varieties.

muscle fiber

Muscle cell.

muscle tissue

Tissue consisting of long muscle cells that are capable of contracting when stimulated by nerve impulses. *See also* skeletal muscle; cardiac muscle; smooth muscle.

mushroom

A fungus in which an aboveground reproductive structure extends from a belowground mycelium.

mutagen

A chemical or physical agent that interacts with DNA and causes a mutation.

mutant trait

The phenotypic trait that is less commonly observed in natural populations; the opposite of the wild-type trait.

mutation

A change in the nucleotide sequence of DNA; a major source of genetic diversity.

mutualism

An interspecific interaction in which both partners benefit.

mycelium (plural, **mycelia**)

The densely branched network of hyphae in a fungus.

mycorrhiza (plural, **mycorrhizae**)

A mutually beneficial symbiotic association of a plant root and fungus.

myofibril

A contractile unit in a muscle cell (fiber) made up of many sarcomeres. Longitudinal bundles of myofibrils make up a muscle fiber.

N

NADPH

An electron carrier (a molecule that carries electrons) involved in photosynthesis. Light drives electrons from chlorophyll to $NADP^+$, forming NADPH, which provides the high-energy electrons for the reduction of carbon dioxide to sugar in the Calvin cycle.

natural killer (NK) cell

A white blood cell that attacks cancer cells and infected body cells as part of internal innate defense.

natural selection

A process in which organisms with certain inherited characters are more likely to survive and reproduce than are organisms with other characters; unequal reproductive success.

negative feedback

A control mechanism in which a chemical reaction, metabolic pathway, or hormone-secreting gland is inhibited by the products of the reaction, pathway, or gland. As the concentration of the products builds up, the product molecules themselves inhibit the process that produced them.

nematode

An animal characterized by a pseudocoelom, a cylindrical, wormlike body form, and a complete digestive tract; also called a roundworm.

nephron

The tubular excretory unit and associated blood vessels of the vertebrate kidney. The nephron extracts filtrate from the blood and refines it into urine.

nerve

A communication line made up of cable-like bundles of neuron fibers (axons and dendrites) tightly wrapped in connective tissue.

nerve cord

An elongated bundle of neurons, usually extending longitudinally from the brain or anterior ganglia. One or more nerve cords and the brain make up the central nervous system in many animals.

nervous system

The organ system that forms a communication and coordination network throughout an animal's body.

nervous tissue
Tissue made up of neurons and supportive cells.

neuron
A nerve cell; the fundamental structural and functional unit of the nervous system, specialized for carrying signals from one location in the body to another.

neurotransmitter
A chemical messenger that carries information from a transmitting neuron to a receiving cell, either another neuron or an effector cell.

neutron
An electrically neutral particle (a particle having no electrical charge), found in the nucleus of an atom.

nondisjunction
An accident of meiosis or mitosis in which a pair of homologous chromosomes or a pair of sister chromatids fail to separate at anaphase.

nonpolar bond
Also called a nonpolar covalent bond; a covalent bond in which electrons are shared equally between two atoms of similar electronegativity; the resulting bond does not have a positive and negative pole; the opposite of a polar bond.

nonsense mutation
A change in the nucleotide sequence of a gene that converts an amino-acid-encoding codon to a stop codon. A nonsense mutation results in a shortened polypeptide.

notochord
A flexible, cartilage-like, longitudinal rod located between the digestive tract and nerve cord in chordate animals, present only in embryos in many species.

nuclear envelope
A double membrane, perforated with pores, that encloses the nucleus and separates it from the rest of the eukaryotic cell.

nuclear pore
A protein-lined opening in the nuclear envelope through which materials (such as messenger RNA) can pass between the nucleus and the cytoplasm.

nuclear transplantation
A technique in which the nucleus of one cell is placed into another cell that already has a nucleus or in which the nucleus has been previously destroyed. The cell is then stimulated to grow, producing an embryo that is a genetic copy of the nucleus donor.

nucleic acid
A polymer consisting of many nucleotide monomers; serves as a blueprint for proteins and, through the actions of proteins, for all cellular structures and activities. The two types of nucleic acids are DNA and RNA.

nucleic acid probe
In DNA technology, a labeled single-stranded nucleic acid molecule used to find a specific gene or other nucleotide sequence within a mass of DNA. The probe hydrogen-bonds to the complementary sequence in the targeted DNA.

nucleotide
An organic monomer consisting of a five-carbon sugar covalently bonded to a nitrogenous base and a phosphate group. Nucleotides are the building blocks of nucleic acids.

nucleus (plural, **nuclei**)
(1) An atom's central core, containing protons and neutrons.
(2) The genetic control center of a eukaryotic cell.

O

obesity
An excessively high body mass index, a ratio of weight to height.

omega-3 fatty acids
A group of essential fatty acids, consumption of which in the diet is required for proper health.

oncogene
A cancer-causing gene; usually contributes to malignancy by abnormally enhancing the amount or activity of a growth factor made by the cell.

oogenesis
The formation of egg cells within the ovaries.

operculum (plural, **opercula**)
A protective flap on each side of a bony fish's head that covers a chamber housing the gills.

optic nerve
A nerve that arises from the retina in each eye and carries visual information to the brain.

oral cavity
An opening through which food is taken into an animal's body; also known as the mouth.

organ
A structure consisting of two or more tissues that coordinate to perform specific functions.

organ system
A group of organs that work together in performing vital body functions.

organelle
A membrane-enclosed structure with a specialized function within a eukaryotic cell.

organic compound
A chemical compound containing the element carbon and usually synthesized by cells.

organism
An individual living thing, such as a bacterium, fungus, protist, plant, or animal.

osmoregulation
The control of the gain or loss of water and dissolved solutes in an organism.

osmosis
The diffusion of water across a selectively permeable membrane.

ovary
(1) In animals, the female gonad, which produces egg cells and reproductive hormones. (2) In flowering plants, the base of a carpel in which the egg-containing ovules develop.

oviduct
The tube that conveys egg cells away from an ovary; also called a fallopian tube.

ovulation
The release of an egg cell from an ovarian follicle.

ovum (plural, **ova**)
A mature reproductive egg.

P

P generation
The parent individuals from which offspring are derived in studies of inheritance. P stands for parental.

pancreas
A gland with dual functions: The nonendocrine portion secretes digestive enzymes and an alkaline solution into the small intestine via a duct; the endocrine portion secretes the hormones insulin and glucagon into the blood.

parasite
An organism that lives in or on another organism (the host) from which it obtains nourishment; an organism that benefits at the expense of another organism, which is harmed in the process.

parathyroid gland
One of four endocrine glands embedded in the surface of the thyroid gland that secrete parathyroid hormone; functions in calcium homeostasis.

pathogen
A disease-causing virus or organism.

pedigree
A family tree representing the occurrence of heritable traits in parents and offspring across a number of generations.

pelagic realm
The open-water region of an ocean.

pepsin
An enzyme present in gastric juice that begins the hydrolysis of proteins.

peptide bond
The covalent linkage between two amino acid units in a polypeptide, formed by a dehydration reaction between two amino acids.

periodic table of the elements
A listing of all the elements ordered by their atomic number (the number of protons found in one atom of the element). Each entry in the periodic table of the elements typically contains the element symbol, its atomic mass, and its atomic number.

peripheral nervous system (PNS)
The network of nerves carrying signals into and out of the central nervous system.

peristalsis
Rhythmic waves of contraction of smooth muscles. Peristalsis propels food through a digestive tube and also enables many animals, such as earthworms, to crawl.

pH scale
A measure of the relative acidity of a solution, ranging in value from 0 (most acidic) to 14 (most basic). pH stands for potential hydrogen and refers to the concentration of hydrogen ions (H^+).

phage
See bacteriophage.

phagocytic cell
A white blood cell that engulfs bacteria, foreign proteins, and the remains of dead body cells as part of internal innate defense.

pharyngeal slit
A gill structure in the pharynx, found in chordate embryos and some adult chordates.

pharynx
The organ in a digestive tract that receives food from the oral cavity; in terrestrial vertebrates, the throat region where the air and food passages cross.

phenotype
The expressed traits of an organism.

phloem
The portion of a plant's vascular system that conveys sugars, nutrients, and hormones throughout a plant. Phloem is made up of live food-conducting cells.

phosphate
A chemical group consisting of a phosphorus atom bonded to four oxygen atoms; found in such molecules as DNA and the phosopholipids that constitute cellular membranes.

phospholipid
A molecule that is a constituent of the inner bilayer of biological membranes, having a hydrophilic head and a hydrophobic tail.

phospholipid bilayer
A double layer of phospholipid molecules (each molecule consisting of a phosphate group bonded to two fatty acids) that is the primary component of all cellular membranes.

photic zone
Shallow water near shore or the upper layer of water away from the shore; region of an aquatic ecosystem where sufficient light is available for photosynthesis.

photosynthesis
The process by which plants, algae, and some bacteria transform light energy to chemical energy stored in the bonds of sugars. This process requires an input of carbon dioxide (CO_2) and water (H_2O) and produces oxygen gas (O_2) as a waste product.

photosystem
A light-harvesting unit of a chloroplast's thylakoid membrane; consists of several hundred molecules, a reaction-center chlorophyll, and a primary electron acceptor.

phylogenetic tree
A branching diagram that represents a hypothesis about evolutionary relationships between organisms.

physiology
The study of the function of an organism's structural equipment.

phytoplankton
Algae and photosynthetic bacteria that drift passively in aquatic environments.

pituitary
An endocrine gland at the base of the hypothalamus. The pituitary produces and secretes many hormones that regulate diverse body functions.

placenta
In most mammals, the organ that provides nutrients and oxygen to the embryo and helps dispose of its metabolic wastes. The placenta is formed of the embryo's chorion and the mother's endometrial blood vessels.

plant
A multicellular eukaryote that carries out photosynthesis and has a set of structural and reproductive terrestrial adaptations, including a multicellular, dependent embryo.

plasma
The yellowish liquid of the blood in which the blood cells are suspended.

plasma membrane
The thin layer of lipids and proteins that sets a cell off from its surroundings and acts as a selective barrier to the passage of ions and molecules into and out of the cell; consists of a phospholipid bilayer in which proteins are embedded.

plasmid
A small ring of self-replicating DNA separate from the chromosome(s). Plasmids are found in prokaryotes and yeasts.

platelet
A piece of cytoplasm from a large cell in the bone marrow; a blood-clotting element.

pleiotropy
The control of more than one phenotypic character by a single gene.

point mutation
A change in a single nucleotide pair of a gene.

polar bond
Also called a polar covalent bond; a covalent bond between atoms that differ in their attraction to electrons. The shared electrons are pulled closer to one atom, making it slightly negative and the other atom slightly positive.

polar ice
A terrestrial biome that includes regions of extremely cold temperature and low precipitation located at high latitudes north of the arctic tundra and in Antarctica.

polygenic inheritance
The additive effect of two or more genes on a single phenotypic character.

polymer
A large molecule consisting of many identical or similar molecular units, called monomers, covalently joined together in a chain.

polymerase chain reaction (PCR)
A technique used to obtain many copies of a DNA molecule or many copies of part of a DNA molecule. A small amount of DNA mixed with the enzyme DNA polymerase, DNA nucleotides, and a few other ingredients replicates repeatedly in a test tube.

polynucleotide
A polymer made up of many nucleotides covalently bonded together.

polyp
One of two types of cnidarian body forms; a stationary (sedentary), columnar, hydra-like body.

polypeptide
A chain of amino acids linked by peptide bonds.

polysaccharide
A carbohydrate polymer consisting of many monosaccharides (sugars) linked by covalent bonds.

population
A group of interacting individuals belonging to one species and living in the same geographic area at the same time.

population density
The number of individuals of a species per unit area or volume of the habitat.

population ecology
The study of how members of a population interact with their environment, focusing on factors that influence population density and growth.

positive feedback
A control mechanism in which the products of a process stimulate the process that produced them.

potential energy
Stored energy; the energy that an object has due to its location and/or arrangement. Water behind a dam and chemical bonds both possess potential energy.

precipitation
The process by which a gas changes to a liquid. For example, water precipitates from clouds and falls as rain.

predation
An interaction between species in which one species, the predator, kills and eats the other, the prey.

prepuce
A fold of skin covering the head of the clitoris or penis.

primary consumer
An organism that eats only autotrophs; an herbivore.

primary growth
Growth in the length of a plant root or shoot produced by an apical meristem.

primary immune response
The initial immune response to an antigen, which includes production of effector cells that respond to the antigen within a few days and also memory cells that will respond to future exposure to the antigen.

primary oocyte
A diploid cell, in prophase I of meiosis, that can be hormonally triggered to develop into an ovum.

primary producer
See producer.

primary production
The amount of solar energy converted to chemical energy (organic compounds) by autotrophs in an ecosystem during a given time period.

primary spermatocyte
A diploid cell in the testis that undergoes meiosis I.

primary succession
A type of ecological succession in which a biological community begins in an area without soil. *See also* secondary succession.

primate
Member of the mammalian group that includes lorises, pottos, lemurs, tarsiers, monkeys, apes, and humans.

primers
Short, artificially created, single-stranded DNA molecules that bind to each end of a target sequence during a PCR procedure.

prion
An infectious form of protein that may multiply by converting related proteins to more prions. Prions cause several related diseases in different animals, including scrapie in sheep, mad cow disease, and Creutzfeldt-Jakob disease in humans.

producer
An organism that makes organic food molecules from carbon dioxide, water, and other inorganic raw materials: a plant, alga, or autotrophic bacterium; the trophic level that supports all others in a food chain or food web.

product
An ending material in a chemical reaction.

prokaryote
An organism characterized by prokaryotic cells.

promoter
A specific nucleotide sequence in DNA, located at the start of a gene, that is the binding site for RNA polymerase and the place where transcription begins.

prostate gland
A gland in human males that secretes an acid-neutralizing component of semen.

protein
A biological polymer constructed from amino acid monomers.

proteomics
The systematic study of the full protein sets (proteomes) encoded by genomes.

protist
Any eukaryote that is not a plant, animal, or fungus.

proto-oncogene
A normal gene that can be converted to a cancer-causing gene.

proton
A subatomic particle with a single unit of positive electrical charge, found in the nucleus of an atom.

protozoan

A protist that lives primarily by ingesting food; a heterotrophic, animal-like protist.

pulmonary circuit

One of two main blood circuits in terrestrial vertebrates; conveys blood between the heart and the lungs. *See also* systemic circuit.

punctuated equilibrium

In the fossil record, long periods of little apparent change (equilibria) interrupted (punctuated) by relatively brief periods of sudden change.

Punnett square

A diagram used in the study of inheritance to show the results of random fertilization.

pupil

The opening in the iris that admits light into the interior of the vertebrate eye. Muscles in the iris regulate its size.

Q

quaternary consumer

An organism that eats tertiary consumers.

R

radial symmetry

An arrangement of the body parts of an organism like pieces of a pie around an imaginary central axis. Any slice passing longitudinally through a radially symmetric organism's central axis divides the organism into mirror-image halves.

radiation therapy

Treatment for cancer in which parts of the body that have cancerous tumors are exposed to high-energy radiation to disrupt cell division of the cancer cells.

radiometric dating

A method for determining the age of fossils and rocks from the ratio of a radioactive isotope to the nonradioactive isotope(s) of the same element in the sample.

radula

A file-like organ found in many molluscs, used to scrape up or shred food.

reactant

A starting material in a chemical reaction.

recessive allele

In heterozygotes, the allele that has no noticeable effect on the phenotype.

recombinant chromosomes

Chromosomes that contain DNA sequences derived from two parents chromosomes, making them a patchwork of regions derived from both parent chromosomes.

recombinant DNA

A DNA molecule carrying genes derived from two or more sources, often from different species.

recombinant plasmid

A bacterial plasmid that contains foreign DNA.

rectum

The terminal portion of the large intestine, where the feces are stored until they are eliminated.

red blood cell

A blood cell containing hemoglobin, which transports O_2; also called an erythrocyte.

relative abundance

The proportional representation of a species in a biological community; one component of species diversity.

reproductive barrier

Anything that prevents individuals of closely related species from interbreeding, even when populations of the two species live together.

reproductive cloning

Using a somatic cell from a multicellular organism to make one or more genetically identical individuals.

reproductive system

The organ system responsible for reproduction.

reptile

Member of the clade of amniotes that includes snakes, lizards, turtles, crocodiles, alligators, birds, and a number of extinct groups (most of the dinosaurs).

respiratory system

The organ system that functions in exchanging gases with the environment, taking in O_2 and disposing of CO_2.

restoration ecology

A field of ecology that develops methods of returning degraded ecosystems to their natural state.

restriction enzyme

A bacterial enzyme that cuts up foreign DNA at one very specific nucleotide sequence, thus protecting bacteria against intruding DNA from phages and other organisms. Restriction enzymes are used in DNA technology to cut DNA molecules in reproducible ways.

restriction fragment

A molecule of DNA produced from a longer DNA molecule cut up by a restriction enzyme; used in genome mapping and other applications.

retina

The light-sensitive layer in an eye, made up of photoreceptor cells and sensory neurons.

retrovirus

An RNA virus that reproduces by means of a DNA molecule. It reverse-transcribes its RNA into DNA, inserts the DNA into a cellular chromosome, and then transcribes more copies of the RNA from the viral DNA. HIV and a number of cancer-causing viruses are retroviruses.

reverse transcriptase

An enzyme that catalyzes the synthesis of DNA on an RNA template.

ribosomal RNA (rRNA)

The type of ribonucleic acid that, together with proteins, makes up ribosomes; the most abundant type of RNA.

ribosome

A cellular structure consisting of RNA and protein organized into two subunits and functioning as the site of protein synthesis in the cytoplasm. The ribosomal subunits are constructed in the nucleolus.

RNA (ribonucleic acid)

A type of nucleic acid consisting of nucleotide monomers, with a ribose sugar, a phosphate group, and the nitrogenous bases adenine (A), cytosine (C), guanine (G), and uracil (U); usually single-stranded; functions in protein synthesis and as the genome of some viruses.

RNA polymerase

An enzyme that links together the growing chain of RNA nucleotides during transcription, using a DNA strand as a template.

RNA splicing
The removal of introns and joining of exons in eukaryotic RNA, forming an mRNA molecule with a continuous coding sequence; occurs before mRNA leaves the nucleus.

rod
A photoreceptor cell in the vertebrate retina, enabling vision in dim light (but only in shades of gray).

root
The underground organ of a plant. Roots anchor the plant in the soil, absorb and transport minerals and water, and store food.

rough ER (rough endoplasmic reticulum)
A network of interconnected membranous sacs in a eukaryotic cell's cytoplasm. Rough ER membranes are studded with ribosomes that make membrane proteins and secretory proteins. The rough ER constructs membrane from phospholipids and proteins.

S

SA (sinoatrial) node
The pacemaker of the heart, located in the wall of the right atrium. At the base of the wall separating the two atria is another patch of nodal tissue called the atrioventricular (AV) node.

sarcomere
The fundamental unit of muscle contraction, composed of thin filaments and thick filaments; the region between two narrow, dark lines in the myofibril.

saturated fat
A fat molecule in which all carbons in the hydrocarbon tails are connected by single bonds and the maximum number of hydrogen atoms are attached to the carbon skeleton. Saturated fats solidify at room temperature and are more commonly found in animal products.

savanna
A terrestrial biome dominated by grasses and scattered trees. Frequent fires and seasonal drought are significant abiotic factors.

scientific method
Scientific investigation involving the observation of phenomena, the formulation of a hypothesis concerning the phenomena, experimentation to demonstrate the truth or falseness of the hypothesis, and results that validate or modify the hypothesis.

sclera
A tough, whitish layer of connective tissue forming the outer surface of the vertebrate eye. The cornea is the front part of the sclera.

scrotum
A pouch of skin outside the abdomen that houses a testis. The scrotum functions in cooling the sperm, thereby keeping them viable.

seaweed
A large, multicellular marine alga.

secondary growth
An increase in a plant's girth, involving cell division in the vascular cambium and cork cambium.

secondary immune response
The immune response elicited by memory cells upon exposure to a previously encountered antigen. The secondary immune response is more rapid, of greater magnitude, and of longer duration than the primary immune response.

secondary oocyte
A haploid cell that results from meiosis I in oogenesis and that will become an ovum after meiosis II.

secondary spermatocyte
A haploid cell that results from meiosis I in spermatogenesis and that will become a sperm cell after meiosis II.

secondary succession
A type of ecological succession that occurs where a disturbance has destroyed an existing biological community but left the soil intact. *See also* primary succession.

seed
A plant embryo packaged with a food supply within a protective covering.

seedless vascular plants
The informal collective name for lycophytes (club mosses and their relatives) and monilophytes (ferns and their relatives).

segmentation
Subdivision along the length of an animal body into a series of repeated parts called segments; allows for greater flexibility and mobility.

semi-conservative
A model for DNA replication in which the replicated double helix consists of one old strand, derived from the old molecule, and one newly made strand.

seminal vesicle
A gland in males that secretes a fluid component of semen that lubricates and nourishes sperm.

seminiferous tubule
A coiled sperm-producing tube in a testis.

sex chromosome
A chromosome that determines whether an individual is male or female; in mammals, for example, the X or Y chromosome.

sex pilus
In bacteria, a structure that links one cell to another at the start of conjugation.

sex-linked gene
A gene located on a sex chromosome.

sexual recombination
The production of offspring with combinations of traits that differ from those found in either parent.

sexual reproduction
The creation of genetically distinct offspring by the fusion of two haploid sex cells (gametes: sperm and egg), forming a diploid zygote.

sexual selection
A form of natural selection in which individuals with certain characteristics are more likely than other individuals to obtain mates.

sexually transmitted disease (STD)
A contagious disease spread by sexual contact.

shoot
The aerial organ of a plant, consisting of stem and leaves. Leaves are the main photosynthetic structures of most plants.

short tandem repeat (STR)
DNA consisting of tandem (in a row) repeats of a short sequence of nucleotides.

signal transduction
The linkage of a stimulus to a specific cellular response.

signal transduction pathway
A series of molecular changes that converts a signal received on a target cell's surface to a specific response inside the cell.

silent mutation

A mutation in a gene that changes a codon to one that codes for the same amino acid as the original codon. The amino acid sequence of the resulting polypeptide is thus unchanged.

single bond

The sharing of one pair of electrons by two atoms to form a covalent bond.

sister chromatid

One of the two identical parts of a duplicated chromosome. While joined, two sister chromatids make up one chromosome; chromatids are eventually separated during mitosis or meiosis II.

skeletal muscle

Striated muscle attached to the skeleton. The contraction of striated muscle produces voluntary movements of the body.

skeletal system

The organ system that provides body support, protects body organs such as the brain, heart, and lungs, and anchors the muscles.

slime mold

A type of protest that has amoeboid cells.

small intestine

The longest section of the alimentary canal. It is the principal site of the enzymatic hydrolysis of food molecules and absorption of nutrients.

smooth ER (smooth endoplasmic reticulum)

A network of interconnected membranous tubules in a eukaryotic cell's cytoplasm. Smooth ER lacks ribosomes. Enzymes embedded in the smooth ER membrane function in the synthesis of certain kinds of molecules, such as lipids.

smooth muscle

Muscle made up of cells without striations, found in the walls of organs such as the digestive tract, urinary bladder, and arteries.

solution

A liquid consisting of a homogeneous mixture of two or more substances: a dissolving agent, the solvent, and a substance that is dissolved, the solute.

somatic cell

Any cell in a multicellular organism except a sperm or egg cell or a cell that develops into a sperm or egg; a body cell.

speciation

An evolutionary process in which one species splits into two or more species.

species

A group of populations whose members possess similar anatomical characteristics and have the ability to interbreed.

species biodiversity

The variety of different species in an ecosystem or in the biosphere as a whole.

species diversity

The variety of species that make up a biological community; the number and relative abundance of species in a biological community.

species richness

The total number of different species in a community; one component of species diversity.

sperm

A male gamete.

spermatogenesis

The formation of sperm cells.

spinal cord

In vertebrates, a jellylike bundle of nerve fibers located within the vertebral column. The spinal cord and the brain together make up the central nervous system.

sponge

An aquatic stationary animal characterized by a highly porous body, choanocytes, and no true tissues.

spore

(1) In plants and algae, a haploid cell that can develop into a multicellular haploid individual, the gametophyte, without fusing with another cell. (2) In fungi, a haploid cell that germinates to produce a mycelium.

sporophyte

The multicellular diploid form in the life cycle of organisms undergoing alternation of generations; results from a union of gametes and meiotically produces haploid spores that grow into the gametophyte generation.

starch

A storage polysaccharide found in the roots of plants and certain other cells; a polymer of glucose.

start codon

On mRNA, the specific three-nucleotide sequence (AUG) to which an initiator tRNA molecule binds, starting translation of genetic information.

stem cell

A relatively unspecialized cell that can give rise to one or more types of specialized cells.

steroid

A type of lipid whose carbon skeleton is in the form of four fused rings: three 6-sided rings and one 5-sided ring. Examples are cholesterol, testosterone, and estrogen.

sticky ends

Single DNA strands located at either end of a segment of DNA that has been chopped by a restriction enzyme. Two pieces of DNA cut by the same restriction enzymes will have complementary sticky ends that will bind together to form a new DNA double helix.

stoma (plural, **stomata**)

A pore surrounded by guard cells in the epidermis of a leaf. When stomata are open, CO_2 enters the leaf, and water and O_2 exit. A plant conserves water when its stomata are closed.

stop codon

In mRNA, one of three triplets (UAG, UAA, UGA) that signal gene translation to stop.

STR analysis

A method of DNA profiling that compares the lengths of STR sequences at specific sites in the genome.

subatomic particles

Protons, neutrons, and electrons; particles smaller than an atom.

substrate

(1) A specific substance (reactant) on which an enzyme acts. Each enzyme recognizes only the specific substrate of the reaction it catalyzes. (2) A surface in or on which an organism lives.

sugar

With respect to DNA, one of the three parts that constitutes a nucleotide. The sugar in DNA is deoxyribose; the sugar in RNA is ribose.

sugar sink

A plant organ that is a net consumer or storer of sugar. Growing roots, shoot tips, stems, and fruits are sugar sinks supplied by phloem.

sugar source
A plant organ in which sugar is being produced by either photosynthesis or the breakdown of starch. Mature leaves are the primary sugar sources of plants.

sugar-phosphate backbone
The alternating chain of sugar and phosphate to which DNA and RNA nitrogenous bases are attached.

survivorship
The chance that an individual organism within a given population will survive a given period of time.

survivorship curve
A plot of the number of individuals that are still alive at each age in the maximum life span; one way to represent age-specific mortality.

sustainable development
The long-term prosperity of human societies and the ecosystems that support them.

swim bladder
A gas-filled internal sac that helps bony fishes maintain buoyancy.

symbiotic
Pertaining to symbiosis, the close physical association of members of different species.

symbol
Also called the atomic symbol, an abbreviation for the name of an element found within the periodic table of the elements.

sympatric speciation
The formation of a new species in populations that live in the same geographic area. *See also* allopatric speciation.

synapse
A junction, or relay point, between two neurons or between a neuron and an effector cell. Electrical and chemical signals are relayed from one cell to another at a synapse.

systematics
A discipline of biology that focuses on classifying organisms and determining their evolutionary relationships.

systemic circuit
One of two main blood circuits in terrestrial vertebrates; conveys blood between the heart and the rest of the body. *See also* pulmonary circuit.

T

T cell
A type of lymphocyte that matures in the thymus and is responsible for the cell-mediated immune response. *See also* B cell.

tail
Extra nucleotides added at the end of an RNA transcript in the nucleus of a eukaryotic cell.

taxonomic hierarchy
An ordered series of categories into which life is classified, running from domain (the most inclusive) to species (the most specific).

taxonomy
The branch of biology concerned with identifying, naming, and classifying species.

tectonic plate
Large plate of Earth's crust that floats on the hot, underlying portion of the mantle. Movements in the mantle cause the tectonic plates, and the continents that reside on them, to move slowly over time.

temperate broadleaf forest
A terrestrial biome located throughout midlatitude regions where there is sufficient moisture to support the growth of large, broadleaf deciduous trees.

temperate grassland
A terrestrial biome located in the temperate zone and characterized by low rainfall and nonwoody vegetation. Tree growth is hindered by occasional fires and periodic severe drought.

tendon
Fibrous connective tissue connecting a muscle to a bone.

terminator
A special sequence of nucleotides in DNA that marks the end of a gene. It signals RNA polymerase to release the newly made RNA molecule, which then departs from the gene.

terrestrial biome
A major land-based life zone characterized by vegetation type.

tertiary consumer
An organism that eats secondary consumers.

testcross
The mating between an individual of unknown genotype for a particular character and an individual that is homozygous recessive for that same character.

testis (plural, **testes**)
The male gonad in an animal. The testis produces sperm and, in many species, reproductive hormones.

tetrapod
A vertebrate with four limbs. Tetrapods include mammals, amphibians, and reptiles (including birds).

theory
A widely accepted explanatory idea that is broad in scope and supported by a large body of evidence.

therapeutic cloning
The cloning of human cells by nuclear transplantation for therapeutic purposes, such as the replacement of body cells that have been irreversibly damaged by disease or injury. *See also* nuclear transplantation; reproductive cloning.

thermal cycler
A laboratory device that performs PCR (polymerase chain reactions).

thermophile
A microorganism that thrives in a hot environment.

thick filament
The thicker of the two types of filaments that make up a sarcomere, consisting of the protein myosin.

thin filament
The thinner of the two types of filaments that make up a sarcomere, consisting of the protein actin.

threatened species
As defined in the U.S. Endangered Species Act, a species that is likely to become endangered in the foreseeable future throughout all or a significant portion of its range.

three-domain system
A system of taxonomic classification based on three basic groups: Bacteria, Archaea, and Eukarya.

thylakoid
One of a number of disk-shaped membranous sacs inside a chloroplast. Thylakoid membranes contain chlorophyll and the enzymes of the light reactions of photosynthesis. A stack of thylakoids is called a granum.

thyroid gland
An endocrine gland, located in the neck, that secretes hormones that increase oxygen consumption and metabolic rate and help regulate development and maturation.

tissue
An integrated group of similar cells that performs a specific function within a multicellular organism.

tissue system
An organized collection of plant tissues. The organs of plants (such as roots, stems, and leaves) are formed from the dermal, vascular, and ground tissue systems.

trace element
An element that is essential for the survival of an organism but is needed in only minute quantities.

trachea (plural, **tracheae**)
(1) The windpipe; the portion of the respiratory tube between the larynx and the bronchi. (2) One of many tiny tubes that branch throughout an insect's body, enabling gas exchange between outside air and body cells.

trait
A variant of a character found within a population, such as purple flowers in pea plants.

trans fats
Unsaturated fatty acids produced by the partial hydrogenation of vegetable oils and present in hardened vegetable oils, most margarines, many commercial baked foods, and many fried foods.

transcription
The synthesis of RNA on a DNA template.

transcription factor
In the eukaryotic cell, a protein that functions in initiating or regulating transcription. Transcription factors bind to DNA or to other proteins that bind to DNA.

transduction
(1) The transfer of bacterial genes from one bacterial cell to another by a phage. (2) See signal transduction pathway.

transfer RNA (tRNA)
A type of ribonucleic acid that functions as an interpreter in translation. Each tRNA molecule has a specific anticodon, picks up a specific amino acid, and conveys the amino acid to the appropriate codon on mRNA.

transformation
The incorporation of new genes into a cell from DNA that the cell takes up from the surrounding environment.

transgenic organism
An organism that contains genes from another organism, typically of another species.

translation
The synthesis of a polypeptide using the genetic information encoded in an mRNA molecule. There is a change of "language" from nucleotides to amino acids.

transpiration
The evaporative loss of water from a plant.

tricep
A muscle of the arm that, when contracted, causes the arm to bend at the elbow.

triglyceride
A dietary fat that consists of a molecule of glycerol linked to three molecules of fatty acids.

triplet code
A set of three-nucleotide-long "words" that specify the amino acids for polypeptide chains.

trisomy 21
See Down syndrome.

trophic structure
The feeding relationships among the various species in a community.

tropical forest
A terrestrial biome characterized by warm temperatures year-round.

tube feet
On the body of an echinoderm, extensions with suction cups that are used to move and to grasp prey.

tumor
An abnormal mass of cells that forms within otherwise normal tissue.

tumor-suppressor gene
A gene whose product inhibits cell division, thereby preventing uncontrolled cell growth.

tundra
A terrestrial biome characterized by bitterly cold temperatures. Plant life is limited to dwarf woody shrubs, grasses, mosses, and lichens. Arctic tundra has permanently frozen subsoil (permafrost); alpine tundra, found at high elevations, lacks permafrost.

U

umbilical cord
A structure containing arteries and veins that connects a developing embryo to the placenta of the mother.

unsaturated fat
Fat with hydrocarbon chains that lack the maximum number of hydrogen atoms and therefore have one or more double covalent bonds. Because of their bent shape, unsaturated fats and fatty acids tend to stay liquid at room temperature.

urethra
A duct that conveys urine from the urinary bladder to outside the body. In the male, the urethra also conveys semen out of the body during ejaculation.

urinary system
The organ system that forms and excretes urine while regulating the amount of water and ions in the body fluids.

uterus
In the reproductive system of a mammalian female, the organ where the development of young occurs; the womb.

V

vaccination
A procedure that presents the immune system with a harmless version of a pathogen, thereby stimulating an adaptive defense when the pathogen itself is encountered.

vacuole
A membrane-enclosed sac, part of the endomembrane system of a eukaryotic cell, having diverse functions.

vagina
Part of the female reproductive system between the uterus and the outside opening; the birth canal in mammals. The vagina accommodates the male's penis and receives sperm during copulation.

vas deferens (plural, **vasa deferentia**)
Part of the male reproductive system that conveys sperm away from the testis; the sperm duct; in humans, the tube that conveys sperm between the epididymis and the common duct that leads to the urethra.

vascular tissue
Plant tissue consisting of cells joined into tubes that transport water and nutrients throughout the plant body. Xylem and phloem make up vascular tissue.

vein
(1) In animals, a vessel that returns blood to the heart.
(2) In plants, a vascular bundle in a leaf, composed of xylem and phloem.

venule
A small vessel that conveys blood between a capillary bed and a vein.

vertebrate
A chordate animal with a backbone. Vertebrates include lampreys, cartilaginous fishes, bony fishes, amphibians, reptiles (including birds), and mammals.

villus (plural, **villi**)
(1) A finger-like projection of the inner surface of the small intestine. (2) A finger-like projection of the chorion of the mammalian placenta. Large numbers of villi increase the surface areas of these organs.

viroid
A plant pathogen composed of molecules of naked, circular RNA several hundred nucleotides long.

virus
A microscopic particle capable of infecting cells of living organisms and inserting its genetic material. Viruses have a very simple structure and are generally not considered to be alive because they do not display all of the characteristics associated with life.

visceral mass
One of the three main parts of a mollusc, containing most of the internal organs.

vitamin
An organic nutrient that an organism requires in very small quantities. Many vitamins serve as coenzymes or parts of coenzymes.

vitreous humor
A jellylike substance that fills the space behind the lens in the vertebrate eye and helps maintain the shape of the eye.

vulva
The outer features of the female reproductive anatomy.

W

wetland
An ecosystem intermediate between an aquatic ecosystem and a terrestrial ecosystem. Wetland soil is saturated with water permanently or periodically.

white blood cell
A blood cell that functions in defending the body against infections; also called a leukocyte.

whole-genome shotgun method
A method for determining the DNA sequence of an entire genome by cutting it into small fragments, sequencing each fragment, and then placing the fragments in the proper order.

wild-type trait
The trait most commonly found in nature.

work
The movement of an object against an opposing force through the expenditure of energy.

X

X chromosome inactivation
In female mammals, the inactivation of one X chromosome in each somatic cell. Once X inactivation occurs in a given cell (during embryonic development), all descendants of that cell will have the same copy of the X chromosome inactivated.

xylem
The portion of a plant's vascular system that provides support and conveys water and inorganic nutrients from the roots to the rest of the plant. Xylem consists mainly of vessel elements and/or tracheids, water-conducting cells.

Z

zygote
The fertilized egg, which is diploid, that results from the union of haploid gametes (sperm and egg) during fertilization.

Credits

PHOTO CREDITS

COVER: Adrian Samson/Getty Images.

DETAILED TOC: Chapter 1: Page xix, Anke van Wyk/Shutterstock; **Chapter 4:** Page xx, Dan Galic/BlueMoon Stock/AGE Fotostock; **Chapter 5:** Page xxi, Dr. Gopal Murti/Science Source; **Chapter 7:** Page xxii, left Andreas Einsiedel/DK Images; right Francois Gohier/Science Source; **Chapter 8:** Page xxii, Michael Just/AGE Fotostock; **Chapter 9:** Page xxiii, Michael P. Gadomski/Science Source; **Chapter 10:** Page xxiii, Anup Shah/Nature Picture Library; **Chapter 12:** Page xxiv, Martin Shields/Science Source.

CHAPTER 1: Module 1.1: Page 2, **elephant mother & baby** Four Oaks/Shutterstock; **elephant eating** Four Oaks/Shutterstock; **elephant eye** karovka/Shutterstock; **elephant walking** NSP-RF/Alamy; Page 3, **virus** Eye of Science/Science Source; **cell** Steve Gschmeissner/Science Source; **elephant spraying** Cucumber Images/Shutterstock; **mammoth** Jon Hughes and Russell Gooday/DK Images; **Module 1.2:** Page 4, **biosphere** Planetary Visions Ltd./Science Source; **ecosystem** Joseph Sohm/AGE Fotostock; **community** Fernando Quevedo de Oliveira/Alamy; **population** Villiers Steyn/Shutterstock; **organism** Anke van Wyk/Shutterstock; Page 5, **tissue** Biology Pics/Science Source; **cells** Ed Reschke/Peter Arnold/Getty Images; **Module 1.3:** Page 6, **TV remote** Ijansempoi/Shutterstock; **scientist** Philippe Psaila/Science Source; Page 7, **plant cell** Biophoto Associates/Science Source; **fat tissue** Science Source; **cookies left** Africa Studio/Shutterstock; **cookies right** M. Unal Ozmen/Shutterstock; **Module 1.4:** Page 9, **jellyfish** Masa Ushioda/AGE Fotostock; **mouse** Steve Gorton/DK Images; **glowing mice** Eye of Science/Science Source; **Module 1.5:** Page 10, **cow** Dudarev Mikhail/Shutterstock; **Module 1.6:** Page 12, **bacteria** Scimat/Science Source; **cell** Steve Gschmeissner/Science Source; Page 13, **sunflowers** Kate Clow, Terry Richardson, Dom/DK Images; **mushrooms** Reinhold Bala/AGE Fotostock; **sheep** blickwinkel/Alamy; **amoeba** Biology Pics/Science Source Biology Pics/Science Source; **hedgehog** Art Wolfe/Science Source/Photo Researchers; **Module 1.7:** Page 14, **rabbits** blickwinkel/Alamy; **water hole** Martin Harvey/Science Source; **lion kill** Christian Heinrich/Imagebroker RF/AGE Fotostock; **ladybirds** PSU Entomology/Science Source; **sea dragon** Peter Scoones/Nature Picture Library; Page 15, **tiger** S. Weber/Blickwinkel/AGE Fotostock; **turtle** David Fleetham/Nature Picture Library; **plant cell** Science Source; **animal cell** Dr. Gopal Murti/Science Source; **Module 1.8:** Page 16, **strawberry plant** Tony Wharton/FLPA; **strawberries** Elena Elisseeva/Shutterstock; **wolf** Corbis/Corbis RF/AGE Fotostock; **great dane** Eric Isselee/Shutterstock; **chihuahua** Dave King/DK

Images; **beagle** Dave King/DK Images; Page 17, **poster** Justin Kase zfivez/Alamy; **bacterium** Dr. Kari Lounatmaa/Science Source; **mosquito** Sinclair Stammers/Science Source; **malaria** Dr. Cecil H. Fox/Science Source; **bottle** Basement Stock/Alamy.

CHAPTER 2: Module 2.1: Page 18, **propane tank** Anthony DiChello/Fotolia; **vinegar** Ivaylo Ivanov/Shutterstock; **aspirin** James Steidl/Shutterstock; Page 19, **sodium** Tim Ridley/DK Images; **chlorine** Leslie Garland Picture Library/Alamy; **salt** Jiri Hera/Shutterstock; **Module 2.2:** Page 20, **calcium** Charles D. Winters/Science Source; **copper** Shchipkova Elena/Shutterstock; **pencil** Gary Ombler/DK Images; Page 21, **oxygen** Phattman/Fotolia; **carbon** Harry Taylor/DK Images; **hydrogen** Agency/Getty Images; **nitrogen** Mark Winword/DK Images; **calcium** Charles D. Winters/Science Source; **phosphorous** Steve Gorton/DK Images; **potassium** Science Photo Library/Alamy; **sulfur** Andy Crawford and Tim Ridley/DK Images; **sodium** James Stevenson/DK Images; **chlorine** Alex Robinson/DK Images; **magnesium** Andrew Paterson/Alamy; **boron** Stephen Oliver/DK Images; **chromium** Atiketta Sangasaeng/Shutterstock; **cobalt** bonchan/Shutterstock; **copper** Shchipkova Elena/Shutterstock; **fluorine** svand/Shutterstock; **iodine** Charles D. Winters/Science Source; **iron** Mike Dunning/DK Images; **manganese** Richard Treptow/Science Source; **molybdenum** Dirk Wiersma/Science Source; **selenium** Charles D. Winters/Science Source; **silicon** Arno Massee/Science Source; **tin** Steve Gorton/DK Images; **vanadium** Charles D. Winters/Science Source; **zinc** GIPhotoStock/Science Source; **Module 2.5:** Page 26, **water background** Nesterov/Shutterstock; **seals** Doug Allan/Science Source; Page 27, **car** Thomas Neubauer/Panther Media/AGE Fotostock; **glass** FoodCollection/AGE Fotostock; **athlete** Joe Fox/Alamy; **spider** Perennou Nuridsany/Science Source; **Module 2.6:** Page 28, **battery** Photosoup/Fotolia; **lemons** Andy Crawford/DK Images; **strawberries** Chris Villano/DK Images; **tomato** Dave King/DK Images; **coffee** Clive Streeter/DK Images; **milk** GorillaAttack/Shutterstock; **water** Mihalec/Shutterstock; **blood** Jane Stockman/DK Images; **forest** James H. Robinson/Science Source; Page 29, **baking soda** Charles D. Winters/Science Source; **ammonia** Garry Watson/Science Source; **bleach** Beth Van Trees/Shutterstock; **lye** Larry Stepanowicz/Fundamental Photographs; **coral reef** David Hall/Science Source; **Module 2.7:** Page 31, **man climbing tree** Wilfried Krecichwost/Getty Images; **Module 2.8:** Page 32, **girl eating** Aaron Amat/Fotolia; Page 33, **arm** Syda Productions/Shutterstock; **Module 2.9:** Page 34, **hospital drip** Dinodia/Dinodia Photo/AGE Fotostock; **fruit** Jules Selmes/DK Images; **ice cream** Neiromobile/Shutterstock; **milk** DK Images; **malt balls** Colin Cooke/AGE Fotostock; **sugar** Danny Smythe/Shutterstock; Page 35, **potatoes** Gary

Holscher/AGE Fotostock; **tree** Mark Goldman/Shutterstock; **runner** mylife photos/AGE Fotostock; **Module 2.10:** Page 36, **oil & vinegar** Martyn F. Chillmaid/Science Source; Page 37, **steak** Spaxiax/Shutterstock; **hands** 3660 Group, Inc. Custom Medical Stock Photo/Newscom; **seals** H. Brehm/Arco Images/AGE Fotostock; **Mark McGwire** AP Photo; **Marion Jones** Peter Foley/EPA/Newscom; **Module 2.11:** Page 38, **beef** Roger Dixon/DK Images; **butter** Stargazer/Shutterstock; **cheese** Ian O'Leary/DK Images; Page 39, **olive oil** David Murray/DK Images; **corn oil** Dino Osmic /Shutterstock; **cupcakes** Anke van Wyk/Fotolia; **omega-3 foods** DK Images; **Module 2.12:** Page 40, **hemoglobin** Kenneth Eward/Science Source; Page 41, **skateboarder** Samot/Blickwinkel/AGE Fotostock; **blood cells** Omikron/Science Source; **Module 2.13:** Page 42, **high bar** Tongro Image Stock/AGE Fotostock; **low bar** Brenda Carson/Shutterstock.

CHAPTER 3: Module 3.1: Page 44, **bacteria** Steve Gschmeissner/Science Source; **archaea** Eye of Science/Science Source; Page 45, **animal** Biophoto Associates/Science Source; **fungus** Biophoto Associates/Science Source; **plant** Biophoto Associates/Science Source; **protist** Biophoto Associates/Science Source; **Module 3.4:** Page 50, **perfume** bobo/Alamy; **beef jerky** Louella938/Shutterstock; **Module 3.5:** Page 52, **plant cell** Biophoto Associates/Science Source; **animal cell** Scott Camazine/Alamy; Page 53, **ribosomes** CNRI/Science Source; **Module 3.8:** Page 58, **plant vacuole** Dr. Jeremy Burgess/Science Source; **fungi vacuole** Thomas Deerinck, NCMIR/Science Source; **protist vacuole** M. I. Walker/Science Source; **cilia** SPL/Science Source; **bacteria** NIBSC/Science Source; **sperm** John Walsh/Science Source; Page 59, **leaf section** Alice J. Belling/Science Source; **cellulose** Biophoto Associates/Science Source; **proteus** Steve Gschmeissner/Science Source; **cytoskeleton** Dr. Torsten Wittmann/Science Source; **matrix** Ioanis Xynos/Science Source.

CHAPTER 4: Module 4.1: Page 60, **slide** Linda Kennedy/Alamy; Page 61, **chameleon** Karl Shone/DK Images; **dishes** Diane Macdonald/Alamy; **Module 4.2:** Page 62, **sunshine** Don Hammond/Design Pics/AGE Fotostock; **kelp** ImageZebra/Shutterstock; **moss** Molodec/Shutterstock; **fern** Zee/DK Images; **bacteria** Michael Abbey/Science Source; **flower** Colin Walton/DK Images; **spruce branch** Matthew Ward/DK Images; Page 63, **elephant** John Lindsay-Smith/Shutterstock; **boy** Glow Images; **caterpillar** Andy Crawford/DK Images; **fungi** Simon Baylis/Shutterstock; **amoeba** Biology Pics/Science Source; **Module 4.3:** Page 64, **plant** AgStock Images/Latitude/Corbis; **micrograph** Eye of Science/Science Source; Page 65, **bark** Galushko Sergey/Shutterstock; **powder** Marekuliasz/Shutterstock; **molecule** Laguna Design/Science Source; **Module 4.6:** Page 70, leaves silver-john/Shutterstock; Page 71, **potato** Christopher Kolaczan/Shutterstock; **tree** Melinda Fawver/Shutterstock; **Module 4.7:** Page 72, **hamburger** Nitr/Shutterstock; Page 73, **woman eating** Dan Galic/BlueMoon Stock/AGE Fotostock; **woman exercising** Dan Galic/BlueMoon Stock /AGE Fotostock; **hamburger** Nitr/Shutterstock; **chocolate** Howard Shooter/DK Images; **orange** Dave King/DK Images; **woman walking** Ruth Jenkinson/DK Images; **Module 4.9:** Page 76, **muscle micrograph** Steve Gschmeissner/ Science Source; **runner** Bob Daemmrich/ PhotoEdit; Page 77, **bacteria** Scimat/Science Source; **pickles** Malyshev Maksim /Shutterstock; **olives** Karina Bakalyan/Shutterstock; **yogurt** Dave King/DK Images; **pepperoni** Sbarabu/ Shutterstock; **yeast** Steve Gschmeissner/Science Source; **beer** Jules Selmes and Debi Treloar/ DK Images; **bread** Anna Sedneva/Shutterstock; **Module 4.10:** Page 78, **oil** Elena Schweitzer/ Shutterstock; **red meat** B. Calkins/Shutterstock; **margarine** Elena Schweitzer/Shutterstock; **butter** Elena Schweitzer/Shutterstock; **bread** Preto Perola/Shutterstock; **potatoes** Iwona Grodzka/Shutterstock; **pasta** Elena Schweitzer/ Shutterstock; **salmon** Volosina/Shutterstock; **beans** Tina Rencelj/Shutterstock; **meat** Margouillat photo/Shutterstock; **eggs** Iwona Grodzka /Shutterstock; Page 79, **tree** Archipoch/ Shutterstock; **man** John Davis/DK Images; **dancer** Ray Moller/DK Images.

CHAPTER 5: Module 5.1: Page 80, **couple** Troels Graugaard/Squaredpixels/iStockphoto; **baby** Jani Bryson/iStockphoto; Page 81, **algal lake** Biophoto Associates/Science Source; **amoeba** Biophoto Associates/Science Source; **algae** M. I. Walker/Science Source; **plant** Michael P. Gadomski/Science Source; **cutting** Lowell Georgia/Science Source; **sea star** D. P. Wilson/ FLPA; **Module 5.2:** Page 82, **woman in blue** Yuri Arcurs/Shutterstock; **cells** Michael Abbey/ Science Source; **chromosomes** James Cavallini/ Science Source; **woman in black** R. Gino Santa Maria/Shutterstock; Page 83, **chromatin top** Biophoto Associates/Science Source; **chromatin bottom** Don. W. Fawcett/Science Source; **Module 5.3:** Page 84, **onions** Lorenzo Vecchia/DK Images; **interphase cells** Warren Rosenberg/Fundamental Photographs; Page 85, **chromosomes** James Cavallini/Science Source; **paramecium** Michael Abbey/Science Source; **Module 5.4:** Page 86, **cell stages** Pr. G. Giménez Martín/Science Source; **Module 5.5:** Page 88, **cleavage furrow** Dr. Gopal Murti/ Science Source; Page 89, **cell plate** Kent Wood/ Science Source; **Module 5.6:** Page 90, **couple** Troels Graugaard/Squaredpixels/ iStockphoto; **baby** Jani Bryson/iStockphoto; Page 91, **karyotypes** Biophoto Associates/ Science Source; **cell** James Cavallini/Science Source; **woman** Yuri Arcurs/Shutterstock; **Module 5.8:** Page 94, **couple** Troels Graugaard/ Squaredpixels/iStockphoto; **male teen** Jackhollingsworthcom/Shutterstock; **woman** Elliot Westacott/iStockphoto; Page 95, **baby left** Jani Bryson/iStockphoto; **baby right** Heidi Brand/Shutterstock; **seated woman** Grady

Reese/iStockphoto; **Module 5.9:** Page 96, **hand** James Steidl/Shutterstock; **sperm** Michael Abbey/Science Source; **egg** Pascal Goetgheluck/ Science Source; **zygote** David M. Phillips/Science Source; **Module 5.10:** Page 99, **boy** R. Gino Santa Maria/Shutterstock; **karyotype** CNRI/Science Source; **chromosomes** Biophoto Associates/ Science Source; **Module 5.11:** Page 100, **portrait** Topham/TopFoto/The Image Works; **purple flower** Mike P. Shepherd/Alamy; **white flower** Nature Alan King/Alamy; **green eye** Ariwasabi/ Shutterstock; **brown eye** Jaimie Duplass/ Shutterstock; **blue eye** Gelpi/Shutterstock; Page 101, **boy** Christopher Robbins/Photodisc/Getty Images; **dog** Eric Isselée/Shutterstock; **flower** Mike P. Shepherd/Alamy; **Module 5.12:** Page 102, **black dog** Eric Isselée/Shutterstock; **brown dog** Erik Lam/Shutterstock; **black puppy** Utekhina Anna/Shutterstock; **brown puppy** Eric Isselée/ Shutterstock; Page 103, **black dog 2** Erik Lam/ Shutterstock; **Module 5.13:** Page 104, **black dog** AMC Photography/Shutterstock; **puppies** Utekhina Anna/Shutterstock; Page 105, **black dog 2** Eric Isselée/Shutterstock; **black dog 3** AMC Photography/Shutterstock; **Module 5.14:** Page 106, **freckles** Stefanie Sudek/Stock4B/Getty Images; **albinism** Ann Marie Kurtz/iStockphoto; **dwarfs** Marmaduke St. John/Alamy; **Module 5.15:** Page 108, **red flowers** Tamara Kulikova/ Shutterstock; **white flowers** Vilor/Shutterstock; **pink flowers** Le Do/Shutterstock; **blood bag** Life in View/Science Source; Page 109, **red blood cells** Sebastian Kaulitzki/Alamy; **people in line** Ljupco Smokovski/Shutterstock; **vial** kizilkayaphotos/ iStockphoto; **babies** Nick White/moodboard/Age Fotostock; **piercings** Wave Royalty Free/Science Source; **Module 5.17:** Page 112, **young man** East/ Shutterstock; **karyotype** Biophoto Associates/ Science Source; **young woman** Kelly Cline/ iStockphoto; Page 113, **sperm egg cell royal family** Mary Evans/John Massey Stewart Russian Collection/The Image Works; **Module 5.18:** Page 114, **nucleus donor** Gordon Clayton/DK Images; **cow egg donor** Margo Harrison/Shutterstock; **cow researcher** Philippe Psaila/Science Source; **cutting plant** Nigel Cattlin/Holt Studios/Science/ Source; **placing plant** Nigel Cattlin/Holt Studios/Science/Source; **seedlings** GardenPhotos .com/Alamy; **adult plant** Sian Irvine/DK Images; Page 115, **surrogate mother cow** Wayne Hutchinson/FLPA/AGE Fotostock; **calf** Gordon Clayton/DK Images; **horses** Mauro Fermariello/ Science Source; **rabbits** Novosti/Science source; **blood cells** Susumu Nishinaga/Science Source; **nerve cells** Juergen Berger/ Science Source; **muscle cells** Eye of Science/Science Source.

CHAPTER 6: Module 6.1: Page 116, **woman** Ariwasabi/Shutterstock; **cell** Biophoto Associates/Science Source; **chromosome** James Cavallini/Science Source; **Module 6.3:** Page 121, **woman** Yuri Arcurs/Shutterstock; **cell** Biophoto Associates/Science Source; **chromosome** James Cavallini/Science Source; **Module 6.6:** Page 126, **ribosomes** CNRI/Science Source; **Module 6.8:** Page 130, **intestinal cell** MedImage/Science Source; **nerve cell** Petit Format/Science Source; **blood cell** Stem Jems/Science Source; **woman** R. Gino Santa Maria/Shutterstock; **Barr body**

Chris Bjornberg/Science Source; **Module 6.9:** Page 133, **embryo** VideoSurgery/Science Source; **baby** Heidi Brand/Shutterstock; **fruit flies** Pascal Goetgheluck/Science Source; **Module 6.10:** Page 134, **sky** djgis/Shutterstock; **X-ray machine** Vereshchagin Dmitry/Shutterstock; **cigarettes** Stillfx/Shutterstock; **fast food** Denis Vrublevski/Shutterstock; **vitamins** Africa Studio/Shutterstock; **sunscreen** Roman Sigaev/ Fotolia; **vegetables** Aprilphoto/Shutterstock; **fruit** Mircea Bezergheaanu/Shutterstock; **Module 6.11:** Page 136, **mammogram** Scott Camazine/Science Source; **Module 6.12:** Page 139, **surgery** Jan Halaska/Science Source; **radiation** Doug Martin/Science Source; **chemotherapy** Will & Deni McIntyre/Science Source; **diet** OtnaYdur/Shutterstock; **smoking** Sergiy Kuzmin/Shutterstock; **sun protection** windu/Shutterstock; **screenings** BURGER/ PHANIE/Science Source; **exercise** Pressmaster/ Shutterstock; **Module 6.13:** Page 140, **bread** andersphoto/Shutterstock; **cow** Christopher Elwell/Shutterstock; **DNA techs** Alexander Raths/Shutterstock; **genetic engineering** Pichi Chuang/Reuters; Page 141, **production** Volker Steger/Science Source; **injection** AJPhoto/ Science Source; **bottle** Leonard Lessin/Science Source; **Module 6.14:** Page 142, **fluorescent chromosomes** Health Protection Agency/ Science Source; Page 143, **DNA synthesis** Laguna Design/Science Source; **microscope** Phanie/ Science Source; **Module 6.15:** Page 144, **plasmid** Vasiliy Koval/Shutterstock; **plant** Science Photo Library/Alamy; **corn** Joerg Beuge/Shutterstock; **rice** Yuttasak Jannarong/Shutterstock; **papayas** George Burba/Shutterstock; Page 145, **petri dishes** Sven Hoppe/Shutterstock; **erythropoietin** BSIP SA/Alamy; **embryo** ktsdesign/Shutterstock; **goat left** Andia/Alamy; **milking** Simon Clay/DK Images; **goat right** Inga Spence/Alamy; **Module 6.16:** Page 146, **enzyme** Laguna Design/Science Source; Page 147, **blood** Korionov/Shutterstock; **primer tubes** luchschen/ Shutterstock; **PCR** Philippe Psaila/Science Source; **Module 6.17:** Page 148, **person A** Wong SzeFei/Fotolia; **cell** Kent Wood/Science Source; **person B** Jason Stitt/Shutterstock; **karyotype** Leonard Lessin/Science Source; Page 149, **crime scene** imago stock&people/Newscom; **suspect** Steliangagiu/Shutterstock; **PCR** Philippe Psaila/ Science Source; **output** Simon Fraser/Science Source; **electrophoresis** Gustoimages/Science Source; **Module 6.18:** Page 151, **woman** Jason Stitt/Shutterstock; **karyotype** Leonard Lessin/ Science Source; **sequencer** Patrick Landmann/ Science Source; **computer** Yuri Arcurs/ Shutterstock; **Module 6.19:** Page 152, **man** Kaarsten/Shutterstock; **doctor & baby** Peter Menzel/Science Source; Page 153, **woman** Yuri Arcurs/Shutterstock.

CHAPTER 7: Module 7.1: Page 154, **Aristotle** Pawel Wojcik/DK Images; **middle ages** North Wind Picture Archives/Alamy; **fossils** Ria Novosti/Science Source; **young Darwin** Science Source; **book page** Paul D. Stewart/ Science Source; **hillside** James Steinberg/Science Source; Page 155, **adult Darwin** Science Source; **older Darwin** SPL/Science Source; **beetles**

Dave King/DK Images, Courtesy of Down House/Natural History Museum, London; **Wallace** New York Public Library/Science Source; **notes** Natural History Museum, London/Science Source; **book page** Science Source; **iguanas** Ashley Toone/Alamy; **tortoises** Natural Visions/Alamy; **Module 7.2:** Page 156, **rabbits** blickwinkel/Alamy; **elephant** Martin Harvey/Science Source; **lion kill** Christian Heinrich/ImageBroker/AGE Fotostock; **ladybirds** PSU Entomology/Science Source; **gazelle** Paul Souders/Corbis; **chromosomes** Biophoto Associates/Science Source; **sea dragon** Peter Scoones/Nature Picture Library; **Module 7.3:** Page 158, **fossil** Colin Keates/DK Images; **trilobite** Colin Keates/DK Images; Page 159, **petrified wood** B. Christopher/Alamy; **duck-billed fossil** Juan Carlos Muñoz/AGE Fotostock; **baby mammoth** Novosti/Science Source; **fossils** Andreas Einsiedel/DK Images; **amber** Francois Gohier/Science Source; **footprints** Francois Gohier/Science Source; **Module 7.4:** Page 160, **globes top to bottom** Christian Darkin/Science Source; Christian Darkin/Science Source; Planetary Visions Ltd./Science Source; **koala** Andrew Harris/DK Images; **kangaroo** DK Images; **planigale** B.G. Thomson/Science Source; **wombat** Ken FindlayDK Images; **opossum** Rick & Nora Bowers/Alamy; Page 161, **forelimbs left to right** Dave King/DK Images; Philip Dowell/DK Images; Dr. Keith Wheeler/Science Source; **chicken embryo** Dr. Keith Wheeler/Science Source; **human embryo** Dr. G. Moscoso/Science Source; **Module 7.5:** Page 162, **islands** Josh McCulloch/All Canada Photos/SuperStock; Page 163, **flowers** Leonardo Viti/Shutterstock; **rabbits** DK Images; **Module 7.6:** Page 164, **camouflage** Roger Powell/Nature Picture Library; Page 165, **elephants** Winfried Wisniewski/FLPA; **pollen** Jim Zipp/Science Source; **cheetahs** AfriPics.com/Alamy; **bird** Phil Savoie/Nature Picture Library; **seals** Morales/AGE Fotostock; **Module 7.7:** Page 166, **bipedal dinosaur** Gary Ombler/DK Images; Page 167, **dinosaurs** Peter Dennis/DK Images; **crater** Mark Pilkington/Geological Survey of Canada/Science Source; **Module 7.8:** Page 168, **Earth formation** Atlantic Digital/DK Images; **lava** Stuart Westmorland/AGE Fotostock; **algae** Michael Abbey/Science Source; **mawsonite** R. Koenig/Blickwinkel/AGE Fotostock; **trilobite** Colin Keates/DK Images; **carboniferous forest** Jon Hughes/DK Images; Page 169, **fault** Design Pics/SuperStock; **volcanoes** Guillaume Soularue/AGE Fotostock; **Jurassic forest** Richard Bizley/Science Source; **stegoceros** Andy Crawford/DK Images, Courtesy of the Royal Tyrrell Museum of Palaeontology, Alberta, Canada; **megazostrodon primate** Martin Shields/Science Source; **hominid** Barbara Strnadova/Science Source; **modern animals** Anup Shah/Nature Picture Library; **Module 7.9:** Page 170, **eastern meadowlark** Bob Steele/DK Images; **western meadowlark** Brian E. Small/DK Images; **dogs** Juniors Bildarchiv/AGE Fotostock; **bacteria** Dr. Tony Brain/Science Source; **frog fossil** Colin Keates/DK Images; Page 171, **peacocks** Ashley Cooper/Specialist Stock/AGE Fotostock; **toad** Vidady/Fotolia; **pine** Dennis

Flaherty/Photographer's Choice/Getty Images; **wasp** McPhoto/Blickwinkel/AGE Fotostock; **sea urchins** Gregory Ochocki/Science Source; **horse & mule** Laurie Noble/DK Images; **donkey** Jerry Young/DK Images; **Module 7.10:** Page 172, **Cambrian sea** The Natural History Museum/The Image Works; Page 173, **desert** Fotosearch/AGE Fotostock; **Grand Canyon** Erik Isakson/AGE Fotostock; **squirrel left** Charlie Ott/Science Source; **squirrel right** Tim Zurowski/All Canada Photos/SuperStock.

CHAPTER 8: Module 8.1: Page 178, **water & rocks** Christophe Sidamon-Pesson & David Allemand/Science Source; Page 179, **early landscape** Chris Butler/Science Source; **cells** David McCarthy/Science Source; **Module 8.2:** Page 180, **helicobacter** A. Barry Dowsett/CAMR/Science Source; **anthrax** Scott Camazine/Science Source; **fission** Dr. Kari Lounatma/Science Source; Page 181, **cocci** Eye of Science/Science Source; **bacilli** Kwangshin Kim/Science Source; **spiral** Juergen Berger/Science Source; **sea vents** B. Murton/Southampton Oceanography Centre/Science Source; **Archaeolobus** Alfred Pasieka/Science Source; **teeth** Alex Bartel/Science Source; **plaque** Steve Gschmeissner/Science Source; **Module 8.3:** Page 182, **landfill** Jim West/ImageBroker/AGE Fotostock; **swamp** Kenneth Murray/Science Source; **bacteria** Dr. M. Rohde/GBF/Science Source; **Antarctic** Michael S Nolan/Specialist Stock/AGE Fotostock; Page 183, **Dead Sea** D. U. Boisberranger Jean/Hemis Sas/AGE Fotostock; **bacteria** Eye of Science/Science Source; **salt crust** NASA image courtesy NASA/GSFC/METI/ERSDAC/JAROS, and U.S./Japan ASTER Science Team; **flamingo** NHPA /SuperStock; **hot spring** Michael Just/AGE Fotostock; **Module 8.4:** Page 184, **sewage plant** Wade H. Massie/Shutterstock; **plant & roots** AgStock Images/Corbis; **bacteria** Steve Gschmeissner/Science Source; **bioremediation** A. Ramey/PhotoEdit, Inc.; **soil** Craig Knowles/DK Images; Page 185, **plague bacteria** Alfred Pasieka/Science Source; **burning** St Bartholomew's Hospital/Science Source; **locker room** Randall Benton/ZUMA Press/Newscom; **streptococcus** BSIP/Science Source; **anthrax attack** Reuters/Corbis; **anthrax bacillus** A. Barry Dowsett/CAMR/Science Source; **rash** James Gathany/CDC; **tick** Scott Camazine/Science Source; **bacteria** Michael Abbey/Science Source; **kitchen** Olaf Doering/Alamy; **salmonella** Dr. Gary Gaugler/Science Source; **Module 8.5:** Page 186, **bacteriophages** Eye of Science/Science Source; Page 187, **conjugation** Dr. Linda M. Stannard, University of Cape Town/Science Source; **plasmids** Dr. Gopal Murti/Science Source; **Module 8.6:** Page 188, **Golgi** Steve Gschmeissner/Science Source; **Module 8.7:** Page 190, **paramecium** Michael Abbey/Science Source; **giardia** Science Source; **plasmodium** Dr. Tony Brain/SPL/Science Source; **foram** Exxon/SPL/Science Source; **amoeba** Gerd Guenther/Science Source; **slime mold** Larry West/FLPA; Page 191, **algae** Michael Abbey/Science Source; **dinoflagellates** Don Paulson/PureStock/AGE Fotostock; **symbiodinium** Mary Beth Angelo/

Science Source; **diatoms** D. P. Wilson/FLPA; **red algae** Marevision/AGE Fotostock; **green algae** Marevision/AGE Fotostock; **brown algae** Mark Conlin/Alamy; **Module 8.8:** Page 192, **volvox** Manfred Kage/Science Source; Page 193, **Earth** Chris Butler /Science Source; **Module 8.9:** Page 195, **flu virus** Hazel Appleton/Health Protection Agency Centre for Infections/Science Source; **hospital** Library of Congress Prints and Photographs Division; **herpes virus** Kwangshin Kim/Science Source; **mouth** Dr P. Marazzi/Science Source; **mumps virus** Hazel Appleton/Health Protection Agency Centre for Infections/Science Source; **face** Dr. P. Marazzi/Science Source; **tobacco mosaic virus** Biophoto Associates/Science Source; **leaf** Norm Thomas/Science Source; **Module 8.10:** Page 196, **AZT** Will & Deni McIntyre/Science Source; Page 197, **cell** Eye of Science/Science Source; **workers** Desirey Minkoh/AFP/Getty Images/Newscom; **Module 8.11:** Page 198, **brain** Simon Fraser/Science Source; **prion** Laguna Design/Science Source; Page 199, **mad cow disease** John Callan/Oxford Scientific/Getty Images; **wasting disease** Karl Gehring/Denver Post/Getty Images; **skulls** Michael Patrick O'Neill/Alamy; **scrapie** Wayne Hutchinson/Alamy; **potato** Nigel Cattlin/Science Source; **peaches** R. Flores/Bugwood Network; **apple** H.J. Larsen/Bugwood Network.

CHAPTER 9: Module 9.1: Page 200, **fungi on log** Michael P. Gadomski/Science Source; Page 201, **petri dish** Biophoto Associates/Science Source; **tablets** Keith Morris/Alamy; **dough** Howard Shooter/DK Images; **yeast cells** Microfield Scientific Ltd/Science Source; **toes** Jane Shemilt/Science Source; **parasitic fungus** Eye of Science/Science Source; **stone** Robert Thompson/Nature Picture Library; **lichen** Eye of Science/Science Source/Photo Researchers; **mold** Cordelia Molloy/Science Source; **mushrooms** Alisafarov/Shutterstock; **corn** David Bleeker Photography/Alamy; **cheese** Roger Phillips/DK Images; **truffles** Trufero/Shutterstock; **Module 9.2:** Page 202, **mushroom** Yon Marsh/Alamy; **mycelium** Blickwinkel/Alamy; **hyphae** Ted Kinsman/Science Source; **mushroom top** Scott Camazine/Science Source; Page 203, **spore print** Scott Camazine/Science Source; **stinkhorn** C. Huetter/Arco Images GmbH/Alamy; **Module 9.3:** Page 204, **charophyte** Michael Abbey/Science Source; **water plant** Marevision/Robert Harding/AGE Fotostock; **land plant** Peter Anderson/DK Images; Page 205, **flowering plant** Steve Shott/DK Images; **mycorrhizae** Dr. Jeremy Burgess/SPL/Science Source; **Module 9.4:** Page 206, **grass** Kim Taylor and Jane Burton/DK Images; **root tip** Dr. Jeremy Burgess/SPL/Science Source; **dandelion** Jeff J. Daly/Alamy; **beet** Andy Crawford/DK Images; **nodules** Inga Spence/Alamy; **mycorrhizae** Eye of Science/Science Source; Page 207, **flowering plant** Tom Viggars/Alamy; **leaf surface** Steve Gschmeissner/Science Source; **stoma** Martin Oeggerli/Science Source; **Module 9.5:** Page 209, **flowering plant** Tom Viggars/Alamy; **Module 9.6:** Page 210, **charophyte** Michael Abbey/Science Source; **liverwort** Adrian Davies/Nature Picture

Library; **hornwort** Daniel Vega/AGE Fotostock; **tasmanian tree** Niels Sloth/DK Images; **tassel fern** Martin Page/DK Images; Page 211, **cypress** Petr Jilek/Shutterstock; **fir** Thorsten Rust/Shutterstock; **columbine** Andrew Lawson/DK Images; **sunflowers** Kate Clow, Terry Richardson, Dominic Whiting/DK Images; **Module 9.7:** Page 212, **bryophyte** Tony Wharton/FLPA; Page 213, **moss** Mari Jensen/Shutterstock; **sphagnum** Nigel Cattlin/FLPA; **liverwort** Larry West/FLPA; **hornwort** Daniel Vega/AGE Fotostock; **Module 9.8:** Page 214, **leaf** William Reavel/DK Images; **beet** Andy Crawford/DK Images; Page 215, **tree stoma** Power and Syred/ Science Source; **Module 9.9:** Page 216, **fern** Philippe Clement/Nature Picture Library; **spore capsules** Roger Smith/DK Images; **gametophyte** Biophoto Associates/Science Source; Page 217, **lycophyte forest** Walter Myers/Science Source; **modern fern** Gerrit Vyn/Nature Picture Library; **Module 9.10:** Page 218, **tree** Steve Taylor ARPS/Alamy; **ovule cone** Michael P. Gadomski/Science Source; **seeds** Peter Chadwick/DK Images; **pollen cone** De Cuveland/ARCO/Nature Picture Library; **pollen** Cheryl Power/Science Source; Page 219, **conifers** Line-of-sight/Fotolia; **bristlecone** S. Sailer/A Sailer/Blickwinkel/AGE Fotostock; **pine redwood** Douglas Peebles/Danita Delimont Photography/Newscom; **Ginkgo** AlcelVision/Fotolia; **leaves** Chungking/Shutterstock; **Module 9.11:** Page 220, **anatomy** Peter Anderson/DK Images; **cactus** Tim Draper/DK Images; **wheat** Nigel Cattlin/Science Source; Page 221, **monocot cotyledon** Peter Anderson/DK Images; **roots** Nigel Cattlin/FLPA; **leaf** Martin Page/DK Images; **flower** Stephen Oliver/DK Images; **dicot cotyledons** Derek Hall/DK Images; **roots** Nigel Cattlin/Alamy; **leaf** Peter Anderson/DK Images; **flower** Neil Fletcher/DK Images; **grass** Martin Page/DK Images; **bean** Peter Anderson/DK Images; **palm** Vibrant Image Studio/Shutterstock; **Module 9.12:** Page 222, **flower** DK Images; Page 223, **fruits** Dave King/DK Images; **pear** Matthew Ward/DK Images; **seeds** Nigel Cattlin/Science Source; **germination** Peter Anderson/DK Images; **Module 9.13:** Page 224, **annuals seedling** David Aubrey/Science Source; **flowering** Michael P. Gadomski/Science Source; **fruiting** Sanddebeautheil/Shutterstock; **biennials seedlings** Peter Anderson/DK Images; **fruiting** Ian Murray/AGE Fotostock; **flowering** David R. Frazier Photolibrary, Inc. /Science Source; **perennials bushes** Modfos/Shutterstock; **flowering** Detail Photography/Alamy; **fruiting** Q-Images/Alamy; Page 225, **root tip** M. I. Walker/Science Source; **growth rings** Blickwinkel/Alamy.

CHAPTER 10: Module 10.1: Page 226, **cheetahs** Paul Sawer/FLPA; **cells chromosomes** Riccardo Cassiani-Ingoni/Science Source; **embryo** Pascal Goetgheluck/Science Source; Page 227, **Cambrian scene** Tom McHugh/Science Source; **trilobite fossils** Sinclair Stammers/Science Source; **Module 10.2:** Page 228, **sponge** Andrew J. Martinez/Science Source; **freshwater sponge** Steve Gschmeissner/Science Source; **sea sponge** Enzo Baradel/AGE fotostock; **tube sponge** Matthew Oldfield/Science Source; **elephant ear sponge** Masa Ushioda/WaterFrame/AGE Fotostock; Page 229, **medusa** Mark Conlin/Image Quest Marine; **polyp** Neil G. McDaniel/Science Source; **tentacles** Image Quest Marine; **sea anemone** Marevision/AGE Fotostock; **jellyfish** Image Quest Marine; **hydra** Blickwinkel/Alamy; **coral** H Goethel/AGE Fotostock; **Module 10.3:** Page 230, **mollusk structure** Andrew J. Martinez/Science Source; **radula** Clouds Hill Imaging Ltd./Science Source; **sea slug** John A. Anderson/Shutterstock; **scallop** Andrew J. Martinez/Science Source; **squid** Marevision/AGE Fotostock; Page 231, **echinoderm structure** Blickwinkel/Alamy; **endoskeleton** Science Photo Library/Alamy; **mouth** Andrew J. Martinez/Science Source; **regeneration** Andrew J. Martinez/Science Source; **sea urchin** Fred McConnaughey/Science Source; **feather star** ESP-Photo/AGE Fotostock; **brittle star** Marevision/AGE fotostock; **sea cucumber** Andrew J. Martinez/Science Source; **Module 10.4:** Page 232, **planarian** Eric Grave/Science Source; **tapeworm model** Geoff Brightling/DK Images; **tapeworm head** SPL/Science Source; **tapeworm in hand** Clouds Hill Imaging Ltd./Science Source; Page 233, **earthworm** Peter Anderson/DK Images; **leech** St Bartholomew's Hospital/Science Source; **polychaete** D. P. Wilson/FLPA; **heartworm** Martin Shields/Science Source; **trichinella** Jessica Wilson/Science Source; **roundworm** Sinclair Stammers/Science Source; **Module 10.5:** Page 234, **arthropod structure** Dave King/DK Images; **scorpion** Stephen Dalton/Science Source; **tarantula** Dave King/DK Images; **tick** Larry West/Science Source; **mite** Andrew Syred/Science Source; Page 235, **shrimp** WaterFrame/AGE Fotostock; **woodlouse** Laurie Noble/DK Images; **barnacles** L. Newman & A. Flowers/Science Source; **crayfish** Tom McHugh/Science Source; **crab** Frank Greenaway/DK Images; **locust** Frank Greenaway/DK Images; **collection** Natural History Museum/Science Source; **centipede** Tom McHugh/Science Source; **millipede** Michael P. Gadomski/Science Source; **Module 10.6:** Page 236, **chordate structure** DK Images; **lancelet** Heather Angel/Natural Visions/Alamy; **tunicates** Georgette Douwma/Science Source; Page 237, **skeleton** Judith Harrington, John Dunlop/DK Images; **lizard** Fivespots/Fotolia; **Module 10.7:** Page 238, **hagfish** Mark Conlin/Alamy; **lamprey** ANP Photo/AGE Fotostock; **mouth** Blickwinkel/Alamy; **shark** Ian Coleman (WAC)/Nature Picture Library; **stingray** Jeff Rotman/Alamy; Page 239, **fish structure** Neil Fletcher/DK Images; **gunnard** Fred McConnaughey/Science Source; **barracuda** Matthew Oldfield/Science Source; **sweetlips** Georgette Douwma/Science Source; **seahorse** Carine Schrurs/Nature Picture Library; **coelacanth** Peter Scoones/Science Source; **lungfish** Tom McHugh/Science Source; **Module 10.8:** Page 240, **frog** Jane Burton/DK Images; **toad** Jan Van Der Voort/DK Images; **caecilian** Dante Fenolio/Science Source; **salamander** PhotoStock-Israel/Alamy; **frogs mating** Francoise Sauze/Science Source; **tadpole** Dr. Keith Wheeler/Science Source; **poison dart frog** Aleksey Stemmer/Shutterstock; Page 241, **snake with eggs** Mike Wilkes/Nature Picture Library; **Galapagos iguana** ashley toone/Alamy; **crocodile** Aditya "Dicky" Singh/Alamy; **emu** Eric Isselee/Shutterstock; **feathers** Cyril Laubscher/DK Images; **bone** Gilbert S. Grant/Science Source; **Module 10.9:** Page 242, **cow and calf** Gordon Clayton/DK Images; **platypus** Jean-Philippe Varin/Science Source; **echidna** Clearviewstock/Shutterstock; Page 243, **opossum** Rolf Nussbaumer/Nature Picture Library; **embryo** John Cancalosi/Alamy; **kangaroo** Will & Deni McIntyre/Science Source; **gorilla** Steve Gorton/DK Images; **amniotic sac** J.-L. Klein & M.-L. Hubert/Science Source; **dog** Jane Burton/DK Images; **bat** Frank Greenaway/DK Images; **dolphin** Jeff Rotman/Science Source; **Module 10.10:** Page 244, **chimp** Anup Shah/Nature Picture Library; Page 245, **lemurs** Edwin Giesbers/Nature Picture Library; **loris** Ian Butler/Alamy; **potto** Jabruson/Nature Picture Library; **tarsier** G. Ronald Austing/Science Source; **macaque** Yukihiro Fukuda/Nature Picture Library; **baboon** Chadden Hunter/Nature Picture Library; **capuchin** Jonathan Irish/Corbis; **spider monkey** José Caldas/Alamy; **gibbon** Jspix/Alamy; **orangutan** Anup Shah/Nature Picture Library; **gorillas** Panda Photo/FLPA; **chimpanzee** Holger Ehlers Naturephoto/Alamy; **human** Mat Hayward/Shutterstock; **Module 10.11:** Page 246, **skull** Alain Beauvilain/Newscom; **footprints** John Reader/Science Source; **skeleton** John Reader/Science Source; **skull & tools** Pascal Goetgheluck/Science Source; **homo habilis model** Harry Taylor/DK Images; Page 247, **stone axe** Pascal Goetgheluck/Science Source; **homo erectus skeleton** lbum/Prisma/Newscom; **Neanderthal model** Tom McHugh/Field Museum Chicago/Science Source; **cave painting** Javier Trueba/Science Source.

CHAPTER 11: Module 11.1: Page 248, **chest X-ray** Innerspace Imaging/Science Source; **heart** SPL/Science Source; **man having EKG** Jochen Tack/AGE Fotostock; **electrocardiogram** Martin M. Rotker/Science Source; Page 249, **man** Asiaselects/Getty Images; **tissue** Susumu Nishinaga/Science Source; **Module 11.2:** Page 250, **bone** Steve Gschmeissner/Science Source; **cartilage** Eric Grave/Science Source; **blood** Scimat/Science Source; **connective** Prof. P. Motta/Dept. of Anatomy/University "La Sapienza", Rome/Science Source; Page 251, **epithelium** Andrew Syred/Science Source; **nervous tissue** Astrid & Hanns-Frieder Michler/Science Source; **smooth muscle** SPL/Science Source; **cardiac muscle** Steve Gschmeissner/Science Source; **skeletal muscle** Steve Gschmeissner/Science Source; **Module 11.3:** Page 252, **cyclist** imagebroker.net/SuperStock; **icy water** Arno Burgi/EPA/Newscom; **sauna** Ferdinando Scianna/Magnum Photos; Page 253, **sweating man** RubberBall/Alamy; **woman** Anatoly_Maltsev/epa/Newscom; **Module 11.4:** Page 254, **girl eating** reka/Jupiter Images; Page 255, **intestinal villi** Steve Gschmeissner/Science Source; **Module 11.5:** Page 256, **girl eating** Aaron Amat/Fotolia; Page 257, **intestinal microvilli** Ami Images/Science Source; **Module 11.6:** Page 258, **tennis player** ImageryMajestic/Shutterstock; Page 259, **food** Ann Cutting/Alamy; **Module 11.7:** Page 260, **girl eating** Aaron Amat/Fotolia; **antacids** Science Source; **bacteria left** Dr. Kari Lounatmaa/Science Source; **bacteria right** A. Dowsett, Health Protection

Agency/Science Source; **medications** Helen Sessions/Alamy; Page 261, **malnutrition** Mauro Fermariello/Science Source; **eating disorders** Dimitri Halkidis/WENN/Newscom; **Module 11.8:** Page 262, **runner** David Madison/Photographer's Choice RF/Getty Images; Page 263, **smoking** Lemoine/AGE Fotostock; **asthma** Science Photo Library/Getty Images; **Module 11.9:** Page 264, **woman** R. Gino Santa Maria/Shutterstock; **Module 11.10:** Page 267, **coronary X-ray** Zephyr/Science Source; **Module 11.11:** Page 268, **red blood cells** Scimat/Science Source; Page 269, **platelets** Eye of Science/Science Source; **white blood cells** Steve Gschmeissner/Science Source; **clotting** Susumu Nishinaga/Science Source; **Module 11.12:** Page 270, **man** Nicholas Piccillo/Shutterstock; **cilia** Susumu Nishinaga/Science Source; Page 271, **phagocyte** Juergen Berger/Science Source; **T cell** CNRI/Science Source; **man** Kaarsten/Shutterstock; **Module 11.13:** Page 272, **man** kaarsten/Shutterstock; Page 273, **vaccination** Nicolas Germain/AFP/Getty Images/Newscom; **Module 11.14:** Page 274, **ragweed** Derrick Ditchburn/Science Source; **pollen** Scott Camazine/Science Source; **woman** Denise Hager/Catchlight Visual Services/Alamy; **X-ray** Zephyr/Science Source; Page 275, **boy** Science Source; **infected cell** Thomas Deerinck/NCMIR/Science Source; **organ rejection** Peter Menzel/Science Source; **stress** PictureIndia/AGE Fotostock; **Module 11.15:** Page 276, **man** Laura Knox/DK Images; Page 277, **monitor** Michael P. Gadomski/Science Source; **Module 11.16:** Page 278, **woman** Anton Gvozdikov/Shutterstock; **dialysis** Picsfive/Shutterstock; **Module 11.19:** Page 284, **bacteria** Juergen Berger/Science Source; **virus** 3D4Medical/Science Source; **protist** Eye of Science/Science Source; **fungi** CNRI/Science Source; **Module 11.20:** Page 286, **man standing** Piccillo/Shutterstock; Page 287, **profile** DK Images; **Module 11.21:** Page 288, **man standing** Piccillo/Shutterstock; Page 289, **soccer player** Alan Dragulin/Alamy; **Module 11.23:** Page 293, **ball & socket** Biophoto Associates/Science Source; **pivot** pisaphotography/Shutterstock; **hinge** Robert Destefano/Alamy; **saddle** itsmejust/Shutterstock; **osteoporosis** Professor Pietro M. Motta/Science Source; **fracture** Zephyr/Science Source; **arthritis** Zephyr/Science Source; **Module 11.24:** Page 294, **arm** Carlos E. Santa Maria/Shutterstock; **sarcomere** Don W. Fawcett/Science Source.

CHAPTER 12: Module 12.1: Page 296, **discovery** Martin Shields/Science Source; **hypothesis** Artyn F. Chillmaid/Science Source; **environmentalism** Aman Rahman/AFP/Getty Images/Newscom; Page 297, **organismal** James Hager/Robert Harding; **population** Juniors Bildarchiv/AGE Fotostock; **community** Rob Daugherty/Flickr/Getty Images; **ecosystem** Galyna Andrushko/Shutterstock; **Module 12.2:** Page 298, **terrestrial** Aflo/Nature Picture Library; **aquatic** Image Quest Marine; **hydrothermal** Verena Tunnicliffe/AFP/Newscom; **corn** Science Photo Library/Alamy; **pitcher plant** Radius Images/Glow Images; **tree** Ashley Cooper/Robert Harding; Page 299, **lizard** Neil Fletcher/DK Images; **cactus** Christian Handl/AGE Fotostock; **fish** Alexander

Raths/Shutterstock; **fire** David R. Frazier Photolibrary, Inc./Alamy; **aftermath** Corbis/AGE Fotostock; **Module 12.3:** Page 300, **fox** Gerard Lacz/FLPA; **bird** Oliver Smart/Alamy; **caterpillar** Premaphotos/Alamy; **trees** Sozaijiten/Imagenavi/Glow Images; **worms** Derek Middleton/FLPA; Page 301, **group** Pavel L Photo and Video/Shutterstock; **individual** Yuri Arcurs/Shutterstock; **cows** jvd-wolf/Shutterstock; **corn** Zeljko Radojko/Shutterstock; **Module 12.4:** Page 302, **volcano** Robert Harding Picture Library Ltd/Alamy; **grass** Alexandr79/Shutterstock; **deer** Guy J. Sagi/Shutterstock; **mushrooms** oksix/Shutterstock; **Module 12.5:** Page 304, **estuary** M-Sat Ltd/Science Source; **sprinklers** Rosenfeld Images Ltd/Science Source; Page 305, **burning forest** George Holton/Science Source; **Module 12.6:** Page 306, **underwater** Wild Wonders of Europe/Munier/Nature Picture Library; **waterfall** Chris Schmid/Nature Picture Library; **wetlands** Nick Upton/2020VISION/Nature Picture Library; **intertidal zone** Image Quest Marine; **birds** David Woodfall/Nature Picture Library; **estuary** Kevin Allen/Alamy; Page 307, **coral reef** Image Quest Marine; **zooplankton** M. I. Walker/Science Source; **deep sea vents** OAR/National Undersea Research Program (NURP)/NOAA; **octopus** B. Murton/Southampton Oceanography Centre/Science Source; **squid** Image Quest Marine; **Module 12.7:** Page 308, **polar** Blickwinkel/AGE Fotostock; **tundra** Wild Wonders of Europe/Munier/Nature Picture Library; **grassland** Jim Parkin/Shutterstock; **tropical forest** blickwinkel/Alamy; Page 309, **chaparral** Verna Johnston/Science Source; **coniferous forest** Igor Shpilenok/Nature Picture Library; **broadleaf forest** Helge Schulz/Premium/AGE Fotostock; **savanna** Elliott Neep/FLPA; **desert** David Wall/Alamy; **Module 12.8:** Page 310, **deer** WILDLIFE GmbH/Alamy; **bear** Biosphoto/SuperStock; **chipmunk** Universal Images Group/AGE Fotostock; Page 311, **coral** Image Quest Marine; **bee** Sergey Lavrentev/Shutterstock; **mycorrhizae** Dr. Jeremy Burgess/Science Source; **predator** Federico Veronesi/Gallo Images/AGE Fotostock; **herbivore** Mark Newman/FLPA; **parasite** Image Quest Marine/Alamy; **pathogen** Nature/Alamy; **Module 12.9:** Page 312, **hawk** Stephen Mcsweeny/Shutterstock; **orca** Gerard Lacz Images/SuperStock; **snake** Wayne Lynch/All Canada Photos/Getty Images; **salmon** Design Pics/Alamy; **chipmunk** Eric Baccega/Nature Picture Library; **flagfin** Kevin Schafer/Alamy; **cricket** Hans Lang/ Image broker RF/AGE Fotostock; **larva** Image Quest Marine; **grass** Derek Croucher/Alamy; **ceratium** Image Quest Marine; Page 313, **horned owl** Eric Isselée/Fotolia; **snake** Rick & Nora Bowers/Alamy; **falcon** Cyril Laubscher/DK Images; **barn owl** Sean Hunter/DK Images; **fox** Jerry Young/DK Images; **bear** Radius Images/Glow Images; **robin** Wild Art/Shutterstock; **blackbird** Steve Byland/Shutterstock; **mantis** Frank Greenaway/DK Images; **whitetail deer** Siddhardha Garige/Alamy; **snail** BSANI/Fotolia; **ants** DK Images; **caterpillar** Valeriy Kirsanov/Fotolia; **cricket** Imagebroker/FLPA; **hare** Sean Hunter/DK Images; **plants** Kim Taylor and Jane Burton/DK Images; **marlin** Colin Newman/DK Images;

mackerel Colin Newman/DK Images; **mercury** DK Images; **Module 12.10:** Page 314, **otter** Michael S. Nolan/AGE Fotostock; Page 315, **Mt. St. Helens 1980** pf/Alamy; **1985** Dr. Robert Spicer/Science Source; **2009** Curved Light USA/Alamy; **2011** Larry Geddis/Alamy; **brushfire** William West/AFP/Getty Images/Newscom; **6 weeks later** Jennifer Hart/Alamy; **1 year later** Jennifer Hart/Alamy; **Module 12.11:** Page 316, **python** Dan Callister/Alamy; **lionfish** WaterFrame/Alamy; **rabbits** SuperStock; **kudzu** M. Timothy O'Keefe/Alamy; **mussels** Wolfgang Pölzer/Alamy; **sign** S. Callahan/Photri Images/Alamy; Page 317, **corn borer mongoose** Steven Lee Montgomery/Photo Resource Hawaii/Alamy; **pest management** Joerg Boethling/Alamy; **Module 12.12:** Page 318, **potato** Nigel Cattlin/Alamy; **memorial** David Grossman/Alamy; **parakeet** Nathan Benn/Alamy; **reef damage** Franco Banfi/WaterF/AGE Fotostock; Page 319, **invasive species** John Mitchell/Science Source; **climate change** Georgette Douwma/Science Source; **habitat destruction** Jacques Jangoux/Alamy; **pollution** U. S. Coast Guard/Science Source; **overharvesting** Onne van der Wal/Bluegreen Pictures/Alamy; **Module 12.13:** Page 320, **Americans** Adrian Weinbrecht/Alamy; **Ethiopians** Gavin Hellier/Robert Harding; Page 321, **man** cristovao/Shutterstock; **birds** Robert Thompson/Nature Picture Library; **fish** Georgette Douwma/Nature Picture Library; **clumped** Norbert Probst/Imagebroker/FLPA; **uniform** Ralph Lee Hopkins/National Geographic Image Collection/Alamy; **random** Imagebroker/FLPA; **Module 12.14:** Page 323, **hyenas** Tony Crocetta/AGE Fotostock; **butterfly** NHPA/SuperStock; **Module 12.15:** Page 324, **New York City 1700s** North Wind Picture Archives/Alamy; **2010** kropic1/Shutterstock; **Module 12.16:** Page 327, **Bolivia 1986** Planet Observer/Science Source; **1998** Planet Observer/Science Source; **depletion California** Robert Brook/Science Source; **Idaho** Shawn Hempel/Alamy; **tagging turtles** Martin Harvey/Science Source; **lynx** Don Johnston/Alamy; **Module 12.17:** Page 328, **manatees** Steven David Miller/Nature Picture Library; **lemurs** Anup Shah/Getty Images; **eagles** Michael Krabs/Imagebroker/FLPA; Page 329, **bridge** Alan Sirulnikoff/Science Source; **bioremediation** Kyodo/AP Images; **tree harvesting** Christophe Vander Eecken/Reporters/Science Source; **greenhouse** GFC Collection/Alamy; **Module 12.18:** Page 331, **bird** Luiz Claudio Marigo/Nature Picture Library; **coral bleaching** Georgette Douwma/Science Source; **polar bear** Jan Martin Will/Shutterstock; **fire** David R. Frazier Photolibrary, Inc./Science Source; **Module 12.19:** Page 332, **goose bumps** Astrid & Hanns-Frieder Michler/Science Source; **feathers** John Devries/Science Source; **hikers** Michael Interisano/AGE Fotostock; **tree** Norbert Probst/Imagebroker/FLPA; **potted plant** Maryann Frazier/Science Source; **marmots** Photoshot/Alamy; Page 333, **clothing** Hana/Datacraft - QxQ images/Alamy; **iguanas basking** Bill Gozansky/Alamy; **migrating herd** Anup Shah/Nature Picture Library.

ILLUSTRATION AND TEXT CREDITS

CHAPTER 1: Module 1.2: Page 5, **DNA molecule** from Simon, Eric J.; Reece, Jane B.; Dickey, Jean L., *Campbell Essential Biology with Physiology*, 3rd Ed. © 2010. Reprinted and electronically reproduced by permission of Pearson Education Inc., Upper Saddle River, NJ 07458; **Module 1.4:** Page 9, **DNA helix**, from Simon, Eric J.; Reece, Jane B.; Dickey, Jean L., *Campbell Essential Biology with Physiology*, 3rd Ed. © 2010. Reprinted and electronically reproduced by permission of Pearson Education Inc., Upper Saddle River, NJ 07458.

CHAPTER 2: Module 2.3: Pages 22-23, **Atom of nitrogen, Nitrogen Isotope, Nitrogen Ion** © Dorling Kindersley.

CHAPTER 5: Module 5.18: Page 118, **Plants cells divide** from Simon, Eric J., *Campbell Essential Biology*, 4th Edition, © 2010. Reprinted by permission of Pearson Education Inc., Upper Saddle River, NJ.

CHAPTER 7: Module 7.4: Page 161, **Bat bone** Dave King © Dorling Kindersley Ltd.; **Porpoise bone** Philip Dowell © Dorling Kindersley; **Human bone** Philip Dowell © Dorling Kindersley, Courtesy of The Natural History Museum, London.

CHAPTER 11: Module 11.5: Page 257, **Salivary glands** Raj Dashi © Dorling Kindersley; **Module 11.10:** Page 267, **Nodes** Zygote Media Group © Dorling Kindersley; **Module 11.17:** Page 280, **Male reproductive anatomy** Zygote Media Group © Dorling Kindersley; **Spermatogenesis diagram** Zygote Media Group © Dorling Kindersley; **Module 11.22:** Page 290, **Sensory receptors** Zygote Media Group © Dorling Kindersley; **Receptor close-up** Zygote Media Group © Dorling Kindersley; Page 291, **Human eye** Zygote Media Group © Dorling Kindersley; **Module 11.23:** Page 292, **Skeletal system** Zygote Media Group © Dorling Kindersley; **Bone structure** Peter Bell © Dorling Kindersley Ltd.; Page 293, **Hinge joint** Zygote Media Group © Dorling Kindersley; **Module 11.24:** Page 295. **Muscle contraction** Zygote Media Group © Dorling Kindersley.

Index

Hemoglobin, 40–41, **269**
Hemophilia, 113
Herbivory, 310–**11**
Heredity, **100**
Heritability, natural selection
	and, 156–57
Herpes, 285
Herpesvirus, 195
Heterotrophs, **226**
Heterozygous alleles, **100**–101
	genetic cross of, 102–3
Hexokinase, 31
High blood pressure, 265
High-density lipoprotein (HDL), 36
Hinge joints, 293
HIV. *See* Human
	immunodeficiency virus
HMS *Beagle* voyage, 155
Hollow nerve cord, **236**
Homeostasis, **252**
	of insulin and glucagon, 277
	negative feedback and, 253
	positive feedback and, 253
Homeotic genes, **133**
Hominins, **246**–47
Homo erectus, 247
Homo habilis, 246–47
Homologous chromosomes, **91**
	crossing over of, 97, 111, 164
	independent assortment of,
		96, 164
Homo neanderthalensis, 247
Homo sapiens, 247
Homozygous alleles, **100**–101
	genetic cross of, 102–3
Honey mushrooms, 200
Hooks, of tapeworms, 232
Hormones, **276**
	fat-soluble, 277
	in female reproductive cycle, 281
	operation of, 277
	regulation with, 276–77
	steroid, 37
	water-soluble, 277
Hornbill bird, 155
Hornworts, 210, 213
Horse, 171
	cloned, 115
	evolution of, 172
Host cells, for viruses, 194–97
Human age structure, 325
Human body, tissues of, 250–51
Human families, pedigrees for
	tracing traits of, 106–7
Human Genome Project, **150**
Human immunodeficiency virus
	(HIV), **196**–97, **275**, 285
	life cycle of, 197
	structure of, 196
Human life cycle, 90–91
Human papillomavirus, 285
Human population growth, 324
Humans, **246**
	development of, 282–83
	evolution of, 244, 246–47

origin of, 193
	as primates, 244–45
Humoral immune response, **272**
Huperzia selago, 217
Hybrid weakness, as
		reproductive barrier, 171
Hydras, 229
Hydrochloric acid, 28
Hydrogen, 21
	in water molecules, 24–26
Hydrogenation, **39**
Hydrogen bonds, **25**
	in DNA, 116–17
	in water, 25–27
Hydrogen ions, acidity and,
	28–29
Hydrolysis reactions, **32**–33
	in food digestion, 255
Hydrophobic molecules, **36**
	lipids as, 36–37
Hydrothermal vents, 181, 183,
	298, 307
Hydroxide ions, acidity and,
	28–29
Hydroxyl group, 30
Hypertension, **265**
Hyphae, **202**–3
Hypothalamus, **276**, 287
Hypothermia, 253
Hypothesis, 6–**7**
	as step in scientific method, 6
Hypothesis-driven science, **296**

I

Ice, floating by, 26
Iguanas, 155
Illnesses, causes of, 260
Immune rejection, **275**
Immune response
	allergies, 274
	cell-mediated, 272
	humoral, 272
	primary, 273
	secondary, 273
Immune system, **270**
	allergies, 274
	antibodies, 272
	autoimmune diseases, 274
	blood cells of, 250
	clonal selection, 272
	complement system, 271
	external defenses, 270
	HIV effects on, 196–97
	immunodeficiencies, 275
	lymphatic system, 271
	malfunctions of, 274–75
	memory cells, 273
	specific attacks of, 272–73
	stress and, 275
	T cells, 273
	white blood cells, 271
Immunization, **273**

Immunodeficiency diseases, **275**
Impotence, 285
Incomplete dominance, **108**–9
Independent assortment, 96–97
	in evolution, 164
	linked genes and, 110–11
	Mendel's law of, 104–5
Indian marsh crocodiles, 241
Induction, **133**
Infant mortality, 324
Infection
	bacteria causing, 185, 285
	of cerebrospinal fluid, 286
	external defenses for, 270
	illnesses with, 260
	prions and viroids causing,
		198–99
	spinal, 286
	STDs, 285
	viral, 194–97
	white blood cells and, 268
Inferior vena cava, 266
Infertility, female, 284
Inflammatory bowel disease, 260
Inflammatory response, 270
Influenza virus, 195
Ingestion, **255**–56
Ingroups, 177
Inhalation, 262–63
Inheritance
	basic principles of, 100–101
	complex patterns of, 108–9
	law of independent
		assortment and, 104–5
	linked genes in, 110–11
	natural selection and, 156
	pedigrees for tracing of, 106–7
	polygenic, 109
	Punnett square for prediction
		of, 102–3
	sex-linked, 112–13
Inhibitors, **43**
Innate white blood cells, 271
Inner electron shell, 22
Inner membrane
	chloroplast, 56, 64
	mitochondria, 57
Insects, **235**
Insertions, 134–**35**
Insulin, 277
	production of, 140–41
Integrated pest management, **317**
Integration, 289
Intermediate biomes, 306
Internal environment, 252–53
Interneuron, 289
Internodes, of plants, 207
Interphase, **84**, 85, **86**
	in meiosis, 92
Interspecies competition, 310
Interspecies interactions, **310**
Interstitial fluid, **265**
Intertidal zones, **306**
Intestinal cells, gene expression
	in, 130

Intestinal villi, 255, **257**
Intestine
	large, 256
	small, 255–56
	swelling of, 260
Intrauterine device (IUD), 284
Introns, **125**, **131**
	in gene expression
		regulation, 131
Invasive species, **316**
	biodiversity loss with, 319
	biological control for, 317
	integrated pest management
		for, 317
Invertebrate chordates, 236
Invertebrates, **228**
Ion, **23**–24
	determination of, 22–23
Ionic bonds, **24**–25
Iris, **291**
Irish Potato Famine, 318
Iron, requirements for, 259
Irrigation, 304
Isomers, **34**
Isotope, **23**
	determination of, 22–23
IUD. *See* Intrauterine device

J

Jacob's syndrome, 99
Japanese macaque, 245
Jaw
	of cartilaginous fishes, 238
	evolution of, 237
Jawless fish, 238
Jellyfish, 9, 229
Jersey calf, 242
Jointed appendages, of
		arthropods, 234
Joints, **293**
	of fetus, 283
	of primates, 244
Jones, Marion, 37
Jurassic period, 169

K

Kangaroo, 160
Karyotype, **91**
Kelp, 191
Keratin, 31, 41
Keystone species, **314**
Kidneys, 278–**79**
	failure of, 278
Kidney tubules, 279
Kilocalories, 73
Kinetic energy, **60**
	conversion and use of, 60–61
Kingdoms, 12–**13**, 174–**75**, 226
King salmon, 312
Klebsiella pneumoniae, 170

Miscarriages, 284
Missense mutations, 134–**35**
Mites, 234
Mitochondria, 46–47, **57**
  cellular respiration in, 258
  endosymbiosis leading to, 189
  endosymbiotic hypothesis
    and, 7
  energy provided by, 56–57, 63,
    71–75
Mitosis, **85–86, 280**
  meiosis compared with,
    94–95
  nucleus duplication during,
    86–87
Mitotic phase, **84**–85
Mitotic spindle, **87**
Moist skin, of amphibians, 240
Molds, 201
Molecules, **5**, **18**
  as level of biological
    organization, 5
  life composed of, 18–19
Mollusks, **230**
  gallery of, 230
  habitats of, 230
  phylogeny of, 227, 230
Mongoose, 317
Monitor lizard, 237
Monkeys, 161, 244–45
Monocots, **221**
Monohybrid cross, **102**–3
Monomers, **32**–33
  first organic, 178–79
  in food digestion, 255
  in proteins, 40
Monosaccharides, **34**
  in carbohydrates, 34–35
Monotremes, 242
Moose, 199
Mosquitoes, 17
Moss, fern, 13
Mosses, 210, 212–13
Motor cortex, 287
Motor neuron, 289, **294**
Motor output, 289
Mountain gorilla, 245
Mountaintop removal, 319
Mouse, 9
  cloned, 115
  genome of, 150
Mouth
  in digestion, 256
  of echinoderms, 231
  of lamprey, 238
Movement
  ATP and energy used for, 79
  of cells, 58–59
  protein role in, 41
mRNA. *See* Messenger RNA
MRSA. *See* Methicillin-resistant
  Staphylococcus aureus
MS. *See* Multiple sclerosis
Mucous membranes, in
  infection defense, 270

Mule, 171
Multicellular life
  animals as, 226
  evolution of, 192–93
Multiple alleles, **108**–9
Multiple sclerosis (MS), 274
Mumps virus, 195
Muscle cells, lactic acid
  fermentation in, 76
Muscle fibers, **251**, **294**
Muscles. *See also* Skeletal muscle
  of animals, 226
  ATP and energy used by, 79
  ciliary, 291
  contraction of, 295
  dehydration synthesis
    reactions in, 33
  of fetus, 283
  pairs of, 295
  of stomach, 256
  structure of, 294
  types of, 251
Muscle tissue, **251**
Musculoskeletal system, of
  amphibians, 240
Mushrooms, 200–201, **202**
  anatomy of, 202
  in reproduction, 203
Mussels. *See* Bivalves
Mutagens, **134**
Mutant trait, **106**
Mutation, **134**, **164**
  in cancer, 136–39
  as mechanism for evolution,
    164–65
  wide-ranging effects of,
    134–35
Mutualism, 310–**11**
Mycelium, **202**–3
Mycorrhizae, **205**–6, 311
Myelin sheath, 288
Myocardial infarction, 267
Myofibril, **294**
Myosin head, 295

N

NADH, in cellular respiration,
  74–75
NADPH, **66**
  in photosynthesis, 66–67,
    69–71
Na^+/K^+ pump. *See* Sodium-
  potassium pump
Nasal cavity, in breathing, 262
Natural gas, 25
Natural killer cell, **271**
Natural selection, **14–15**, **156**
  in action, 17
  argument for, 156
  Darwinian fitness in, 164
  in Darwin's theory of
    evolution, 154

  sexual selection in, 165
  unequal reproductive success
    in, 14, 156–57
Natural world, evidence of
  evolution in, 160–61
Neanderthals, 247
Needles, 218
Negative feedback, **253**
Neisseria gonorrhoeae, 187
Nematodes. *See* Roundworms
Nephrons, **279**
Nerve, **288**
Nerve cells, gene expression
  in, 130
Nerve cord, **236**
Nervous system, **286**, **288**
  of animals, 226
  brain as hub of, 286–87
  human, 286
  organization of, 289
Nervous tissue, 251
Neuron membrane, 288
Neurons, **251**, **288**
  anatomy of, 288
  motor, 289
  sensory, 289–90
Neurotransmitter, **289**
Neutrons, **22**–23
Nitrogen, 21
  atom of, 22–23
Nitrogen-15 isotope, 23
Nitrogen cycle, 303
Nitrogen fixation, 303
Nitrogen-fixing bacteria, 184
Nitrogen ion, 23
Node, of plants, 207
Nonanthropoids, 245
Nonbranching evolution,
  166–67
Noncompetitive inhibitors, 43
Nondisjunction, **98**–99
Nonpolar bonds, **25**
Nonsense mutations, 134–**35**
Nonvascular plants, 210–11
North American opossum, 160
Nose
  cartilage of, 250
  URIs of, 263
Notochord, **236**
Novel features, 166
Nuclear envelope, 46–47,
  52–**53**, 188
  during mitosis, 86–87
Nuclear pore, 52–**53**, 122, 125
Nuclear transfer, 114–15
Nuclear transplantation, **114**–15
Nucleic acid probes, **142**–43
Nucleic acids, 31, **116**, 120–21
  of viruses, 194
Nucleoid, 44
Nucleolus, 52–53
Nucleotides, **116**–17
  in DNA, 116–17, 120
  in human and animal
    genomes, 150

  mutations affecting, 134–35
  in RNA, 120–21
Nucleus, **22**, **46**, **52**
  of atoms, 22
  of cells, 45–47, 52–53, 59
  DNA and chromosomes in,
    52–53, 82–83
  duplication of, 86–87
  of eukaryotes, 226
Nutrients
  as abiotic factor, 298
  blood transport of, 264, 266
  essential, 258
  small intestine absorption of,
    255–57
Nutrition
  animal obtainment of, 227
  eating disorders, 261
  energy and building materials
    from, 258–59
  imbalances in, 261
  malnutrition, 261
  plant, 205, 214–15
  prokaryote, 181

O

Obesity, **261**
  extreme, 261
  in United States, 261
Observations, 6–7
  as step in scientific method, 6
  in theory of evolution, 14, 156
Ocean acidification, 29
Ocellated lizard, 299
Octopi. *See* Cephalopods
Oil gland, 270
Old world monkey, 161
Omega-3 fatty acids, **39**
Omega-6 fatty acid, 259
Oncogenes, **136**–38
Onions, 84
Oocyte
  primary, 281
  secondary, 281
Oogenesis, **281**
Operculum, **239**
Opossums, 160
Optic nerve, **291**
Oral cavity, **256**
Oral contraceptives, 284
Orange elephant ear
  sponge, 228
Orange sea sponge, 228
Orangutans, 245
Orca, 312
Orchids, 171
Order, as property of life, 2
Orders, 174
Ordovician period, 168
Organ, **5**, **209**, **249**
  exercise and, 249
  formation of, 283